# PHYSICS IN FOCUS

**YEAR 12**

**Robert Farr**
Kate Wilson
Philip Young
Darren Goossens
2ND EDITION

**Contributing author**
Neil Champio

Physics In Focus Year 12
2nd Edition
Robert Farr
Kate Wilson
Philip Young
Darren Goossens
Contributing author: Neil Champion
9780170409131

Publisher: Eleanor Gregory
Project editor: Felicity Clissold
Editor: Elaine Cochrane (Gillespie & Cochrane)
Proofreader: Jane Fitzpatrick
Permissions researcher: Debbie Gallagher
Indexer: Don Jordan (Antipodes Indexing)
Text design: Leigh Ashforth (Watershed Design)
Cover design: Chris Starr (MakeWork)
Cover image: iStock.com/loveguli
Production controller: Karen Young
Typeset by: MPS Limited

Any URLs contained in this publication were checked for currency during the production process. Note, however, that the publisher cannot vouch for the ongoing currency of URLs.

For product information and technology assistance,
in Australia call **1300 790 853**;
in New Zealand call **0800 449 725**

For permission to use material from this text or product, please email
**aust.permissions@cengage.com**

**National Library of Australia Cataloguing-in-Publication Data**
A catalogue record for this book is available from the National Library of Australia.

**Cengage Learning Australia**
Level 7, 80 Dorcas Street
South Melbourne, Victoria Australia 3205

**Cengage Learning New Zealand**
Unit 4B Rosedale Office Park
331 Rosedale Road, Albany, North Shore 0632, NZ

For learning solutions, visit **cengage.com.au**

Printed in China by 1010 Printing International Limited.
3 4 5 6 7 22 21 20 19

# CONTENTS

## MODULE FIVE » ADVANCED MECHANICS · 30

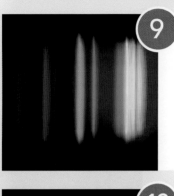
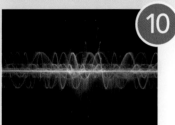

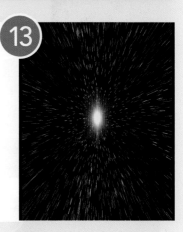

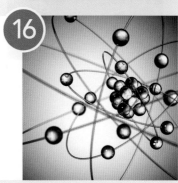

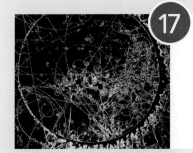

## 17 Deep inside the atom 412

# INTRODUCTION

*Physics in Focus Year 12 (2nd edition)* has been written to meet the requirements of the NESA NSW Physics Stage 6 Syllabus (updated January 2018). The text has been written to enable students to meet the requirements of achieving a Band 6 in the Higher School Certificate. It also allows all students to maximise their learning and results.

Physics deals with the wonderfully interesting and sometimes strange Universe. Physicists investigate space and time (and space–time), from the incredibly small to the incredibly large, from the nucleus of atoms to the origin of the Universe. They look at important, challenging and fun puzzles and try to work out solutions.

Physicists deal with the physical world where energy is transferred and transformed, where things move, where electricity and magnetism affect each other, where light and matter interact. As a result, physics has been responsible for about 95% of the world's wealth – including electricity supply and distribution, heating and cooling systems, computers, diagnostic and therapeutic health machines, telecommunications and safe road transport.

Physicists are not just concerned with observing the Universe. They explain these observations, using models, laws and theories. Models are central to physics. Physicists use models to describe, explain, relate and predict phenomena. Models can be expressed in a range of ways – via words, images, mathematics (numerical, algebraic, geometric, graphical), or physical constructions. Models help physicists to frame physical laws and theories, and these laws and theories are also models of the world. Models are not static – as scientific understanding of concepts or physical data or phenomena evolves, so too do the models scientists use to describe, explain, relate and predict these. Thus, the text emphasises the observations and quantitative data from which physicists develop the models they use to explain the data. Central to this is the rigorous use of mathematical representations as a key element of physics explanations.

*Physics in Focus Year 12 (2nd edition)* is written by academic and classroom teaching experts. They were chosen for their comprehensive knowledge of the physics discipline and best teaching practice in physics education at secondary and tertiary levels. They have written the text to make it accessible, readable and appealing to students. They have included numerous current contexts to ensure students gain a wide perspective on the breadth and depth of physics. This mathematically rigorous and methodological approach is designed to ensure students can reach the highest possible standard. The intention is to ensure all students achieve the level of depth and interest necessary to pursue tertiary studies in physics, engineering, technology and other STEM-related courses. Physics taken for the Higher School Certificate provides opportunities for students to arrive at a deeper understanding of their world, whether they are intending to pursue STEM-related careers or take a different pathway.

Each chapter of the *Physics in Focus* text follows a consistent pattern. Learning outcomes from the syllabus appear on the opening page. The text is then broken into manageable sections under headings and sub-headings. Question sets are found at the end of each section within the chapter. Relevant diagrams that are easy to interpret and illustrate important concepts support the text. New terms are bolded and defined in a glossary at the end of the book. Important concepts are summarised to assist students to take notes.

Worked examples, written to connect important ideas and solution strategies, are included throughout the text. Solutions are written in full, including algebraic transformations with substitution of values with units and significant figures. In order to consolidate learning, students are challenged to try similar questions on their own.

There is a comprehensive set of review questions at the end of each chapter which expands on the questions sets for further revision and practice. Questions have been set to accommodate the abilities of all students. Complete worked answers appear on the teacher website.

Investigations demonstrate the high level of importance the authors attach to understanding-by-doing physics. These activities introduce, reinforce and enable students to practise first-hand investigation skills, especially experimental design, data collection, analysis and conclusions. Chapter 1 explores in detail the concepts of reliability, validity and the nature of scientific investigation using the scientific method, and provides valuable information for performing and analysing investigations. This is designed to enhance students' experiences and to provide them with information that will maximise their marks in this fundamental area, and is reinforced throughout the course.

Système Internationale d'Unités (SI) units and conventions, including accuracy, precision, uncertainty and error, are also introduced in the first chapter. This invaluable chapter supports student learning through questions and investigations.

*Physics in Focus Year 12 (2nd edition)* provides students with a comprehensive study of modern physics that will fully prepare them for exams and any future studies in the area.

**Robert Farr (lead author)**

9780170409131

# AUTHOR AND REVIEWER TEAMS

## Author team

**Rob Farr** has taught Science for over 30 years, 20 of those as Head of Department. He has extensive experience as an HSC marker in Physics and Chemistry, and is a past Supervisor of Marking. Rob has co-authored the very successful *Physics in Focus* series and is a contributing author to the *iScience for NSW* series and the *Nelson Physics for the Australian Curriculum* books. He writes trial HSC examinations for Physics, used in over 120 schools across NSW, and leads workshops for the Broken Bay Diocese Science teachers to help improve their HSC results. Rob maintains his passion for Science teaching through active engagement with bodies such as the CSIRO and the STANSW, as well as sitting on the experienced teacher accreditation assessment panel for the NSW Association of Independent Schools (AIS). He is a NESA Board Curriculum Committee (BCC) member for the Stage 6 Science syllabuses, representing the NSW AIS.

**Dr Kate Wilson** is a senior lecturer and Scientia Education Fellow at UNSW Canberra (ADFA) in the School of Engineering and IT. She has a PhD in physics from Monash University and a GradDipEd (Secondary Teaching) from the University of Canberra. Kate has been first year coordinator in physics at the ANU and Director of the Australian Science Olympiads Physics Program. She is a past member of the Sydney University Physics Education Research Group and has held an Innovative Teaching and Educational Technology Fellowship at UNSW (Kensington). Kate has published more than 30 research papers, including more than 20 in physics education research, and is an author of the first-year university text *Physics* by Serway, Jewett, Wilson and Wilson. Kate also runs a primary school science enrichment program.

**Philip Young** is a former director of the National Space Society in Washington DC, and former President of the National Space Society of Australia. He was Coordinator for the Australian Space Network run by the Australian Centre for Astrobiology, a NASA-affiliated organisation hosted at Macquarie University. For the last decade he has been teaching high school Science, specialising in Physics, and writing textbook materials at both senior and junior levels for the Australian Curriculum in several states. He holds a Bachelor of Science (Physics and Mathematics) from the University of Sydney.

**Dr Darren Goossens** has a PhD in Physics from Monash University and a GradCert in Professional Writing (Editing) from the University of Canberra. He is currently a science writer, editor and educator with Biotext in Canberra. Darren has worked as a research scientist at the Australian Nuclear Science and Technology Organisation, and as an educator and researcher at the University of New South Wales and the Australian National University, where he won several awards for his teaching. He has published over 100 research papers, including work in education research. In 2012 he won the inaugural Sandy Mathieson Medal of the Society of Crystallographers in Australia and New Zealand for distinguished contributions to crystallography.

# ACKNOWLEDGEMENTS

## Author acknowledgements

Rob Farr would like to thank his wife **Elisa** and children **Josh** and **Lauren** for the use of their kitchen table, study and other rooms in the house during the writing of this book. Without their calming support it would not be possible to produce a work such as this.

Kate Wilson would like to thank **David Low** for valuable suggestions and feedback, and her students who have very patiently been guinea pigs for her teaching experiments.

Philip Young would like to thank his wife **Jennie** and children **Sophie** and **Mark** for their forbearance at his distractedness.

Darren Goossens would like to thank his co-authors, particularly **Dr Kate Wilson**, for their guidance and advice.

## Publisher acknowledgements

Eleanor Gregory sincerely thanks **Rob**, **Kate**, **Philip** and **Darren** for their perseverance and dedication in writing this manuscript. Thanks also to **Dr Xiao L. Wu** for co-writing the first edition of *Physics in Focus* and for allowing use of some of his material in this edition. She also thanks **Neil Champion**, **Bill Matchett** and **Megan Mundy** for reviewing the manuscript to ensure that it was of the best quality.

Also thanks to **Bill Matchett**, **Neil Champion**, **Gillian Dewar** and **Philip Young** for authoring NelsonNet material.

# USING *PHYSICS IN FOCUS*

*Physics in Focus* has been purposely crafted to enable you, the student, to achieve maximum understanding and success in this subject. The text has been authored and reviewed by experienced Physics educators, academics and researchers to ensure up-to-date scientific accuracy for users. Each page has been carefully considered to provide you with all the information you need without appearing cluttered or overwhelming. You will find it easy to navigate through each chapter and see connections between chapters through the use of margin notes. Practical investigations have been integrated within the text so you can see the importance of the interconnectedness between the conceptual and practical aspects of Physics.

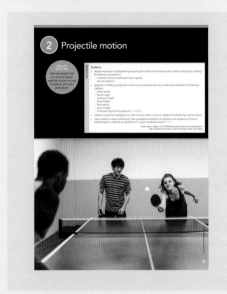

The content is organised under four modules as set out in the NESA Stage 6 Physics syllabus. Each module begins with a **Module opener**.

Each chapter begins with a **Chapter opener**. This presents the learning outcomes from the NESA Stage 6 Physics syllabus that will be covered in the chapter and also gives you the opportunity to monitor your own progress and learning.

To assist comprehension, a number of strategies have been applied to the preparation of our text to improve literacy and understanding. One of these is the use of shorter sentences and paragraphs. This is coupled with clear and concise explanations and real-world examples. New terms are bolded as they are introduced and are consolidated in an end-of-book glossary.

Throughout the text, important ideas, concepts and theories are summarised in **Concept boxes**. This provides repetition and summary for improved assimilation of new ideas.

**Learning across the curriculum content** has been identified by NESA as important learning for all students. This content provides you with the opportunity to develop general capabilities beyond the Physics course, as well as

Critical and creative thinking

links into areas that are important to Australia and beyond. This content has been identified by a margin icon.

Mathematical relationships are presented in context. Step-by-step instructions on how to perform mathematical calculations are shown in the **Worked examples**. The logic behind each step is explained and you can practise these steps by attempting the related problems presented at the end of the worked example.

KEY CONCEPTS

- The trajectory of a particle is a parabolic arc.
- The highest point of the arc is the maximum height. The maximum height can be found from $h = y_0 - \dfrac{u_y^2}{2g}$, where $y_0$ is the launch height and $u_y = u\sin\theta$.
- The time of flight of a projectile is the time between launch and landing, and is given by $t_{flight} = \dfrac{2u_y}{-g}$ when the launch and landing heights are the same. When the launch and landing

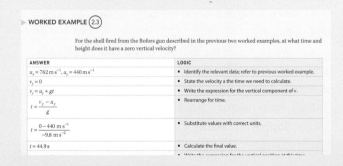

► WORKED EXAMPLE (2.3)

For the shell fired from the Bofors gun described in the previous two worked examples, at what time and height does it have a zero vertical velocity?

| ANSWER | LOGIC |
|---|---|
| $u_x = 762\,\mathrm{m\,s^{-1}}$, $u_y = 440\,\mathrm{m\,s^{-1}}$ | • Identify the relevant data; refer to previous worked example. |
| $v_y = 0$ | • State the velocity a the time we need to calculate. |
| $v_y = u_y + gt$ | • Write the expression for the vertical component of $v$. |
| $t = \dfrac{v_y - u_y}{g}$ | • Rearrange for time. |
| $t = \dfrac{0 - 440\,\mathrm{m\,s^{-1}}}{-9.8\,\mathrm{m\,s^{-2}}}$ | • Substitute values with correct units. |
| $t = 44.9\,\mathrm{s}$ | • Calculate the final value. |

9780170409131

Physics is a science, and you need to be given the opportunity to explore and discover the physical world through practical investigations. **Investigations** introduce and reinforce the Working scientifically skills listed in the NESA Stage 6 Physics syllabus. In some cases, the investigations are open-ended. These provide you with the opportunity to design and carry out your own scientific investigation, either individually or in a group. At times you are prompted to consider ideas for improvement to illustrate that science is constantly undergoing review and improvement. At other times investigations are secondary-sourced, meaning that you need to research the subject using data and information gained by other people. Further information on how to conduct a scientific investigation can be found in the **Working scientifically and depth study** chapter on page 1.

Full understanding of a concept is often constructed from many pieces of information. Due to the sequential nature of a book, this information cannot always be presented together as it is best placed in other chapters. Links between concepts that occur on other pages and chapters are indicated using the **Margin notes**.

You may want to review chapter 4 of *Physics in Focus Year 11* to remind yourself how to work with vectors.

Regular opportunities to recall new terms and review recent concepts are provided as short **Check your understanding** question sets throughout each chapter.

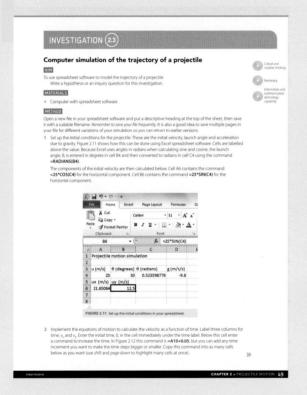

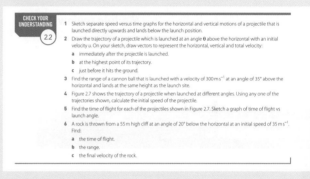

The end-of-chapter review provides:

- a **Summary** of the important concepts that have been covered in the chapter. This will be a valuable tool when you are revising for tests and exams.

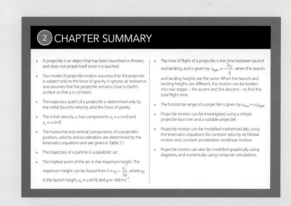

The **Risk assessment** table occurs within the investigations. The table highlights the risks of the investigation and provides suggestions on how to minimise these risks – they are not to be considered comprehensive. Teachers are expected to amend this table in the case of substitutions or in the case of any additional risks. This may mean obtaining and following Safety Data Sheets (SDS) for certain chemicals. All teachers are required to follow the safety guidelines of their specific school and associated government legislation when students are in their care.

- **Chapter review questions** that review understanding and provide opportunities for application and analysis of concepts and how they interrelate.

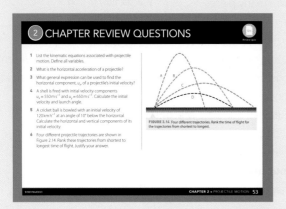

Each module concludes with a **Module review**. This contains short-answer questions that provide you with the opportunity to assimilate content from across the chapters that fall within that module.

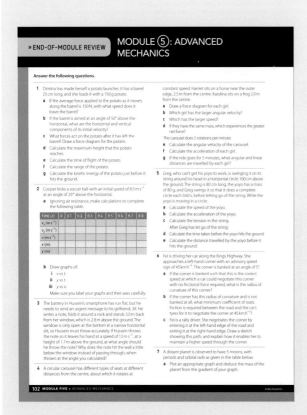

The **Depth study** provides you with the opportunity to pursue a topic of interest from within the course. It enables you to study a topic in more depth and present your findings in a format of your choice. Advice and support to assist you in undertaking your depth study can be found in chapter 1, and there are suggestions for topics provided at the end of each module review. Refer to the NESA Stage 6 Physics syllabus for the full details on scoping and completion of your depth study.

## NelsonNet

NelsonNet is your protected portal to the premium digital resources for Nelson textbooks, located at www.nelsonnet.com.au. Once your registration is complete you will have access to a helpful suite of digital resources for each chapter to further enhance and reinforce learning.

Each chapter will be supplemented with the following digital resources:

- Worksheets to review concepts and to practise applying understanding to new examples

- A review quiz containing 20 auto-correcting multiple-choice questions to review understanding

- Links to websites that contain extra information. These are hotspotted within the ebook and they can also be accessed at http://physicsinfocus12nelsonnet.com.au.

Please note that complimentary access to NelsonNet and the NelsonNetBook is only available to teachers who use the accompanying student textbook as a core educational resource in their classroom. Contact your sales representative for information about access codes and conditions.

# OUTCOME GRID

## Working Scientifically mapping

Content statements from the NESA Stage 6 Physics syllabus are shown in full on the chapter opening pages of the chapters where they are dealt with. A full mapping of chapters and content statements can be found on the NelsonNet Teacher website. Below is a mapping of the outcome statements for Working scientifically across all the chapters of *Physics in Focus Year 12*.

| OUTCOME STATEMENTS STUDENTS: | CHAPTER | | | | | | | | | | | | | | | | |
|---|---|---|---|---|---|---|---|---|---|---|---|---|---|---|---|---|---|
| | 1 | 2 | 3 | 4 | 5 | 6 | 7 | 8 | 9 | 10 | 11 | 12 | 13 | 14 | 15 | 16 | 17 |
| **PH11/12-1** develops and evaluates questions and hypotheses for scientific investigation | ✓ | ✓ | ✓ | ✓ | ✓ | ✓ | ✓ | ✓ | ✓ | ✓ | ✓ | ✓ | | | | ✓ | |
| **PH11/12-2** designs and evaluates investigations in order to obtain primary and secondary data and information | ✓ | | | | | | | ✓ | | ✓ | ✓ | | | ✓ | | | |
| **PH11/12-3** conducts investigations to collect valid and reliable primary and secondary data and information | ✓ | ✓ | ✓ | ✓ | ✓ | ✓ | ✓ | ✓ | ✓ | ✓ | ✓ | ✓ | ✓ | ✓ | ✓ | ✓ | ✓ |
| **PH11/12-4** selects and processes appropriate qualitative and quantitative data and information using a range of appropriate media | ✓ | ✓ | ✓ | ✓ | ✓ | ✓ | ✓ | ✓ | ✓ | ✓ | ✓ | ✓ | ✓ | ✓ | ✓ | ✓ | ✓ |
| **PH11/12-5** analyses and evaluates primary and secondary data and information | ✓ | ✓ | ✓ | ✓ | ✓ | ✓ | ✓ | ✓ | ✓ | ✓ | ✓ | ✓ | ✓ | ✓ | ✓ | ✓ | ✓ |
| **PH11/12-6** solves scientific problems using primary and secondary data, critical thinking skills and scientific processes | ✓ | ✓ | ✓ | ✓ | ✓ | ✓ | ✓ | ✓ | ✓ | ✓ | ✓ | ✓ | ✓ | ✓ | ✓ | ✓ | ✓ |
| **PH11/12-7** communicates scientific understanding using suitable language and terminology for a specific audience or purpose | ✓ | ✓ | ✓ | ✓ | ✓ | ✓ | ✓ | ✓ | ✓ | ✓ | ✓ | ✓ | ✓ | ✓ | ✓ | ✓ | ✓ |

Physics Stage 6 Syllabus © 2017 NSW Education Standards Authority (NESA) for and on behalf of the Crown in right of the State of New South Wales.

# 1 Working scientifically and depth studies

OUTCOMES

## Knowledge and understanding

A student:

- describes and analyses qualitatively and quantitatively circular motion and motion in a gravitational field, in particular, the projectile motion of particles (PH12-12)
- explains and analyses the electric and magnetic interactions due to charged particles and currents and evaluates their effect both qualitatively and quantitatively (PH12-13)
- describes and analyses evidence for the properties of light and evaluates the implications of this evidence for modern theories of physics in the contemporary world (PH12-14)
- explains and analyses the evidence supporting the relationship between astronomical events and the nucleosynthesis of atoms and relates these to the development of the current model of the atom (PH12-15)

## Skills

A student:

- develops and evaluates questions and hypotheses for scientific investigation (PH12-1)
- designs and evaluates investigations in order to obtain primary and secondary data and information (PH12-2)
- conducts investigations to collect valid and reliable primary and secondary data and information (PH12-3)
- selects and processes appropriate qualitative and quantitative data and information using a range of appropriate media (PH12-4)
- analyses and evaluates primary and secondary data and information (PH12-5)
- solves scientific problems using primary and secondary data, critical thinking skills and scientific processes (PH12-6)
- communicates scientific understanding using suitable language and terminology for a specific audience or purpose (PH12-7)

Science is the systematic study, by observation and experiment, of the natural and physical world (Figure 1.1). Science is characterised by a way of thinking and working, and, most fundamentally, by questioning. Physics asks questions about matter and energy, and the interactions between matter and energy. As you have already seen in *Physics in Focus Year 11*, the theories and models of physics, which are based on matter and energy, give us powerful explanatory and predictive power. In this book we will continue to explore the interactions of matter and energy, using the same fundamental concepts of forces, motion, energy and conservation principles. You will extend your knowledge and understanding of mechanical systems and electricity and magnetism, and begin to learn about atomic and subatomic particles and their interactions. By undertaking investigations and depth studies you will develop your scientific problem-solving skills. These skills include questioning and developing hypotheses, planning and carrying out investigations, analysing the results and evaluating both the outcomes and the process itself.

**FIGURE 1.1** Science is the systematic study of the natural and physical world.

## 1.1  Physics knowledge and understanding

The knowledge that has arisen from answering the questions asked by physicists can be broadly categorised into five areas. These are mechanics, waves, thermodynamics, electromagnetism and quantum physics. You have already begun your study of the first four of these areas in *Physics in Focus Year 11*. The first half of this book (chapters 2 to 8) describes more applications of mechanics and electromagnetism. The second half (chapters 9 to 17) introduces quantum mechanics.

While at first glance these may seem to be distinct topics, they are all related, and they are all based on the same central ideas introduced in chapter 1 of *Physics in Focus Year 11*: energy, forces and conservation principles.

Recall from *Physics in Focus Year 11* that mechanics describes the motion and interaction of objects and uses the ideas of force and energy to explain phenomena. We use mechanics to describe and predict the behaviour of small numbers of macroscopic (bigger than atomic-sized) objects. In *Physics in Focus Year 11* the core ideas of motion, forces, energy and conservation principles were introduced. The mathematical models that we use to describe mechanical systems, including kinematics and Newton's laws, were applied to objects accelerating in straight lines.

In chapters 2, 3 and 4 of this book we extend the classical mechanics that was first introduced in *Physics in Focus Year 11*. We apply Newton's laws of motion to two special cases where there is a constant magnitude net force.

The first of these special cases is projectile motion; the motion of an object close to, but not touching, the surface of Earth and subject only to a gravitational force that is constant in both magnitude and direction. The model we present is an approximation – it ignores the interaction of the air with the projectile, so that drag and lift forces are neglected. So this model, *like all models*, is limited, and gives useful predictive and descriptive power only in limited situations. But it is a useful starting model, to which more detail can be added as needed.

The second special case is uniform circular motion. Uniform circular motion occurs when there is a constant magnitude net force that always acts perpendicular to the direction of motion. Such a force is called a centripetal force. The net force may be a single force, like the tension in a string, or a combination of forces, like contact forces and gravity. It causes an object to travel in a circle with constant speed and radius.

In chapter 4 we look at the gravitational force, and see how Newton's laws apply for this fundamental force. We also see an example of circular motion due to the gravitational force; the orbital motion of satellites. The concept of energy is very important here, because potential energy is stored in the gravitational field.

Figure 1.2 shows how these forces and types of motion are related to the fundamental concepts of motion, force, energy and conservation that were introduced in *Physics in Focus Year 11*.

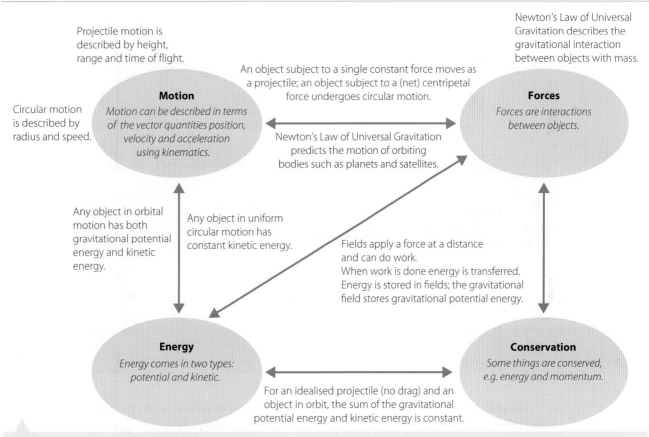

**FIGURE 1.2** Concept map for mechanics (chapters 2 to 4)

Chapters 5 to 8 extend the description of electromagnetism that was begun in chapters 12 to 14 of *Physics in Focus Year 11*. In chapter 5 of this book we will look at how charged particles interact with electric and magnetic fields. As you will see, a charged particle in a uniform electric field behaves exactly as a projectile in a gravitational field, as described in chapter 2. A charged particle in a magnetic field follows a curved path, and may undergo uniform circular motion, as described in chapter 3. The models that were introduced in mechanics apply to situations in electromagnetism also. The underlying mathematical models are the same. This is why it is important not to treat different topics in physics separately – they are really just different examples of the application of the same fundamental ideas. We will also use the idea of potential energy stored in the electric field to understand how work can be done by a field.

In chapters 5 to 8 we look at useful applications of the interaction between charged particles and magnetic fields; motors, generators and transformers. The concept of energy is central to our understanding of these devices. The core ideas are again that energy is stored in fields, in this case the magnetic field, and energy can be transformed from one form into another but it is always conserved.

Figure 1.3 summarises the main ideas introduced in chapters 5 to 8 and how they relate to the core ideas of force, energy and conservation.

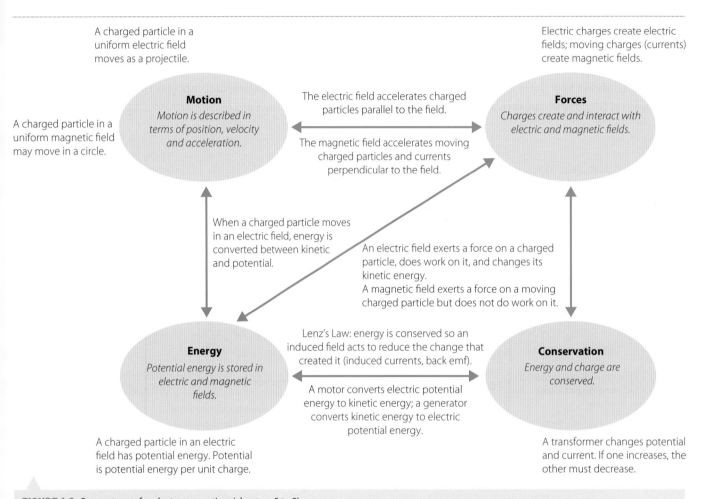

**FIGURE 1.3** Concept map for electromagnetism (chapters 5 to 8)

Chapters 9 to 12 describe light. Two different models are presented: the electromagnetic wave model and the particle (photon) model. Light is a pair of coupled, oscillating electric and magnetic fields that travel together. Energy is stored in the fields, and as one field gets smaller, the other gets bigger, so that the energy of the two fields is constant. Light travels as a wave, and we can describe it in terms of wave properties including frequency and wavelength.

But when light interacts with matter it also behaves like a particle. When light is emitted or absorbed by matter it is always in discrete amounts, with a particular energy. The quantisation of the energy of light into discrete amounts called photons is the beginning of your study of quantum mechanics and particle physics. Because light is a pair of coupled, oscillating fields, it does not need a medium to travel through – the fields are both the wave and its medium. As a result, the speed of light in vacuum is always the same, regardless of the motion of an observer. This is the basis of special relativity, which is described in chapter 12.

Figure 1.4 shows how the ideas in chapters 9 to 12 are related to each other, and to the core ideas of mechanics and electromagnetism. You will see that a fifth core idea has been added – the idea of quantisation.

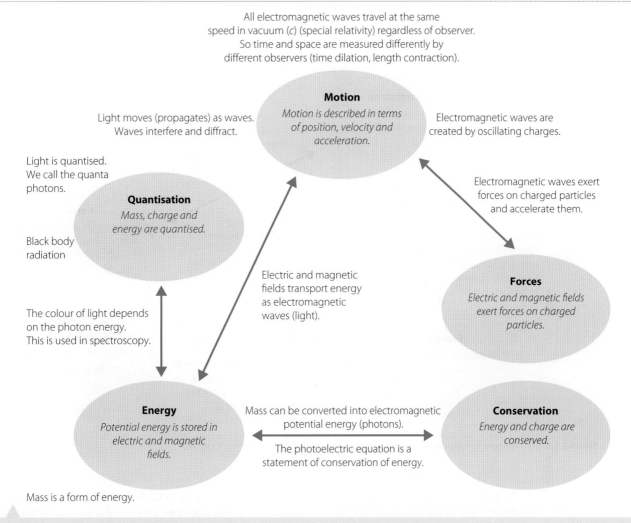

All electromagnetic waves travel at the same speed in vacuum (*c*) (special relativity) regardless of observer. So time and space are measured differently by different observers (time dilation, length contraction).

Light moves (propagates) as waves. Waves interfere and diffract.

**Motion**
*Motion is described in terms of position, velocity and acceleration.*

Electromagnetic waves are created by oscillating charges.

Light is quantised. We call the quanta photons.

**Quantisation**
*Mass, charge and energy are quantised.*

Electromagnetic waves exert forces on charged particles and accelerate them.

Black body radiation

Electric and magnetic fields transport energy as electromagnetic waves (light).

**Forces**
*Electric and magnetic fields exert forces on charged particles.*

The colour of light depends on the photon energy. This is used in spectroscopy.

**Energy**
*Potential energy is stored in electric and magnetic fields.*

Mass can be converted into electromagnetic potential energy (photons).

The photoelectric equation is a statement of conservation of energy.

**Conservation**
*Energy and charge are conserved.*

Mass is a form of energy.

**FIGURE 1.4** Concept map for light (chapters 9 to 12). Note that a new core idea is introduced – quantisation.

The final section of this book, chapters 13 to 17, deals primarily with particle physics. Particle physics describes the fundamental interactions of matter and energy. It is based on quantum mechanics, and so quantisation is central to particle physics.

Conservation principles are particularly important in particle physics, and we introduce a new type of potential energy – the potential energy associated with mass. This allows us to model nuclear and other subatomic reactions in which the conversion of mass-energy to or from other forms is important.

The application of particle physics to cosmology is explored in chapter 13. While the structure of the Universe may seem a very different topic from the interactions of fundamental particles, in fact they are intimately related, as is every part of physics with every other part. Figure 1.5 summarises the main ideas presented in chapters 13 to 17.

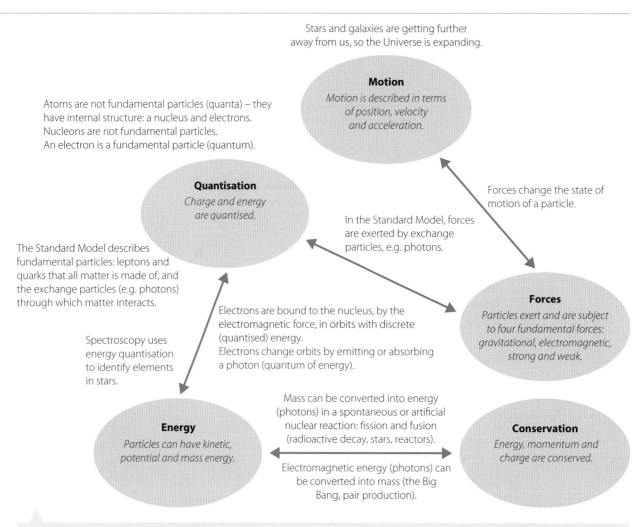

Stars and galaxies are getting further away from us, so the Universe is expanding.

**Motion**
*Motion is described in terms of position, velocity and acceleration.*

Atoms are not fundamental particles (quanta) – they have internal structure: a nucleus and electrons.
Nucleons are not fundamental particles.
An electron is a fundamental particle (quantum).

**Quantisation**
*Charge and energy are quantised.*

Forces change the state of motion of a particle.

In the Standard Model, forces are exerted by exchange particles, e.g. photons.

The Standard Model describes fundamental particles: leptons and quarks that all matter is made of, and the exchange particles (e.g. photons) through which matter interacts.

**Forces**
*Particles exert and are subject to four fundamental forces: gravitational, electromagnetic, strong and weak.*

Spectroscopy uses energy quantisation to identify elements in stars.

Electrons are bound to the nucleus, by the electromagnetic force, in orbits with discrete (quantised) energy.
Electrons change orbits by emitting or absorbing a photon (quantum of energy).

Mass can be converted into energy (photons) in a spontaneous or artificial nuclear reaction: fission and fusion (radioactive decay, stars, reactors).

**Energy**
*Particles can have kinetic, potential and mass energy.*

**Conservation**
*Energy, momentum and charge are conserved.*

Electromagnetic energy (photons) can be converted into mass (the Big Bang, pair production).

**FIGURE 1.5** Concept map for particle physics (from the Universe to the atom) (chapters 13 to 17).

As you learn more of the content knowledge of physics, you need to create your own mental models to help you understand it. Concept maps are a useful way of representing your mental models. They help to remind you that physics is not simply a collection of facts and formulae. Every idea in physics is connected to other ideas, and always to one of the fundamental ideas of force, energy and conservation. *All of the theories, laws and equations of physics are fundamentally interconnected*, none of them stands alone.

These concept maps (Figures 1.2 to 1.5) summarise some of our knowledge and understanding of physics. This knowledge and understanding was arrived at by physicists who asked questions and then tried to answer those questions by conducting investigations, making measurements and evaluating their results. Working scientifically is more characteristic of, and more important to, the study of physics than any particular collection of content knowledge. You will practise working scientifically, working like a physicist, when you undertake investigations and depth studies.

# 1.2 Solving scientific problems: depth studies

You have already had some experience of performing depth studies. You will further hone the skills that you developed by doing more depth studies, and develop new skills. In particular, your prior experience will help you to evaluate your own work and that of others.

Depth studies provide you with an opportunity to:

» use the research methods that scientists use

» analyse and evaluate works for scientific relevance and validity

» broaden your range of reading in a field of interest

» extend your depth of thinking and understanding

» ask questions and investigate areas that do not have definite answers

» investigate contentious issues and use critical thinking skills to evaluate views expressed in a variety of sources

» use inquiry-based learning to develop your creative thinking.

When performing a depth study you pose questions, develop a **hypothesis** to answer your questions, and then seek evidence to support or disprove your hypotheses. The evidence may come from the existing scientific literature, or from your own experiments. You need to analyse data to determine whether your hypotheses are supported. Analysing data usually requires you to represent it in some way, often mathematically or graphically. Finally, as scientists do, you need to communicate your findings to others. There are many ways that you can do this, and you need to choose the method most appropriate to the audience you wish to communicate to.

Depth studies can take different forms, and over the year you may undertake several different types of depth study.

## Types of depth study

There are two broad types of depth study. In first-hand practical investigations, you design and perform experiments or make observations to gather primary data. Investigations based on secondary sources require you to research and evaluate information and data collected by other people.

First-hand investigations include experiments in a laboratory, field work at home, school or elsewhere, or the creation and testing of a model or device.

Secondary-sourced depth studies may include undertaking a literature review, investigating emerging technologies, evaluating a hypothesis, or developing and evaluating an evidence-based argument.

Depth studies may be presented in different forms, some of which include:

» written texts (reports, summaries, essays)

» visual presentations (diagrams, flow charts, posters, portfolios)

» multimedia presentations

» physical models

» a blend of the above.

All depth studies will involve the analysis of data, either primary data that you collect or other people's research. Looking for patterns and trends in data will involve analysing and constructing graphs, tables, flow charts and diagrams.

All depth studies involve evaluation – evaluation of the hypothesis or question posed, evaluation of the method as **valid** and appropriate, and evaluation the data as **reliable** and valid.

## Stages in a depth study

The summary below outlines four main stages of conducting a depth study, as well as the *Working scientifically* skills that you will need to develop and apply at each stage.

1 Initiating and planning involves:

*Questioning and predicting* (PH12-1): formulating and evaluating questions or hypotheses

*Planning* (PH12-2): researching background information; assessing risks and considering ethical issues; evaluating methods and secondary sources to plan valid experiments

2 Implementation and recording

*Conducting investigations* (PH12-3): safely carrying out valid investigations using appropriate technology and measuring instruments

*Processing data and information* (PH12-4): collecting, organising, recording and processing data/information.

3 Analysing and interpreting

*Analysing data* (PH12-5): looking for trends or patterns, finding mathematical relationships and evaluating data

*Problem-solving* (PH12-6): drawing and justifying conclusions, testing hypotheses and answering questions, evaluating the study.

4 *Communicating* (PH12-7): using appropriate language, visualisations and technologies to communicate scientific ideas, procedures and results.

Critical and creative thinking

## Posing questions and formulating hypotheses

**FIGURE 1.6** Brainstorm as many ideas as you can in your group.

The first step to beginning any investigation or depth study is deciding on a question.

A good way to start is by 'brainstorming' for ideas (Figure 1.6). This works whether you are working on your own or in a group. Write down as many ideas as you can think of. Don't be critical at this stage. Get everyone involved to contribute, and write every idea down.

After you have run out of ideas, it is time to start being critical. You need to evaluate each of the ideas.

Decide which questions or ideas are the most interesting. Think about which of these it is actually possible to investigate given the time and resources available. Make a shortlist of questions, but keep the long list too for the moment. Once you have your shortlist it is time to start refining and evaluating your ideas.

Information and communication technology capability

Your depth study will be based on one of the areas described in Figures 1.2 to 1.5, which are discussed in the remaining chapters of this book.

Because the purpose of a depth study is to *extend* your knowledge, while at the same time building your skills at working scientifically, you will need to go beyond the basic syllabus content. The next step is therefore to find out what is already known about the ideas on your list. You need to do a literature review. If your depth study is a secondary-sourced investigation, then the literature review may be the investigation itself. A formal, written literature review includes the information you have found, and complete references to the sources of information. It also includes interpretation and critique of what you have read. This is particularly important for a secondary-sourced investigation.

Literature reviews are important because they help you to increase your breadth of knowledge and learn from others, and stimulate new ideas. They are necessary to identify gaps in current knowledge that you may wish to research and to identify methods that you could use. You may also find that there are a variety of views (sometimes opposing views) in an area of research. A literature review will help you evaluate your questions and hypotheses by showing you whether they have already been posed or investigated by someone else and, if so, what the results of other investigations to answer them have been.

## Your literature review

 Literacy

A literature review is a search and evaluation of available literature in a particular subject area. It has a particular focus that is defined by your research question or hypothesis.

The process of conducting a literature review involves researching, analysing and evaluating the literature. It is *not* merely a descriptive list of the information gathered on a topic, or a summary of one piece of literature after another. It outlines any opposing points of view in the research and also expresses the *writer's perspective* of the strengths and weaknesses of the research being reviewed. A literature review brings together results of different studies, pointing out areas where researchers or studies agree, where they disagree, and where major questions remain. By identifying gaps in research, literature reviews often indicate directions for future research.

Your literature review will give you an idea of past findings and procedures, techniques and research designs that have already been used. This will help you to evaluate which methods are worth copying, which need modifying and which to avoid (those that have been inconclusive or invalid). You may plan your investigation to target a gap in research or try to replicate an investigation to test or validate it.

The length of your literature review will depend on its purpose. If it is a depth study in itself, it will need to be more detailed and draw conclusions about the research and evaluate existing studies. If it is used as an introduction to inform your own research, it will be shorter and more focused.

To write a literature review you first need to define the topic. It may help to formulate a literature review question. Then write a list of key words that will help you search for information.

To find articles you can use library catalogues, databases and the internet. Refine your search technique by using specific words that narrow your search. Record search words that are successful and, if necessary, modify your search strategy.

When you write your literature review for your report it should have an introduction that defines the topic and gives your specific focus. For a lengthy secondary-sourced depth study it may also explain the structure of the review.

The main body of the review will then group the literature according to common themes and provide an explanation of the relationship between the research question and the literature reviewed. It should proceed from the general, wider view of the research to the specific area you are targeting. Include information about the usefulness, recency and major authors or sources of the literature. This information is necessary to support your evaluation of the sources and of your results.

The literature review concludes by summarising the major contributions of the literature and explaining the link between your investigation and the literature reviewed. It should also evaluate the current state of knowledge in the area and point out major flaws or gaps in research if appropriate.

**Literature review**
More information about literature reviews, and how to complete them.

## Evaluating sources

Always be critical of what you read. Be wary of pseudoscience and any material that has not been peer reviewed. Apply the CRAAP (currency, relevance, authority, accuracy, purpose) test to websites that you find. The most reliable sites are from educational institutions, particularly universities, and government and scientific organisations such as the CSIRO and NASA, and professional organisations such as the Australian Institute of Physics and international equivalents. You can narrow your search to particular types of sites by including in your search terms "site:edu" or "site:gov" so that you only find sites from educational or government sources.

**FIGURE 1.7** Start researching your topic and make sure you keep a record of all your references. Good record-keeping is important in scientific research, and it begins at this stage of the investigation.

Make sure you keep a record of the information that you find as well as the sources, so you can reference them correctly later (Figure 1.7). You should start a logbook at this stage. You can write in references, or attach printouts to your logbook. This can save you a lot of time later on! Your logbook may be hardcopy or electronic, but either way, *begin keeping it now*.

Finally, talk to your teacher about your ideas. They may be able to suggest sources of information. They will also be able to tell you whether your ideas are likely to be possible given the equipment available. They may have had students with similar ideas in the past and will make suggestions.

After you have researched and evaluated your questions and ideas, you will hopefully be able to narrow the shortlist down to the one question that you want to tackle. If none of the questions or ideas looks possible (or still interesting), then you need to go back to the long list.

## Proposing a research question or hypothesis

You need to evaluate your question or hypothesis. An appropriate research question is one that can be answered by performing experiments or making observations. A good hypothesis is a prediction of the results of an experiment that can be tested by performing experiments or making observations. It should be able to be clearly disproved and, ideally, be clearly supported. The predictions of your hypothesis should be different enough from those of other hypotheses that experiments can be performed to distinguish between them.

**FIGURE 1.8** You need to frame your research question carefully. These students are investigating the launch angle at which their water rocket will achieve maximum range.

You need to frame a research question carefully. A good research question should define the investigation, set boundaries and provide some direction to the investigation. It needs to be specific enough that it guides the design of the investigation (Figure 1.8). A specific question rather than a vague one will make the design of your investigation much easier. Asking 'what volume of water gives the maximum height for a water rocket?' tells you what you will be varying and what you will be measuring. It also gives a criterion for judging whether you have answered the question. Use these criteria to evaluate your proposed research question, and reframe it if needed.

Asking 'How can we make a water rocket fly the best?' is not a good question. This question does not say what will be varied, nor does it tell you when you have answered the question. 'Best' is a vague term. What you mean by 'best' may not be what someone else means.

A hypothesis is a tentative explanation or prediction, such as 'The height attained by a water rocket will increase with the amount of water contained in the rocket'. Your hypothesis should give a prediction that you can test, ideally quantitatively. A hypothesis needs to be falsifiable, that is, able to be proved false. Remember that for a hypothesis or theory to be considered scientific, it must be able to be disproved (but no theory can ever be proved to be true).

A hypothesis is usually based on some existing model or theory. It is a prediction of what will happen in a specific situation based on that model. For example, kinematics describes the trajectory of projectiles. A hypothesis based on the kinematics model predicts the range of a specific projectile launched at a given angle and speed.

Often in physics there is more than one model that can be used to generate a hypothesis, so there may be multiple hypotheses to explain a given phenomenon. A good hypothesis for an investigation is one that gives a prediction sufficiently different from those of other hypotheses. The prediction is sufficiently different if the experiment can disprove one or more hypotheses, and ideally provide support for only one.

You can use these criteria to evaluate your hypothesis: is it based on an existing model, is it testable, and will any test distinguish between your hypothesis and competing hypotheses.

A good research question or hypothesis identifies the variables that will be investigated. Usually you will have one dependent variable and one independent variable. For a depth study you may have two or more independent variables that you control.

If your experiments agree with predictions based on your hypothesis, then you can claim that they support your hypothesis. This *increases your confidence* in your model, but *it does not prove that it is true*. Hence an aim for an experiment should *never* start 'To prove ...', because it is not possible to actually prove a hypothesis, only to disprove it.

If your experimental results disagree with your hypothesis, then you may have disproved it. This is not a bad thing! Often the most interesting discoveries in science start when a hypothesis based on an existing model is disproved, because this raises more questions.

Even if you evaluate your question or hypothesis and it meets these criteria, do not be surprised if you change or modify the question during the course of your investigation or depth study. In scientific research, the question you set out to answer is often only a starting point for more questions.

## 1.3 Planning your depth study

Critical and creative thinking

There are many things to consider when planning an investigation. You need to think about how much time you will have, what space and equipment you will need and where you will go if you want to make measurements or observations outside. If you are doing a secondary-sourced investigation or some other type of depth study, such as a creative work, you still need to plan ahead to make sure you have the resources you need.

You may be working in a group or on your own. Most scientists work in groups. If you can choose who you work with, think about it carefully. It is not always best to work with friends. Think about working with people who have skills that are different from your own.

Having a plan allows you to ensure that you collect the data, whether primary- or secondary-sourced, that you need to test your hypothesis. The longer the investigation, the more important it is that you have a clear plan. There are several things to consider, as listed in Table 1.1.

**TABLE 1.1** Factors to consider when planning your depth study

| PRIMARY-SOURCED INVESTIGATION | SECONDARY-SOURCED INVESTIGATION |
|---|---|
| What data will you need to collect? | What information will you need to gather? |
| What materials and equipment will you need? | What sources will you use? |
| When and where will you collect the data? | When and where will you gather the information? |
| If you are working in a group, what tasks are assigned to which people? | If you are working in a group, what tasks are assigned to which people? |
| Who will collect the data? | Who will collect what information? |
| Who will be responsible for record-keeping? | How will record-keeping be done to avoid plagiarism? |
| How will the data be analysed? | How will the information be analysed? |
| How will sources be referenced? | How will sources be referenced? |

The most common problem that students have is time management. It is important to plan to have enough time to perform the experiments, including repeat measurements, to analyse them and to report on them.

A good plan will help you keep on track. Your teacher may ask you to hand in a plan of your depth study before you begin the implementation stage. Table 1.2 gives an idea of the types of things you should think about.

**TABLE 1.2** Depth study plan

| 1 INTRODUCTION | |
|---|---|
| Title<br>What? | Choose a title for your depth study. |
| Rationale<br>Why? | Explain why you have chosen this area of research.<br>Describe what you are hoping to achieve through this investigation. Include any applications. |
| Type of depth study<br>Which? | State the type of depth study you intend conducting (e.g. literature review / practical investigation).<br>Where applicable, describe any theoretical models (e.g. kinematics) that you will use. |

| 2 TIMELINE | |
|---|---|
| **Action and time frame – When?** | **Working scientifically skills – How?** |
| 1 Initiating and planning<br>When? (e.g. week 1 and 2) | *Questioning and predicting:* Formulate questions and/or a hypothesis.<br>*Planning:* Wide reading – research background information; assess risks and ethical issues; plan methods; design experiments. |
| 2 Implementation and recording<br>When? | *Conducting investigations:* Carry out experiments safely; make observations and/or measurements; use appropriate technology and measuring instruments.<br>*Process and record data and information:* Collect, organise, record and process information and/or data *as you go.* |
| 4 Analysing and interpreting<br>When? | *Analyse data and information:* Begin looking for trends or patterns or mathematical relationships.<br>*Problem-solve:* Evaluate the adequacy of data (relevance, accuracy, validity and reliability) from primary and/or secondary sources; answer your research question; draw and justify conclusions. |

9780170409131

| 5 | Communicating When? | *Present your depth study:* Write the report or other presentation, using appropriate language, visualisations, and technologies. |
| | Final presentation | Due date: *Allow time for proofreading and editing.* |

| 3 | DATA COLLECTION | |
|---|---|---|
| **Variables**<br>What will you measure and what will you hold constant?<br><br>Identify dependent and independent variables. | | **Measurements and uncertainties**<br>How will you make measurements?<br><br>What equipment will you need?<br><br>How will you minimise uncertainties? |

| 4 | DATA ANALYSIS AND PROBLEM-SOLVING | |
|---|---|---|
| **Data analysis**<br>What method(s) will you use to analyse the data and how will you represent the trends and patterns? | | **Conclusions**<br>How will you judge whether the experiment was valid?<br><br>How will your data allow you to test your hypothesis or answer your question? |

Keep a record of your planning. This should go in your logbook. Recording what you plan to do, and why, will help you stay focused. This is particularly important for a depth study. If you are working in a group, then keep a record of what each person agrees to do. But remember, the plan may need to be adjusted as you go.

## Designing your depth study

When designing your depth study or investigation you should be aiming for reliable and valid measurements with good accuracy and precision. This is also true if you are doing a secondary-sourced investigation – in this case you should be trying to find resources that have these characteristics. Remember that when you have completed your depth study you will need to evaluate it, so you should consider this at the design stage.

Some good questions to ask to evaluate the reliability, validity, accuracy and precision of your design are given in the table below.

**TABLE 1.3** Evaluating a depth study for reliability, accuracy and validity in investigations

| | PRIMARY INFORMATION AND DATA | SECONDARY INFORMATION AND DATA |
|---|---|---|
| **Reliability** | Have I tested with repetition? | How consistent is the information with other reputable sources?<br><br>Are the data presented based on repeatable processes? |
| **Accuracy and precision** | Have I designed my experiments to minimise uncertainties?<br><br>Have I used repeat measurements to estimate random errors?<br><br>Have I used the best measuring equipment available, and used it correctly? | Is this information similar to information presented in peer-reviewed scientific journals?<br><br>Are the data given with uncertainties, and are these uncertainties small compared to the measured values? |
| **Validity** | Does my experiment actually test the hypothesis that I want it to?<br><br>Have all variables apart from those being tested been kept constant? | Do the findings relate to the hypothesis or problem?<br><br>Are the findings accurate and the sources reliable? |

## Selecting equipment

A well-framed question or hypothesis will help you choose the equipment that you need. For example, if your hypothesis predicts a temperature change of 0.5°C, then you will need a thermometer that can measure to at least this precision (precision and accuracy are discussed later). You also need to know how to use the equipment correctly. Always ask if you are unsure. Reading the user manual is also a good idea. It will usually specify the precision of the device, and let you know of any potential safety risks.

**Minimising uncertainty**

Find out how Foucault measured the speed of light so precisely.

Ethical understanding

Personal and social capability

You need to think about how you can minimise uncertainties. Minimising uncertainty is not just about using the most precise equipment you can find, it is also about clever experimental technique. Very precise measurements are possible using simple equipment. For example, in 1862 Léon Foucault measured the speed of light with an uncertainty of 0.2%, without a computer, data logger or even a digital stopwatch. Remember that it is a poor workman who blames his tools!

## Working safely: risk assessment

You may be required to complete a risk assessment before you begin your depth study. You need to think about three factors.

1   What are the possible risks to you, to other people, to the environment or property?
2   How likely is it that there will be an injury or damage?
3   How serious are the consequences likely to be if there is an injury or damage to property or environment?

A 'risk matrix', such as that shown in Table 1.4, can be used to assess the severity of a risk associated with an investigation. The consequences are listed across the top from negligible to catastrophic. Negligible may be getting clothes dirty. Marginal might be a bruise from falling off a bike, or a broken branch in a tree. Severe could be a more substantial injury or a broken window. Catastrophic would be a death or the release of a toxin into the environment. You need to ensure that your investigation is low risk.

**TABLE 1.4** Risk matrix for assessing for severity of risk

| CONSEQUENCES → LIKELIHOOD ↓ | NEGLIGIBLE | MARGINAL | SEVERE | CATASTROPHIC |
| --- | --- | --- | --- | --- |
| Rare | Low risk | Low risk | Moderate risk | High risk |
| Unlikely | Low risk | Low risk | High risk | Extreme risk |
| Possible | Low risk | Moderate risk | Extreme risk | Extreme risk |
| Likely | Moderate risk | High risk | Extreme risk | Extreme risk |
| Certain | Moderate risk | High risk | Extreme risk | Extreme risk |

Once you have considered what the possible risks are, you need to think about what you will do about them. What you will do to minimise them, and what you will do to deal with the consequences if something does happen? You can use a risk assessment table like Table 1.5.

**TABLE 1.5** Example risk assessment table

| WHAT ARE THE RISKS IN DOING THIS EXPERIMENT? | HOW CAN YOU MANAGE THESE RISKS TO STAY SAFE? |
| --- | --- |
| Water from the rocket may be spilled and someone might slip. | Clean up all spills immediately. |

**Stay safe online!**

Think about what you could do to keep yourself safe online.

Consider where you will perform your experiments or observations. Will you need to consider the convenience or safety of others? Talk to your teacher about what space is available.

In a secondary-sourced investigation, take precautions with cyber safety and remember to keep your personal information private.

**KEY CONCEPTS**
- In primary-sourced investigations you collect and analyse your own data. In secondary-sourced investigations you analyse someone else's data.
- Investigations need to be planned carefully so that they answer your research question. You also need to consider safety and possible environmental impacts of your investigation.

# Implementation and collecting data

Scientists keep a logbook for each project that they work on. The logbook is the primary source of information when a scientist writes up their work for publication. A logbook is a legal document for a working scientist. If the work is called into question, then the logbook acts as important evidence. Logbooks are even provided as evidence in court cases sometimes; for example, in patent disputes, claims of academic misconduct or scientific fraud. The logbook is necessary for the evaluation of the investigation, or for any subsequent investigation of the research.

Every entry in a scientist's logbook is dated, records are kept in indelible form (pen, not pencil), and entries may even be signed. Scientists' logbooks include details of experiments, ideas and analysis. They frequently include printouts of data, photocopies of relevant information, photos and other items.

## Your logbook records

You need to begin keeping a logbook as soon as you begin planning your depth study. Your logbook may be paper or electronic. Either way, your logbook is a detailed record of *what you did* and *what you found out* during your investigation. Make an entry in the logbook *every* time you work on your depth study (Figure 1.9). At the start of each session you should record the date and the names of all the people with whom you are working at the time. A logbook is particularly vital for primary-sourced investigations, but is also important for secondary-sourced investigations.

*Always write down what you do as you do it.* It is easy to forget what you did if you do not write it down immediately.

Record the results of all measurements immediately and directly into your logbook, in pen if using hardcopy. Never record data onto bits of scrap paper instead of your logbook!

**FIGURE 1.9** Make sure you keep an accurate record of what you do *as you do it.*

Results must be recorded in indelible form. Never write your results in pencil or use white-out. If you want to cross something out, just put a line through it and make a note explaining why it was crossed out. If you are using an electronic logbook then do not delete data or working; label it appropriately and keep it.

A good logbook contains:

- notes taken during the planning of your investigation
- a record of when, where and how you carried out each experiment
- diagrams showing the experimental setups, circuit diagrams, etc.
- all your raw results
- all your derived results, analysis and graphs
- all the ideas you had while planning, carrying out experiments and analysing data
- printouts, file names and locations of any data not recorded directly in the logbook.

It is not a neat record, but it is a *complete* record. It needs to be complete to enable you to evaluate your work thoroughly.

## Data collection

If you are conducting a secondary-sourced investigation then your literature review will be the basis of your investigation, and your data will come from the existing literature. Remember that a literature review is not simply a summary of what you have read; you need to add meaning and evaluate both the sources and the state of knowledge in the field overall.

Adding meaning may come from comparing and contrasting competing models and constructing an argument, or by analysing and presenting secondary-sourced data. When using secondary sources, remember to make comparisons between data and claims in a number of reputable sources, including science texts, scientific journals and reputable internet sites, and to reference these appropriately.

When evaluating sources, apply the CRAAP test, and also consider aspects such as limitations of the experiments, reliability (have others repeated the experiments?) and ability of the experiments to distinguish between competing hypotheses. When evaluating the state of knowledge in the area, ask what existing models there are, what hypotheses based on these models have already been tested, and what the unanswered questions are.

If you are doing a primary-sourced investigation then you will be performing measurements to gather data yourself. You can collect data by performing experiments or making observations in the field. You will gain practice at making measurements if you do some of the investigations in the following chapters. These investigations can form a basis for your depth study.

## Data collection for primary-sourced investigations

When doing experiments you need to decide which variables you will measure and which variables you will control. Typically in an experiment we have an independent variable, which we control and vary, and a dependent variable, which is what we measure. We assume that the dependent variable is in some way dependent on the independent variable. There may also be **controlled variables**, which are kept constant so that they do not interfere with the results.

Whenever possible you should make repeat measurements. This allows you to check that your measurements are reliable. Your results are reliable if repeat measurements give the same results within experimental uncertainty. If a result is not **reproducible**, it is not a reliable result. It may be that a variable other than the ones you are controlling is affecting its value. If this is the case, you need to determine what this other variable is, and control it if possible. When you evaluate your depth study, reliability is an important factor.

You also need to consider how many data points to collect. In general, it is better to have more data than less. However, you will have limited time to collect your data, and you need to allow time for analysis and communicating your results. A minimum of 6 to 10 data points is usually required to establish a relationship between variables if the relationship is linear. A linear relationship is one which, if you plot one variable against the other, gives a straight line. If you think the relationship might not be linear then take more data points, and think carefully about how they will be spaced. You should try to collect more data in the range where you expect the dependent variable to be changing more quickly. For example, if you are measuring the temperature of a hot object as it cools, then you should collect more data early when cooling is more rapid.

Draw a table to record data in. Label the columns in the table with the name and units of the variables. If you know that the uncertainty in all your measurements is the same, then you can record this at the top of the column as well. Otherwise, each data entry should have its uncertainty recorded in the cell with it.

It is a good idea to start your analysis while you are collecting your data. If you spot an outlier while you are still making measurements then you have the opportunity to repeat that measurement. If you made a mistake, then put a line through the mistake, write in the new data, and make a comment in your logbook.

If you have not made a mistake, then plotting and analysing as you go allows you to spot something interesting early on. You then have a choice between revising your hypothesis or question to follow this new discovery, or continuing with your plan. Many investigations start with one question and end up answering a completely different one. These are often the most fun, because they involve something new and exciting.

## Accuracy and precision of measurements

When making measurements, your aim is to be as precise and accurate as possible.

An accurate measurement result is one that represents the 'true value' of the measured quantity as closely as possible. When we take repeated measurements we assume that the mean of the measurements will be close to the 'true value' of the variable. However, this may not always be the case. For example, if you have ever been a passenger in a car with an analogue speedometer and tried to read it, your reading will be consistently different from what the driver reads. This is because of parallax error. The needle sits above the scale, and when viewed from the side it does not line up correctly with the true speed. Beware of parallax error with any equipment that uses a needle indicator. This is an example of a systematic error, in which measurements differ from the true value by a consistent amount. Note that often we do not know what the 'true value' is. Figure 1.10 compares accurate measurements (a) with inaccurate measurements due to a systematic error (b).

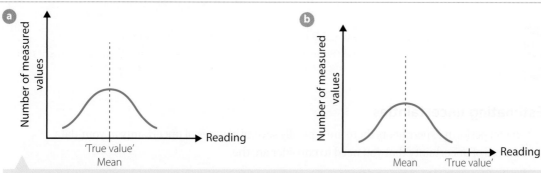

**FIGURE 1.10** In a plot of number of measured values versus reading, results may: **a** be accurate and cluster close to the 'true value', or **b** cluster about some other value.

Precision is a measure of the variability of the measurements, so it affects the spread of the repeated measurements about the mean value. The smaller the spread, the greater the precision. This is shown in Figure 1.11. Figure 1.11a shows precise measurements, Figure 1.11b shows less precise measurements. Note that both data sets are centred about the same average so they have the same accuracy.

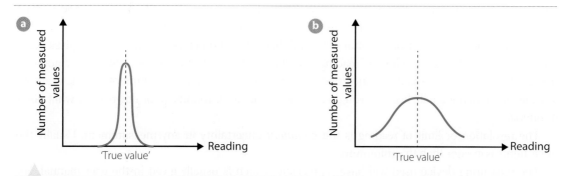

**FIGURE 1.11** In a plot of number of measured values versus reading, results may: **a** be precise and have a small spread about the mean; **b** have a large spread about the mean.

Figure 1.12 shows the difference between accuracy and precision for archery. Accuracy is how close to the centre the arrows hit, and precision is how closely the arrows are grouped.

All of these factors – reliability, precision and accuracy – need to be considered when planning and evaluating your investigation.

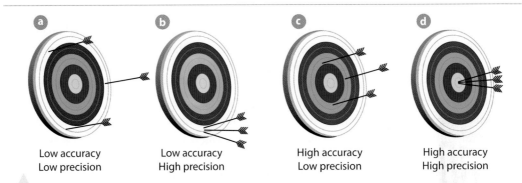

| Low accuracy<br>Low precision | Low accuracy<br>High precision | High accuracy<br>Low precision | High accuracy<br>High precision |

**FIGURE 1.12** On a target, accuracy is determined by how close to the centre (bullseye) your arrows land. Precision is how closely you can group your arrows.

**KEY CONCEPTS**

- Precision is a measure of the variability or spread in repeat measurements.
- Accuracy is a measure of the difference between the average value of repeat measurements and the 'true value' of the thing being measured. Often we do not know the true value.

## Estimating uncertainties

When you perform experiments there are typically several sources of uncertainty in your data.

Sources of uncertainty that you need to consider are the:

- limit of reading of measuring devices
- precision of measuring devices
- variation of the measurand (the variable being measured).

For *all* devices there is an uncertainty due to the limit of reading of the device. The limit of reading is different for analogue and digital devices.

Analogue devices have continuous scales and include swinging needle multimeters and liquid-in-glass thermometers. For an analogue device, the limit of reading, sometimes called the resolution, is half the smallest division on the scale. We take it as half the smallest division because you will generally be able to see which division mark the indicator (needle, fluid level, etc.), is closest to. So, for a liquid-in-glass thermometer with a scale marked in degrees Celsius, the limit of reading is 0.5°C.

Digital devices such as digital multimeters and digital thermometers have a scale that gives you a number. A digital device has a limit of reading uncertainty of a whole division. So a digital thermometer that reads to whole degrees has an uncertainty of 1°C. For a digital device the limit of reading is always a whole division, not a half, because you do not know whether it rounds up or down, or at what point it rounds.

The resolution or limit of reading is the *minimum* uncertainty in any measurement. Usually the uncertainty is greater than this minimum.

The measuring device used will have a precision, which is usually given in the user manual. For example, a multimeter such as that shown in Figure 1.13 may have a precision of 1.5% on a voltage scale. This means if you measure a potential difference of 12.55 V on this scale, the uncertainty due to the precision of the meter is $0.015 \times 12.55\,V = 0.19\,V$. This is greater than the limit of reading uncertainty, which is 0.01 V in this case. The precision is a measure of repeatability. If you measured a potential difference of 12.55 V on this multimeter, you could expect the reading to vary by as much as 0.19 V, even if the potential difference stayed the same.

9780170409131

Many students think that digital devices are more precise than analogue devices. This is often not the case. A digital device may be easier for you to read, but *this does not mean it is more precise.* The uncertainty due to the limited precision of the device is generally greater than the limit of reading.

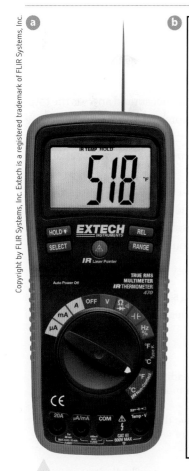

| Function | Range | Resolution | Accuracy | |
|---|---|---|---|---|
| DC voltage | 400 mV | 0.1 mV | ±(0.3% reading + 2 digits) | |
| | 4 V | 0.001 V | | |
| | 40 V | 0.01 V | ±(0.5% reading + 2 digits) | |
| | 400 V | 0.1 V | | |
| | 1000 V | 1 V | ±(0.8% reading + 3 digits) | |
| AC voltage | | | 50 to 400 Hz | 400 Hz to 1 kHz |
| | 400 mV | 0.1 mV | ±(1.5% reading + 15 digits) | ±(2.5% reading + 15 digits) |
| | 4 V | 0.001 V | | |
| | 40 V | 0.01 V | ±(1.5% reading + 6 digits) | ±(2.5% reading + 8 digits) |
| | 400 V | 0.1 V | | |
| | 750 V | 1 V | ±(1.8% reading + 6 digits) | ±(3% reading + 8 digits) |
| Frequency | 5.000 Hz | 0.001 Hz | ±(1.5% reading + 5 digits) | |
| | 50.00 Hz | 0.01 Hz | | |
| | 500.0 Hz | 0.1 Hz | | |
| | 5.000 kHz | 0.001 kHz | ±(1.2% reading + 2 digits) | |
| | 50.00 kHz | 0.01 kHz | | |
| | 500.0 kHz | 0.1 kHz | | |
| | 5.000 MHz | 0.001 MHz | ±(1.5% reading + 4 digits) | |
| | 10.00 MHz | 0.01 MHz | | |
| | Sensitivity: 0.8 V rms min. @20% to 80% duty cycle and <100 kHz; 5 V rms min. @20% to 80% duty cycle and >100 kHz | | | |
| Duty cycle | 0.1 to 99.9% | 0.1% | ±(1.2% reading + 2 digits) | |
| | Pulse width: 100 μs – 100 ms, Frequency: 5 Hz to 150 kHz | | | |

**Note:** Accuracy specifications consist of two elements:
• (% reading) – This is the accuracy of the measurement circuit.
• (+ digits) – This is the accuracy of the analog to digital converter.

**FIGURE 1.13** **a** A typical small digital multimeter; **b** A page from the user manual giving the precision on various scales. Note that instruction manuals often refer to 'accuracy' when the correct term is precision.

Finally, the measurand itself may vary. For example, the flight of a water rocket depends strongly on initial conditions, wind and other factors. Even keeping launch conditions as close to identical as possible, it is unlikely that in repeat experiments you will be able to get a rocket to attain the same height within the limit of reading or equipment precision. Making repeat measurements allows you to estimate the size of the variation, using the maximum and minimum values.

Sometimes you will be able to see how the measurand varies during a measurement by watching a needle move or the readings change on a digital device. Watch and record the maximum and minimum values.

The difference between the maximum and minimum value is the range:

$$\text{Range} = \text{maximum value} - \text{minimum value}$$

The value of the measurand is the average value for repeated measurements, or the centre of the range for a single varying measurement:

$$\text{Measurand} = \text{minimum value} + \frac{1}{2}(\text{range})$$

The uncertainty in the measurement is half the range:

$$\text{Uncertainty} = \frac{1}{2}(\text{range}) = \frac{1}{2}(\text{maximum value} - \text{minimum value})$$

Accuracy, precision and resolution – what is the difference?

For example, if you are using a multimeter and you observe that the reading fluctuates between 12.2 V and 12.6 V then your measurement should be recorded as $(12.4 \pm 0.2)$ V. Note that the measurement and uncertainty are together in the brackets, indicating that the unit applies to both the measurement and its uncertainty.

When you take repeat measurements, the best estimate of the measurand is the average value. If you have taken fewer than ten measurements then the best estimate of the uncertainty is half the range. If you have more than ten measurements, the best estimate of the uncertainty is the standard deviation, given by:

$$\text{standard deviation} = \left[ \frac{\Sigma_i (x_i - \overline{x})^2}{n} \right]^{\frac{1}{2}}$$

where $x_i$ is an individual value of the measurand, $\overline{x}$ is the average value of the measurand and $n$ is the total number of measurements. The sum is over all values of $x_i$. Most calculators and spreadsheet software have built-in statistical functions such as standard deviation.

*Remember that repeat measurements means repeating under the same conditions.* It is not the same as collecting lots of data points under different conditions.

Uncertainties can be expressed in two ways: absolute and fractional (also called relative or proportional) uncertainty. The absolute uncertainty has the same units as the measurement, and is the uncertainty we have been describing above. The relative or fractional uncertainty is the absolute uncertainty as a fraction of the measurement, often expressed as a percentage. For example, if you measure your height to be 164 cm with an absolute uncertainty of 1 cm, then the fractional uncertainty is $\dfrac{1\,\text{cm}}{164\,\text{cm}} = 0.006$ or 0.6%.

The uncertainty in the measurements will determine whether you can say that your experiment has disproved or supported your hypothesis. So you must calculate uncertainties. Knowing the uncertainties in the data is also required for your evaluation of your investigation.

**KEY CONCEPTS**

- The uncertainty in any measurement depends upon the limit of reading of the measuring device, the precision of the device and the variation of the measurand. The uncertainty in the measurand is whichever is the greatest of these.
- Uncertainties can be recorded as absolute uncertainties, which have the same units as the measurand, or fractional uncertainties, which have no units and are usually expressed as percentages.

# 1.5 Analysing and interpreting your results

Critical and creative thinking

Numeracy

Information and communication technology capability

The first step in analysing data, whether primary or secondary, is to organise it. This will usually involve tabulating it. Tables of data need to have headings with units for each column, and a caption telling you what the data means, or how it was collected. Tables are used for recording raw data, and also for organising derived data.

## Calculating derived data from raw data

Raw data is what you actually measured (with units and uncertainties). Derived data is data that you have calculated using your raw data. For example, your raw data may be time and distance measurements, from which you derive average speed and acceleration data.

When you record your data, write down the units for all your measurements. You may need to convert these to SI units, for example cm to m (see Appendix 1). Include the units with all numbers as you do your calculations. In this way you will make sure you have the correct units on all derived data. It also allows you to check that any equations you are using are dimensionally correct.

It is good practice *in general*, not just in investigations, to include units at each step in all your calculations.

Your raw data should be recorded with uncertainties. All your derived results should also have uncertainties.

Whenever you add or subtract raw data, then you simply add the absolute uncertainties.

Whenever you multiply or divide with raw data, you add the fractional uncertainties in all the values used. For example, if calculating $a$ using the equation $a = \dfrac{b^2}{2c}$, then the relative uncertainty in $a$ will be the relative uncertainty in $c$ plus the relative uncertainty in $b$ plus the relative uncertainty in $b$ again:

$$\frac{\Delta a}{a} = \frac{\Delta b}{b} + \frac{\Delta b}{b} + \frac{\Delta c}{c}$$

We add the relative uncertainty of $b$ twice because we multiply by $b^2$, which is the same as multiplying by $b$ twice. Note that we do not include the relative uncertainty in $\dfrac{1}{2}$ because we assume it to be an exact number with no uncertainty.

If you have more complicated calculations, then you should refer to a guide such as the Guide to Uncertainties in Measurement or a book on experimental techniques.

## Drawing and using graphs

If you look at any physics journal, you will see that almost every article contains graphs. Graphs are not only a useful way of representing data, they are also commonly used to analyse relationships between variables. You should have lots of graphs in your logbook as part of your exploration of the data. It is often useful to plot your data in different ways, especially if you are unsure what relationship to expect between your dependent and independent variables.

Graphs should be large and clear. The axes should be labelled with the names of the variables and their units. Choose a scale so that your data takes up most of the plot area. This will often mean that the origin is not shown in your graph. Usually there is no reason why it should be. Figure 1.14 shows an example of (a) a poor graph, and (b) a good one.

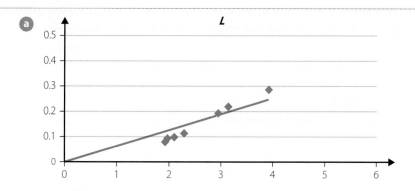

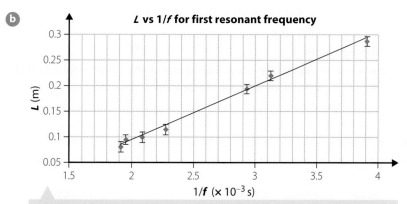

**FIGURE 1.14** **a** An example of a poor graph; **b** An example of a good graph of the same data. How many problems can you identify on graph **a**?

When you are looking for a relationship between variables, plot a scatter plot (also called a scatter graph). This is a graph showing your data as points. Do not join them up as in a dot-to-dot picture. Usually the independent variable is plotted on the *x*-axis and the dependent variable goes on the *y*-axis.

To determine a relationship you need to have enough data points, and the range of your data points should be as large as possible. A minimum of six data points is generally considered adequate if the relationship is expected to be linear (a straight line), but always collect as many as you reasonably can, given the available time.

For non-linear relationships you need more data points than this. Try to collect more data in regions where you expect rapid variation. Imagine you are measuring an interference pattern from two slits, as shown in Figure 1.15a. You may need more than a hundred data points to clearly see the sinusoidal variation in intensity (Figure 1.15b) due to the two-slit interference.

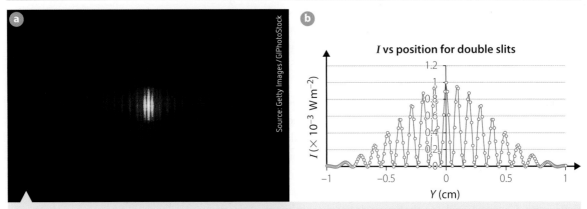

**FIGURE 1.15** **a** Interference pattern from two slits; **b** Plot of intensity as a function of position for this experiment

**Data points**

Some helpful advice on deciding the number of data points

A useful graph to start with is simply a graph of the raw data. You will usually be able to tell by looking whether the graph is linear. If it is, then fit a straight line either by hand or using graphing software. Graphing software has a linear regression tool which calculates an $R^2$ number, which is a measure of 'goodness of fit'. The closer $R^2$ is to 1, the better the fit. If it is not very close to 1, typically better than 0.95, then the relationship is probably not linear. Alternatively, you can calculate the uncertainty in the gradient by using lines of maximum and minimum gradient. If the uncertainty is large, then the relationship may not be linear.

If you have a hypothesised equation then use it to generate a fit on a graph of your data. *Do not substitute your data into your hypothesised equation and try to show that it fits.* Note that a line of best fit is not the same as joining the dots. It is rarely useful or appropriate to join the dots, even though this is often the default setting in spreadsheet software.

If it is a linear relationship, then finding the equation for the line of best fit will be useful. Remember that a linear relationship is of the form $y = mx + c$ where $y$ is the dependent variable plotted on the vertical axis, $x$ is the independent variable on the horizontal axis, $m = \dfrac{\Delta y}{\Delta x}$ is the gradient of your graph and $c$ is the *y*-intercept from your graph.

*Never force a line of best fit through the origin.* The intercept gives you useful information. It may even indicate a systematic error, such as a zero error in calibration of your equipment.

When you plot your raw data you may find that one or two points are outliers. These are points that do not fit the pattern of the rest of the data. These points may be mistakes; for example, they may have been incorrectly recorded or a mistake may have been made during measurement. They may also be telling you something important. For example, if they occur at extreme values of the independent variable then it might be that the behaviour of the system is linear in a certain range only. You may choose to ignore outliers when fitting a line to your data, but you should be able to justify why.

## Non-linear data and linearising

Relationships between variables are often not linear. If you plot your raw data and it is a curve, then *do not draw a straight line through it*. In this case you need to think a little harder. If your hypothesis predicts the shape of the curve, then try fitting a theoretical curve to your data. If it fits well, then your hypothesis is supported.

If possible, you should linearise your data based on your hypothesis. This means you should write it in the form $y = mx + c$. For example, if your hypothesis is that $h = \frac{1}{2}gt^2$, try plotting your data as a function of $t^2$. Here $h$ is the initial height of a falling object, $g$ is the acceleration due to gravity and $t$ is the time taken for it to fall:

$$h = (\tfrac{1}{2}g)(t^2) + 0$$
$$\uparrow \qquad \uparrow \uparrow \qquad \uparrow$$
$$y = \quad m \ x \ + c$$

Hence a plot of $h$ vs $t^2$ should be a straight line with gradient $\frac{1}{2}g$ and a $y$-intercept of zero. So if you plot $h$ vs $t^2$ and get a straight line of gradient $\frac{1}{2}g$ with a $y$-intercept of zero, then your hypothesis is supported.

Log–log graphs are useful for power laws. A log–log graph will give you a straight line if there is a power law relationship between the variables. For example, if the relationship is of the form $y = ax^n$, then if we take logarithms of both sides we get $\log y = \log a + n \log x$. A plot of $\log y$ vs $\log x$ then has gradient $n$ and intercept $\log a$.

Consider the data shown in Table 1.6. If we plot the raw data in Table 1.6 as shown in Figure 1.16a, we see that the relationship between power and temperature is clearly *not* linear. *Do not* draw a straight line through data like this. We can test whether power radiated and temperature are related by a power law by plotting a log–log graph, as shown in Figure 1.16b.

**TABLE 1.6** Power radiated by a hot filament as a function of filament temperature.

| TEMPERATURE (K) | POWER (W) |
| --- | --- |
| 1000 | 23 |
| 1100 | 34 |
| 1200 | 48 |
| 1300 | 67 |
| 1400 | 89 |
| 1500 | 117 |
| 1600 | 152 |
| 1700 | 194 |
| 1800 | 243 |
| 1900 | 301 |
| 2000 | 370 |

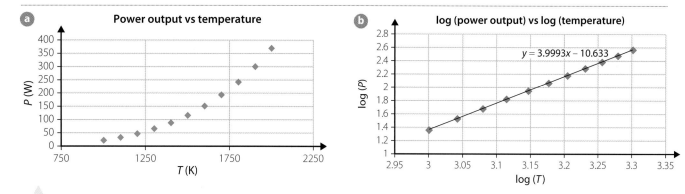

FIGURE 1.16 **a** Direct plot of the data in Table 1.6; **b** Log–log graph of the same data, with line of best fit displayed.

We can see from the equation for the line of best fit that the gradient is 3.9993 and the intercept is –10.633. Hence we deduce that the relationship between $P$ and $T$ is:

$$P = (2.7 \times 10^{-11} \, \text{W K}^{-4}) T^4$$

This is a powerful technique, and one well worth practising.

Sometimes you do not know what relationship to expect between variables, or you may have tried to fit your data and it has not worked. With spreadsheet software it is very quick to generate log–linear, log–log graphs and other plots, so it is worth trying a few different types of relationship.

It is always better to linearise your data rather than try to fit a curve to non-linear data. Often a curve for an exponential relationship can look very much like a curve for a power law. Linearising your data allows you to distinguish between the two. Finding the mathematical relationship that best fits your data is important if you are trying to distinguish between hypotheses that predict different relationships. You need to be able to do this when you evaluate your investigation.

**KEY CONCEPTS**

- Data is usually recorded in tables. Graphs are used to represent and analyse data.
- Linear graphs are useful for analysing data. If your data is not linear, then a log–log graph may tell you what the relationship between the variables is.
- If you predict a particular mathematical relationship then you should linearise your data and then graph it to test your prediction.

## Interpreting your results

Once you have analysed your results you need to interpret them. This means being able to either answer your research question or state whether your results support your hypothesis.

You need to take into account the uncertainties in your results when you decide whether they support your hypothesis. For example, suppose you have hypothesised that the maximum range of a rocket occurs at a launch angle of 45°. But your results show that the maximum range occurs at an angle of 47°. You may think that this result does not support your hypothesis. To say whether the result agrees with the prediction, you need to consider the uncertainty. If the uncertainty is 1°, then the results disagree with the hypothesis. If the uncertainty is 2° or more, then the results do agree and the hypothesis is supported.

If your hypothesis is not supported it is not enough to simply say 'our hypothesis is wrong'. If the hypothesis is wrong, what is wrong with it?

It may be that you have used a model that is too simple. For example, if you have based your hypothesis on the kinematics model and ignored the effect of air resistance, the range is likely to be shorter and the maximum height lower than you predicted. For many projectiles, including rockets, air resistance is not negligible. If you find that this is the case, then you may conclude that your situation is better described by a model that includes air resistance.

9780170409131

Before you decide that the model is at fault, however, it is a good idea to check carefully that you have not made any mistakes.

It is *never* good enough to conclude that 'the experiment didn't work'. Either a mistake was made or the model used was not appropriate for the situation. It is your job to work out which.

Experiments that do not support predictions based on existing models are crucial in the progress of science. It is these experiments that tell us that there is more to find out, and inspire our curiosity as scientists.

- You must know the uncertainty in your results to be able to test your hypothesis.
- When your hypothesis is not supported, try to figure out why.

## Evaluating your investigation

After you have completed your investigation you need to evaluate it. Your logbook records are central to this.

Your evaluation needs to consider the validity and reliability of your experiment. Is your experiment reproducible? Is your data accurate?

If you started with a hypothesis, is your data precise enough that your hypothesis is either clearly disproved or clearly supported? Does your data clearly support (or disprove) your hypothesis but not competing hypotheses – does it distinguish between models?

If you started with a question, have you answered it? How certain can you be? You need to know the precision of your data to answer this question; this is why you must calculate uncertainties.

You need to address the limitations of your investigation. Do your findings apply to only a specific situation, or are they more broadly applicable? For example, if you found the optimum volume of water in a water rocket to achieve maximum height, does this volume apply only to water rockets of the same shape? Does it apply only to water but not to other liquids?

Finally, it is a good idea to consider how your investigation could have been improved. Do not focus only on limitations in your equipment. Instead, consider how you might have improved your technique, or the design of the experiment, or even your question or hypothesis. If you had to do the investigation again, what would you do differently?

## 1.6  Communicating your understanding

If research is not reported on, then no-one else can learn from it. An investigation is not complete until the results have been communicated. Most commonly a report is written. Scientists also use other means to communicate their research to each other, such as posters and talks. School science fairs may also include posters and oral presentations. Science shows and demonstrations, websites, videos and blogs may also be used. You need to select the mode that best suits the *content* you wish to communicate and the *audience* you wish to communicate to.

Literacy

Numeracy

Information and communication technology capability

### Writing reports

A report is a formal and carefully structured account of your investigation or depth study. It is based on the data and analysis in your logbook. However, the report is a summary. It contains only a small fraction of what appears in the logbook. Your logbook contains all your ideas, rough working and raw data. The report typically contains very little of this.

A report consists of distinct sections, each with a particular purpose.

» Abstract

» Introduction

» Method

» Results and analysis

» Discussion

» Conclusion

» Acknowledgements

» References

» Appendices

Reports are *always* written in the past tense, because they describe what you have done.

The *abstract* is a very short summary of the entire report, typically between 50 and 200 words long. It appears at the start of the report, but is always the last thing that you write. Try writing just one sentence to summarise each part of your report.

The *introduction* tells the reader why you did this investigation or depth study and what your research question or hypothesis is. This is the place to explain why this research is interesting. The introduction also includes the literature review, which gives the background information needed to be able to understand the rest of the report. The introduction for secondary-sourced reports is similar to that for a primary-sourced investigation. In both cases it is important to reference all your sources correctly.

The *method* summarises what you did. It says what you measured and how you measured it. *It is not a recipe for someone else to follow.* It is *always* written in past tense, for example 'We measured the temperature every 10 s' *not* 'measure the temperature every 10 s'. It also explains, briefly, why you chose a particular method or technique.

For a primary-sourced investigation the method describes how you carried out your experiments or observations in enough detail that someone with a similar knowledge level could repeat your experiments. It should include large, clear diagrams of equipment setup, circuits and so on. You should have diagrams in your logbook, but these are generally rough sketches. Diagrams should be redrawn neatly for a report, as in Figure 1.17.

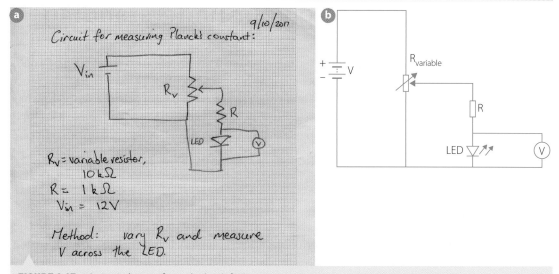

**FIGURE 1.17**  **a** A circuit diagram from a logbook; **b** The same circuit diagram redrawn in a formal report

The method section for a secondary-sourced investigation is generally shorter. If you are doing a review of the current literature on a topic then your method will say what literature searches you carried out, and how you decided which sources to use.

The *results* section is a summary of your results. It is usually combined with the analysis section, although they may be kept separate.

Tables comparing the results of different experiments or secondary sources are useful, but avoid including long tables of raw data in your report. Wherever possible use a graph instead of a table. If you need to include a lot of raw data, then put it in an appendix attached to the end of the report.

Think about what sort of graph is appropriate. If you want to show a relationship between two variables then use a scatter plot. Display your data as points with uncertainty bars and clearly label any lines you have fitted to the data. Column and bar charts are used for comparing different data sets. Do not use a column or bar chart to try to show a mathematical relationship between variables. Examples of these two types of graphs are shown in Figure 1.18.

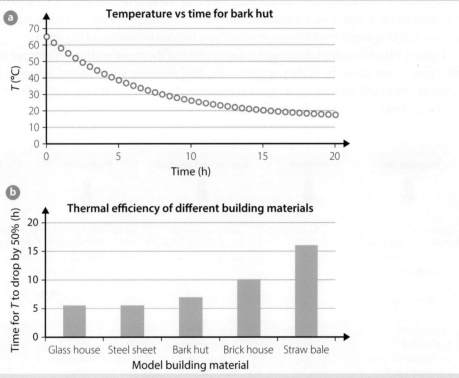

FIGURE 1.18 **a** A scatter plot demonstrating a mathematical relationship; **b** A column graph comparing results from different experiments

Any data and derived results should be given in correct SI units with their uncertainties. If you performed calculations then show the equations you used. You might want to show one sample calculation, but do not show more than one if the procedure used is repeated.

The *discussion* should summarise what your results mean. If you began with a research question, give the answer to the question here. If you began with a hypothesis, state whether or not your results supported your hypothesis. If not, explain why. If your investigation led you to more questions, as is often the case, say what further work could be done to answer those questions. Your discussion also includes the evaluation of the investigation (step 3 in Figure 1.19).

The *conclusion* is a very brief summary of the results and their implications. Say what you found out and what it means. A conclusion should only be a few sentences long.

Scientific reports often include *acknowledgements* thanking people and organisations that helped with the investigation. This includes people who supplied equipment or funding, as well as people who gave you good ideas or helped with the analysis. In science, as in other aspects of your life, it is always polite to say thank you.

The final section of a report is the *reference list*. It details the sources of all information that were actually used to write the report. This will generally be longer for secondary-sourced investigations. Wherever a piece of information or quotation is used in your report it must be referenced at that point. This is typically done either by placing a number in brackets at the point, for example [2], or the author and year of publication, for example (Smith, 2014). The reference list is then either provided in a footnote at the end of each page or as a single complete list at the end of the report. There are different formats for referencing, so check with your teacher what format they prefer. There are several good online guides to referencing.

Note that a reference list is not the same as a bibliography. A bibliography is a list of sources that are useful to understanding the research. They may or may not have actually been used in writing the report. You should have a bibliography in your logbook from the planning stage of your investigation. The references will be a subset of these sources. A primary-sourced investigation does not include a bibliography. A secondary-sourced investigation may include a bibliography as well as references, to demonstrate the scope of your literature search. For some secondary-sourced investigations, such as an annotated bibliography, the bibliography itself may be a major section of the report.

Figure 1.19 is a flow chart showing the steps in each of the major sections of a formal report. Note that this figure shows the steps as they appear in the final report to give a logical sequence for your reader. Typically you would not write your report in this order. For example, students often find it easiest to write the method section first and the abstract last.

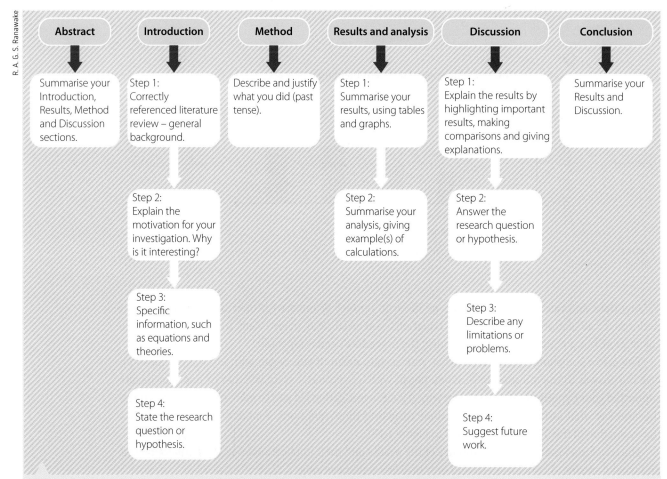

**FIGURE 1.19** Steps in each of the major sections of a formal report

- A formal report has the same form as an article written by a scientist.
- It begins with an abstract briefly summarising the entire work.
- It includes an introduction with a literature review, a method, results and analysis, discussion and conclusion.
- All sources need to be referenced correctly.

## Other ways of communicating your work

You may want to present the results of your investigation in some other way. Look at examples of science investigations reported on websites, in the newspaper, on the TV and so on. This will give you an idea of the different styles used in the different media. Think about the purpose. Is it to inform, to persuade or both? What sort of language is used?

Think about your audience and purpose, and use appropriate language and style. A poster is not usually as formal as a report. A video or webpage may be more or less formal, depending on your audience.

Posters and websites use a lot of images. Images are usually more appealing than words and numbers, but they need to be relevant. Make sure any images you use communicate the information you want them to.

Consider accessibility if you are creating a website. Fonts need to be large enough and digital images should have tags. You can follow the weblink for more information on accessibility and webpage design.

If you make a video then consider who your audience is and what will appeal to them. Think about how you will balance content with entertainment.

A formal report uses referencing to show where you found information. Other means of communicating about your depth study or investigation also need to acknowledge the sources of information that you used. You also need to be very careful about using copyright content – for example, you cannot copy images from other people's websites without permission unless the site gives that permission. Talk to your teacher about how they would like you to acknowledge your sources.

However you communicate your work, make sure you know what the message is and who the audience is. Once you have established that, you will be able to let other people know about the interesting things you have discovered in your investigation.

**Website accessibility**

The Royal Society for the Blind has information on making websites accessible.

KEY CONCEPTS

- There are many ways of communicating your findings. Choose a method that is appropriate to your investigation and your intended audience.

## Ideas for depth studies

As you progress through this book you will find suggestions for investigations in each chapter. These investigations may give you ideas for primary investigations which can be part of a depth study.

At the end of each module there is a short section called Depth Study Suggestions. Here you will find ideas for primary- and secondary-sourced investigations that build on the content of the preceding modules. These suggestions are sourced from experienced teachers and university academics, and from physics education literature. Your teacher will also have ideas and suggestions and you can generate your own ideas by reading about topics you are interested in.

Consider what skills from other areas you might bring to a depth study, particularly if you are artistic or musical or good at making things. Many physicists combine their love of science with other creative pursuits.

By carrying out depth studies you will extend your knowledge and understanding in physics, but more importantly, you will learn how to work scientifically – you will learn how to *do* physics.

# ADVANCED MECHANICS

Getty Images / Abeleao

9780170409131

# 2 Projectile motion

OUTCOMES

**Students:**

- analyse the motion of projectiles by resolving the motion into horizontal and vertical components, making the following assumptions:
  - a constant vertical acceleration due to gravity
  - zero air resistance
- apply the modelling of projectile motion to quantitatively derive the relationship between the following variables:
  - initial velocity
  - launch angle
  - maximum height
  - time of flight
  - final velocity
  - launch height
  - horizontal range of the projectile (ACSPH099)
- conduct a practical investigation to collect primary data in order to validate the relationships derived above
- solve problems, create models and make quantitative predictions by applying the equations of motion relationships for uniformly accelerated and constant rectilinear motion ICT, N

Physics Stage 6 Syllabus © 2017 NSW Education Standards Authority (NESA) for and on behalf of the Crown in right of the State of New South Wales.

Shutterstock.com/Rawpixel.com

FIGURE 2.1 A projectile is launched. As it moves we model it as being subject only to the gravitational force.

A thrown or kicked ball (Figure 2.1), a bullet fired from a gun and the water spat by an archer fish at its prey are all projectiles. A projectile is an object that has been projected, which means launched, thrown or fired in some way. A projectile is not self-propelled, like a bird or an aeroplane; it does not provide energy for its flight. It has some initial momentum and energy from the force or forces that were used to launch it but, after launching, its path is determined by the local gravitational field and its initial velocity only.

For simplicity, we will make the assumption that air resistance is negligible, and so gravity is the only force we need to consider. But remember that this is a simplified model and, in reality, air resistance is often significant.

## 2.1 Projectile motion

**Projectile motion game**

Many games include projectile motion, and calculate projectile paths. This one shows you how the calculations work.

We call the path of a projectile its trajectory. Figure 2.2 shows the trajectory of a thrown ball. The trajectory of a projectile is determined by two things: its initial velocity and the forces that act on it. The initial velocity is the velocity a projectile has immediately after being launched.

In this chapter, we will ignore the effects of friction on projectiles and model their motion as if the only force applied to them after being launched is the gravitational force. This force is always directed vertically downwards and has magnitude $F = mg$, where $m$ is the mass of the projectile and $g$ is the acceleration due to gravity. (As we shall see in chapter 4, $g$ is also the gravitational field strength.) We shall only consider projectiles close to the surface of Earth, so we can make the approximation that the gravitational force is constant.

### Acceleration of a projectile

As the gravitational force close to Earth's surface is given by $F = mg$, from Newton's second law,

$$a = \frac{F}{m} = \frac{mg}{m} = g$$

So the acceleration of a projectile, once launched, has magnitude $g$, and is in the direction of the gravitational force. We will use $g = -9.8\,\mathrm{m\,s^{-2}}$, where the negative sign indicates that the acceleration is downwards.

The gravitational force due to Earth acts vertically downwards, towards the centre of Earth. Remember that acceleration is a vector. Vectors can always be broken into perpendicular

FIGURE 2.2 The trajectory of a thrown basketball. The acceleration at all points on the path is directly downwards. The velocity changes with time. At the top of the path the velocity is horizontal.

components, usually (but not always) horizontal and vertical components. So if we break the acceleration into vertical and horizontal components, we get:

$a_x = 0$ for the horizontal component because there is no horizontal force, and

$a_y = g$ for the vertical component.

So the acceleration of a projectile is directly downwards. This does *not* mean that the velocity is directly downwards. Remember that acceleration is the *change* in velocity over time; velocity and acceleration do not need to be in the same direction. The force on any projectile is always downwards, but the direction of motion can be in any direction, including upwards.

## Velocity of a projectile

Recall from chapter 2 of *Physics in Focus Year 11* that the velocity of an object that is subject to a constant acceleration is given by $v = u + at$, where $v$ is the velocity vector at some time $t$, and $u$ is the initial velocity at time $t = 0$. We can use this equation to find the velocity of a projectile at any time, if we know initial velocity.

### Initial velocity

Projectiles can be launched vertically, horizontally or at any angle. The angle, $\theta$, at which the projectile is launched is called the launch angle. The initial velocity of a projectile, $u$, is determined by an initial force such as an explosion, a throw or a kick. Remember that velocity is a vector: it has both magnitude and direction. Like all vectors, velocity can be decomposed into perpendicular components.

Always start any problem by drawing a diagram. In this case we want to find the horizontal and vertical components of the initial velocity. So we draw a vector diagram, as shown in Figure 2.3.

Using trigonometry, we can see that the horizontal component of the initial velocity is:

$$u_x = u \cos \theta$$

and the vertical component is

$$u_y = u \sin \theta$$

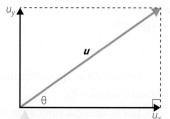

FIGURE 2.3 The horizontal and vertical components of the initial velocity.

You may want to review chapter 4 of *Physics in Focus Year 11* to remind yourself how to work with vectors.

---

▶ **WORKED EXAMPLE** (2.1)

A Bofors gun, as shown in Figure 2.4, fires a shell with a muzzle velocity of $880 \, \mathrm{m \, s^{-1}}$. If it has an elevation (angle above the horizontal) of 30°, what are the vertical and horizontal components of the initial velocity?

FIGURE 2.4 A 40 mm Bofors anti-aircraft gun fires a shell with an initial velocity of around $1 \, \mathrm{km \, s^{-1}}$.

| ANSWER | LOGIC |
|---|---|
| $u = 880\,\text{m s}^{-1}, \theta = 30°$ | • Identify the relevant data. |
| $u_x = u \cos\theta$ | • Write the expression for the horizontal component of $u$. |
| $u_x = 880\,\text{m s}^{-1} \cos 30°$ | • Substitute values with correct units. |
| $u_x = 762\,\text{m s}^{-1}$ | • Calculate final value. |
| $u_y = u \sin\theta$ | • Write the expression for the vertical component of $u$. |
| $u_y = 880\,\text{m s}^{-1} \sin 30°$ | • Substitute values with correct units. |
| $u_y = 440\,\text{m s}^{-1}$ | • Calculate the final value. |
| $u_x = 760\,\text{m s}^{-1}, u_y = 440\,\text{m s}^{-1}$ | • State final answer with correct units and appropriate significant figures. |

**TRY THESE YOURSELF**

1   If the gun was aimed with an elevation of 45°, what would the horizontal and vertical components of the initial velocity be?

2   At what angle is the vertical component of the initial velocity greatest? What is this maximum vertical component?

## Velocity of a projectile during flight

Recall from chapter 2 of *Physics in Focus Year 11* that the velocity of an object with constant acceleration is given by:

$$v = u + at$$

We can decompose this into two independent equations, one for the vertical velocity and one for the horizontal velocity.

The horizontal velocity is given by:

$$v_x = u_x + a_x t$$

and the vertical by:

$$v_y = u_y + a_y t$$

In the horizontal direction, the acceleration is zero. So:

$$v_x = u_x + a_x t = u_x + 0 = u_x$$

This tells us that the horizontal component of a projectile's velocity is constant and is equal to its initial velocity. This is because we have assumed that there are no horizontal forces acting.

In the vertical direction, the acceleration is $g$. So:

$$v_y = u_y + a_y t = u_y + gt$$

This tells us that the vertical component of a projectile's velocity changes with time because of the gravitational force. The acceleration due to gravity is constant, and directed downwards.

▶ **WORKED EXAMPLE** (2.2)

Consider the shell fired by a Bofors gun as described in Worked example 2.1. Find the vertical and horizontal components of the shell's velocity after 15 s. What is the magnitude and angle of the shell velocity at this time?

| ANSWER | LOGIC |
|---|---|
| $t = 15\,s$ | ▪ Identify the relevant data. |
| $u_x = 762\,m\,s^{-1}$, $u_y = 440\,m\,s^{-1}$ | ▪ Refer to the previous worked example. |
| $v_x = u_x$ | ▪ Write the expression for the horizontal component of $v$. |
| $v_x = 762\,m\,s^{-1}$ | ▪ Substitute values with correct units to obtain the final value. |
| $v_y = u_y + a_y t$ | ▪ Write the expression for the vertical component of $v$. |
| $v_y = 440\,m\,s^{-1} + (-9.8\,m\,s^{-2})(15\,s)$ | ▪ Substitute values with correct units. |
| $v_y = 293\,m\,s^{-1}$ | ▪ Calculate the final value. |
| $v = \sqrt{v_x^2 + v_y^2}$ | ▪ Write the expression for the magnitude of $v$ using Pythagoras theorem. |
| $v = \sqrt{(762\ m\,s^{-1})^2 + (293\ m\,s^{-1})^2}$ | ▪ Substitute values with correct units. |
| $v = 816\,m\,s^{-1}$ | ▪ Calculate the final value. |
| $\tan\theta = \dfrac{v_y}{v_x}$ | ▪ Relate the angle to the velocity components. |
| $\theta = \tan^{-1}\left(\dfrac{v_y}{v_x}\right)$ | ▪ Rearrange for angle. |
| $\theta = \tan^{-1}\left(\dfrac{293\ m\,s^{-1}}{762\ m\,s^{-1}}\right)$ | ▪ Substitute values with correct units. |
| $\theta = 21.0°$ | ▪ Calculate the final value. |
| $v_x = 760\,m\,s^{-1}$, $v_y = 290\,m\,s^{-1}$ <br><br> $v = 820\,m\,s^{-1}$, $\theta = 21°$ | ▪ State the final answer with correct units and appropriate significant figures. |

**TRY THESE YOURSELF**

1  How long would it take for the vertical component of the velocity to reach 0?

2  When the vertical velocity is 0, what is the horizontal velocity?

## Position of a projectile

The position of a projectile at any moment can be found from the kinematic equations in chapter 2 of *Physics in Focus Year 11*. The position of an object that has a constant acceleration is given by $s = ut + \dfrac{1}{2}at^2$, where $s$ is the position at time $t$. Note that $s$ is a vector, and has two components, a horizontal component, $x$, and a vertical component, $y$.

If we separate the position into vertical and horizontal components we have:

$$x = x_0 + u_x t + \frac{1}{2} a_x t^2 = x_0 + u_x t$$

and

$$y = y_0 + u_y t + \frac{1}{2} a_y t^2 = y_0 + u_y t + \frac{1}{2} g t^2$$

Often we take the starting position of a projectile to be the origin, so that $x_0 = 0$ and $y_0 = 0$, which simplifies the equations to:

$$x = u_x t$$

and

$$y = u_y t + \frac{1}{2} g t^2$$

▶ **WORKED EXAMPLE** (2.3)

For the shell fired from the Bofors gun described in the previous two worked examples, at what time and height does it have a zero vertical velocity?

| ANSWER | LOGIC |
|---|---|
| $u_x = 762\,\mathrm{m\,s^{-1}}$, $u_y = 440\,\mathrm{m\,s^{-1}}$ | ▪ Identify the relevant data; refer to previous worked example. |
| $v_y = 0$ | ▪ State the velocity a the time we need to calculate. |
| $v_y = u_y + gt$ | ▪ Write the expression for the vertical component of $v$. |
| $t = \dfrac{v_y - u_y}{g}$ | ▪ Rearrange for time. |
| $t = \dfrac{0 - 440\ \mathrm{m\,s^{-1}}}{-9.8\ \mathrm{m\,s^{-2}}}$ | ▪ Substitute values with correct units. |
| $t = 44.9\,\mathrm{s}$ | ▪ Calculate the final value. |
| $y = u_y t + \dfrac{1}{2} g t^2$ | ▪ Write the expression for the vertical position at this time. |
| $y = (440\,\mathrm{m\,s^{-1}})(44.9\,\mathrm{s}) + \dfrac{1}{2}(-9.8\,\mathrm{m\,s^{-1}})(44.9\,\mathrm{s})^2$ | ▪ Substitute values with correct units. |
| $y = 9877\,\mathrm{m}$ | ▪ Calculate the final value. |
| $t = 45\,\mathrm{s}$, $y = 9900\,\mathrm{m}$ | ▪ State the final answer with correct units and appropriate significant figures. |

**TRY THESE YOURSELF**

1 What is the horizontal position of the shell at this time?

2 Repeat the example above, with the gun aimed at an angle of 60°. You will need to start by repeating the calculations for the initial horizontal and vertical components of the velocity.

Projectile motion: some basic ideas and calculations

In the worked examples we have used the equations for projectile motion assuming that no air resistance was acting. For a fast-moving projectile like a shell or bullet this is not a good assumption, and the height that we calculated will be an overestimate.

The physical significance of the height at which the vertical velocity of the projectile is zero is that this is the **maximum height** that it obtains. When the vertical component of the velocity has dropped to zero, the projectile is no longer rising. It has reached its maximum height and will now begin to fall. Note that while the vertical velocity is zero at this point, the horizontal velocity is still the same as the initial horizontal velocity and the projectile continues to move in the same horizontal direction. The path of the projectile is symmetric, with the second half of the path, after the peak, being a mirror image of the first half.

**Golf on the Moon**

A projectile's trajectory depends only on the initial velocity and gravitational acceleration. So a golf ball travels much further on the Moon.

**Applications of projectile motion**

The physics of projectile motion is used in solving crimes.

**KEY CONCEPTS**

- A projectile is an object that has been launched or thrown, and does not propel itself once it is launched.
- Our model of projectile motion ignores air resistance and assumes that the projectile remains close to Earth's surface so that $g$ is constant.
- The trajectory (path) of a projectile is determined only by the initial (launch) velocity, and the force of gravity.
- The initial velocity has components $u_x = u\cos\theta$ and $u_y = u\sin\theta$.
- The horizontal and vertical components of a projectile's position, velocity and acceleration are determined by the kinematics equations and are given in Table 2.1.

**TABLE 2.1** The horizontal and vertical components of a projectile's position, velocity and acceleration

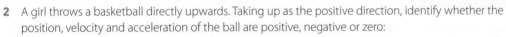

| | HORIZONTAL COMPONENT | VERTICAL COMPONENT |
|---|---|---|
| position | $x = u_x t = (u\cos\theta)\,t$ | $y = y_0 + u_y t + \frac{1}{2}gt^2 = y_0 + (u\sin\theta)\,t + \frac{1}{2}gt^2$ |
| velocity | $v_x = u_x = u\cos\theta$ | $v_y = u_y + gt = u\sin\theta + gt$ |
| acceleration | $a_x = 0$ | $a_y = g = -9.8\,\text{m s}^{-2}$ |

**CHECK YOUR UNDERSTANDING**

**2.1**

1 On a vector diagram, show the horizontal and vertical components of the initial velocity of a projectile. Why do we assume that there is no horizontal component of acceleration for a projectile?

2 A girl throws a basketball directly upwards. Taking up as the positive direction, identify whether the position, velocity and acceleration of the ball are positive, negative or zero:

   a just after the ball has left her hand.

   b at the very top of its trajectory.

   c just before she catches the ball.

3 A potato is fired at an angle of 30° to the horizontal with a speed of 120 m s$^{-1}$.

   a Find the horizontal component of the potato's initial velocity.

   b Some time later, the potato hits the ground. With what horizontal speed does this occur? Explain your answer.

4 Sketch the following situations, then find the horizontal and vertical components of the initial velocity of projectiles fired:

   a at an angle of 65° above the horizontal and a speed of 45 m s$^{-1}$.

   b from a high cliff at an angle of 25° below the horizontal at 50 m s$^{-1}$.

5 An arrow is fired at an angle of 45° above the horizontal with a speed of 86 m s$^{-1}$.

   a Find the vertical component of the initial velocity of the projectile.

   b Find the maximum height (vertical position) reached by the projectile.

   c What is the time taken for the projectile to reach its maximum height?

6 A golf ball is hit at an angle of 60° above the horizontal and reaches a maximum height of 25 m. What was its initial speed?

# 2.2 The trajectory of a projectile

A projectile can be fired at any angle, $\theta$, and from any initial height. The initial height is called the launch height.

Figure 2.5a shows the trajectory of a projectile, which is launched from ground level and stops at ground level. This projectile has a launch height of $y_0 = 0$. Figure 2.5b shows the trajectory when the projectile is launched from above ground level. The final vertical position is lower than the launch height. If we define the launch height as $y_0 = 0$ then the final vertical position will be negative. Alternatively, we can define the final height as zero, and then the launch height is positive. It doesn't matter what you choose, as long as you are consistent.

Figure 2.5 shows the highest point, the maximum height, reached by the projectile, and the total horizontal distance travelled, called the horizontal range, or just the range.

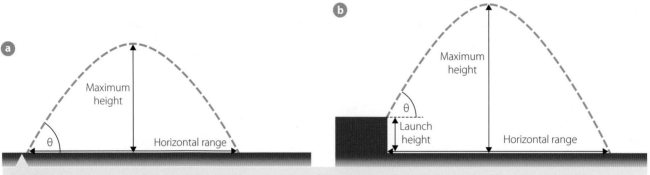

**FIGURE 2.5** **a** The trajectory of a projectile launched from ground level; **b** The trajectory of a projectile launched from above ground level

## Maximum height

The maximum height, $h$, that a projectile will reach depends only on the vertical component of the initial velocity and the launch height. The maximum height occurs when the vertical velocity is zero.

The vertical position of a projectile at any time is given by:

$$y = u_y t + \frac{1}{2} g t^2$$

if we take launch height to be zero. Then the maximum height, $h$, is when

$$v_y = u_y + gt = 0 \text{ and } y = h$$

We can rearrange this equation to find the time at which the maximum height occurs:

$$t = -\frac{u_y}{g}$$

Note that, while the expression for time is negative, because the value for $g$ is also negative the value calculated for time will be positive.

We can now substitute this time into the expression for vertical position:

$$h = u_y t + \frac{1}{2} g t^2 = -\frac{u_y^2}{g} + \frac{u_y^2}{2g} = -\frac{u_y^2}{2g}$$

If the launch height is not zero, then the maximum height is simply the launch height plus the value for maximum height found above:

$$h = y_0 + u_y t + \frac{1}{2} g t^2 = y_0 - \frac{u_y^2}{2g}$$

Alternatively, if we start with the equation $v^2 = u^2 + 2as$ from chapter 2 of *Physics in Focus Year 11* and write this in the vertical direction as:

$$v_y^2 = u_y^2 + 2gy$$

then recognising that at the peak of the trajectory the vertical velocity is zero, and the vertical position is the maximum height $h$, we can write:

$$h = -\frac{u_y^2}{2g}$$

which is exactly the result we arrived at above.

▶ **WORKED EXAMPLE** (2.4)

In a moment of frustration, Jen (who is very strong and very angry) throws her maths textbook out the window, 2.4 m above the ground. She throws it with initial velocity $12\,\text{m s}^{-1}$ at an angle of 45° above the horizontal. Calculate the maximum height that the book reaches.

| ANSWER | LOGIC |
|---|---|
| $u = 12\,\text{m s}^{-1}, \theta = 45°, y_0 = 2.4\,\text{m}$ | ▪ Identify the relevant data. |
| 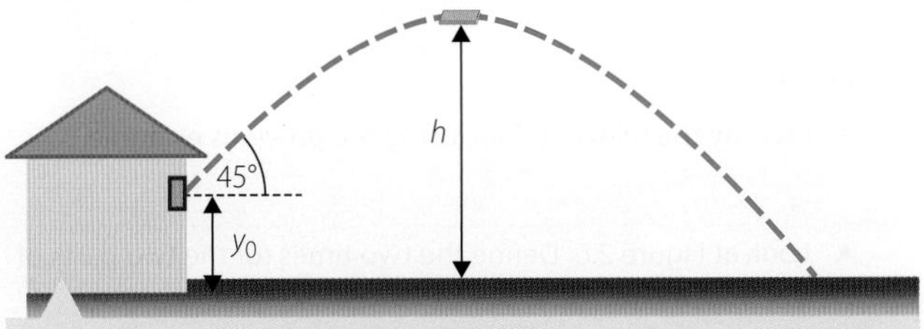  FIGURE 2.6 Trajectory of the maths book | ▪ Draw a diagram (Figure 2.6). |
| $h = y_0 - \dfrac{u_y^2}{2g}$ | ▪ Write the expression for the maximum height. |
| $u_y = u \sin\theta$ | ▪ Write the expression for the initial vertical velocity. |
| $h = y_0 - \dfrac{(u\sin\theta)^2}{2g}$ | ▪ Substitute the expression for initial velocity into the expression for height. |
| $h = 2.4\,\text{m} - \dfrac{(12\text{ m s}^{-1}\sin 45°)^2}{2 \times -9.8\text{ m s}^{-2}}$ | ▪ Substitute values with correct units. |
| $h = 6.07\,\text{m}$ | ▪ Calculate the final value. |
| $h = 6.0\,\text{m}$ | ▪ State the final answer with correct units and appropriate significant figures. |

**TRY THIS YOURSELF**
What would be the maximum height if Jen threw the book directly upwards with the same initial velocity?

## Time of flight

The total time between when a projectile is launched and when it hits the ground is called the **time of flight**. Look again at Figure 2.5a. Note that the path is symmetric, and is parabolic in shape. This is the case when there is no significant air resistance. When the launch height is the same as the final height, the time taken to reach the maximum height is the same as the time taken to go from the maximum height down to the ground again.

The time taken to reach the maximum height was found to be:

$$t = -\frac{u_y}{g}$$

so the time of flight, $t_{flight}$, must be twice this:

$$t_{flight} = -\frac{2u_y t}{g}$$

When the launch and final heights are not the same, it is simplest to consider the two parts of the flight separately; the ascent to the maximum height and the descent from the maximum height to the landing point.

► **WORKED EXAMPLE** (2.5)

Jen has thrown her maths text book out the window, 2.4 m above the ground. She throws it with initial velocity $12\,\text{m s}^{-1}$ at an angle of 45° above the horizontal. How long does it take for the book to reach the ground?

| ANSWER | LOGIC |
|---|---|
| $u = 12\,\text{m s}^{-1}, \theta = 45°, y_0 = 2.4\,\text{m}$ <br> $h = 6.07\,\text{m}$ | ▪ Identify the relevant data, using the previous example. |
| $t_1$ = time taken to reach maximum height <br> $t_2$ = time taken to fall from maximum height to the ground | ▪ Look at Figure 2.6. Define the two times for the two parts of the trajectory. |
| $t_1 = -\dfrac{u_y}{g} = -\dfrac{u\sin\theta}{g}$ | ▪ Write the expression for $t_1$. |
| $t_1 = -\dfrac{12\ \text{m s}^{-1}\sin 45°}{-9.8\ \text{m s}^{-2}}$ | ▪ Substitute values with correct units. |
| $t_1 = 0.866\,\text{s}$ | ▪ Calculate the value for $t_1$. |
| $-h = u_y t_2 + \dfrac{1}{2}g t_2^{\ 2}$ | ▪ Relate the maximum height to the time $t_2$. Note that we use $-h$ for the change in vertical position because the book is now going down. |
| $-h = \dfrac{1}{2}g t_2^{\ 2}$ | ▪ Simplify the expression, noting that $u_y = 0$ for this part of the path. |
| $t_2 = \sqrt{\dfrac{2(-h)}{g}}$ | ▪ Rearrange for $t_2$. |
| $t_2 = \sqrt{\dfrac{2(-6.07\ \text{m})}{-9.8\ \text{m s}^{-2}}}$ | ▪ Substitute values with correct units, noting that the change in vertical position is negative because the book is falling. |
| $t_2 = 1.113\,\text{s}$ | ▪ Calculate the value for $x_{max}$. |
| $t_{flight} = t_1 + t_2 = 0.866\,\text{s} + 1.113\,\text{s} = 1.979\,\text{s}$ | ▪ Calculate the final answer. |
| $t_{flight} = 2.0\,\text{s}$ | ▪ State the final answer with correct units and appropriate significant figures. |

**TRY THIS YOURSELF**

What would the time of flight be if Jen threw her maths book with the same initial velocity from ground level?

## Final velocity

For a projectile that lands at the same height from which it was launched the final velocity, the velocity at which the projectile lands, has the same magnitude as the initial velocity. This is because the horizontal velocity does not change and the vertical velocity increases by the same amount during the descent as it decreased during the ascent. The path is symmetrical, so the velocity is also symmetrical. The only difference is the angle. The angle of the final velocity is the same below the horizontal as the initial velocity was above the horizontal.

When the launch and landing heights are different, then it is easiest to consider just the descent phase of the trajectory. If we know the time of flight for this phase, we can calculate the final vertical velocity. If we know the initial (and hence also final) horizontal velocity we can calculate the total final velocity.

The trajectory of a projectile

▶ **WORKED EXAMPLE** (2.6)

What is the final velocity of the maths textbook in the previous example?

| ANSWER | LOGIC |
|---|---|
| $u = 12\,\mathrm{m\,s^{-1}}, \theta = 45°, t_2 = 1.113\,\mathrm{s}$ | ▪ Identify the relevant data, using previous examples. |
| $v_x = u_x = u\cos\theta$ | ▪ Write the expression for the horizontal velocity. |
| $v_x = 12\,\mathrm{m\,s^{-1}}\cos 45°$ | ▪ Substitute values with correct units. |
| $v_x = 8.49\,\mathrm{m\,s^{-1}}$ | ▪ Calculate the value for $v_x$. |
| $v_y = u_y + gt$ | ▪ Write the general expression for the vertical component of $v$. |
| $v_y = 0 + gt$ | ▪ Simplify the expression by considering only the descent stage. |
| $v_y = (-9.8\,\mathrm{m\,s^{-2}})(1.113\,\mathrm{s})$ | ▪ Substitute values with correct units. |
| $v_y = -10.91\,\mathrm{m\,s^{-1}}$ | ▪ Calculate the value for $v_y$. |
| $v = \sqrt{v_x^2 + v_y^2}$ | ▪ Write the expression for the magnitude of $v$. |
| $v = \sqrt{(8.49\ \mathrm{m\,s^{-1}})^2 + (-10.91\ \mathrm{m\,s^{-1}})^2}$ | ▪ Substitute values with correct units. |
| $v = 13.8\,\mathrm{m\,s^{-1}}$ | ▪ Calculate the final value. |
| $\tan\theta = \dfrac{v_y}{v_x}$ | ▪ Relate the angle to the velocity components. |
| $\theta = \tan^{-1}\left(\dfrac{v_y}{v_x}\right)$ | ▪ Rearrange for angle. |
| $\theta = \tan^{-1}\left(\dfrac{-10.91\ \mathrm{m\,s^{-1}}}{8.49\ \mathrm{m\,s^{-1}}}\right)$ | ▪ Substitute values with correct units. |
| $\theta = -52.1°$ | ▪ Calculate the final value. |
| $v_{\text{final}} = 14\,\mathrm{m\,s^{-1}}$ at 52° below the horizontal. | ▪ State the final answer with correct units and appropriate significant figures. |

**TRY THIS YOURSELF**

What would the final velocity be if Jen threw her maths book from ground level?

## Horizontal range of a projectile

The horizontal range of a projectile (see Figure 2.5) can be found from the time of flight. Recall that while the vertical component of the velocity varies with time, the horizontal component is constant. This is because we assume no horizontal forces are acting.

From the kinematics equations, the range, $x_{max}$, is:

$$x_{max} = u_x t_{flight}$$

▶ **WORKED EXAMPLE** (2.7)

When Jen threw her maths book out the window in the previous examples, at what horizontal distance from her window did it land?

| ANSWER | LOGIC |
|---|---|
| $u = 12\,\mathrm{m\,s^{-1}}$, $\theta = 45°$, $t_{flight} = 1.979\,\mathrm{s}$ | ▪ Identify the relevant data, using previous example. |
| $x_{max} = u_x t_{flight} = u\cos\theta\, t_{flight}$ | ▪ Write the expression for $x_{max}$. |
| $x_{max} = (12\,\mathrm{m\,s^{-1}})\cos 45°\,(1.979\,\mathrm{s})$ | ▪ Substitute values with correct units. |
| $x_{max} = 16.9\,\mathrm{m}$ | ▪ Calculate the value for $t_1$. |
| $x_{max} = 17\,\mathrm{m}$ | ▪ State the final answer with correct units and appropriate significant figures. |

**TRY THIS YOURSELF**
What would the range be if Jen threw her maths book from ground level?

Starting with the expression for the range, $x_{max} = u_x t_{flight}$, and substituting in the expression for time of flight, $t_{flight} = -\dfrac{2u_y}{g}$, we can write the expression for the range as:

$$x_{max} = u_x t_{flight} = \frac{2u_x u_y}{g} = \frac{2u^2 \sin\theta\,\cos\theta}{g}$$

This is for a projectile that lands at the launch height.

The maximum possible horizontal range of a projectile occurs for a launch angle of 45° when there is no air resistance. Figure 2.7 compares the trajectories of a projectile launched with the same initial velocity at different angles.

**Projectile motion calculators**

Use the calculators to model the trajectories of projectiles fired vertically, horizontally and at any arbitrary angle.

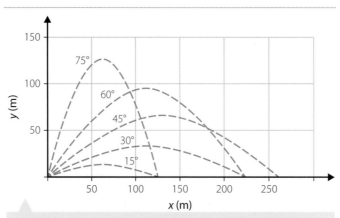

**FIGURE 2.7** The trajectory of a particle for different launch angles

## Horizontal range of a projectile

Critical and creative thinking

Numeracy

RISK ASSESSMENT

### AIM

To find the launch velocity and time of flight of a projectile
Write a hypothesis for this investigation.

### MATERIALS

- Curved ramp such as a toy car track
- Ball bearing
- Tape measure
- Sand tray

| WHAT ARE THE RISKS IN DOING THIS INVESTIGATION? | HOW CAN YOU MANAGE THESE RISKS TO STAY SAFE? |
| --- | --- |
| The ball bearing may hit someone. | Keep the area at the end of the ramp clear. |

What other risks are associated with your investigation, and how can you manage them?

### METHOD

1 Set up the curved track on the edge of a desk as shown in Figure 2.8. Ensure the end of the track is horizontal.

2 Measure the height $h_{desk}$ and the height $h_{track}$.

3 Release the ball bearing from the top of the track. If the ball bearing does not land in the sandbox, move the sand box so that the ball bearing will land in it next time.

4 Release the ball bearing from the top of the track.

5 Measure the distance, $x$, from the edge of the desk to the point at which the ball bearing landed in the sand.

6 Smooth the sand and repeat steps 4 and 5 *at least* five times.

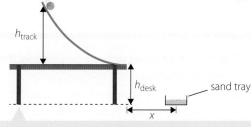

**FIGURE 2.8** Experimental set up

### RESULTS

Record your results as you measure them.
You should have a record of the heights $h_{desk}$ and $h_{track}$, and several measurements of the horizontal range.

### ANALYSIS OF RESULTS

1 Use the height $h_{desk}$ to calculate the time of flight of the ball bearing. Write down any assumptions that you make.

2 Calculate the average value of the horizontal range and its uncertainty.

3 Use the average value of the range and the time of flight to calculate the initial velocity of the ball bearing. Write down any assumptions that you make and calculate the uncertainty in the initial velocity.

4 Using conservation energy (see chapter 5 of *Physics in Focus Year 11*) calculate the expected velocity of the ball bearing as it leaves the ramp. Write down any assumptions that you make.

1 Did the expected value of the initial velocity, based on the height of the ramp, agree with the calculated value from the range? Don't forget to consider the uncertainty in the value.

2 If the values for the initial velocity did not agree, which of your assumptions do you think was not a valid approximation? Can you think of any way of testing which assumptions were valid?

3 State whether your hypothesis was supported or not.

**CONCLUSION**

Write a conclusion summarising the outcomes of your investigation.

**KEY CONCEPTS**

- The trajectory of a particle is a parabolic arc.
- The highest point of the arc is the maximum height. The maximum height can be found from $h = y_0 - \dfrac{u_y^2}{2g}$, where $y_0$ is the launch height and $u_y = u\sin\theta$.
- The time of flight of a projectile is the time between launch and landing, and is given by $t_{\text{flight}} = -\dfrac{2u_y}{g}$ when the launch and landing heights are the same. When the launch and landing heights are different, the motion can be broken into two stages, the ascent and the descent, to find the total flight time.
- The final velocity of a projectile that lands at the same height from which it was launched has the same magnitude as the initial velocity. The angle of the final velocity is as far below the horizontal as the initial velocity was above the horizontal.
- The horizontal range of a projectile is given by $x_{\text{max}} = u_x t_{\text{flight}}$.

**CHECK YOUR UNDERSTANDING**

**2.2**

1 Sketch separate speed versus time graphs for the horizontal and vertical motions of a projectile that is launched directly upwards and lands below the launch position.

2 Draw the trajectory of a projectile that is launched at an angle $\theta$ above the horizontal with an initial velocity $u$. On your sketch, draw vectors to represent the horizontal, vertical and total velocity:

a immediately after the projectile is launched.

b at the highest point of its trajectory.

c just before it hits the ground.

3 Find the range of a cannon ball that is launched with a velocity of $300\,\text{m s}^{-1}$ at an angle of $35°$ above the horizontal and lands at the same height as the launch site.

4 Figure 2.7 shows the trajectory of a projectile when launched at different angles. Using any one of the trajectories shown, calculate the initial speed of the projectile.

5 Find the time of flight for each of the projectiles shown in Figure 2.7. Sketch a graph of time of flight vs launch angle.

6 A rock is thrown from a 55 m high cliff at an angle of $20°$ below the horizontal at an initial speed of $35\,\text{m s}^{-1}$. Find:

a the time of flight.

b the range.

c the final velocity of the rock.

# 2.3 Investigating projectile motion

Projectile motion can be investigated experimentally using simple equipment. The biggest challenge is reproducibility, as it is very hard to throw a ball consistently at the same speed and initial angle. Usually some sort of projectile launcher is required. In the investigation described below a tennis ball launcher, such as is used for tennis or cricket practice, is used. Other projectile launchers such as nerf guns can also be used, and it is fairly simple to make your own.

*Projectile launchers can be very dangerous, and you should only use low-powered launchers and soft projectiles.*

In the investigation below you will make measurements and then use those measurements to reconstruct the trajectory of the projectile. In the next section we shall see how the trajectory can be modelled using a simple computer simulation.

## INVESTIGATION (2.2)

### Trajectory of a projectile

**AIM**

To plot the trajectory of a projectile

Write a hypothesis or an inquiry question for this investigation.

○ Critical and creative thinking

○ Numeracy

**MATERIALS**

- Tennis ball launcher
- Tennis ball
- Chalk dust
- Stopwatch
- Large tape measure (at least 30 m)
- Large protractor
- Extension cord and power source for tennis ball launcher

RISK ASSESSMENT

| WHAT ARE THE RISKS IN DOING THIS INVESTIGATION? | HOW CAN YOU MANAGE THESE RISKS TO STAY SAFE? |
|---|---|
| The ball may hit someone or break a window. | Perform the investigation in an open area away from other people and stand clear of the launcher when firing. |

What other risks are associated with your investigation, and how can you manage them?

**METHOD**

1. Load and fire a tennis ball a few times to check the range. You will need the ball to land somewhere where it will leave a chalk mark, such as on concrete.
2. Measure the launch angle with the protractor. This will be the angle between the launch tube and the ground.
3. Roll the ball in chalk dust so it has a thin coating.
4. Fire the ball from the launcher.
5. Use the stopwatch to measure the time of flight. Record the time and its uncertainty. Remember that the uncertainty will be more than the limit of reading error (see page 18).

»

6    Measure the horizontal range of the projectile from the launcher to the chalk mark made when the ball first hit the ground. Record the uncertainty in this measurement.

7    Adjust the angle of the launcher and repeat steps 1 to 7. Do this twice so that you have three sets of data.

If you have a video camera with a high frame rate you can also record the flight and make measurements from your recordings.

## RESULTS

Record your results for angle, range and time of flight as you measure them. Write them in a table such as the one below.

| | | | |
|---|---|---|---|
| Launch angle (°) | | | |
| Range (m) | | | |
| Time of flight (s) | | | |
| Maximum height (m) | | | |
| Initial horizontal velocity ($m\,s^{-1}$) | | | |
| Initial vertical velocity ($m\,s^{-1}$) | | | |
| Total initial velocity ($m\,s^{-1}$) | | | |
| Horizontal velocity at top of arc ($m\,s^{-1}$) | | | |
| Vertical velocity at top of arc ($m\,s^{-1}$) | | | |
| Total velocity at top of arc ($m\,s^{-1}$) | | | |
| Final horizontal velocity ($m\,s^{-1}$) | | | |
| Final vertical velocity ($m\,s^{-1}$) | | | |
| Total final velocity ($m\,s^{-1}$) | | | |

## ANALYSIS OF RESULTS

Copy the results table and complete it by making the following calculations.

1    Use the measured time of flight to calculate the maximum height of the tennis ball's trajectory and the initial vertical velocity. State any assumptions that you make.

2    Use the measured values of the horizontal range and the angle to calculate the initial horizontal velocity and the initial vertical velocity. State any assumptions that you make.

3    Draw a vector diagram and calculate the total initial velocity.

4    Calculate the horizontal, vertical and total velocity at the top of the trajectory and at the end, just before the tennis ball hits the ground. State any assumptions that you make.

5    Draw scale diagrams of the trajectory for each launch angle, like that shown in Figure 2.9. Note the scale and show the maximum height and the horizontal range on your diagrams. Write the values for the velocity and its components at the start, top and end of the trajectory. Draw vector diagrams for the velocity at these points on your diagram, as shown in Figure 2.9.

»

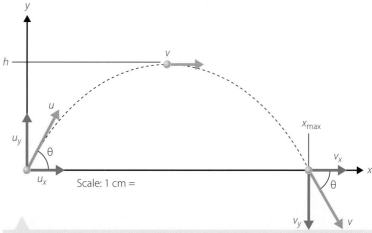

**FIGURE 2.9** Make a scale drawing for each of the three trajectories. Note the scale, and write the values for the velocity and its components at the start, peak and end of the trajectory.

## DISCUSSION

1 Compare and contrast the three trajectories. Comment on how the velocity components change during the flight of the projectile.

2 Did the two calculated values of the initial vertical velocity agree? Do not forget to consider the uncertainties in the values.

3 Do you think the assumptions that you made were valid?

4 How could this investigation be improved or extended?

5 Give the answer to your inquiry question or state whether your hypothesis was supported or not.

## CONCLUSION

Write a conclusion summarising the outcomes of your investigation.

**KEY CONCEPTS**

- Projectile motion can be investigated using a simple projectile launcher and a suitable projectile.
- Care must be taken as projectiles can move fast and be very dangerous.
- The velocity of a projectile and its horizontal and vertical components can be calculated from the range, time of flight and launch angle of a projectile using the kinematics equations.
- The model for projectile motion we are using does not include air resistance, so it is an approximation.

**CHECK YOUR UNDERSTANDING**

**2.3**

1 Why do we assume that a projectile travels the same horizontal distance before and after the peak of its trajectory? If we do not make this assumption, in which part of its trajectory does it travel the greater horizontal distance?

2 In a projectile motion experiment, Ali measures the time of flight to be $3.52\,\text{s} \pm 0.02\,\text{s}$. Calculate the maximum height of the trajectory and the uncertainty in this value. State any assumptions that you make.

3 Calculate the initial vertical velocity of the projectile in Question **2** above. Calculate the uncertainty in this value.

4 A projectile reaches a maximum height of 15 m and has a horizontal range of 25 m. Calculate the magnitude and angle of the initial velocity.

5 Copy the sketch of the projectile trajectory shown in Figure 2.10. At each of points B, C, D and E, draw a vector diagram showing the velocity and its components. In a different colour, draw a vector representing the acceleration at each of these points.

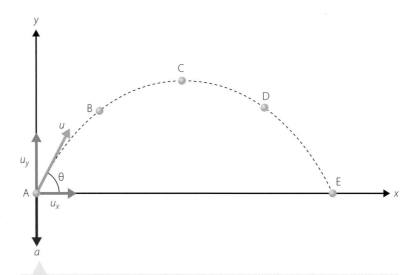

FIGURE 2.10 A projectile's trajectory. Add vector diagrams showing the velocity, velocity components and acceleration at each point.

6 In a projectile motion experiment, Umut measures the launch angle to be $45° \pm 1°$ and the horizontal range to be $(15.5 \pm 0.1)$ m. Calculate the initial velocity and its uncertainty. Use the range method to calculate the uncertainty.

## 2.4 Modelling projectile motion

Physicists use different ways of modelling a system to help them understand it and to solve problems and make predictions.

So far we have modelled projectile motion mathematically by treating it as a combination of two rectilinear motions. Rectilinear motion is motion in a straight line, or along a single direction. Projectile motion is a combination of uniformly accelerated motion in the vertical direction and constant (zero acceleration) motion in the horizontal direction. The *mathematical model* that we have used is shown in Table 2.1. It is the constant acceleration kinematics equations for the vertical motion and the constant velocity equations for the horizontal motion. In the first two sections of this chapter we used this mathematical model to solve problems and make predictions. As we shall see in chapter 5, this model also applies to charged particles moving in constant electric fields.

In Investigation 2.2, you used experimental data, combined with the mathematical model, to make a *graphical* or *visual model* of projectile motion. Visual representations such as that shown in Figure 2.9 enable us to see relationships; in this case the relationship between launch angle and range.

A third type of model, which also draws on the mathematical model we have been using, is a *numerical model* or computer simulation.

**Projectile motion simulations**

Experiment with the simulations to see how initial conditions affect the trajectory.

# Computer simulation of the trajectory of a projectile

Critical and creative thinking

Numeracy

Information and communication technology capability

**AIM**

To use spreadsheet software to model the trajectory of a projectile

Write a hypothesis or an inquiry question for this investigation.

**MATERIALS**

▪ Computer with spreadsheet software

**METHOD**

Open a new file in your spreadsheet software and put a descriptive heading at the top of the sheet, then save it with a suitable filename. *Remember to save your file frequently.* It is also a good idea to save multiple pages in your file for different variations of your simulation so you can return to earlier versions.

1 Set up the initial conditions for the projectile. These are the initial velocity, launch angle and acceleration due to gravity. Figure 2.11 shows how this can be done using Excel spreadsheet software. Cells are labelled above the value. Because Excel uses angles in radians when calculating sine and cosine, the launch angle, θ, is entered in degrees in cell B4 and then converted to radians in cell C4 using the command **=RADIANS(B4)**.

The components of the initial velocity are then calculated below. Cell A6 contains the command **=25*COS(C4)** for the horizontal component. Cell B6 contains the command **=25*SIN(C4)** for the horizontal component.

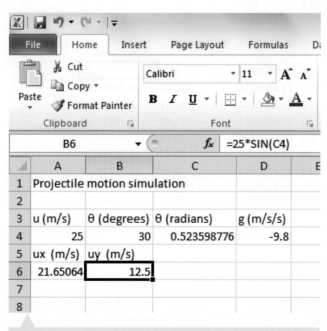

**FIGURE 2.11** Set up the initial conditions in your spreadsheet.

2 Implement the equations of motion to calculate the velocity as a function of time. Label three columns for time, $v_x$ and $v_y$. Enter the initial time, 0, in the cell immediately under the time label. Below this cell enter a command to increase the time. In Figure 2.12 this command is **=A10+0.05**, but you can add any time increment you want to make the time steps bigger or smaller. Copy this command into as many cells below as you want (use *shift* and *page down* to highlight many cells at once).

»

>> Set up the horizontal velocity in the next column. This is constant, and always equal to the initial horizontal velocity. So cell B10 and all those below it contain the command **=$A$6**. Note that the $ signs tell the software that the cell reference is not to be changed, unlike in the command for calculating the time, which updates the reference in each cell.

Set up the vertical velocity in the third column. This is the constant acceleration part of the motion. The command used in cell C10 and copied to those below it is **=$B$6+A10*$D$4**. This command takes the initial velocity and adds *at* to it, where $a = g$.

Add a fourth column to give the magnitude of the total velocity. This is found using Pythagoras theorem, and is calculated in cell D10 using the command **=SQRT(B10*B10+C10*C10)**, which is then copied into the column of cells below.

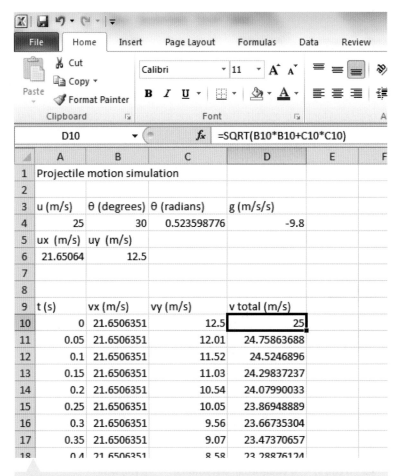

**FIGURE 2.12** Implement the equations of motion to calculate the velocity as a function of time.

3  Implement the equations of motion to calculate the position as a function of time, as shown in Figure 2.13. Label the next two columns x and y. The *x* position is calculated in the cell beneath the x label using the command **=$A$6*A10**, which is then copied into the cells below.

The *y* position is calculated in the cell beneath the y label using the command **=$B$6*A10+0.5*$D$4*A10*A10**, which is then copied into the cells below.

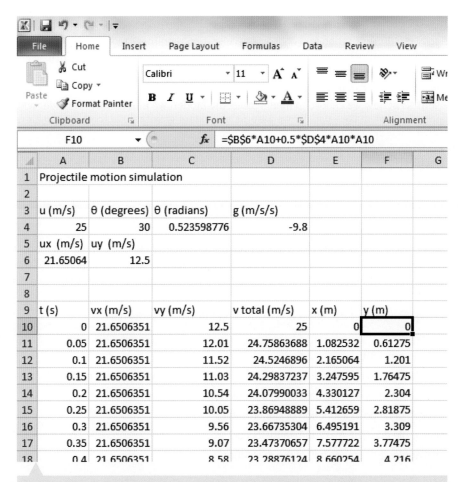

Cell F10 formula bar: `=$B$6*A10+0.5*$D$4*A10*A10`

**Projectile motion simulation**

| | A | B | C | D | E | F |
|---|---|---|---|---|---|---|
| 1 | Projectile motion simulation | | | | | |
| 2 | | | | | | |
| 3 | u (m/s) | θ (degrees) | θ (radians) | g (m/s/s) | | |
| 4 | 25 | 30 | 0.523598776 | -9.8 | | |
| 5 | ux (m/s) | uy (m/s) | | | | |
| 6 | 21.65064 | 12.5 | | | | |
| 7 | | | | | | |
| 8 | | | | | | |
| 9 | t (s) | vx (m/s) | vy (m/s) | v total (m/s) | x (m) | y (m) |
| 10 | 0 | 21.6506351 | 12.5 | 25 | 0 | 0 |
| 11 | 0.05 | 21.6506351 | 12.01 | 24.75863688 | 1.082532 | 0.61275 |
| 12 | 0.1 | 21.6506351 | 11.52 | 24.5246896 | 2.165064 | 1.201 |
| 13 | 0.15 | 21.6506351 | 11.03 | 24.29837237 | 3.247595 | 1.76475 |
| 14 | 0.2 | 21.6506351 | 10.54 | 24.07990033 | 4.330127 | 2.304 |
| 15 | 0.25 | 21.6506351 | 10.05 | 23.86948889 | 5.412659 | 2.81875 |
| 16 | 0.3 | 21.6506351 | 9.56 | 23.66735304 | 6.495191 | 3.309 |
| 17 | 0.35 | 21.6506351 | 9.07 | 23.47370657 | 7.577722 | 3.77475 |
| 18 | 0.4 | 21.6506351 | 8.58 | 23.28876124 | 8.660254 | 4.216 |

Spreadsheet file

**FIGURE 2.13** Implement the equations to calculate the position as a function of time.

You now have a numerical model of projectile motion that can be easily modified to investigate the effects of changing launch angle, initial velocity and acceleration due to gravity. When you decide what you want to investigate, copy your starting worksheet so you have an original and a copy. Modify the copy by changing the variable of interest. You can then compare the two models.

## RESULTS AND ANALYSIS OF RESULTS

*Make sure that you have saved your file!*

Plot scatter graphs of your data. Create graphs of:

1 horizontal, vertical and total velocity as a function of time

2 horizontal, vertical and total velocity as a function of position

3 trajectory (vertical position as a function of horizontal position).
Save and/or print these graphs.

## DISCUSSION

1 Comment on the shape of your graphs. Are they what you expect, given the physical experiments you have done with projectiles and the equations used?

2 What happened when you varied the initial conditions? For example, if you varied the launch angle, what happened to the range and maximum height? Can you generate a graph like that shown in Figure 2.7? If you varied *g*, what happened to the time of flight, range and maximum height?

3 Give the answer to your inquiry question or state whether your hypothesis was supported or not.

## CONCLUSION

Write a conclusion summarising the outcomes of your investigation.

● Projectile motion can be modelled mathematically, using the kinematics equations for constant velocity rectilinear motion and constant acceleration rectilinear motion.

● Projectile motion can be modelled graphically, using visual representations such as graphs and diagrams.

● Projectile motion can be modelled numerically, using computer simulations.

● Regardless of how we choose to model projectile motion we must always consider what assumptions we make – usually we assume a constant gravitational force and no air resistance or other friction forces.

● We can use models of projectile motion to solve problems and to predict the motion of projectiles.

---

**CHECK YOUR UNDERSTANDING**

**2.4**

1 Describe three different ways in which projectile motion can be modelled.

2 A simulation of projectile motion is used to investigate the effects of changing the gravitational force acting on a projectile. Describe how the trajectory of a projectile in a small gravitational field would be different from the trajectory in a large gravitational field. Draw a diagram showing the two different trajectories.

3 In the investigation above, what effect does increasing the initial velocity have on:

 a the range.

 b the maximum height.

4 Sketch a graph showing how the horizontal, vertical and total velocity vary with position for a projectile.

5 Minh has collected the following data using a simulation of projectile motion.

 a Plot a graph of range as a function of launch angle.

 b For what angle does the maximum range occur?

 c For what angles is the range half of the maximum?

**Physclips interactive projectile motion tutorials**

Work along with the calculations and try the experiments for yourself.

| θ (°) | 0 | 5 | 10 | 15 | 20 | 25 | 30 | 35 | 40 | 45 | 50 | 55 | 60 | 65 | 70 | 75 | 80 | 85 | 90 |
|---|---|---|---|---|---|---|---|---|---|---|---|---|---|---|---|---|---|---|---|
| Range (m) | 0 | 44 | 87 | 128 | 164 | 195 | 221 | 240 | 251 | 255 | 251 | 240 | 221 | 195 | 164 | 128 | 87 | 44 | 0 |

6 Use a spreadsheet to create a numerical model of the time of flight of a projectile as a function of launch angle, for an initial speed of $50\,\text{m s}^{-1}$.

 a Plot a graph of time of flight as a function of launch angle.

 b Describe the shape of the graph.

- A projectile is an object that has been launched or thrown, and does not propel itself once it is lauched.

- Our model of projectile motion assumes that the projectile is subject only to the force of gravity. It ignores air resistance and assumes that the projectile remains close to Earth's surface so that $g$ is constant.

- The trajectory (path) of a projectile is determined only by the initial (launch) velocity, and the force of gravity.

- The initial velocity, $u$, has components $u_x = u \cos \theta$ and $u_y = u \sin \theta$.

- The horizontal and vertical components of a projectile's position, velocity and acceleration are determined by the kinematics equations and are given in Table 2.1.

- The trajectory of a particle is a parabolic arc.

- The highest point of the arc is the maximum height. The maximum height can be found from $h = y_0 - \dfrac{u_y^2}{2g}$, where $y_0$ is the launch height, $u_y = u \sin \theta$, and $g = -9.8\,\mathrm{m\,s^{-1}}$.

- The time of flight of a projectile is the time between launch and landing, and is given by $t_{flight} = -\dfrac{2u_y}{g}$ when the launch and landing heights are the same. When the launch and landing heights are different, the motion can be broken into two stages – the ascent and the descent – to find the total flight time.

- The horizontal range of a projectile is given by $x_{max} = u_x t_{flight}$.

- Projectile motion can be investigated using a simple projectile launcher and a suitable projectile.

- Projectile motion can be modelled mathematically using the kinematics equations for constant velocity rectilinear motion and constant acceleration rectilinear motion.

- Projectile motion can also be modelled graphically using diagrams, and numerically using computer simulations.

# 2 CHAPTER REVIEW QUESTIONS

Review quiz

1. List the kinematic equations associated with projectile motion. Define all variables.

2. What is the horizontal acceleration of a projectile?

3. What general expression can be used to find the horizontal component, $u_x$, of a projectile's initial velocity?

4. A shell is fired with initial velocity components $u_x = 550\,\mathrm{m\,s^{-1}}$ and $u_y = 650\,\mathrm{m\,s^{-1}}$. Calculate the initial velocity and launch angle.

5. A cricket ball is bowled with an initial velocity of $120\,\mathrm{km\,h^{-1}}$ at an angle of $10°$ below the horizontal. Calculate the horizontal and vertical components of its initial velocity.

6. Four different projectile trajectories are shown in Figure 2.14. Rank these trajectories from shortest to longest time of flight. Justify your answer.

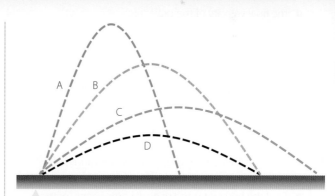

**FIGURE 2.14** Four different trajectories. Rank the time of flight for the trajectories from shortest to longest.

**7** A ball is thrown vertically upwards at $35\,\text{m s}^{-1}$.

    **a** What is the maximum height reached by the ball?

    **b** How long will the ball take to reach this height?

**8** Are there any points on the trajectory for a projectile that is launched directly upwards at which the acceleration and velocity vectors are parallel? Are there any points at which they are perpendicular? If so, identify these points on a sketch of a trajectory.

**9** Are there any points on the trajectory for a projectile that is not launched directly upwards at which the acceleration and velocity vectors are parallel? Are there any points at which they are perpendicular? If so, identify these points on a sketch of a trajectory.

**10** A plane is flying over a house that has been cut off by flood waters. To drop a package of food on to the roof, should the package be dropped before the plane is over the house, when it is directly overhead, or after it has passed? Explain your answer, and include a diagram showing the path of the package.

**11** A ball is thrown directly upwards at an initial speed of $45\,\text{m s}^{-1}$.

    **a** What maximum height does it reach?

    **b** At what heights is its speed half that of its initial speed?

**12** A projectile is launched at an initial speed of $45\,\text{m s}^{-1}$ at an angle of 60° above the horizontal, from a height of 10 m above the ground.

    **a** Sketch the path of the trajectory.

    Calculate, and mark on your sketch:

    **b** the maximum height.

    **c** the horizontal range.

    **d** the final velocity of the projectile.

**13** Harry is throwing a tantrum. He shoves his dinner plate and it slides across the table and over the edge, to land 1.5 m from the edge of the table, which is 1.2 m high.

    **a** At what speed did the plate leave the table?

    **b** What was the time of flight of the plate?

    **c** At what speed (total velocity) does the plate hit the floor?

**14** A projectile is fired from ground level at an angle of 45° above the horizontal with a speed of $38\,\text{m s}^{-1}$.

    **a** Find the horizontal and vertical components of the projectile's initial velocity.

    **b** Find the time of flight for the projectile, assuming it is travelling over level ground.

    **c** Find the horizontal distance (the range) of this projectile.

**15** A projectile is launched at an initial speed of $45\,\text{m s}^{-1}$ at an angle of 60° above the horizontal, from a height of 10 m above the ground. Copy and complete the table by calculating the horizontal and vertical position of the projectile.

| $t$ (s) | 1 | 2 | 3 | 4 | 5 | 6 | 7 | 8 | 9 |
|---|---|---|---|---|---|---|---|---|---|
| $x$ (m) | | | | | | | | | |
| $y$ (m) | | | | | | | | | |

**16** A golf ball is struck at $45\,\text{m s}^{-1}$ and an angle of 15° above the horizontal.

    **a** What is the vertical component of the ball's initial velocity?

    **b** Will the ball pass over the head of a 2.0 m high golfer standing 50 m away?

**17** In February 1971, astronaut Alan Shepherd hit a golf ball on the Moon. If he hit a ball with an initial speed of $50\,\text{m s}^{-1}$ at an angle of 20° above the horizontal, what was its range? With what speed would he have needed to hit the ball, at that same angle, to get the same range on Earth? $g_{Moon} = -1.6\,\text{m s}^{-2}$

**18** An inexperienced archer fires an arrow at a target 100 m away. She fires along a line pointing directly at the centre of the target, with an initial speed of $55\,\text{m s}^{-1}$ at an angle of 15° above the horizontal. By how much does the arrow miss the centre of the target?

**19** How does the trajectory of a real projectile differ from that of the model of projectile motion used in this chapter?

**20** A projectile is fired at an angle of 20° above the horizontal. Another projectile is fired at the same speed at an angle of 70° above the horizontal. Show that the horizontal distance travelled by each projectile is the same.

**21** A projectile is fired such that its maximum height is equal to its range. At what launch angle was the projectile fired?

# 3 Circular motion

## INQUIRY QUESTION
Why do objects move in circles?

**OUTCOMES**

**Students:**

- conduct investigations to explain and evaluate, for objects executing uniform circular motion, the relationships that exist between:
  - centripetal force
  - mass
  - speed
  - radius
- analyse the forces acting on an object executing uniform circular motion in a variety of situations, for example:
  - cars moving around horizontal circular bends
  - a mass on a string
  - objects on banked tracks (ACSPH100) CCT, ICT
- solve problems, model and make quantitative predictions about objects executing uniform circular motion in a variety of situations, using the following relationships:

  - $a_c = \dfrac{v^2}{r}$

  - $v = \dfrac{2\pi r}{T}$

  - $F_c = \dfrac{mv^2}{r}$   ICT, N

  - $\omega = \dfrac{\Delta\theta}{t}$

- investigate the relationship between the total energy and work done on an object executing uniform circular motion
- investigate the relationship between the rotation of mechanical systems and the applied torque
  $-\tau = r_\perp F = rF\sin\theta$ ICT, N

Physics Stage 6 Syllabus © 2017 NSW Education Standards Authority (NESA) for and on behalf of the Crown in right of the State of New South Wales.

FIGURE 3.1 A car on a horizontal road turns as a result of the friction force applied to the tyres by the road.

Source: Shutterstock.com/Restuccia Giancarlo

In chapter 2 we saw that an object can move in a parabolic path as a result of the gravitational force. In this chapter we look at how different forces can result in motion along a circular path.

Newton's first law states that an object travelling in a straight line at a constant speed has no net force acting on it. What happens when the object travels at constant speed going around a corner? Because the direction of motion is changing, clearly a net force must be acting on the object. For the car in Figure 3.1, the road surface pushes on the car through the tyres. It is the friction force of the road on the tyres that makes the car turn the corner.

We will see that circular motion can be the result of all sorts of forces, singly or in combination, but we can apply the same mathematical model to describe the motion in each case.

## 3.1 Uniform circular motion

Uniform circular motion is motion in a circular path at constant speed. The motion may be that of a whole object, such as a totem tennis ball moving around on its string, or a single point on an object, such as a person sitting on a Ferris wheel. The motion may be for one or many circles or only a fraction of a complete circle. We can model cornering of a car as circular motion for some fraction of a circle. Uniform circular motion gives us a mathematical model that we can use to describe and predict these motions.

### Velocity and acceleration in circular motion

An object going around a circle of radius $r$ at a constant speed $v$ takes a **period** of time $T$ to complete one revolution as it moves around the circumference. Speed is the distance travelled per unit time interval. In circular motion the object travels one circumference in one time period, so the speed is:

$$v = \frac{\text{circumference}}{\text{time}} = \frac{2\pi r}{T}$$

Although the object's speed is constant, its direction is changing, so its velocity is changing. The velocity at any point on a circle is the speed in the direction along the tangent to the circle at that point. This is shown in Figure 3.2.

FIGURE 3.2 The velocity of the object at any point on the circle is tangential to the circle at that point.

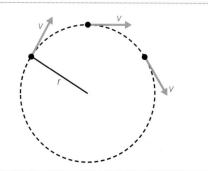

9780170409131

The velocity is constantly changing, so the object is accelerating. Recall from chapter 2 of *Physics in Focus Year 11* that acceleration is the rate of change of velocity. In Figure 3.3a, $v_1$ and $v_2$ are velocity vectors separated by a time interval $\Delta t$. The average acceleration, $a_{ave}$, over the time interval is:

$$a_{ave} = \frac{\Delta v}{\Delta t}$$

In the time interval $\Delta t$, the distance travelled by the object is $v\Delta t$ and the change in velocity is $\Delta v$. Figure 3.3b shows the vector subtraction, $\Delta v = v_2 - v_1$. The direction of this change in velocity is towards the centre of the circle. Because acceleration, $a$, is in the same direction as $\Delta v$, the acceleration also points towards the centre of the circle, as shown in Figure 3.3c.

This is called a **centripetal acceleration**, because it points towards the centre. Centripetal means centre-seeking.

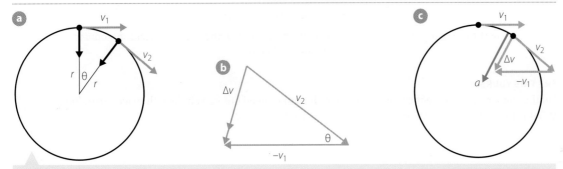

**FIGURE 3.3 a** The velocity vectors for a particle undergoing circular motion at two times $\Delta t$ apart; **b** The vector subtraction of the two velocities gives the change in velocity, $\Delta v$; **c** The acceleration vector, $a_{ave} = \dfrac{\Delta v}{\Delta t}$, is parallel to $\Delta v$ and points radially inwards towards the centre of the path.

To find the magnitude of the centripetal acceleration we can use similar triangles. Figure 3.4 shows two triangles. The first shows the positions of the object at times $t_1$ and $t_2$, and the displacement over the time $\Delta t$. The second triangle is the same one as shown in Figure 3.3b, but rotated so that the angle $\theta$ is in the same place as the angle $\theta$ in Figure 3.4a.

Using the two triangles in Figure 3.4, we can say that:

$$\frac{\Delta v}{v} = \frac{v\Delta t}{r}$$

$$\frac{\Delta v}{\Delta t} = \frac{v^2}{r}$$

$$a_c = \frac{v^2}{r}$$

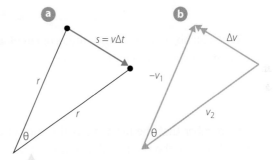

**FIGURE 3.4 a** Triangle showing radial vectors at two positions, separated by a displacement $s = v\Delta t$; **b** A similar triangle showing the velocity vectors. This is the same triangle as in Figure 3.3a, rotated to place the angle $\theta$ in the same position.

where $a_c$ is the magnitude of the centripetal acceleration, which points towards the centre of the circular path and $v$ is the magnitude of the velocity vectors, $v_1$ and $v_2$.

Ben is on a ride at the Royal Easter Show. The carriage he is in moves in a circle of radius 3.5 m at a speed of $10 \, \text{m s}^{-1}$. What is Ben's centripetal acceleration?

| ANSWER | LOGIC |
|---|---|
| $v = 10 \, \text{m s}^{-1}, r = 3.5 \, \text{m}$ | ▪ Identify the relevant data. |
| $a = \dfrac{v^2}{r}$ | ▪ Write the expression for the centripetal acceleration. |
| $a = \dfrac{(10 \, \text{m s}^{-1})^2}{3.5 \, \text{m}}$ | ▪ Substitute values with correct units. |
| $a = 28.6 \, \text{m s}^{-2}$ | ▪ Calculate the final value. |
| $a = 29 \, \text{m s}^{-2}$ towards the centre of the circular path. | ▪ State the final answer with correct units and appropriate significant figures. |

**TRY THIS YOURSELF**

On the same ride, a different carriage moves at the same speed but in a circle with twice the radius. What is the acceleration experienced by people in that carriage?

## Centripetal force

Newton's second law (see chapter 4 of *Physics in Focus Year 11*) tells us that when an object accelerates, it does so because there is a net force acting on it. The net force is:

$$F_{\text{net}} = \Sigma F = ma$$

When an object is moving in uniform circular motion we call the net force the centripetal force. Its direction is towards the centre of the circle, and its magnitude is given by:

$$F_{\text{c}} = \frac{mv^2}{r}$$

Remember that the net force is the sum of all forces acting. So the centripetal force is a sum of forces. It will be the result of different combinations of forces that depend on the particular situation. For example, when a car drives around a corner the forces acting on it are the normal force of the road, the gravitational force of Earth and the friction force of the road. It is the sum of all these that gives the centripetal force, although on a flat road it is only the friction force that acts horizontally and so provides a centripetal acceleration. In other situations, the forces involved may include a tension force, and electric and magnetic forces.

Ben is on a ride at the Royal Easter Show. The carriage he is in moves in a circle of radius 3.5 m at a speed of 10 m s$^{-1}$. Ben has a mass of 65 kg. What net force is applied to Ben?

| ANSWER | LOGIC |
|---|---|
| $v = 10$ m s$^{-1}$, $r = 3.5$ m, $m = 65$ kg | ▪ Identify the relevant data. |
| $\Sigma F = \dfrac{mv^2}{r}$ | ▪ Write the expression for the net (centripetal) force. |
| $\Sigma F = \dfrac{65 \text{ kg}(10 \text{ m s}^{-1})^2}{3.5 \text{ m}}$ | ▪ Substitute values with correct units. |
| $\Sigma F = 1857$ kg m s$^{-2}$ | ▪ Calculate the final value. |
| $\Sigma F = 1900$ N towards the centre of the circular path. | ▪ State the final answer with correct units and appropriate significant figures. |

**TRY THIS YOURSELF**

If Ben is to experience a net force equal to twice the gravitational force that acts on him, at what speed does his carriage need to move?

## Angular position and velocity

So far we have described motion in terms of linear distances and speeds, with units of m and m s$^{-1}$ respectively. However, when describing circular motion or rotational motion it is often more useful to use angular units.

### Angular displacement

The angular displacement of an object or a point on an object that is moving in a circle can be described by the angle, $\Delta\theta$, through which it has moved (Figure 3.5).

Figure 3.5 shows the path of an object that has moved through an arc of a circle of length $s$. The angular displacement of the object is $\Delta\theta$. Using the definition of radians we can write the angular displacement as $\Delta\theta = \dfrac{s}{r}$.

The units of $\Delta\theta$ are radians, *not degrees*. The radian is an unusual unit in that it is dimensionless, it is a pure number because it is a ratio of two lengths.

To convert between radians and degrees, remember that $2\pi$ rad $= 360°$.

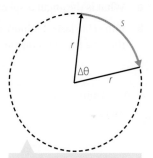

**FIGURE 3.5** An object moving in a circular path through a distance $s$ has an angular displacement $\Delta\theta$.

Circular motion: Some basic ideas and calculations

## Angular velocity

The angular velocity, $\omega$, is the angular displacement per unit time:

$$\omega = \frac{\Delta\theta}{\Delta t}$$

$\omega$ has units of radians per second, $\mathrm{rad\,s^{-1}}$.

If we assume the change in time, $\Delta t$, is from $t = 0$ to some later time $t$ then we can write this equation as:

$$\omega = \frac{\Delta\theta}{t}$$

We can relate the angular velocity to the linear velocity, $v$, by remembering the definition of radians:

$$\omega = \frac{\Delta\theta}{\Delta t} = \frac{\Delta s}{r\Delta t} = \frac{v}{r}$$

so $v = \omega r$.

Remember that $v$ has units of $\mathrm{m\,s^{-1}}$ and $\omega$ has units of $\mathrm{rad\,s^{-1}}$. The unit of radians comes from the ratio m/m.

▶ **WORKED EXAMPLE** (3.3)

Ben is on a ride at the Royal Easter Show. The carriage he is in moves in a circle of radius 3.5 m at a speed of $10\,\mathrm{m\,s^{-1}}$.

**a** What is the angular speed, $\omega$, at which Ben moves?

**b** If the ride lasts 3 minutes, what is Ben's angular displacement in this time? Make the approximation that the speed is constant over this time.

| ANSWER | LOGIC |
|---|---|
| $v = 10\,\mathrm{m\,s^{-1}}$, $r = 3.5\,\mathrm{m}$, $\Delta t = 3\,\mathrm{min}$ | ▪ Identify the relevant data. |
| $t = 3\,\mathrm{min} \times 60\,\mathrm{s\,min^{-1}} = 180\,\mathrm{s}$ | ▪ Convert to SI units. |
| **a** $\quad \omega = \dfrac{v}{r}$ | ▪ Write the expression for the angular velocity. |
| $\omega = \dfrac{10\ \mathrm{m\,s^{-1}}}{3.5\ \mathrm{m}}$ | ▪ Substitute values with correct units. |
| $\omega = 2.85\,\mathrm{rad\,s^{-1}}$ | ▪ Calculate the final value. |
| $\omega = 2.9\,\mathrm{rad\,s^{-1}}$ | ▪ State the final answer with correct units and appropriate significant figures. |
| **b** $\quad \omega = \dfrac{\Delta\theta}{t}$ | ▪ Relate the angular velocity to the angular displacement. |
| $\Delta\theta = \omega t$ | ▪ Rearrange for the angular displacement. |
| $\Delta\theta = (2.85\,\mathrm{rad\,s^{-1}})(180\,\mathrm{s})$ | ▪ Substitute values with correct units. |
| $\Delta\theta = 513\,\mathrm{rad}$ | ▪ Calculate the final value. |
| $\Delta\theta = 510\,\mathrm{rad}$ | ▪ State the final answer with correct units and appropriate significant figures. |

**TRY THIS YOURSELF**

How many revolutions does Ben do in this time?

# Measuring friction with a turntable

## AIM

To measure the maximum static friction force between an object and a turntable

Write an inquiry question for this investigation.

Critical and creative thinking

Numeracy

Information and communication technology capability

## MATERIALS

- Ruler
- Stopwatch
- Turntable, for example a two-speed record player or a pottery wheel
- Eraser
- Coin
- Weighing scale

RISK ASSESSMENT

| WHAT ARE THE RISKS IN DOING THIS INVESTIGATION? | HOW CAN YOU MANAGE THESE RISKS TO STAY SAFE? |
|---|---|
| Long hair can get caught in rotating equipment. | Always tie back long hair when doing experiments. |

What other risks are associated with your investigation, and how can you manage them?

## METHOD

1 Measure the mass of the eraser and the coin.

2 Place the eraser at a point close to the centre of the turntable. Measure its distance, $r$, from the centre.

3 Turn on the turntable at the lower speed setting and measure the time taken for ten rotations.

4 Repeat step 3 with the turntable at its higher speed setting.

5 Move the eraser outwards approximately 1 cm further away from the centre. Measure the new radial distance $r$.

6 Repeat steps 2 and 3, noting whether the eraser begins to slide.

7 Repeat the experiment, replacing the eraser with the coin.

If you have a turntable with continuously adjustable speed, such as a pottery wheel, you can instead hold the radius constant and vary the speed. Finding the speed at which the object (eraser or coin) begins to slide will then allow you to calculate the maximum friction force between the object and the turntable.

## RESULTS

Record your results for $r$ and $10T$ in a table like the one below as you measure them.

| Object | $r$ (m) | $10T$ (s) | $T$ (s) | $\omega$ (rad s$^{-1}$) | $v$ (m s$^{-1}$) | $a$ (m s$^{-2}$) | $F$ (kg m s$^{-2}$) | Slide? |
|---|---|---|---|---|---|---|---|---|
| | | | | | | | | |
| | | | | | | | | |
| | | | | | | | | |
| | | | | | | | | |
| | | | | | | | | |

Note in the 'Slide?' column whether the object began to slide at this speed.

»

1 Complete the table by calculating $T$, $\omega$, $v$, $a$ and $F$.

If you create a data table in a spreadsheet program you can use the spreadsheet to calculate all the required values.

2 Draw a force diagram for the object on the turntable. Note that the net force (the centripetal force) is the friction force. This is what holds the object to the turntable and stops it sliding.

The maximum friction force between the object and the turntable has a value between the largest force at which the object did not slide and the smallest force at which it did slide. This gives you an upper and lower bound on the value of the maximum friction force between the objects and the turntable.

## DISCUSSION

1 Give an estimated value and range for the maximum friction force between each object and the turntable. For which object was it greater?

2 Summarise the relationships between centripetal force, radius of path and speed. Explain why an object is more likely to slide off the turntable at large angular speeds and large radius.

3 Give the answer to your inquiry question.

## CONCLUSION

Write a conclusion summarising the outcomes of your investigation.

---

**KEY CONCEPTS**

- An object travelling in a circular path at constant speed is undergoing uniform circular motion.
- An object in uniform circular motion has a speed given by $v = \dfrac{2\pi r}{T}$, where $r$ is the radius of the circular path and $T$ is the period of the motion.
- The acceleration of an object in circular motion is $a_c = \dfrac{v^2}{r}$, which points radially towards the centre of the circular path.
- The net force that causes this acceleration is called the centripetal force, $F_c = \dfrac{mv^2}{r}$.
- The angular displacement of an object is $\Delta\theta = \dfrac{s}{r}$ where $s$ is the arc length of the displacement.
- The angular velocity is $\omega = \dfrac{\Delta\theta}{\Delta t} = \dfrac{s}{r\Delta t} = \dfrac{v}{r}$.

---

**CHECK YOUR UNDERSTANDING**

**3.1**

1 An object undergoing uniform circular motion has a constant speed, so how can it have a non-zero acceleration?

2 A particle is moving in a circle with constant speed $v$. Its speed then increases to twice this value. By how much has the acceleration increased?

3 For an object undergoing uniform circular motion, sketch a graph of acceleration as a function of:

  a speed.

  b radius.

4 A dog is running in a circle, chasing its own tail. If it moves in a circle of radius 1.0 m, turning through a complete circle every second, what is:

  a its speed?

  b its acceleration?

  c its angular velocity?

5 What is the maximum speed allowed for a rotating ride of radius 2.5 m if the maximum net (centripetal) force exerted on a 70 kg person is not allowed to exceed 1000 N?

6 What is the radius of a rotor ride that is exerting a 400 N net force on a 30 kg child when the speed of the child is 4.5 m s$^{-1}$?

# Applications of the uniform circular motion model

The mathematical model for uniform circular motion described above can be applied to many systems. These include planets, moons and satellites in circular orbits (chapter 4), electrons orbiting atoms (chapter 14) and charged particles travelling in circles in magnetic fields (chapter 5). In this section we shall look at three common examples: an object on a string and a car moving around a corner on a flat road and on a banked road.

## Motion in a horizontal circle

Consider an object being whirled in a horizontal circle, such as the glider in Figure 3.6. Two forces act on this object: the gravitational force, $mg$, downwards, and the tension force, $F_T$, due to the string pulling on the object.

A string can provide a **tension** (a pull) in the direction of the string only. Because strings are 'floppy' they cannot exert a force in any other direction. From Newton's third law, the tension force *exerted* by the string is also *experienced* by the string, and is referred to as the tension *in* the string.

The string makes an angle $\theta$ with the horizontal so that the tension has both a horizontal and a vertical component. The horizontal component is $F_{T,x} = F_T \cos\theta$ and the vertical component is $F_{T,y} = F_T \sin\theta$.

It is always a good idea to start your analysis of any system with a force diagram. Figure 3.6b shows a force diagram for the glider with the tension broken into horizontal and vertical components.

The vertical component of the tension is equal to the gravitational force, as there is zero vertical acceleration.

$$\Sigma F_y = F_{T,y} + mg$$

$$\text{So } F_{T,y} = F_T \sin\theta = -mg$$

$T_x$ is the only force acting in the horizontal direction, so:

$$\Sigma F_x = F_{T,x} = F_T \cos\theta$$

The net force in the horizontal direction is the centripetal force, so:

$$\Sigma F_x = F_{T,x} = F_T \cos\theta = \frac{mv^2}{r}$$

The tension in this case has a constant magnitude, but its direction varies so that it always points towards the centre of the circle.

**FIGURE 3.6 a** A boy whirls a glider in a horizontal circle; **b** A force diagram for the glider

A 250 g glider on the end of a string is swung in a horizontal circle with a radius of 1.2 m. The glider makes one revolution every 2.0 s.

   **a**  What is the speed of the glider?

   **b**  What is the horizontal component of the tension in the string?

| ANSWER | LOGIC |
|---|---|
| $T = 2.0\,\text{s}, r = 1.2\,\text{m}, m = 0.25\,\text{kg}$ | ▪ Identify the relevant data and convert to SI units. |
| **a**  $v = \omega r$ | ▪ Relate velocity to angular velocity. |
| $\omega = \dfrac{2\pi}{T}$ | ▪ Relate angular velocity to period. |
| $v = \dfrac{2\pi r}{T}$ | ▪ Substitute the expression for $\omega$ into the expression for $v$. |
| $v = \dfrac{2\pi \times 1.2\ \text{m}}{2.0\ \text{s}}$ | ▪ Substitute values with correct units. |
| $v = 3.777\ \text{m s}^{-1}$ | ▪ Calculate the final value. |
| $v = 3.8\ \text{m s}^{-1}$ | ▪ State the final answer with correct units and appropriate significant figures. |
| **b**  $F_{\text{T},x} = \dfrac{mv^2}{r}$ | ▪ Relate the horizontal force to the velocity. |
| $F_{\text{T},x} = \dfrac{0.25\ \text{kg}(3.777\ \text{m s}^{-1})^2}{1.2\ \text{m}}$ | ▪ Substitute values with correct units. |
| $F_{\text{T},x} = 2.96\ \text{kg m s}^{-2}$ | ▪ Calculate the final value. |
| $F_{\text{T},x} = 3.0\ \text{N}$ | ▪ State the final answer with correct units and appropriate significant figures. |

**TRY THESE YOURSELF**

**1**  Find:

   **a**  the vertical component of the tension.

   **b**  the total tension.

   **c**  the angle θ.

**2**  The period of the motion is halved.

   **a**  What is the new tension in the string?

   **b**  What is the new angle θ?

## Motion in a vertical circle

When an object is whirled in a vertical circle, as in Figure 3.7, the gravitational force, $F_g = mg$, and the vertical component of the tension add to give the net force towards the centre. Take the simple case of an object being whirled in a vertical circle with the cable also vertical. At any point on the circle the net force is:

$$\Sigma F = F_g + F_T = mg + F_T$$

9780170409131

At the bottom of the circle, the net force is upwards. Hence, at this point the tension must be larger than the gravitational force, as they act in opposite directions. At the top of the circle, where $mg$ is acting in the same direction as $F_T$, the total force must still be the same, so the tension must be less and may even drop to zero. In this case the tension varies not only in direction but also in magnitude.

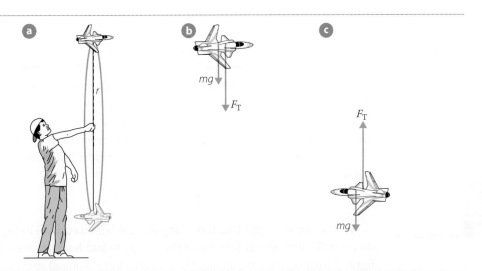

**FIGURE 3.7 a** A glider is whirled in a vertical circle; **b** A force diagram for the glider at the top of the circular path; **c** A force diagram for the glider at the bottom of the path

## WORKED EXAMPLE (3.5)

If the glider in Worked example 3.4 is to be whirled in a vertical circle of the same radius at the same constant speed, what must be the tension in the string when the glider is at the bottom of its path?

| ANSWER | LOGIC |
|---|---|
| $r = 1.2\,\text{m}, m = 0.25\,\text{kg}, v = 3.777\,\text{m s}^{-1}$ | ▪ Identify the relevant data, using the previous example. |
| $\Sigma F = F_T - mg = \dfrac{mv^2}{r}$ | ▪ Relate the net force to the velocity. |
| $F_T = \dfrac{mv^2}{r} + mg$ | ▪ Rearrange for the tension. |
| $F_T = \dfrac{0.25\ \text{kg}(3.777\ \text{m s}^{-1})^2}{1.2\,\text{m}} + (0.25\,\text{kg})(9.8\,\text{m s}^{-2})$ | ▪ Substitute values with correct units. |
| $F_T = 5.42\,\text{kg m s}^{-2}$ | ▪ Calculate the final value. |
| $F_T = 5.4\,\text{N}$ | ▪ State the final answer with correct units and appropriate significant figures. |

**TRY THIS YOURSELF**
What is the tension at the top of the path?

## A car turning a horizontal corner

To travel around a circular path, such as running around the bend in a 200 m race or driving a car around a corner, you must push outwards on the ground. The ground will push back on you (Newton's third law) towards the centre of the circular path you are taking. Remember from *Physics in Focus Year 11* chapter 4 that the contact force has two components: the parallel friction component, and the perpendicular normal component. When you drive a car, it is the friction force that acts on the tyres in the direction of the car's motion that makes the car go forwards.

When you want to drive around a corner on a flat road, the tyres must push down at an angle to the horizontal. The ground then pushes back with an equal and opposite force. It is friction, or the parallel component of this reaction force, that supplies the net (centripetal) force needed to make the car go around the corner. Figure 3.8 shows a force diagram for a car rounding a corner on a horizontal road. Note that in reality the normal and friction forces are distributed among all the tyres.

Hence when a car corners on a flat road, we model the corner as part of a circle, and so:

$$\Sigma F = F_{\text{friction}} = \frac{mv^2}{r}$$

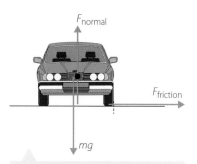

**FIGURE 3.8** Forces acting on a car on a horizontal road, turning towards the car's left (your right)

We can see from this that the ability of a car (and its driver) to negotiate a corner successfully depends on how sharp the corner is and how fast the car is going. The faster the car is going, the greater the frictional force required, and the force increases with the square of the velocity. There is a maximum friction force that the road can exert on the tyres. Therefore, slowing down for a corner can make a big difference between staying on the road and having an accident. The maximum friction force is significantly reduced by water, mud or oil on the road, and by worn tyres.

## Cornering on a banked road

Some roads have corners that are **banked**. A banked corner is sloped so that the edge of the road on the outside of the corner is slightly higher than the inside edge. Recall that we looked at cars on inclined planes in chapter 5 of *Physics in Focus Year 11*. A banked corner is an inclined plane, but in this case it is angled across the road, rather than in the direction of the road. The result is a net force that accelerates the car in the direction of the corner.

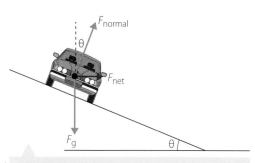

**FIGURE 3.9** Forces acting on a car on a banked road (ignoring friction)

There are at least two forces acting on a car as it turns a banked corner. These are the gravitational force, $F_g$, and the contact force due to the road acting on the tyres. Air resistance or drag may also be important if the car is going fast. Consider the contact force as two components: the normal force, $F_N$, and the friction force, $F_{\text{friction}}$. For this analysis we begin by making the simplifying assumption that only the gravitational and normal forces are significant, as shown in Figure 3.9.

The net force acting on the car (ignoring the friction force) is:

$$\Sigma F = F_{\text{net}} = F_g + F_N$$

The gravitational force acts in the vertical direction only, while the normal force has components in both the vertical and horizontal directions. If the car is to stay at the same vertical height, so it doesn't slide across the road, then the net force in the vertical direction must be:

$$\Sigma F_y = F_{\text{net},y} = F_g + F_{N,y} = F_g + F_N \cos\theta = 0$$

so $F_N \cos\theta = mg$ or $F_N = \dfrac{mg}{\cos\theta}$.

In the horizontal direction we have:

$$\Sigma F_x = F_{net,x} = F_{N,x} = F_N \sin\theta$$

Using the expression for $F_N$ derived by considering the forces in the vertical direction, we can say that:

$$F_{net} = F_N \sin\theta = \frac{mg\sin\theta}{\cos\theta} = mg\tan\theta$$

Because the car is moving around a corner, which we model as an arc of a circle, this net force is the centripetal force:

$$F_{net} = mg\tan\theta = \frac{mv^2}{r}$$

We can use this equation to estimate the angle at which a road should be banked to assist cars to move around the corner at a particular speed.

This analysis assumes no friction force. The friction force is of course never really zero. The friction force may act parallel to and up the slope, or down the slope, depending on the speed of the car. If the car is going slowly, the friction force acts up the slope, preventing (or slowing) the car from sliding down the slope. If the car is going very quickly, the friction force may act in the opposite direction, acting against the car's tendency to slide up the slope at high speeds.

Applications of the uniform circular motion model

▶ **WORKED EXAMPLE** (3.6)

The car in Figure 3.9 has a mass of 1500 kg. It is travelling horizontally at $80\,km\,h^{-1}$ around a bend that is banked at $10°$ to the horizontal.

**a**  What is the net force acting on the car?

**b**  What radius of curvature must the road have so that the car can turn the corner with no friction force acting?

| ANSWER | LOGIC |
|---|---|
| $\theta = 10°$, $m = 1500\,kg$, $v = 80\,km\,h^{-1} \times 1000\,m\,km^{-1} \times \dfrac{s}{3600\,h} = 22.2\,m\,s^{-1}$ | ▪ Identify the relevant data, and convert to SI units. |
| **a**  $F_{net} = mg\tan\theta$ | ▪ Write the expression for the net force. |
| $F_{net} = (1500\,kg)(9.8\,m\,s^{-2})\tan 10°$ | ▪ Substitute values with correct units. |
| $F_{net} = 2592\,kg\,m\,s^{-2}$ | ▪ Calculate the final value. |
| $F_{net} = 2600\,N$ | ▪ State the final answer with correct units and appropriate significant figures. |
| **b**  $F_{net} = \dfrac{mv^2}{r}$ | ▪ Relate the net force to the radius. |
| $r = \dfrac{mv^2}{F_{net}}$ | ▪ Rearrange for the radius. |
| $r = \dfrac{(1500\,kg)(22.2\,m\,s^{-1})^2}{2592\,N}$ | ▪ Substitute values with correct units. |
| $r = 286\,m$ | ▪ Calculate the final value. |
| $r = 290\,m$ | ▪ State the final answer with correct units and appropriate significant figures. |

**TRY THIS YOURSELF**

A car moving at $60\,km\,h^{-1}$ enters a curve with a radius of 150 m. What is the ideal angle of banking for this road so that the horizontal component of the normal force on the car is equal to the centripetal force required to maintain the car's circular motion? Why is the mass of the car not needed for this calculation?

Critical and creative thinking

Numeracy

Information and communication technology capability

## Designing a race track

### AIM

To design a race track so that each corner is banked at an appropriate angle, using spreadsheet software to calculate the angles

Write a design brief for your investigation.

### MATERIALS

- Computer with spreadsheet software
- Paper, pens, compass, ruler

### METHOD

1 Sketch your race track carefully to scale on a large sheet of paper. Use a compass for drawing the corners, and measure the radius of each corner as you draw it. Mark the radius of curvature on each corner.

Open a new file in your spreadsheet software and put a descriptive heading at the top of the sheet, then save it with a suitable filename. *Remember to save your file frequently.*

2 Set up the values that remain constant for the first corner: the radius, and the acceleration due to gravity. Figure 3.10 shows how this can be done using spreadsheet software. The example commands given here are for Excel. Cells are labelled above the value. Cells A6 and B6 contain the radius and the value for *g* for Earth, but you could instead use a value for the Moon or a different planet and design a racetrack for a new space colony.

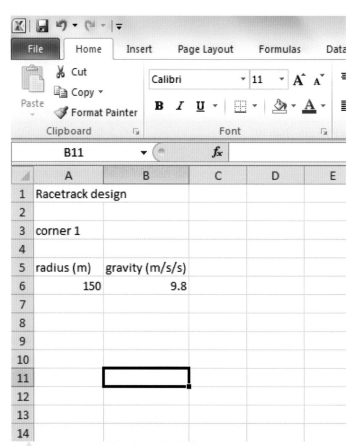

**FIGURE 3.10** Set up the constant values in your spreadsheet.

**» 3** Implement the equations to calculate the banking angle required as a function of speed. Label three columns for speed, angle, and angle in degrees.

**4** In the cell under the label 'speed', enter 0. Below this cell enter a command to increase the speed. In Figure 3.11 this command is **=A9+0.5**, but you can add any speed increment you want to make the steps bigger or smaller. Copy this command into as many cells below as you want.

**5** Calculate the required angle in the next column. This is calculated from the equation $mg \tan\theta = \dfrac{mv^2}{r}$. You need to rearrange this equation for $\theta$.

To implement this equation in Excel, type the formula **=ATAN((A9\*A9/(B\$6\*A\$6)))** in the first cell under the 'angle' label (Figure 3.11). This requires that A9 contains the speed, A6 contains the radius and B6 contains the acceleration due to gravity. Note that the $ signs tell Excel not to increment this part of the cell reference. Copy this to the cells in the column below.

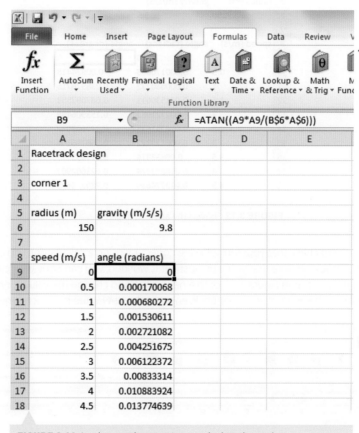

**FIGURE 3.11** Implement the equation to calculate the angle.

**6** Convert the angle to degrees.

Excel calculates angles in radians. If you want to convert the angle to degrees, type the command **=DEGREES(B9)** under the label 'angle (degrees)'. Copy this to the cells below. See Figure 3.12.

»

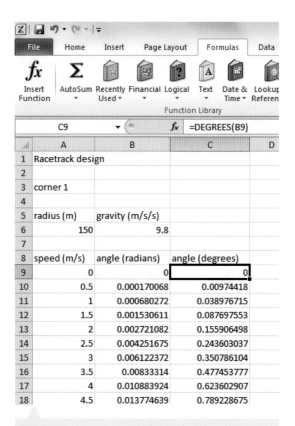

FIGURE 3.12 Convert the angle to degrees.

**7** It is a good idea to check that your formulae are correct by using a calculator to do at least one calculation manually. When you are sure that your spreadsheet is working, you can copy your calculations either to new cells to the right, as in Figure 3.13, or to a new sheet, and change the value for *r* for the second corner. Repeat this for each corner of your race track, entering the new value for *r* each time.

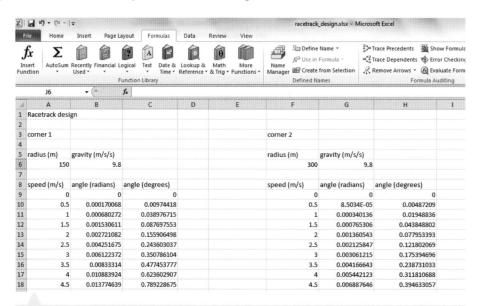

FIGURE 3.13 Copy the calculations for a second corner and change the value for radius.

Racetrack design_Excel

## RESULTS AND ANALYSIS OF RESULTS

**1** Make sure that you have saved your file!

**2** Create a plot of required banking angle as a function of speed for at least one corner.

>> 3  Decide on a suitable banking angle for each corner, given how fast you want a vehicle to be able to go around that corner. Mark the speed and angle on your racetrack.

4  Estimate how long it would take to do a lap of your racetrack.

## DISCUSSION

1  Comment on the shape of your graph. How does the required banking angle vary with speed? What effect does radius of curvature have?

2  Compare your racetrack design to those of others in the class. Whose is the fastest? Whose looks the most exciting or dangerous? Why?

3  How does your racetrack compare to a real racetrack design?

4  Have you met the requirements of your design brief?

## CONCLUSION

Write a conclusion summarising the outcomes of your investigation.

KEY CONCEPTS

- The mathematical model for uniform circular motion can be applied to many situations.
- When an object is whirled on a string in a horizontal circle, the horizontal component of the tension provides the net (centripetal) force:
$$F_{net} = F_{T,x} = F_T \cos \theta = \frac{mv^2}{r}$$
- When an object is whirled on a string in a vertical circle, the tension and the gravitational forces provide the net (centripetal) force:
$$F_{net} = F_T + mg = \frac{mv^2}{r}$$
- When a car turns a corner on a horizontal road we can model the corner as a circular arc and the motion as uniform circular motion. In this case it is the friction force that provides the net (centripetal) force:
$$F_{net} = F_{friction} = \frac{mv^2}{r}$$
- When a car turns a corner on a banked road it is the horizontal component of the contact force of the road on the car that provides the net (centripetal) force. In the absence of friction, this is the normal force:
$$F_{net} = F_{N,x} = mg \tan \theta = \frac{mv^2}{r}$$

### CHECK YOUR UNDERSTANDING

**3.2**

1  Explain why it is more important to slow down for corners in wet weather than in dry weather. Use a diagram and an appropriate equation to support your answer.

2  A car undergoing uniform circular motion going around a horizontal corner has a net (centripetal) force $F_{net}$ acting on it.
   a  In which direction is this force?
   b  What is the origin of this force?
   c  In terms of $F_{net}$, what is the magnitude of the centripetal force if the car enters the same curve moving at twice the speed? Show your working.

3  A rock with mass 1.5 kg is being whirled in a horizontal circle on a string 0.80 m long. The speed of the rock is a constant 5.0 m s$^{-1}$. Find the magnitude and direction of the tension in the string.

4  A rock with mass 1.5 kg is being whirled in a vertical circle on a string 0.80 m long. The speed of the rock is constant. At the top of the circle, the tension $T$ in the string is zero.
   a  Find the speed of the rock.
   b  Find the tension in the string when the rock is at the bottom of the circle.

**5** A car with a mass of 1500 kg is travelling horizontally at $20\,\mathrm{m\,s^{-1}}$ around a bend that is banked at 10° to the horizontal.

    **a** What is the normal force acting on the car?

    **b** What do you need to consider when calculating the net force acting on the car?

    **c** What radius of curvature must the road have so that the car can turn the corner with no friction force acting?

**6** A car moving at $20\,\mathrm{m\,s^{-1}}$ enters a curve with a radius of 150 m. What is the ideal angle of banking for this road so that the horizontal component of the normal force on the car is equal to the centripetal force required to maintain the car's circular motion?

**7** A girl is whirling a poi (a weight on a string) in a vertical circular path. Assume the poi has approximately constant speed. Sketch a graph of the magnitude of:

    **a** the gravitational force acting on the poi as a function of time.

    **b** the tension force acting on the poi as a function of time.

    **c** the net force acting on the poi as a function of time.

# 3.3 Energy and work in uniform circular motion

Energy and work in uniform circular motion

Remember from chapter 5 of *Physics in Focus Year 11* that there are two types of energy: kinetic and potential. The kinetic energy, $K$, of an object (sometimes written $E_k$) is due to its motion, and is given by:

$$K = \frac{1}{2}mv^2$$

When an object is undergoing uniform circular motion its speed is constant, so its kinetic energy is also constant.

The potential energy of a system is due to the forces acting between objects in the system, and depends on the positions of the objects. For the gravitational force, the potential energy of an Earth–object system depends on how high above Earth's surface the object is:

$$U_g = mgh$$

For an object undergoing uniform circular motion in a horizontal plane the potential energy is also constant.

Remember that the work done on an object is equal to the energy transferred to the object. So for an object moving in uniform circular motion in a horizontal plane there is no work being done because the total energy is constant.

We can also see that this is the case by using the definition of work: $W = Fs$ where $s$ is the displacement in the direction of the force.

Figure 3.14 shows an object in uniform circular motion and the net force acting on it. Because the direction of displacement at any moment is in the same direction as the velocity, the force and displacement are always perpendicular. This means that $s$ in the direction of the force is always zero, so:

$$W = Fs = 0$$

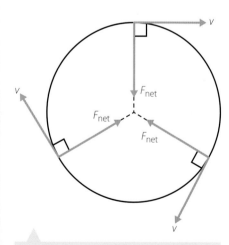

**FIGURE 3.14** For uniform circular motion the net force always points towards the centre of the path, and is perpendicular to the velocity.

For an object that is undergoing uniform motion and changing potential energy as it does so, there will be a change in energy; hence, work done over any part of a cycle. Because the object returns to its initial position in any cycle, and therefore to its initial potential energy, the total work done in a complete circle is still zero.

9780170409131

## WORKED EXAMPLE (3.7)

The 250 g glider in Figure 3.7 is being whirled in a vertical circle of radius 1.2 m at a constant speed of 3.8 m s$^{-1}$. At the bottom of the path the glider is 0.2 m above the ground. Calculate the kinetic, potential and total energy of the glider at the top and bottom of its path.

| ANSWER | LOGIC |
|---|---|
| $m = 0.25$ kg, $r = 1.2$ m, $v = 3.8$ m s$^{-1}$ <br> $h_{bottom} = 0.2$ m | ▪ Identify the relevant data, and convert to SI units. |
| At the bottom: | |
| $K = \dfrac{1}{2} mv^2$ | ▪ Write the expression for the kinetic energy. |
| $K = \dfrac{1}{2} (0.25 \text{ kg})(3.8 \text{ m s}^{-1})^2$ | ▪ Substitute values with correct units. |
| $K = 1.81$ kg m$^2$ s$^{-2}$ | ▪ Calculate the value. |
| $K = 1.8$ J | ▪ State the final answer with correct units and appropriate significant figures. |
| $U_g = mgh_{bottom}$ | ▪ Write the expression for the potential energy. |
| $U_g = (0.25 \text{ kg})(9.8 \text{ m s}^{-2})(0.2 \text{ m})$ | ▪ Substitute values with correct units. |
| $U_g = 0.49$ kg m$^2$ s$^{-2}$ | ▪ Calculate the value. |
| $U_g = 0.49$ J | ▪ State the final answer with correct units and appropriate significant figures. |
| $E_{total} = K + U_g$ | ▪ Write the expression for the total energy. |
| $E_{total} = 1.81 \text{ J} + 0.49 \text{ J}$ | ▪ Substitute values with correct units. |
| $E_{total} = 2.30$ J | ▪ Calculate the value. |
| $E_{total} = 2.3$ J | ▪ State the final answer with correct units and appropriate significant figures. |
| At the top: | |
| $K = \dfrac{1}{2} mv^2 = 1.8$ J | ▪ Recognise that the kinetic energy is constant. |
| $U_g = mgh_{top} = mg(h_{bottom} + 2r)$ | ▪ Write the expression for the potential energy, recognising that the height at the top is the height at the bottom plus the diameter of the circle. |
| $U_g = (0.25 \text{ kg})(9.8 \text{ m s}^{-2})(0.2 \text{ m} + 2.4 \text{ m})$ | ▪ Substitute values with correct units. |
| $U_g = 6.37$ kg m$^2$ s$^{-2}$ | ▪ Calculate the value. |
| $U_g = 6.4$ J | ▪ State the final answer with correct units and appropriate significant figures. |
| $E_{total} = K + U_g$ | ▪ Write the expression for the total energy. |
| $E_{total} = 1.81 \text{ J} + 6.37 \text{ J}$ | ▪ Substitute values with correct units. |
| $E_{total} = 8.18$ J | ▪ Calculate the value. |
| $E_{total} = 8.2$ J | ▪ State the final answer with correct units and appropriate significant figures. |

### TRY THIS YOURSELF

For the example above, calculate the change in energy as the glider moves between the top and the bottom of its path, the change in energy as it moves from the bottom to the top, and the change in energy over a complete loop.

- Kinetic energy, $K = \frac{1}{2}mv^2$, is constant for an object undergoing uniform circular motion.
- If the potential energy is also constant, for example if the motion is horizontal, then the total energy is constant.
- The work done by the net force acting on an object undergoing uniform circular motion is always zero because the net force is perpendicular to the velocity.

1   Explain why the kinetic energy of an object undergoing uniform circular motion is constant.

2   Explain why the work done over any complete cycle of uniform circular motion must be zero.

3   A 1000 kg car turns a horizontal corner with radius of curvature 200 m, travelling at a speed of $10\,\mathrm{m\,s^{-1}}$. Calculate:

    a   the net force acting on the car.

    b   the kinetic energy of the car.

    c   the work done on the car if it moves through an angle of 90°.

4   A girl is swinging a 0.5-kg rock on a string. It moves in a horizontal circle of radius 1.0 m, 1.7 m above the ground, at a speed of $8.0\,\mathrm{m\,s^{-1}}$. Calculate the total energy of the rock.

5   A girl is swinging a 0.5- rock on a string. It moves in a vertical circle of radius 1.0 m, with the lowest point 0.1 m above the ground, at a speed of $8.0\,\mathrm{m\,s^{-1}}$. Calculate the total energy of the rock at the top and bottom of its path.

6   Refer to the situation described in Question 5.

    a   What is the work done by the gravitational force between the bottom and top of the path?

    b   What is the work done by the gravitational force between the top and bottom of the path?

    c   What is the total work done by the gravitational force in a complete loop?

# 3.4 Rotation and torque

When a force acts on an object, the object accelerates in a straight line in the direction of the net force (chapter 4, *Physics in Focus Year 11*). Straight-line motion is also called *translational motion*. We have now also looked at uniform circular motion, which is a type of *rotational motion*. To make an object rotate, you need to apply a torque. Torque is the rotational equivalent of force.

Imagine using a spanner to undo a bolt. You want the bolt to turn, but you don't want to move it sideways or up or down. You want it to rotate but not translate, so you need to apply a torque to the bolt. A torque is due to a force acting on an object at some distance from a pivot point or axis of rotation.

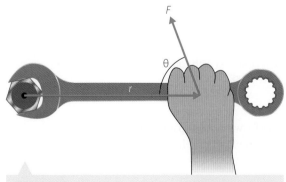

The torque, $\tau$, acting on an object due to a force, $F$, applied at a distance $r$ from the pivot point is given by:

$$\tau = rF_{\perp} = rF\sin\theta$$

where $F_{\perp}$ is the component of the force in the direction of the vector $r$. The vector $r$ points from the pivot point to the point at which the force is applied. The angle, $\theta$, is the angle between the line joining the pivot point to the point of application of the force and the force vector. This is shown in Figure 3.15.

Torque has units of N m. These are the same units as work, which we usually write in units of J because work is an energy. However torque *is not* a type of energy, so we never write the units of torque as J.

FIGURE 3.15 The torque applied to the spanner by the hand is $\tau = rF\sin\theta$.

Like force, torque is a vector. Imagine if the hand shown in Figure 3.15 was applying the force in the opposite direction, angled down instead of up. This would result in rotation in the other direction.

To find the direction of torque, point the fingers of your right hand in the direction of the line joining the pivot to point of application of the force, as shown in Figure 3.16. Now curl your fingers in the direction of the force. Your thumb points in the direction of the torque vector, and your fingers show the direction of the resulting rotation.

It may seem odd that the torque vector is perpendicular to the direction of rotation. However, every point on a rotating object is changing the direction of its velocity constantly, as we saw in circular motion. So it makes sense to define the torque vector as perpendicular to the plane in which rotation occurs. This is the only way that we can assign it a unique direction for the whole rotating object.

To apply a large torque, you need to apply a large force, at a large distance from the pivot point. The distance from the pivot point to the point of application of the force is called the lever arm. A lever is a device that applies a torque.

The angle is also important. To get the maximum torque, you apply the force perpendicular to $r$, as shown in Figure 3.17. When $\theta = 90°$ the torque has its maximum value, $\tau_{max} = rF$. If you pull or push on the handle of the spanner along the line of the handle, then $\theta = 0$ and the torque is also zero, and no rotation is achieved.

The pivot point is the point about which the object rotates, such as the hinges of a door, or the centre of the bolt in Figure 3.17. If you try to balance a pencil on its end on your desk (Figure 3.18a), it will rotate about its tip and fall over. The tip is the pivot point. Sometimes we talk about an axis of rotation rather than a pivot point. This just means we are thinking about a three-dimensional object rotating about a particular axis. For example, a bolt (Figure 3.18b) rotates about a line through its own long axis when it is turned with a spanner.

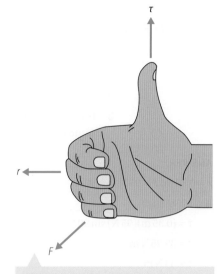

FIGURE 3.16 The right-hand rule for finding the direction of a torque

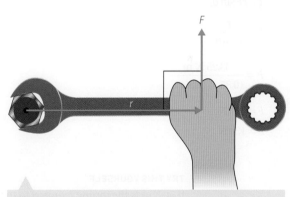

FIGURE 3.17 Torque is a maximum when the lever arm ($r$) and the force ($F$) are perpendicular.

Rotation and torque

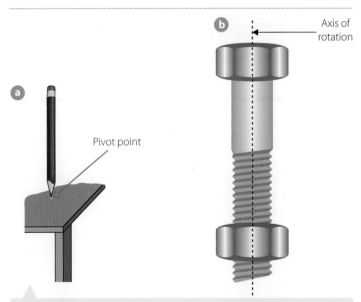

FIGURE 3.18 **a** The tip of the pencil is the pivot about which the pencil will rotate when it falls. **b** The bolt rotates about a line through its own long axis.

▶ **WORKED EXAMPLE** (3.8)

A mechanic uses a spanner to tighten a bolt in a car. She uses a spanner with a handle 35 cm long, and applies a force of 95 N perpendicular to the handle.

**a** What torque does she apply?

**b** If she applied the force at an angle of 45°, how much force would she need to apply to produce the same torque?

| ANSWER | LOGIC |
|---|---|
| $r = 0.35$ m, $F = 95$ N, $\theta = 90°$ | ▪ Identify the relevant data, and convert to SI units. |
| **a** $\tau = rF \sin \theta$ | ▪ Write the expression for the torque. |
| $\tau = (0.35\,\text{m})(95\,\text{N}) \sin 90°$ | ▪ Substitute values with correct units. |
| $\tau = 33.25$ N m | ▪ Calculate the value. |
| $\tau = 33$ N m | ▪ State the final answer with correct units and appropriate significant figures. |
| **b** Now $\theta = 45°$ | ▪ Identify changed variables. |
| $\tau = rF \sin \theta$ | ▪ Write the expression for the torque. |
| $F = \dfrac{\tau}{r \sin \theta}$ | ▪ Rearrange for force. |
| $F = \dfrac{33.25 \text{ N m}}{(0.35 \text{ m}) \sin 45°}$ | ▪ Substitute values with correct units. |
| $F = 134$ N | ▪ Calculate the value. |
| $F = 130$ N | ▪ State the final answer with correct units and appropriate significant figures. |

**TRY THIS YOURSELF**

If 50 N m is the torque required to turn the bolt, and the mechanic can exert a maximum force of 150 N, what is the minimum length handle she needs on her spanner?

**KEY CONCEPTS**

● Torque, $\tau$, is the rotational equivalent of force, and results in rotational motion.
● $\tau = rF \sin \theta$

**CHECK YOUR UNDERSTANDING**

**3.4**

1 Write the equation for torque and define all the terms used.

2 Explain why is it so much harder to hold a heavy bag with your arm outstretched horizontally than to hold it close to your body.

3 Ivan pushes on a door at a horizontal distance of 0.7 m from the hinges. If he applies a force of 90 N at an angle of 75° to the plane of the door, what torque does he exert on the door?

4 How much force must be applied at a distance of 35 cm from a bolt to produce a torque of 250 N m on the bolt?

5 You are having difficulty undoing a screw. Should you use a screwdriver with a longer handle or a thicker handle? Draw a diagram and explain your answer.

6 Ivan pushes on a door with a force of 90 N, at a distance of 0.7 m from the hinges.

　a Calculate the maximum and minimum torques that he can apply.

　b Sketch a graph of the applied torque as a function of the angle at which he pushes.

9780170409131

- An object travelling in a circular path at constant speed is undergoing uniform circular motion.

- An object in uniform circular motion has a speed given by $v = \dfrac{2\pi r}{T}$ where $r$ is the radius of the circular path and $T$ is the period of the motion.

- The (centripetal) acceleration of an object in circular motion is $a_c = \dfrac{v^2}{r}$, and points towards the centre of the circular path.

- The net force that causes this acceleration (called the centripetal force) is $F_c = \dfrac{mv^2}{r}$.

- The angular displacement of an object in circular motion is $\Delta\theta = \dfrac{s}{r}$, where $s$ is the arc length of the displacement.

- The angular velocity is $\omega = \dfrac{\Delta\theta}{\Delta t} = \dfrac{s}{r\Delta t} = \dfrac{v}{r}$.

- For an object whirled on a string in a horizontal circle, $F_{net} = F_{T,x} = F_T \cos\theta = \dfrac{mv^2}{r}$.

- For an object whirled on a string in a vertical circle, $F_{net} = F_T + mg = \dfrac{mv^2}{r}$.

- When a car turns a corner on a horizontal road, $F_{net} = F_{friction} = \dfrac{mv^2}{r}$.

- When a car turns a corner on a banked road, with no friction, $F_{net} = F_{N,x} = mg \tan\theta = \dfrac{mv^2}{r}$.

- Kinetic energy, $K = \dfrac{1}{2}mv^2$, is constant for an object undergoing uniform circular motion.

- The work done by the net force acting on an object undergoing uniform circular motion is always zero because the net force is perpendicular to the velocity.

- Torque, $\tau$, is the rotational equivalent of force, and results in rotational motion.

- $\tau = r_\perp F = rF \sin\theta$.

# 3 CHAPTER REVIEW QUESTIONS

Qz
Review quiz

1 Define 'centripetal acceleration'.

2 Why are long-handled tools generally more effective than short-handled tools?

3 Explain why it is that an object undergoing uniform circular motion is accelerating.

4 Why are the speed limits on winding roads lower than those on straight roads?

5 Explain why curved railway tracks are usually banked towards the inside of the curve. Use a force diagram to help explain your answer.

6 The Australian Synchrotron has a ring-shaped tunnel of circumference 130 m through which electrons travel at close to the speed of light. What is the centripetal acceleration of an electron moving through the tunnel at $2.8 \times 10^8\,\mathrm{m\,s^{-1}}$?

7 Mai is driving her car (mass 1200 kg) around a roundabout at $40\,\mathrm{km\,h^{-1}}$. The roundabout has a radius of 8.0 m.
   a Calculate the acceleration of Mai's car.
   b Calculate the net force acting on Mai's car.
   c What force provides the net force?

8 Explain how banking a corner can make a corner safer, particularly in wet weather. Use a diagram to help explain.

9 How many times more centripetal force is required to act on a vehicle moving around a curve with a radius of curvature of 200 m compared with a curve with a radius of curvature of 400 m if the vehicle is travelling safely at the same speed?

10 The bottom of a rollercoaster ride has a track with a radius of curvature of 28 m. The carriage passes this point with a speed of $12\,\mathrm{m\,s^{-1}}$. What is the normal force on a 50 kg person by the carriage? Hint: start by drawing a force diagram.

11 Kevin is lost in Canberra and is driving around and around a large roundabout, trying to work out which exit to take. Kevin is driving at $60\,\mathrm{km\,h^{-1}}$ and the net force acting on his car (mass 1800 kg) is 20 kN.
   a What is the radius of the roundabout?
   b What is the period of Kevin's motion?
   c What is Kevin's angular velocity, $\omega$?

**12** Mahmud is driving around a banked corner on the Hume Highway. The corner is banked at an angle of 5.0° and has a radius of curvature of 300 m. Ignoring friction, at what maximum speed should Mahmud drive around this corner?

**13** A spanner is used to apply a torque of 500 N m to a nut.

   **a** What minimum force must be applied to achieve this torque if the handle of the spanner is 45 cm long?

   **b** What minimum force must be applied if it is applied at an angle of 25° to the handle?

**14** A train goes around a bend of radius 350 m that is banked at an angle of 8.0°. At what speed should the train travel around the bend?

**15** A girl is swinging on a maypole in a playground as shown in Figure 3.19.

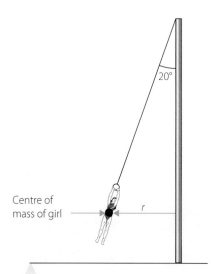

**FIGURE 3.19** A girl swings around a maypole.

The girl has a mass of 36 kg, and when she is moving with a speed of 2.0 m s$^{-1}$ the light rope makes an angle of 20° with the vertical. Consider the motion of the girl's centre of mass, which moves in a horizontal circle of radius $r$.

   **a** What is the vertical component of the tension in the rope?

   **b** What is the horizontal component of the tension in the rope?

   **c** What is the tension in the rope?

   **d** What is the net force acting on the girl?

   **e** What is the radius of the circle?

**16** A hammer uses a torque to effectively hammer a nail into a piece of wood. Explain, with the aid of diagrams, why this is the case. Hint: Consider what you do with your wrist and arm when you use a hammer.

**17** A street performer is whirling a ball on the end of a rope in a horizontal circle at constant speed. The ball has a mass of 0.5 kg and the radius of the path is 1.5 m. If the ball is moving at 10 m s$^{-1}$, what is the tension in the rope? Give the magnitude and direction.

**18** The street performer in the previous question gradually changes the way he whirls the ball until it is moving in a vertical circle. Assuming it still moves at 10 m s$^{-1}$, what is the maximum and minimum magnitude of the tension in the rope now?

**19** A 2.0 kg rock is being whirled around in a horizontal circle held by a 0.70 m long string that will break if the tension in it exceeds 38 N. Find the shortest period, $T$, for the motion of the rock that will not break the string. Model the string as approximately horizontal.

**20** A proton is undergoing circular motion in a magnetic field with an angular speed of 1500 rad s$^{-1}$. If the net force acting on the proton is $3.8 \times 10^{-22}$ N, what is the radius of its path?

# 4 Motion in gravitational fields

**OUTCOMES**

**INQUIRY QUESTION**

How does the force of gravity determine the motion of planets and satellites?

**Students:**

- apply qualitatively and quantitatively Newton's Law of Universal Gravitation to:
  - determine the force of gravity between two objects $F = \dfrac{GMm}{r^2}$
  - investigate the factors that affect the gravitational field strength $g = \dfrac{GM}{r^2}$
  - predict the gravitational field strength at any point in a gravitational field, including at the surface of a planet (ACSPH094, ACSPH095, ACSPH097)
- investigate the orbital motion of planets and artificial satellites when applying the relationships between the following quantities: CCT, ICT, N
  - gravitational force
  - centripetal force
  - centripetal acceleration
  - mass
  - orbital radius
  - orbital velocity
  - orbital period
- predict quantitatively the orbital properties of planets and satellites in a variety of situations, including near the Earth and geostationary orbits, and relate these to their uses (ACSPH101) ICT, N
- investigate the relationship of Kepler's Laws of Planetary Motion to the forces acting on, and the total energy of, planets in circular and non-circular orbits using: (ACSPH101)
  - $v = \dfrac{2\pi r}{T}$
  - $\dfrac{r^3}{T^2} = \dfrac{GM}{4\pi^2}$ ICT, N
- derive quantitatively and apply the concepts of gravitational force and gravitational potential energy in radial gravitational fields to a variety of situations, including but not limited to: ICT, N
  - the concept of escape velocity $v_{esc} = \sqrt{\dfrac{2GM}{r}}$
  - total potential energy of a planet or satellite in its orbit $U = -\dfrac{GMm}{r}$
  - total energy of a planet or satellite in its orbit $U + K = -\dfrac{GMm}{2r}$
  - energy changes that occur when satellites move between orbits (ACSPH096)
  - Kepler's Laws of Planetary Motion (ACSPH101)

**FIGURE 4.1** It is the gravitational force that keeps the Moon in orbit about Earth, and the planets in orbit about the Sun.

The idea of universal, fundamental forces is central to modern physics. Newton's development of the Law of Universal Gravitation was a crucial step in the development of modern physics because it was the first mathematical model to describe a fundamental force that was universally applicable (Figure 4.1). Newton's Law of Universal Gravitation can be applied to all objects with mass, from atoms to planets to galaxies. It provided a theoretical framework to explain the many empirical observations made by astronomers such as Kepler and Brahe.

Physicists now believe that there are only four fundamental forces: the electromagnetic, strong and weak forces, and gravity. You have already met the two aspects of the electromagnetic force – electricity and magnetism – in *Physics in Focus Year 11*, and we shall revisit them in more detail in the following four chapters. In this chapter the gravitational force, as described by Newton, is discussed. This model gives us excellent descriptive and predictive power. Only very recently, new discoveries in particle physics, such as the identification of the Higg's boson, have given us further insight into the underlying mechanisms by which gravity works. This is discussed later, in the chapters dealing with quantum mechanics and particle physics, where you will also meet the strong and weak forces. There is energy associated with all forces, and gravitational waves carry gravitational energy, in much the same way that electromagnetic waves (chapter 9) carry electromagnetic energy. Gravitational waves are very difficult to detect and have only been detected a few times from massive events like the collision of neutron stars.

## 4.1 The gravitational force and the gravitational field

By the late 17th century the orbits of the planets and the Moon had been well described, but there was no single mathematical model that could be universally applied to describe and predict them all until Newton developed his model for universal gravitation.

Newton's model of gravity was built on the measurements of Tycho Brahe and other astronomers, and the mathematical interpretation of these data by Johannes Kepler.

The most important aspect of Newton's mathematical model was its universality – it did not just apply to celestial motion, it applied to earthly motion as well. This was a major philosophical, as well as physical, advance. That a mathematical equation could describe both 'heavenly' and 'earthly' motion was new and, at the time, shocking to many.

**The history of gravity**

**Measuring the gravitational constant, G**

The Cavendish experiment was the first to measure the strength of the gravitational constant *G*. Find out how it worked.

### Newton's Law of Universal Gravitation

Newton's Law of Universal Gravitation gives the gravitational force of attraction that any object with mass $M$ exerts on a second object with mass $m$, where the distance between the two objects is $r$:

$$F = \frac{GMm}{r^2}$$

The constant $G$ is the universal gravitational constant and has the value $G = 6.67 \times 10^{-11}\ \text{N kg}^{-2}\,\text{m}^2$.

The gravitational force points back towards the object that exerts the force, as shown in Figure 4.2. The force acts towards the object that exerts it and hence it is an attractive force. Each object exerts an attractive force on the other. Hence the force pulls the object of mass $m$ towards the object with mass $M$, and vice versa.

The form of this equation should look familiar to you. It is the same form as Coulomb's Law, which was used in chapter 12 of *Physics in Focus Year 11* to describe the interaction between charged objects.

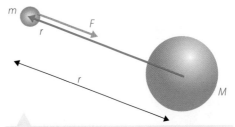

Strength of gravity

**FIGURE 4.2** The object with mass $M$ exerts a gravitational force on the object with mass $m$.

> ## WORKED EXAMPLE (4.1)

Calculate the gravitational force that the Sun exerts on Earth. You will need the following data:
Mass of Earth $= 5.97 \times 10^{24}$ kg, mass of the Sun $= 2.00 \times 10^{30}$ kg, mean radius of Earth's orbit around the Sun $= 1.50 \times 10^{11}$ m.

| ANSWER | LOGIC |
|---|---|
| $M_E = 5.97 \times 10^{24}$ kg, $M_S = 2.00 \times 10^{30}$ kg, $r = 1.50 \times 10^{11}$ m | • Identify the relevant data. |
| $F = \dfrac{GM_S M_E}{r^2}$ | • Write the expression for the force. |
| $F = \dfrac{(6.67 \times 10^{-11}\,\text{N kg}^{-2}\text{m}^2)(2.00 \times 10^{30}\,\text{kg})(5.97 \times 10^{24}\,\text{kg})}{(1.50 \times 10^{11}\,\text{m})^2}$ | • Substitute values with correct units. |
| $F = 3.54 \times 10^{22}$ N | • Calculate the final value. |
| $F = 3.54 \times 10^{22}$ N directed towards the Sun. | • State the final answer with correct units, direction, and appropriate significant figures. |

**TRY THIS YOURSELF**
Given the mass of the Moon is $7.35 \times 10^{22}$ kg and its mean orbital radius about Earth is $3.84 \times 10^8$ m, find the gravitational force exerted by Earth on the Moon.

## Gravity and Newton's third law

Recall that Newton's third law tells us that all forces are interactions. Whatever force one object exerts on another object, it will experience an equal and opposite force due to the interaction. We write this as $F_{a\,on\,b} = -F_{b\,on\,a}$. The forces are equal in magnitude but opposite in direction.

Newton's Law of Universal Gravitation shows this clearly when written in vector form.

The gravitational force of the Sun on Earth is:

$$F_{S\,on\,E} = -\frac{GM_S M_E}{r^2_{Sun\,to\,Earth}}$$

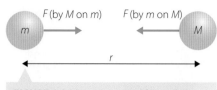

FIGURE 4.3 The gravitational force of $M$ on $m$ is the Newton's third law pair to the gravitational force of $m$ on $M$.

The negative sign indicates that the force points back towards the Sun. The gravitational force of Earth on the Sun is:

$$F_{\text{E on S}} = \frac{GM_{\text{E}}M_{\text{S}}}{r^2_{\text{Sun to Earth}}}$$

This has the same magnitude as $F_{\text{S on E}}$, but points in the opposite direction, towards Earth, so we can say that $F_{\text{S on E}} = -F_{\text{E on S}}$ and both have magnitude $F = \frac{GM_{\text{E}}M_{\text{S}}}{r^2}$. This is shown in Figure 4.3.

## WORKED EXAMPLE 4.2

What is the gravitational force that Earth exerts on the Sun?

| ANSWER | LOGIC |
|---|---|
| $F_{\text{S on E}} = -F_{\text{E on S}}$ | ▪ Recognise that the force of Earth on the Sun is the Newton's third law pair to the force of the Sun on Earth. |
| $F_{\text{S on E}} = -F_{\text{E on S}} = -3.54 \times 10^{22}\,\text{N}$ | ▪ Find the value previously calculated for the force of the Sun on Earth. |
| $F_{\text{E on S}} = 3.54 \times 10^{22}\,\text{N}$ directed towards Earth | ▪ State the final answer with correct units, direction, and appropriate significant figures. |

### TRY THIS YOURSELF

What is the gravitational force exerted by the Moon on Earth?

## The gravitational field

Recall from chapter 12 of *Physics in Focus Year 11* that the electric field is defined as the force per unit charge that acts on a small test charge: $E = \frac{F}{q}$. When we apply this to Coulomb's Law for point charges $Q$ and $q$, we get $E = \frac{1}{4\pi\varepsilon_0}\frac{Q}{r^2}$.

Like the electrostatic and magnetic forces, gravity also acts at a distance. So we say the gravitational force is mediated by a gravitational field. We can define the gravitational field, $g$, due to a mass in the same way: as the force per unit mass acting on a small test mass:

$$g = \frac{F}{m}$$

The units of the gravitational field are hence $\text{N kg}^{-1}$.

When we combine this definition of gravitational field with the Law of Universal Gravitation, we get:

$$g = \frac{F}{m} = \frac{GM}{r^2}$$

All objects with mass are surrounded by a gravitational field (Figure 4.4) that depends on the mass of the object and decreases with the square of the distance from the object. This equation is valid for point-like objects, and for spherically symmetric objects. The distance, $r$, is measured from the centre of the object, *not the surface*. For more complex objects the field is calculated by adding the fields due to each particle of the object. Note that $g$ is a vector, so the rules of vector addition apply.

The gravitational field can be represented using gravitational field lines, in the same way as an electric field is represented using electric field lines. The gravitational field lines show the direction of $g$, which is the direction of the gravitational force acting on a small test mass. The density of the field lines is

The gravitational force and the gravitational field

9780170409131

an indication of the field strength. As you can see in Figure 4.4, the gravitational field of a spherical mass looks like the electric field of a negative point charge.

Unlike electric field lines, gravitational field lines always point towards an object with mass, and never away, because the gravitational force is always attractive.

We can use this field equation to find the gravitational field at a distance $r$ from any object of mass $M$. For example, given the mass and radius of Earth, we can use it to find the gravitational field at Earth's surface:

$$g_{\text{Earth surface}} = \frac{GM_{\text{Earth}}}{r_{\text{Earth}}^2} = \frac{(6.67 \times 10^{-11}\,\text{N kg}^{-2}\,\text{m}^2)(5.97 \times 10^{24}\,\text{kg})}{(6.37 \times 10^3\,\text{m})^2} = 9.8\,\text{N kg}^{-1}$$

The unit $\text{N kg}^{-1}$ can be written as $\text{m s}^{-2}$, because $1\,\text{N} = 1\,\text{kg m s}^{-2}$. The gravitational field is equal to the acceleration due to gravity, which is due to the gravitational force. At the surface of Earth, the gravitational field strength is $9.8\,\text{N kg}^{-1} = 9.8\,\text{m s}^{-2}$.

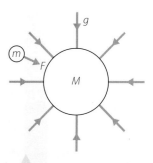

**FIGURE 4.4** The gravitational field due to a spherical mass, and the force it exerts on a small nearby object.

▶ **WORKED EXAMPLE** (4.3)

Calculate the acceleration due to gravity on the surface of the Moon, which has mass $7.3 \times 10^{22}\,\text{kg}$ and radius 1740 km.

| ANSWER | LOGIC |
|---|---|
| $M_M = 7.3 \times 10^{22}\,\text{kg}$, $r = 1740\,\text{km} = 1.74 \times 10^6\,\text{m}$ | ▪ Identify the relevant data and convert to SI units. |
| $g_{\text{Moon}} = \dfrac{GM_M}{r^2}$ | ▪ Write the expression for the field. |
| $g_{\text{Moon}} = \dfrac{(6.67 \times 10^{-11}\,\text{N kg}^{-2}\,\text{m}^2)(7.3 \times 10^{22}\,\text{kg})}{(1.74 \times 10^6\,\text{m})^2}$ | ▪ Substitute values with correct units. |
| $g_{\text{Moon}} = 1.61\,\text{N kg}^{-1}$ | ▪ Calculate the final value. |
| $g_{\text{Moon}} = 1.6\,\text{m s}^{-2}$ | ▪ State the final answer with correct units and appropriate significant figures. |

**TRY THESE YOURSELF**

1  Calculate the gravitational field strength for a satellite orbiting Earth at an altitude of 36 000 km.

2  If Earth's gravitational field strength is $4.9\,\text{m s}^{-2}$ at a satellite, how far away from Earth's centre must the satellite be?

Close to the surface of Earth, or other very large objects, the gravitational field is approximately constant. As long as the changes in positions that we are considering are small compared to the radius of the object, the changes in field will also be small and we can model the field as constant. This is what we do whenever we use $F = mg = $ constant for projectiles and other objects moving close to Earth's surface. This is called the 'near Earth' or 'surface' approximation. In this approximation, the field lines are parallel to each other, which means that the field is constant and perpendicular to Earth's surface, as in Figure 4.5.

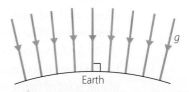

**FIGURE 4.5** The 'near-Earth' field approximation: gravitational field lines close to Earth's surface are parallel to each other, which means the field is constant and perpendicular to Earth's surface.

- Newton's Law of Universal Gravitation provided a mathematical model that explained the empirical observations of generations of astronomers.
- Newton's Law of Universal Gravitation, $F = \dfrac{GMm}{r^2}$, states that the force one object with mass exerts on another is proportional to the product of the masses and inversely proportional to the square of the distance between the objects.
- Newton's third law applies to the gravitational force, as it does to all other forces.
- The gravitational field at any point is the gravitational force per unit mass at that point: $g = \dfrac{F}{m}$, which is the local acceleration due to gravity.
- The units of the gravitational field are $N\,kg^{-1}$, which is the same as $m\,s^{-2}$.
- The gravitational field near a point mass or spherical mass is $g = \dfrac{F}{m} = \dfrac{GM}{r^2}$.
- Close to the surface of Earth, the field is uniform and has magnitude $g = 9.8\,N\,kg^{-1} = 9.8\,m\,s^{-2}$.

**CHECK YOUR UNDERSTANDING**

**4.1**

1 Define 'gravitational field'. How does gravitational field differ from gravitational force?

2 A schoolbag sits on the ground. Identify the Newton's third law force pair to the gravitational force of Earth acting on the bag. Draw a force diagram for the bag. Note that you cannot show the Newton's third law force to the gravitational force *on the bag* on the force diagram *for the bag*.

3 What two variables determine the acceleration due to gravity at the surface of a planet?

4 Two masses, $m$ and $M$, are in an isolated system. The gravitational forces, $F$ (by $M$ on $m$) and $F$ (by $m$ on $M$), are equal and opposite. Why do they not add to a zero net force? Explain your answer.

5 Two masses, $m$ and $M$, have a gravitational force of attraction $F$ when they are a distance $r$ apart. What is the force when this distance is increased to $4r$?

6 What is the gravitational force of attraction due to Earth on a 500 kg satellite with an orbital radius of 7000 km? (Mass of Earth $= 5.97 \times 10^{24}$ kg)

7 What is the acceleration of a rock dropped onto the surface of Mars? The radius of Mars is 3390 km and its mass is $6.39 \times 10^{23}$ kg.

8 Given $m_{proton} = 1.7 \times 10^{-27}$ kg, $m_{electron} = 9.1 \times 10^{-31}$ kg, $e = 1.6 \times 10^{-19}$ C, calculate:

   a the gravitational force that a proton exerts on an electron $10^{-10}$ m away.

   b the electrostatic (Coulomb force) that a proton exerts on an electron $10^{-10}$ m away.

## 4.2 The orbits of planets and satellites

The gravitational field model explains how objects such as planets and satellites maintain their orbits. It allows us to calculate information such as stellar masses from observed planetary orbits, and work out how fast a satellite must be moving to be placed in a stable orbit about Earth.

To do this we combine our model of uniform circular motion from the previous chapter with Newton's Law of Universal Gravitation. If we model the orbit of a planet as circular then the net force acting on the planet is equal to the centripetal force:

$$\Sigma F = \frac{mv^2}{r}$$

This net force is provided by the gravitational force of the central object (star or planet) about which the planet or satellite orbits:

$$F = \frac{GMm}{r^2} = \frac{mv^2}{r}$$

The centripetal acceleration of an orbiting body such as a planet or satellite is hence given by:

$$a = \frac{v^2}{r} = \frac{GM}{r^2}$$

which can be rearranged to give $v = \sqrt{\dfrac{GM}{r}}$.

This equation allows us to deduce the mass of distant stars by observing the orbital period and radius of any planets that orbit them. Recall that the period of uniform circular motion is given by $T = \dfrac{2\pi r}{v}$, which can be rearranged to give $v = \dfrac{2\pi r}{T}$, where $v$ is the orbital velocity.

These equations allow us to relate the period to the orbital radius and the orbital velocity, and hence to the mass of the central object (Figure 4.6). Remember that the velocity and force are perpendicular at all points in a circular orbit.

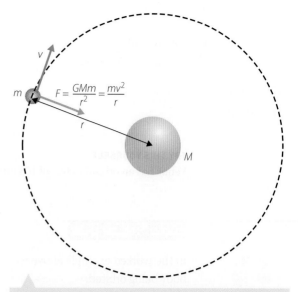

FIGURE 4.6 A planet in a circular orbit about a star experiences a net (centripetal) force towards its star, due to the gravitational force.

▶ WORKED EXAMPLE (4.4)

Jupiter has an orbital period of 12 years and an average orbital radius of 780 million km. Use this information to calculate the mass of the Sun.

| ANSWER | LOGIC |
|---|---|
| $T = 12 \text{ years} = 12 \times \dfrac{365 \text{ days}}{\text{year}} \times \dfrac{24 \text{ h}}{\text{day}} \times \dfrac{60 \text{ min}}{\text{h}} \times \dfrac{60 \text{ s}}{\text{min}} = 3.78 \times 10^8 \text{ s}$ <br> $r = 780 \times 10^6 \text{ km} = 7.80 \times 10^{11} \text{ m}$ | ▪ Identify the relevant data and convert to SI units. |
| $a = \dfrac{v^2}{r} = \dfrac{GM}{r^2}$ | ▪ Use the expression for acceleration to relate the mass to the radius and velocity. |
| $M = \dfrac{rv^2}{G}$ | ▪ Rearrange for mass. |
| $T = \dfrac{2\pi r}{v}$ | ▪ Relate the period to the velocity. |
| $v = \dfrac{2\pi r}{T}$ | ▪ Rearrange for $v$. |
| $M = \dfrac{rv^2}{G} = \dfrac{r}{G}\left(\dfrac{2\pi r}{T}\right)^2 = \dfrac{4\pi^2 r^3}{GT^2}$ | ▪ Substitute into the expression for mass and simplify. |
| $\text{kg} = \dfrac{(\text{m})^3}{(\text{N kg}^{-2} \text{m}^2)(\text{s})^2}$ <br> $= \dfrac{(\text{m})^3}{(\text{kg m s}^{-2} \text{kg}^{-2} \text{m}^2)(\text{s})^2}$ <br> $= \dfrac{(\text{m})^3}{(\text{kg}^{-1} \text{m}^3 \text{s}^{-2})(\text{s})^2}$ <br> $= \text{kg}$ | ▪ As this is a complicated expression, it is worth doing a dimension check to make sure we have not made a mistake before we substitute numbers. |

| $M = \dfrac{4\pi^2 (7.8 \times 10^{11}\,\text{m})^3}{(6.67 \times 10^{-11}\,\text{N kg}^{-2}\,\text{m}^2)(3.78 \times 10^8\,\text{s})^2}$ | ▪ Substitute values with correct units. |
|---|---|
| $M = 1.96 \times 10^{30}\,\text{N}^{-1}\,\text{kg}^2\,\text{m s}^{-2}$ | ▪ Calculate the final value. |
| $M = 2.0 \times 10^{30}\,\text{kg}$ | ▪ State the final answer with correct units and appropriate significant figures. |

**TRY THIS YOURSELF**

Venus has an orbital radius of 108 million km. How long, in Earth days, is a year on Venus?

## Satellite orbits

The orbits of planets and satellites

In the worked example above we derived a relationship between the orbital period and the mass of the body being orbited:

$$M = \frac{4\pi^2 r^3}{GT^2}$$

This can also be written as $T^2 = \dfrac{4\pi^2 r^3}{GM}$.

We can use this expression to calculate the altitude at which a satellite must be placed for it to have a given period. Alternatively, if we know what altitude we want a satellite to have, we can use this equation to work out what period, and hence velocity, it must have.

This expression also applies to any object orbiting a central large mass, including natural satellites such as moons.

### Geostationary and geosynchronous satellites

A geostationary satellite stays above the same point on Earth's surface over the equator. They are used for a range of purposes including communications and global positioning (GPS).

For a satellite to stay above the same point on Earth's surface it needs to have the same orbital period as the rotation of Earth; that is, one day.

▶ **WORKED EXAMPLE** (4.5)

What is the orbital radius of a geostationary satellite?

| ANSWER | LOGIC |
|---|---|
| $T = 1\,\text{day} = 1\,\text{day} \times \dfrac{24\,\text{h}}{\text{day}} \times \dfrac{60\,\text{min}}{\text{h}} \times \dfrac{60\,\text{s}}{\text{min}} = 8.64 \times 10^4\,\text{s}$ | ▪ Identify the relevant data and convert to SI units. |
| $M = M_{\text{Earth}} = 5.97 \times 10^{24}\,\text{km}$ | ▪ Look up other required data. |
| $T^2 = \dfrac{4\pi^2 r^3}{GM}$ | ▪ Relate period to radius. |
| $r = \left( \dfrac{GMT^2}{4\pi^2} \right)^{\frac{1}{3}}$ | ▪ Rearrange for radius. |

| | |
|---|---|
| $r = \left( \dfrac{(6.67 \times 10^{-11}\,\text{N kg}^{-2}\,\text{m}^2)(5.97 \times 10^{24}\,\text{kg})(8.64 \times 10^4\,\text{s})^2}{4\pi^2} \right)^{\frac{1}{3}}$ | ▪ Substitute values with correct units. |
| $r = 4.22 \times 10^7\,(\text{N kg}^{-1}\,\text{m}^2\text{s}^2)^{\frac{1}{3}}$ | ▪ Calculate the final value. |
| $r = 4.22 \times 10^7\,\text{m}$ | ▪ Note that $\text{N kg}^{-1}\,\text{m}^2\,\text{s}^2 = \text{m}^3$, and state the final answer with correct units and appropriate significant figures. |
| This $r$ is the orbital radius, so the distance is measured from the centre of Earth. To find the altitude above Earth's surface we need to subtract the radius of Earth, which is 6370 km or $6.37 \times 10^6\,\text{m}$. | |
| $h = r - R_{\text{Earth}} = 4.22 \times 10^7\,\text{m} - 6.37 \times 10^7\,\text{m} = 3.59 \times 10^7\,\text{m}$ | ▪ State the final answer with correct units and appropriate significant figures. |

**TRY THESE YOURSELF**

Calculate:

**a** the speed at which a geostationary satellite moves.

**b** its centripetal acceleration. This is the acceleration due to gravity, and hence the gravitational field strength at that altitude.

Geosynchronous satellites travel above any great circle. A great circle is any circle on Earth's surface whose radius extends from Earth's centre (Figure 4.7), so that it has the same circumference as the equator. Both geostationary and geosynchronous orbits have 24-hour periods and altitudes of approximately 36 000 km. A geostationary satellite is a special case of a geosynchronous satellite that is above the equator and so always stays above the same point.

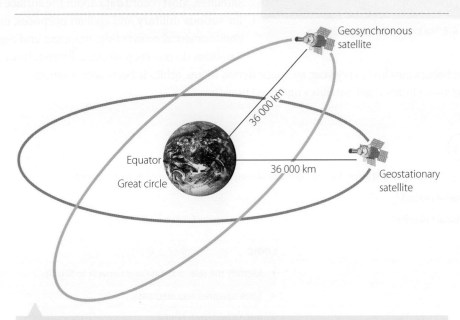

**FIGURE 4.7** A geostationary satellite stays at an altitude of 36 000 km above a point on the equator on Earth's surface. A geosynchronous satellite travels in an orbit above a great circle. (Not to scale)

## Near-Earth orbits

Near-Earth or low-Earth orbit satellites typically have orbits less than 1200 km above Earth's surface. They have orbital periods much less than geostationary satellites, typically between 80 and 130 minutes. These near-Earth satellites, including the International Space Station (Figure 4.8), which are in orbit close to Earth's surface, have different purposes. These include monitoring of weather, military surveillance and mapping. Their low altitude allows these satellites to collect high resolution images, and because they move rapidly over Earth's surface they can map large areas rapidly.

The first artificial satellite, Sputnik 1, was a sphere only about 60 cm in diameter, launched by the Soviet Union in 1957. Politically this was a very important event because it occurred during the Cold War, and it triggered the 'space race'. Sputnik 1 had an elliptical orbit that took it over almost all of Earth's surface.

The International Space Station (ISS) is the largest near-Earth satellite. The first components of the ISS were launched in 1998 and it has been expanded with many additional modules since then. It orbits at an altitude of approximately 400 km in an almost circular orbit. The gravitational field at the ISS is approximately 90% of the gravitational field on Earth's surface. The ISS is used for many scientific purposes, including experiments in 'microgravity' that are made possible because the ISS and all its contents are constantly in free fall about Earth.

**The ISS**

Find out more about the ISS, and watch live video from onboard, or the view of Earth.

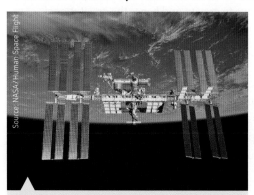

Source: NASA/Human Space Flight

**FIGURE 4.8** The International Space Station

The Hubble Space Telescope is also a near-Earth orbit satellite, at an altitude of 540 km. This puts it well above Earth's atmosphere and so enables it to collect astronomical data that is not available to Earth-based telescopes. Observations from the Hubble Space Telescope have contributed to accurate calculations of the expansion rate of the Universe (see chapter 13).

There are more than 500 near-Earth orbit satellites. Most record data about the surface of Earth for various military and civilian purposes, including environmental monitoring, mapping and espionage. Satellites do not keep working forever, however, and more are being launched every year, so space debris in low orbits is becoming a significant problem. Space debris includes 'dead' satellites and used boosters.

**Current positions of satellites**

Is there a satellite passing over you right now?

> **WORKED EXAMPLE (4.6)**

The International Space Station has an altitude above Earth's surface of 400 km. What is:

**a**  its orbital period?

**b**  its orbital velocity?

| ANSWER | LOGIC |
|---|---|
| $h = 400\,\text{km} = 4 \times 10^5\,\text{m}$ | ▪ Identify the relevant data and convert to SI units. |
| $M = M_{\text{Earth}} = 5.97 \times 10^{24}\,\text{kg}$ | ▪ Look up other required data. |
| **a**  $T^2 = \dfrac{4\pi^2 r^3}{GM}$ | ▪ Relate period to radius. |
| $T = \sqrt{\dfrac{4\pi^2 r^3}{GM}}$ | ▪ Rearrange for $T$. |

| | |
|---|---|
| $r = R_{Earth} + h$ | ■ Relate radius to altitude. |
| $T = \sqrt{\dfrac{4\pi^2 (R_{Earth} + h)^3}{GM}}$ | ■ Substitute expression for $r$ into equation for $T$. |
| $T = \sqrt{\dfrac{4\pi^2 (6.37 \times 10^6\,\text{m} + 4.00 \times 10^5\,\text{m})^3}{(6.67 \times 10^{-11}\,\text{N kg}^{-2}\,\text{m}^2)(5.97 \times 10^{24}\,\text{kg})}}$ | ■ Substitute values with correct units. |
| $T = 5.55 \times 10^3 (\text{N}^{-1}\text{kg}^{-1}\text{m})^{\frac{1}{2}}$ <br><br> $T = 5.55 \times 10^3\,\text{s}$ <br> This is approximately $1\frac{1}{2}$ hours. | ■ Calculate the final value and state the final answer with correct units and appropriate significant figures. |
| **b** $T = \dfrac{2\pi r}{v}$ | ■ Relate period to velocity. |
| $v = \dfrac{2\pi r}{T}$ | ■ Rearrange for velocity. |
| $v = \dfrac{2\pi(6.37 \times 10^6\,\text{m} + 4.00 \times 10^5\,\text{m})}{5.55 \times 10^3\,\text{s}}$ | ■ Substitute values with correct units. |
| $v = 7.66 \times 10^3\,\text{m s}^{-1}$ <br> $v = 766\,\text{km s}^{-1}$ | ■ Calculate the final value and state the final answer with correct units and appropriate significant figures. |

**TRY THIS YOURSELF**

Calculate the centripetal acceleration of the International Space Station. This is the acceleration due to gravity, and hence the gravitational field strength at that altitude. What fraction of the surface value of $g$ is this equal to?

# INVESTIGATION (4.1)

## Planetary orbits

Critical and creative thinking

Numeracy

Information and communication technology capability

### AIM

In this investigation, you will use an orbit simulator to model the orbit of an additional hypothetical Planet X in our solar system.

Write an inquiry question for this investigation.

### MATERIALS

■ Computer with internet access

### METHOD

1 Open the orbit simulator at the weblink.

The tutorial will open automatically the first time you open the simulator. You can also open it at any time using the help tab on the right.

**Orbit Simulator**
Use this simulator to model a planetary orbit.

Work through the tutorial so you know what each control does, and what the information panel is telling you.

2    Change the average distance of Planet X from the Sun. What happens to its period as the orbit gets larger? What about its velocity?

3    Change the eccentricity of Planet X's orbit and observe how the shape of the orbit changes. How does the average velocity change as you increase the eccentricity?

Note the range of eccentricities of the orbits of the other planets, and how these relate to their shapes.

4    Set the eccentricity of Planet X's orbit to be extremely large, at least 0.5. Pause the simulation when Planet X is at its closest point to the Sun. Note the velocity at this point. Now restart the simulation, then pause it again when Planet X is at its furthest point from the Sun. Note your observations – how does the velocity change?

5    Return the eccentricity of Planet X to a less extreme value, less than 0.2.

6    Open the graphing panel. Experiment with plotting different graphs, including linear graphs of eccentricity as a function of average distance, period as a function of distance and velocity and distance as a function of period. Try using different scales including linear–linear, log–linear and log–log.

### RESULTS

Make notes of your observations as you go.

For each graph that you plot, make a sketch or take a screen shot for your records.

### ANALYSIS OF RESULTS

Open the notebook tab and work through the questions. You can open and close it as many times as you like, and make further observations to answer the questions.

Keep a record of your work by printing a copy when you are done.

### DISCUSSION

1    How does the period of an orbit vary with average distance from the Sun?

2    How did the velocity of Planet X vary with its position when it had a high eccentricity?

3    What relationships between variables did you observe?

4    Give the answer to your inquiry question.

### CONCLUSION

Write a conclusion summarising the outcomes of your investigation.

---

**KEY CONCEPTS**

- The gravitational force provides a centripetal (net) force to keep planets and satellites in orbit. If we model orbital motion as uniform circular motion then $F = \dfrac{GMm}{r^2} = \dfrac{mv^2}{r}$.

- The centripetal acceleration of an orbiting body such as a planet or satellite is given by $a = \dfrac{v^2}{r} = \dfrac{GM}{r^2}$.

- The period of an object in orbit is $T = \dfrac{2\pi r}{v}$, and $T^2 = \dfrac{4\pi^2 r^3}{GM}$.

- Geostationary satellites stay in orbit above a specific point on Earth's equator, at an altitude of 36 000 km.

- Geosynchronous satellites have the same altitude and period as geostationary satellites and travel above a great circle.

- Low-Earth orbit satellites have much lower altitudes and shorter periods.

1  What is a geostationary satellite? Draw a diagram to help explain your answer.

2  Explain how geostationary and near-Earth orbit satellites differ, and give some examples of the uses of each.

3  Explain why planets that are further from the Sun have longer orbital periods.

4  If you wanted to move a satellite into higher orbit, you would also need to change its speed. Would you need to increase or decrease its speed? Use the expression for the centripetal acceleration to justify your answer.

5  Find the altitude of a satellite in orbit around Earth if its orbital speed is $5.0 \, \text{km s}^{-1}$.

6  Mars has an orbital period of 687 days. Calculate its orbital radius.

7  Mars has a mass of $6.39 \times 10^{23}$ kg, and two moons, Phobos and Deimos. Phobos has an orbital radius of 9400 km and Deimos has an orbital radius of 23 000 km. Calculate:

   a  the orbital periods of Phobos and Deimos.

   b  their orbital velocities.

8  Mars has a mass of $6.39 \times 10^{23}$ kg, a day length of 1 (Earth) day and 40 minutes, and a radius of 3390 km. If you wished to place a stationary satellite above Mars, at what altitude would it need to be?

9  a  Given the mass of the Sun is $2.00 \times 10^{30}$ kg, calculate Earth's:

      i  orbital radius.

      ii  orbital velocity.

   b  Compare Earth's orbital speed to the rotational speed at a point on the equator, given that Earth has a radius of 6370 km.

# 4.3 Kepler's laws

Before Newton published his Law of Universal Gravitation, there was no theory that could explain the orbits of all the planets and moons in terms of a single force.

However, various astronomers did propose models based on astronomical data that gave good predictive power. One of the most important of these was Johannes Kepler. While Kepler's model has been replaced by the simpler and more complete model of Newtonian gravity, it is still of historical interest.

Kepler inherited the volumes of Tycho Brahe's meticulous observations of the motions of the planets, the Moon and the stars. Built up over many years, Brahe's measurements and recordings, all made before the invention of the telescope, enabled Kepler to propose a model for the motion of the planets.

## Kepler's first law: the Law of Orbits

Before Kepler, it had been assumed that the planets orbited Earth, and in later models the Sun, in perfectly circular orbits. This was, in part, due to the belief that the heavens were perfect and that circles are a perfect shape. Kepler found that if the planets were considered as moving in elliptical orbits, then their observed positions in the sky could be predicted almost perfectly. While most planets' orbits are very close to circular, very precise measurements such as those made by Brahe demonstrated that they were not *perfectly* circular.

Kepler's first law states that all planets move in elliptical orbits with the Sun at one focus (Figure 4.9).

An ellipse is a curved shape. It has a major, or long, axis that is the longest line between two points on the edge drawn through the geometric centre. The minor axis of an ellipse is the shortest line joining two points on the edge drawn through the geometric centre. An ellipse has two foci. These are special points such that lines drawn from one focal point to any point on the edge and then to the other focal point all have the same length.

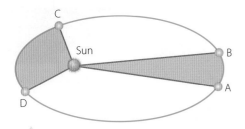

**FIGURE 4.9** Planets move in elliptical orbits (Kepler's first law), and a line joining the planet to the Sun sweeps out equal areas in equal times (Kepler's second law). Note that the eccentricity of this orbit is greatly exaggerated compared to that of any planets in our solar system.

A circle is a special case of an ellipse in which the two axes are equal and the two foci are at the same position. The orbits of most planets, moons and satellites are very close to circular, with some exceptions. The orbit shown in Figure 4.9 is not representative of that of any planets in our solar system, most of which would not be distinguishable from a circle by eye on such a diagram. If you worked through Investigation 4.1 above, you would have noticed that only the orbits of Mercury and Pluto have significant eccentricity. Eccentricity is a measure of how far from circular an orbit is, and varies between 0 and 1, where 0 is circular. Pluto and Mercury both have eccentricities of approximately 0.2. All other planets in our solar system have eccentricities less than 0.1, and Earth's is 0.02.

## Kepler's second law: the Law of Areas

Kepler noticed that the speeds of the planets changed during their orbits. Nearer to the Sun their speeds increased; further away their speeds decreased. He was able to conclude that the areas swept out in equal time intervals were the same. If the planet shown in Figure 4.9 moves from A to B in a time $\Delta t$, and it takes the same time, $\Delta t$, for it to move from C to D, then the two shaded sections have the same area.

A planet in orbit and the star it orbits about can be modelled as a closed system with no work done by external forces. Hence the mechanical energy, the sum of the kinetic and potential energies, is constant. As the planet gets closer to the star the gravitational potential energy of the system decreases. If the gravitational potential energy decreases, then the kinetic energy must increase, and so the planet speeds up.

## Kepler's third law: the Law of Periods

**WS**

Kepler's laws

Kepler's third law states that the square of the period of a planet's orbit is proportional to the cube of the mean radius of its orbit: $T^2 \propto r^3$, or $\dfrac{T^2}{r^3} = \text{constant}$.

We have already seen how this relationship can be derived from Newton's Law of Universal Gravitation, combined with circular motion:

$$\frac{T^2}{r^3} = \frac{4\pi^2}{GM}$$

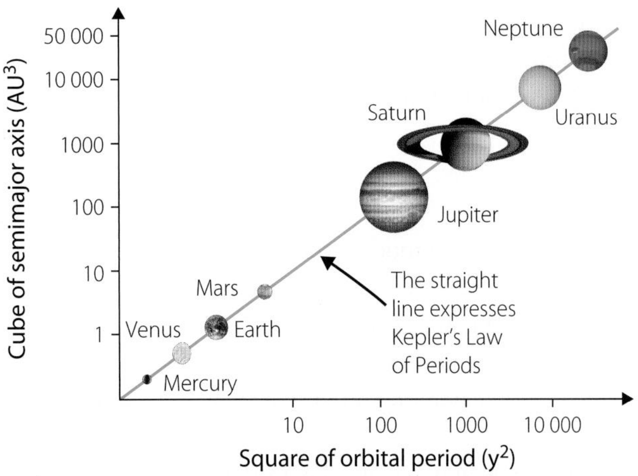

Kepler was able to deduce his third law, based on careful analysis of precise observations, before Newton formulated his Law of Universal Gravitation. He had no theoretical basis to support the relationship. The agreement between Kepler's empirically derived relationship and the theoretical one based on Newton's Law of Universal Gravitation was important in providing support for Newton's work.

Figure 4.10 shows a plot of orbital radius cubed as a function of period squared ($r^3$ vs $T^2$) for the planets of our solar system. You can see that the data gives a straight line.

The gradient of this line is $\dfrac{GM}{4\pi^2}$, and from such a graph we can deduce the mass of our Sun. We can also use similar graphs to deduce the mass of other stars from observations of the motions of their planets.

**FIGURE 4.10** A graph of $r^3$ versus $T^2$ for the planets in our solar system, giving a straight line. Note that 1 AU (astronomical unit) = $1.5 \times 10^{11}$ m.

▶ **WORKED EXAMPLE** (4.7)

A small planet is observed to orbit a star every 30 days. A second planet orbits the same star at a distance that is nine times the orbital radius of the first planet. What is the period of the second planet?

| ANSWER | LOGIC |
|---|---|
| $T_1 = 30$ days, $r_2 = 9r_1$ | ▪ Identify the relevant data. |
| $\dfrac{T^2}{r^3} = \dfrac{4\pi^2}{GM}$ | ▪ Relate period to radius. |
| $\dfrac{T^2}{r^3} = $ constant <br><br> $\dfrac{T_1^2}{r_1^3} = \dfrac{T_2^2}{r_2^3}$ | ▪ Recognise that the relationship between $T$ and $r$ is the same for all planets in a system. |
| $T_2 = \sqrt{\dfrac{r_2^3\, T_1^2}{r_1^3}}$ | ▪ Rearrange for $T_2$. |
| $T_2 = \sqrt{\dfrac{(9r_1)^3 (30\text{ days})^2}{r_1^3}} = \sqrt{(9)^3 (30\text{ days})^2}$ | ▪ Substitute values and simplify. |
| $T_2 = 810$ days | ▪ Calculate the final answer. |
| $T_2 = 810$ days | ▪ State the final answer with correct units and appropriate significant figures. |

**TRY THIS YOURSELF**

A spacecraft orbits the Moon with a period of 4.0 hours. It is then lowered into an orbit with half the radius. What is its new orbital period?

**KEY CONCEPTS**

- Kepler's laws are empirical. They were arrived at through careful mathematical analysis and modelling of precise astronomical observations.
- Kepler's first law states that all planets move in elliptical orbits with the Sun at one focus.
- Kepler's second law states that a line joining a planet to the Sun sweeps out equal areas in equal times. Hence planets move faster when they are closer to the Sun.
- Kepler's third law states that the square of the period of a planet's orbit is proportional to the cube of the mean radius of its orbit: $T^2 \propto r^3$, or $\dfrac{T^2}{r^3} = $ constant.

**Planetary orbits lab**

Work through the planetary orbits lab, and answer the questions about orbits and Kepler's laws.

**CHECK YOUR UNDERSTANDING**

4.3

1 State Kepler's three laws. Why are Kepler's laws referred to as empirical laws?

2 Describe the difference between an ellipse and a circle. Draw a diagram of an ellipse and mark on it the major and minor axes and the two foci.

3 Would a graph like that shown in Figure 4.10 for a different solar system with a more massive star have a smaller or larger gradient? Explain your answer.

4 Calculate the value of the constant in Kepler's second law, or $\dfrac{T^2}{r^3} = $ constant , for our solar system. $M_{Sun} = 2.00 \times 10^{30}$ kg.

5 Titan, a moon of Saturn, has an orbital radius of $1.22 \times 10^6$ km. It takes Titan 15 days and 22 hours to revolve around Saturn. Find the mass of Saturn.

6 Find the orbital period of an asteroid around the Sun, given that its mean orbital radius is twice Earth's orbital radius. Do not use the mass of the Sun in this calculation.

# 4.4 Energy in gravitational fields

Recall that in chapter 5 of *Physics in Focus Year 11* we derived the expression for the gravitational potential energy near the surface of Earth: $U = mgh$. This is the potential energy of the system of Earth and an object of mass $m$ that is at a height $h$ above Earth's surface. We derived this expression using the definition of work, $W = Fs$, and applying it to an object moving through a distance $h$ in the direction of the gravitational field $g$. We made the assumption in deriving this equation that the gravitational force was constant.

This model for gravitational potential energy works well for objects close to the surface of Earth, but it is *not* appropriate when the gravitational field cannot be treated as constant. We can, however, use the same process to derive the gravitational potential energy for any two objects with mass that are a distance $r$ apart.

## Potential energy in a gravitational field

The gravitational force experienced by mass $m$ due to mass $M$ is:

$$F = \frac{GMm}{r^2}$$

The force that $M$ exerts on $m$ is zero only when the two objects are infinitely far apart. So we take that to be the zero of gravitational potential energy for a system of masses. So $U = 0$ when $r = \infty$.

Imagine bringing one mass, $m$, gradually closer to a second mass $M$, starting with the two masses very, very far apart and at rest. The work done on mass $m$ due to the force by mass $M$, as it moves from infinitely far away ($r = \infty$) to its final position ($r$), is:

$$W = K = \frac{GMm}{r}$$

This is the change in kinetic energy of the object, and it is positive. By conservation of energy, the total energy, which is $E = K + U = 0$, stays constant. So:

$$U = -\frac{GMm}{r}$$

Note that it may look like we have simply taken the expression for force and multiplied it by a distance $-r$. However this is *not* how this equation arises. The force varies with distance so we must use calculus and integrate the force with respect to $r$, from the starting point with $r = \infty$ to the final configuration. If you are studying calculus, you can do the integration yourself to confirm that this is the correct expression.

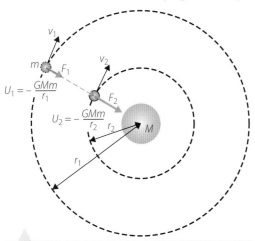

The equation $U = -\frac{GMm}{r}$ tells us that the gravitational potential energy of two masses is proportional to the product of the masses and inversely proportional to their separation. It is also negative. This makes sense because we chose our zero for potential energy as the configuration where the masses are infinitely far apart. At any separation less than infinite the potential energy is less than zero, and hence negative. This is because the gravitational force is attractive. If you want to move two massive objects further apart, you have to do positive work (add energy) by applying a force in the opposite direction to the field. If you allow an object to move in a gravitational field, it will accelerate in the direction of the field and its kinetic energy will increase. Because energy is conserved, this increase in kinetic energy comes from a decrease in potential energy (Figure 4.11).

**FIGURE 4.11** The potential energy of two massive objects decreases as they get closer together. At the same time the force increases and the velocity of the orbiting object increases.

Table 4.1 summarises the energy changes when one object moves within the gravitational field of a second object.

**TABLE 4.1** Summary of potential and kinetic energy changes for two objects with mass

| SYSTEM | OBJECT MOVES | WORK IS DONE | POTENTIAL ENERGY | KINETIC ENERGY |
|---|---|---|---|---|
| Isolated | with the field (objects get closer together) | by the field | decreases | increases |
| | against the field (objects get further apart) | on the field | increases | decreases |
| Open or closed | with the field (objects get closer together) | by external agent and by the field | decreases | increases |
| | against the field (objects get further apart) | by external agent and on the field | increases | either (depends which does more work) |

This expression for gravitational potential energy applies to any two objects of masses $M$ and $m$, for example a planet and a star, a satellite and a planet, even an electron and a nucleus. Remember that it is the Law of *Universal* Gravitation.

This gravitational potential energy is a type of binding energy: it is the energy that you would need to add to a system of two massive objects to move them infinitely far apart. To escape from Earth's gravitational field, if you have mass $m$ you need to have an amount of work done on you equal to or

greater than $W = \dfrac{GM_{\text{Earth}}m}{R_{\text{Earth}}}$.

▶ **WORKED EXAMPLE** (4.8)

What is the change in potential energy of a satellite with a mass of 75 kg when it is in near-Earth orbit with an initial altitude of 300 km and its orbit decays to an altitude of 290 km?

| ANSWER | LOGIC |
|---|---|
| $h_i = 300\,\text{km} = 3.00 \times 10^5\,\text{m}$, $h_f = 290\,\text{km} = 2.90 \times 10^5\,\text{m}$, $m = 75\,\text{kg}$ $M_{\text{Earth}} = 5.97 \times 10^{24}\,\text{kg}$, $R_{\text{Earth}} = 6.37 \times 10^6\,\text{m}$ | ▪ Identify the relevant data and convert to SI units. ▪ Find additional data needed. |
| $U = -\dfrac{GMm}{r}$ | ▪ Relate potential energy to position. |
| $\Delta U = U_f - U_i$ | ▪ Write the expression for change in potential energy. |
| $\Delta U = -\dfrac{GMm}{r_f} + \dfrac{GMm}{r_i}$ | ▪ Expand the expression. |
| $r_f = R_{\text{Earth}} + h_f$ $r_i = R_{\text{Earth}} + h_i$ | ▪ Note that the orbital radius is the altitude plus Earth's radius. |
| $\Delta U = -\dfrac{GMm}{R_{\text{Earth}} + h_f} + \dfrac{GMm}{R_{\text{Earth}} + h_i}$ | ▪ Substitute the expression for $r$ into the equation for $U$. |
| $\Delta U = -\dfrac{(6.67 \times 10^{-11}\,\text{N kg}^{-2}\text{m}^2)(5.97 \times 10^{24}\,\text{kg})(75\,\text{kg})}{6.37 \times 10^6\,\text{m} + 2.90 \times 10^5\,\text{m}}$ $+ \dfrac{(6.67 \times 10^{-11}\,\text{N kg}^{-2}\text{m}^2)(5.97 \times 10^{24}\,\text{kg})(75\,\text{kg})}{6.37 \times 10^6\,\text{m} + 3.00 \times 10^5\,\text{m}}$ $\Delta U = -4.4842 \times 10^9\,\text{N m} + 4.4775 \times 10^9\,\text{N m}$ | ▪ Substitute values with correct units. |
| $\Delta U = -6.723 \times 10^6\,\text{N m}$ $\Delta U = -6.72\,\text{MJ}$ | ▪ Calculate the final value and state the final answer with correct units and appropriate significant figures. |

**TRY THIS YOURSELF**

Calculate the change in potential energy of this satellite when it is first launched from ground level to an altitude of 300 km.

To raise a satellite into orbit requires a great deal of energy because the potential energy of the Earth–satellite system is much greater when the satellite is in orbit than when it is on the ground. But the potential energy is only one component of the total energy of the Earth–satellite system. We also need to consider kinetic energy.

## Total orbital energy

An object in orbit has both kinetic and potential energy. (Remember that the potential energy belongs to the system – both the orbiting mass and the mass being orbited.)

The potential energy is given by $U = -\dfrac{GMm}{r}$

The kinetic energy is given by $K = \dfrac{1}{2}mv^2$ if we take the central (orbited) object as stationary.

We can use the expression for $v$ that we derived earlier, $v = \sqrt{\dfrac{GM}{r}}$, to write the kinetic energy as:

$$K = \frac{1}{2}mv^2 = \frac{GMm}{2r}$$

The total energy of the system is then the sum of the kinetic and potential energies:

$$E_{\text{total}} = U + K = -\frac{GMm}{r} + \frac{GMm}{2r} = -\frac{GMm}{2r}$$

From this equation we can also derive a simple relationship between the potential, kinetic and total energies of an object in orbit:

$$E_{\text{total}} = \frac{U}{2} = -K$$

The kinetic energy is positive, and is larger for objects in closer orbits. The potential energy is negative, and has a magnitude twice that of the kinetic energy.

---

▶ **WORKED EXAMPLE** (4.9)

Find the potential, kinetic and total energy of Earth due to its orbit about the Sun. The orbit has an average radius of 150 million km.
$M_{\text{Sun}} = 2.00 \times 10^{30}\,\text{kg};\ M_{\text{Earth}} = 5.97 \times 10^{24}\,\text{kg}$

| ANSWER | LOGIC |
|---|---|
| $M_{\text{Sun}} = 2.00 \times 10^{30}\,\text{kg},\ M_{\text{Earth}} = 5.97 \times 10^{24}\,\text{kg}$ <br><br> $r = 150 \times 10^6\,\text{km} = 1.50 \times 10^{11}\,\text{m}$ | ▪ Identify the relevant data and convert to SI units. |
| $U = -\dfrac{GMm}{r}$ | ▪ Write the expression for the potential energy. |
| $U = -\dfrac{(6.67 \times 10^{-11}\,\text{N kg}^{-2}\text{m}^2)(5.97 \times 10^{24}\,\text{kg})(2.00 \times 10^{30}\,\text{kg})}{1.5 \times 10^{11}\,\text{m}}$ | ▪ Substitute values with correct units. |
| $U = -5.31 \times 10^{33}\,\text{N m}$ | ▪ Calculate the final value. |
| $U = -5.31 \times 10^{33}\,\text{J}$ | ▪ State the answer with correct units and appropriate significant figures. |
| $K = -\dfrac{U}{2}$ | ▪ Relate kinetic energy to the potential energy. |

| | |
|---|---|
| $K = -\dfrac{-5.31 \times 10^{33}\,\text{J}}{2}$ | ▪ Substitute values with correct units. |
| $K = 2.65 \times 10^{33}\,\text{J}$ | ▪ Calculate the final value. |
| $E_{\text{total}} = \dfrac{U}{2}$ | ▪ Relate the total energy to the potential energy. |
| $E_{\text{total}} = \dfrac{-5.31 \times 10^{33}\,\text{J}}{2}$ | ▪ Substitute values with correct units. |
| $E_{\text{total}} = -2.65 \times 10^{33}\,\text{J}$ | ▪ Calculate the final value. |
| $U = -5.31 \times 10^{33}\,\text{J},\ K = 2.65 \times 10^{33}\,\text{J},\ E_{\text{total}} = -2.65 \times 10^{33}\,\text{J}$ | ▪ State the final answers with correct units and appropriate significant figures. |

**TRY THIS YOURSELF**

Calculate the total energy of a 50 kg geostationary satellite. Use results from previous worked examples.

## Escape velocity

Recall that the gravitational potential energy of a system of two objects is zero when the objects are infinitely far apart. At this distance, they no longer experience any force due to each other's gravitational field. Imagine launching a spacecraft that is to travel from the surface of Earth to a distant galaxy, so that it must escape completely from Earth's gravitational field. For an object to escape Earth's gravitational field, starting at the surface of Earth, it must have enough kinetic energy that can be converted into potential energy (which is negative), so the final potential energy is zero.

For the limiting case in which the object starts with just enough kinetic energy, so that when it has escaped Earth's gravitational field all of the kinetic energy has been converted to potential energy:

$$E_{\text{total}} = U + K = 0$$

because the potential energy at infinite separation (zero field) is zero.

As the object moves further from Earth, conservation of energy tells us that $\Delta U = -\Delta K$, or

$$-\frac{GMm}{r_{\text{f}}} + \frac{GMm}{r_{\text{i}}} = -\frac{1}{2}mv_{\text{f}}^2 + \frac{1}{2}mv_{\text{i}}^2$$

If the object starts at Earth's surface, ends up infinitely far away, and has a final speed of zero, then:

$$-\frac{GMm}{\infty} + \frac{GMm}{R_{\text{Earth}}} = -\frac{1}{2}m0^2 + \frac{1}{2}mv_{\text{i}}^2$$

which we can simplify to

$$\frac{GMm}{R_{\text{Earth}}} = \frac{1}{2}mv_{\text{i}}^2$$

We can rearrange this equation to find the necessary initial speed:

$$v_{\text{i}} = \sqrt{\frac{2GM}{R_{\text{Earth}}}}$$

This is the minimum speed required of an object at the surface of Earth so that it can escape into space and not be pulled back. Note that it must also be moving in the right direction – radially away from Earth. This speed is called escape velocity, and at the surface of Earth it has the value $11.2\,\text{km s}^{-1}$. So for an object to be able to escape from Earth and into deep space it must leave Earth's surface with a speed of at least $11.2\,\text{km s}^{-1}$. Of course, it must also be pointing away from Earth's surface.

WS

Energy in gravitational fields

This expression applies for any object at any distance from a large body such as a planet or star, and can be more generally written as:

$$v_{esc} = \sqrt{\frac{2GM}{r}}$$

## WORKED EXAMPLE (4.10)

Calculate the escape velocity for the International Space Station, in orbit 400 km above the surface of Earth.

| ANSWER | LOGIC |
|---|---|
| $M_{Earth} = 5.97 \times 10^{24}$ kg <br><br> $r = R_{Earth} + 400$ km $= 6370$ km $+ 400$ km $= 6.77 \times 10^6$ m | ■ Identify the relevant data and convert to SI units. |
| $v_{esc} = \sqrt{\frac{2GM}{r}}$ | ■ Write the expression for escape velocity. |
| $v_{esc} = \sqrt{\frac{2(6.67 \times 10^{-11}\,\text{N kg}^{-2}\,\text{m}^2)(5.97 \times 10^{24}\,\text{kg})}{6.77 \times 10^6\,\text{m}}}$ | ■ Substitute values with correct units. |
| $v_{esc} = 1.0846 \times 10^4\,\text{m s}^{-1}$ | ■ Calculate the final value. |
| $v_{esc} = 10.8\,\text{km s}^{-1}$ <br><br> Note that this is only slightly less than the escape velocity from the surface of Earth. | ■ State the answer with correct units and appropriate significant figures. |

### TRY THIS YOURSELF

Calculate the escape velocity on the surface of Mars, which has a mass of $6.39 \times 10^{23}$ kg and a radius of 3400 km. Would it be easier to launch a spacecraft from Mars or from Earth?

**Satellites**

Watch the video clip of this satellite launch. Observe the angle of the launch vehicle shortly after lift-off.

Satellites are typically launched from close to the equator so that the rotation of Earth contributes to their kinetic energy. They are also launched at an angle to take advantage of this, rather than directly upwards.

**KEY CONCEPTS**

- The gravitational potential energy of a system of two objects with masses $M$ and $m$, a distance $r$ apart, is $U = -\frac{GMm}{r}$. This is negative because we define the zero of gravitational potential energy as occurring when $r = \infty$.
- The kinetic energy of an object in orbit is $K = \frac{1}{2}mv^2 = \frac{GMm}{2r}$.
- The total energy of an object in orbit is: $E_{total} = U + K = -\frac{GMm}{2r}$, so $E_{total} = \frac{U}{2} = -K$.
- The minimum speed required for an object to escape the gravitational field of an object of mass $M$ is the escape velocity, $v_{esc} = \sqrt{\frac{2GM}{r}}$, when the object is a distance $r$ from the mass $M$.

1 Distinguish between orbital velocity and escape velocity.

2 Two spaceships are to be launched from Earth's surface to travel into interstellar space. One has mass $m$ the other has mass $2m$.

   a For which satellite is the escape speed greater?

   b Which will require more kinetic energy at launch?

3 A satellite is in an unstable orbit such that its altitude is gradually decreasing. Describe how its potential and kinetic energies change as its orbit decays.

4 For a satellite with a mass of 5.0 kg that is in a geostationary orbit, calculate the:

   a gravitational potential energy.

   b kinetic energy.

   c total energy.

5 The Moon has a mass of $7.35 \times 10^{22}$ kg and an orbital radius of 384 000 km.

   a Using energy considerations, calculate the orbital speed of the Moon.

   b Calculate the orbital period of the Moon.

6 Kepler's second law says that planets in elliptical orbits move faster when closer to the Sun. Using energy conservation, explain why this must be the case.

7 The escape velocity on Jupiter is 59.5 km s$^{-1}$. Jupiter has a mass of $1.60 \times 10^{27}$ kg. Calculate the radius of Jupiter.

8 Sketch graphs, on a single set of axes, of the kinetic, potential and total energy as a function of orbital radius, $r$.

- Newton's Law of Universal Gravitation, $F = \dfrac{GMm}{r^2}$, states that the force one object with mass exerts on another object with mass is proportional to the product of the masses and inversely proportional to the square of the distance between the objects.

- The gravitational field at any point is the gravitational force per unit mass at that point: $g = \dfrac{F}{m}$, which is the local acceleration due to gravity. The units of $g$ are $N\,kg^{-1}$ or $m\,s^{-2}$.

- The gravitational field strength near a point mass or spherical mass is $g = \dfrac{GM}{r^2}$.

- Close to the surface of Earth, the gravitational field is uniform and has magnitude $g = 9.8\,N\,kg^{-1} = 9.8\,m\,s^{-2}$.

- The gravitational force provides the centripetal (net) force that keep planets and satellites in orbit. For circular orbits $F = \dfrac{GMm}{r^2} = \dfrac{mv^2}{r}$.

- The centripetal acceleration of an orbiting body such as a planet or satellite is given by $a = \dfrac{v^2}{r} = \dfrac{GM}{r^2}$.

- The period of an object in orbit is $T = \dfrac{2\pi r}{v}$, and $T^2 = \dfrac{4\pi^2 r^3}{GM}$.

- Geostationary satellites stay in orbit above a specific point on Earth's equator, at an altitude of 36 000 km. Geosynchronous satellites have the same altitude and travel above a great circle.

- Low-Earth orbit satellites have much lower altitudes and shorter periods.

- Kepler's first law states that all planets move in elliptical orbits with the Sun at one focus.

- Kepler's second law states that a line joining a planet to the Sun sweeps out equal areas in equal times. Hence planets move faster when they are closer to the Sun.

- Kepler's third law states that the square of the period of a planet's orbit is proportional to the cube of the mean radius of its orbit: $T^2 \propto r^3$, or $\dfrac{T^2}{r^3} = \text{constant}$.

- We define the zero of gravitational potential energy as occurring when $r = \infty$.

- The gravitational potential energy of a system of two objects is $U = -\dfrac{GMm}{r}$.

- The kinetic energy of an object in orbit is $K = \dfrac{1}{2}mv^2 = \dfrac{GMm}{2r}$.

- The total energy of an object in orbit is: $E_{total} = U + K = -\dfrac{GMm}{2r}$, so $E_{total} = \dfrac{U}{2} = -K$.

- The minimum speed required for an object to escape the gravitational field of an object of mass $M$ is $v_{esc} = \sqrt{\dfrac{2GM}{r}}$.

# 4 CHAPTER REVIEW QUESTIONS

Review quiz

1 Write the expression for gravitational field strength according to Newton's Law of Universal Gravitation. Give the units for gravitational field strength.

2 Why is the net force acting on a satellite considered to be a centripetal force?

3 Explain how it is that no work is done by Earth's gravitational field on a satellite in a stable orbit. Use a diagram to help explain your answer.

4 What happens to the gravitational force between two objects when they are moved ten times further apart? Explain your answer.

5 Explain why an antenna for satellite TV does not need to move (to 'track') to continue receiving a signal.

6 Draw sketches to distinguish between geostationary, geosynchronous and low-Earth orbits. Explain why these different orbits are used for different types of satellites.

7 What is the orbital speed of the moon of a distant planet, given that the planet has a mass of $4.5 \times 10^{24}\,kg$ and the moon's mean orbital radius is $3.8 \times 10^8\,m$?

8 A meteor is at a distance above Earth's surface equal to twice the radius of Earth, and is falling towards the surface. Calculate its acceleration.

**9** Calculate the orbital period in Earth years for an asteroid that is orbiting the Sun with a mean orbital radius that is three times the orbital radius of Earth.

**10** What is the mass of the central body being orbited by a planet if the planet's orbital period is 8.0 hours and its distance from the central body is $4.5 \times 10^{10}$ m?

**11** Explain why the acceleration of a satellite with an orbital radius of 42 000 km is less than that for a satellite with 7000 km orbital radius.

**12** A satellite is orbiting the Sun with a radius of $1.5 \times 10^{10}$ m. What is its orbital speed, given the mass of the Sun is $2.0 \times 10^{30}$ kg?

**13** A 20 kg satellite is in orbit 400 km above the surface of Earth.

　**a** What is the force that Earth exerts on the satellite? Make two calculations – first using the near-Earth approximation and then using Newton's Law of Universal Gravitation.

　**b** Repeat part **a** for the same satellite in a geosynchronous orbit.

**14** A 20 kg satellite is initially in orbit 400 km above the surface of Earth. It is then raised to a geostationary orbit.

Calculate the change in gravitational potential energy of the Earth–satellite system. Make two calculations – first using the near-Earth approximation and then using the expression for potential energy derived from Newton's Law of Universal Gravitation.

**15** Explain how Kepler's third law can be used by astronomers to deduce the mass of a star.

**16** A space probe is launched from the surface of Earth with an initial speed twice the escape speed at Earth's surface. What will be its speed when it is very, very far from Earth?

**17** A 5.0 kg rock is dropped from 10 m above the surface of Ganymede, a moon of Jupiter. Ganymede has a mass of $1.5 \times 10^{23}$ kg and radius of 2600 km.

　**a** What is the gravitational field near Ganymede's surface?

　**b** What is the acceleration of the rock?

　**c** How much gravitational potential energy was transformed as the rock fell to the surface?

　**d** What was the maximum amount of kinetic energy gained by the rock?

**18** What is the speed of any satellite orbiting Earth at an altitude of 630 km?

**19** A 5000 kg spacecraft is to be launched from Earth to travel into deep space. What minimum kinetic energy must it have?

**20** A 10 kg synchronous satellite is to be placed in orbit above the equator of Jupiter. Jupiter has a day length of 9 hours and 56 minutes, a mass of $1.9 \times 10^{27}$ kg and a radius of 69 900 km.

　**a** At what altitude should the satellite be placed?

　**b** What will its orbital velocity be?

　**c** What will be the total energy of the Jupiter–satellite system?

**21** A distant planet is observed to have a moon orbiting it with a period of 39 days at a distance of 1.1 million km. Calculate the mass of this planet.

**22** A satellite in low-Earth orbit is subject to a small frictional force due to Earth's atmosphere. This results in a decrease in the total mechanical energy of the Earth–satellite system.

　**a** Explain how it is possible for the speed of the satellite to increase as a result of this. Hint: assume it moves to a new circular orbit and think about the energies involved.

　**b** Describe the energy changes in terms of the work done by the different forces acting on the satellite. Use a diagram to help explain your answer.

**Answer the following questions.**

1   Dimitra has made herself a potato launcher. It has a barrel 25 cm long, and she loads it with a 150 g potato.

   **a**   If the average force applied to the potato as it moves along the barrel is 150 N, with what speed does it leave the barrel?

   **b**   If the barrel is aimed at an angle of 50° above the horizontal, what are the horizontal and vertical components of its initial velocity?

   **c**   What forces act on the potato after it has left the barrel? Draw a force diagram for the potato.

   **d**   Calculate the maximum height that the potato reaches.

   **e**   Calculate the time of flight of the potato.

   **f**   Calculate the range of the potato.

   **g**   Calculate the kinetic energy of the potato just before it hits the ground.

2   Cooper kicks a soccer ball with an initial speed of $8.5\,\mathrm{m\,s^{-1}}$ at an angle of 25° above the horizontal.

   **a**   Ignoring air resistance, make calculations to complete the following table.

| TIME (s) | 0 | 0.1 | 0.2 | 0.3 | 0.4 | 0.5 | 0.6 | 0.7 | 0.8 |
|---|---|---|---|---|---|---|---|---|---|
| $v_x\,(\mathrm{m\,s^{-1}})$ | | | | | | | | | |
| $v_y\,(\mathrm{m\,s^{-1}})$ | | | | | | | | | |
| $v\,(\mathrm{m\,s^{-1}})$ | | | | | | | | | |
| $x\,(\mathrm{m})$ | | | | | | | | | |
| $y\,(\mathrm{m})$ | | | | | | | | | |

   **b**   Draw graphs of:

     **i**   $v$ vs $t$.

     **ii**   $y$ vs $t$.

     **iii**   $y$ vs $x$.

     Make sure you label your graphs and their axes carefully.

3   The battery in Hussein's smartphone has run flat, but he needs to send an urgent message to his girlfriend, Jill. He writes a note, folds it around a rock and stands 3.0 m back from her window, which is 2.8 m above the ground. The window is only open at the bottom in a narrow horizontal slit, so Hussein must throw accurately. If Hussein throws the note so it leaves his hand at an angle of 30° at a height of 1.7 m above the ground, at what speed should he throw the note? Why does the note hit the wall a little below the window instead of passing through, when thrown at the speed you calculated?

4   A circular carousel has different types of seats at different distances from the centre, about which it rotates at

constant speed. Harriet sits on a horse near the outer edge, 2.5 m from the centre. Karolina sits on a frog 2.0 m from the centre.

   **a**   Draw a force diagram for each girl.

   **b**   Which girl has the larger angular velocity?

   **c**   Which has the larger speed?

   **d**   If they have the same mass, which experiences the greater net force?

   The carousel does 5 rotations per minute.

   **e**   Calculate the angular velocity of the carousel.

   **f**   Calculate the acceleration of each girl.

   **g**   If the ride goes for 5 minutes, what angular and linear distances are travelled by each girl?

5   Greg, who can't get his yoyo to work, is swinging it on its string around his head in a horizontal circle 100 cm above the ground. The string is 80 cm long, the yoyo has a mass of 80 g, and Greg swings it so that it does a complete circle each 0.60 s, before letting go of the string. While the yoyo is moving in a circle:

   **a**   Calculate the speed of the yoyo.

   **b**   Calculate the acceleration of the yoyo.

   **c**   Calculate the tension in the string.

   After Greg has let go of the string:

   **d**   Calculate the time taken before the yoyo hits the ground.

   **e**   Calculate the distance travelled by the yoyo before it hits the ground.

6   Fei is driving her car along the Kings Highway. She approaches a left-hand corner with an advisory speed sign of $45\,\mathrm{km\,h^{-1}}$. The corner is banked at an angle of 5°.

   **a**   If the corner is banked such that this is the correct speed at which a car could negotiate this corner with no frictional force required, what is the radius of curvature of this corner?

   **b**   If the corner has this radius of curvature and is not banked at all, what minimum coefficient of static friction is required between the road and the car's tyres for it to negotiate the corner at $45\,\mathrm{km\,h^{-1}}$?

   **c**   Fei is a rally driver. She negotiates the corner by entering it at the left-hand edge of the road and exiting it at the right-hand edge. Draw a sketch showing this path, and explain how it enables her to maintain a higher speed through the corner.

7   A distant planet is observed to have 5 moons, with periods and orbital radii as given in the table below.

   **a**   Plot an appropriate graph and deduce the mass of the planet from the gradient of your graph.

**b** Some cosmological theories suggest that as the Universe is expanding the value of $G$ will change. If the value of $G$ decreases, what happens to the gradient of your graph?

| MOON | PERIOD (DAYS) | ORBITAL RADIUS ($10^6$ km) |
|---|---|---|
| Alpha | 39 | 1.1 |
| Beta | 110 | 2.2 |
| Gamma | 230 | 3.6 |
| Delta | 343 | 4.7 |
| Epsilon | 519 | 6.2 |

**8** The distance between Earth and the Moon is 384 000 km. Earth has a mass of $5.97 \times 10^{24}$ kg and radius of 6370 km; the Moon has a mass of $7.35 \times 10^{22}$ kg.

**a** Calculate the gravitational field due to Earth at:

**i** the International Space Station at an altitude of 400 km.

**ii** a geostationary satellite at an altitude of 36 000 km.

**iii** the Moon.

**b** Calculate the gravitational force that the Moon exerts on Earth.

**c** Find the distance from the centre of Earth, along a line from Earth to the Moon, at which the gravitational forces of the Moon and Earth are equal, so the net gravitational force is zero.

**9** The Sun has a mass of $2.00 \times 10^{30}$ kg and rotates once every 24 days. If you want to put a heliostationary satellite (that stays over the same point on the Sun's equator) with mass 120 kg into orbit about the Sun.

**a** What must its orbital radius be?

**b** What must its orbital speed be?

**c** What is the total energy of the satellite in this orbit?

**d** What is the escape speed from this orbit?

**10** A 15 000 kg spacecraft is to be launched for travel into deep space from the equator of Mars. Mars has a day length of 24 hours and 40 minutes, a radius of 3390 km, and a mass of $6.39 \times 10^{23}$ kg.

**a** Calculate the minimum kinetic energy that must be given to the spacecraft, given that it already has some kinetic energy due to the rotation of Mars, to ensure that it can reach deep space.

**b** Repeat your calculations for the spacecraft if it is to be launched from Earth, which has mass $5.97 \times 10^{24}$ kg, radius 6370 km.

## DEPTH STUDY SUGGESTIONS

→ Perform a first-hand investigation to measure the trajectory of a projectile using data loggers or a webcam.

→ Perform a first-hand investigation to compare the trajectories of projectiles of different shapes and densities, to explore the effect of air resistance on projectiles.

→ Extend the numerical modelling investigation of projectile motion in chapter 2 to include the effect of air resistance.

→ Investigate projectile motion in one or more sports. Are different sports characterised by different typical trajectories? Is air resistance an important factor in any of them?

→ Investigate circular motion on carnival rides – what typical speeds, accelerations and forces are involved?

→ Perform a first-hand investigation to measure the force required to keep an object moving in a circular path on a string. Investigate how that force varies with speed and radius.

→ Investigate levers and the role of torque in tool design to provide mechanical advantage.

→ Create a numerical model (computer simulation) to calculate the orbital speed, period, and energies of satellites at different radii about different planets.

→ Develop a lesson plan or website to teach younger students about the orbit, rotation and phases of the Moon.

→ Do a literature review and make a timeline of the history of the development of our understanding of gravity, including early theories, classical mechanics, general relativity and particle physics.

# ELECTROMAGNETISM

9780170409131

# 5 Charged particles, conductors and electric and magnetic fields

## INQUIRY QUESTION

What happens to stationary and moving charged particles when they interact with an electric or magnetic field?

**OUTCOMES**

## Students:

- investigate and quantitatively derive and analyse the interaction between charged particles and uniform electric fields, including: **(ACSPH083) ICT, N**
  - electric field between parallel charged plates $E = \dfrac{V}{d}$
  - acceleration of charged particles by the electric field $\vec{F}_{net} = m\vec{a}$, $\vec{F} = q\vec{E}$
  - work done on the charge $W = qV$, $W = qEd$, $K = \dfrac{1}{2}mv^2$
- model qualitatively and quantitatively the trajectories of charged particles in electric fields and compare them with the trajectories of projectiles in a gravitational field **CCT, ICT, L, N**
- analyse the interaction between charged particles and uniform magnetic fields, including: **(ACSPH083)**
  - acceleration, perpendicular to the field, of charged particles
  - the force on the charge $F = qv_\perp B = qvB\sin\theta$ **N**
- compare the interaction of charged particles moving in magnetic fields to: **CCT**
  - the interaction of charged particles with electric fields
  - other examples of uniform circular motion **(ACSPH108)**

Physics Stage 6 Syllabus © 2017 NSW Education Standards Authority (NESA) for and on behalf of the Crown in right of the State of New South Wales.

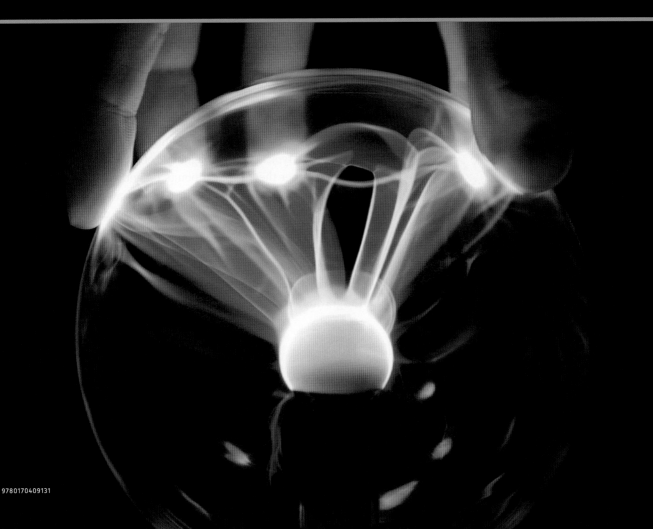

Getty Images/whiterabbit83

FIGURE 5.1 Part of the accelerator mass spectrometer in the Oxford Radiocarbon Accelerator Unit, used for dating samples of organic material.

In chapter 4 we explored how objects interact with the gravitational field. In this chapter we examine how objects interact with electric and magnetic fields. Any charged object experiences a force in an electric field, and a moving charged object experiences a force in a magnetic field. The mass spectrometer shown in Figure 5.1 makes use of that effect.

Earth has a gravitational field that holds us to Earth's surface and keeps the Moon in orbit. Earth also has an electric field and a magnetic field, and these exert forces on charged particles. Over small distances both these fields can be modelled as uniform. We can also create uniform electric fields using charged parallel plates, and uniform magnetic fields using current-carrying solenoids. This was described in chapters 12 and 14 of *Physics in Focus Year 11*. Here we build on that material and investigate the interaction of uniform electric and magnetic fields with charged particles.

## 5.1 Charged particles in uniform electric fields

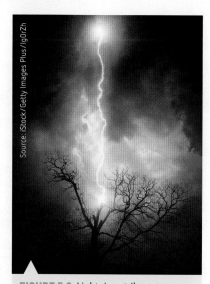

FIGURE 5.2 Lightning strikes a tree when the electric field between the cloud and the ground becomes so large that air particles ionise, producing free charged particles that can form a current.

Electric fields are created by charged objects, such as charged particles, and exert forces on charged objects. Charge is a fundamental property of matter.

You will be familiar with the local effects of Earth's electric field in stormy weather, such as that shown in Figure 5.2. The electric field varies over the surface of Earth but can be treated as uniform, that is, constant in magnitude and direction, over distances of a few hundred metres or so.

The size and direction of Earth's electric field varies with weather conditions (Figure 5.3). In sunny weather it points downwards and typically has a magnitude of around $100\,\mathrm{V\,m^{-1}}$. In stormy weather the local field points up and can be much stronger. Electric field lines can be used to represent electric fields. The direction of the field lines shows the direction of force on a positively charged particle, and the density of the field lines is proportional to the field strength. The force due to Earth's electric field on a small charged particle such as an electron is many orders of magnitude greater than that due to Earth's gravitational field.

FIGURE 5.3 Electric field line diagram for Earth's electric field in **a** fair weather and **b** stormy weather

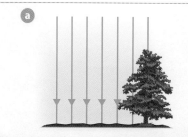

If we want to make a strong, uniform electric field we can do so by using two parallel plates with opposite charges. The field is uniform between the plates, and only becomes non-uniform very close to the edges. Figure 5.4 shows the electric field created by a pair of parallel plates, one with charge $Q$ and the other with charge $-Q$. The plates are spaced a distance $d$ apart and the **potential difference** between the plates is $\Delta V$. Potential difference is the difference in **electric potential** between the two plates. Electric potential is the **electric potential energy** per unit charge at a point in a field, and is measured in units of volts, V, where $1\,\mathrm{V} = 1\,\mathrm{J\,C^{-1}}$.

The electric field between the plates is $E = -\dfrac{\Delta V}{d}$.

The potential difference, $\Delta V$, is measured in V and $d$ is measured in m, so the units of $E$ are $\mathrm{V\,m^{-1}}$. Recall from chapter 12 of *Physics in Focus Year 11* that the electric field is the force per unit charge at a point, so its units can also be written as $\mathrm{N\,C^{-1}}$.

Field, like force, is a vector. The negative sign in the equation above tells us the direction of the field. The field points in the direction of decreasing potential.

You may sometimes see this equation written as $E = \dfrac{V}{d}$, but remember that it is the potential *difference*, the *change* in potential, not the value of the potential at a single point that determines the field. It is not enough that the potential is non-zero, *it has to be changing* for there to be an electric field. So, in this version of the equation, the $V$ really means the change in potential over the distance $d$. The $E$ is the magnitude of the field; the direction is not given by this equation.

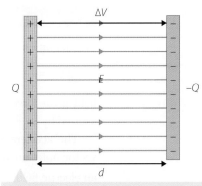

**FIGURE 5.4** An electric field line diagram for a pair of oppositely charged parallel plates

Charged particles in uniform electric fields

▶ **WORKED EXAMPLE** (5.1)

The pair of charged plates in Figure 5.4 are at potentials of +6.0 V and −6.0 V. The distance between the plates is 1.0 mm. What is the magnitude and direction of the electric field between the plates?

| ANSWER | LOGIC |
|---|---|
| $d = 1.0 \times 10^{-3}\,\mathrm{m}$ <br><br> $V_{\text{left plate}} = +6.0\,\mathrm{V}$, $V_{\text{right plate}} = -6.0\,\mathrm{V}$ | • Identify the relevant data in the question, noting that the plate with positive charge is at the higher potential. |
| $E = \dfrac{V}{d}$ | • Relate the field to the change in potential. |
| $V = \Delta V = V_{\text{right plate}} - V_{\text{left plate}}$ | • Relate the potential difference to the potentials of the two plates, working left to right. |
| $E = \dfrac{\Delta V}{d} = \dfrac{V_{\text{right plate}} - V_{\text{left plate}}}{d}$ | • Substitute the expression for $\Upsilon V$. |
| $E = \dfrac{-6.0\,\mathrm{V} - +6.0\,\mathrm{V}}{1.0 \times 10^{-3}\,\mathrm{m}}$ | • Substitute correct values with units. |
| $E = -12\,000\,\mathrm{V\,m^{-1}}$ | • Calculate the answer. |
| $E = 12\,\mathrm{kV\,m^{-1}}$ pointing to the right | • State the final answer with correct units, direction and appropriate significant figures. Note that the field points from high to low potential, so to the right. |

**TRY THIS YOURSELF**

On a stormy day (Figure 5.3) the electric field has a magnitude of $2.0\,\mathrm{kV\,m^{-1}}$. What is the potential difference between the ground and the clouds 2.5 km above? Which is at higher potential, the clouds or the ground?

**Millikan's oil-drop experiment**

Find out how the charge of the electron was measured using a uniform electric field.

## Acceleration of charged particles by an electric field

The definition of electric field is the electrostatic force per unit charge exerted on a small positive test charge, $E = \dfrac{F}{q}$ (chapter 12 of *Physics in Focus Year 11*). This is analogous to the definition of gravitational field as the force per unit mass on a test mass, $g = \dfrac{F}{m}$ (see chapter 4). The electrostatic force is the force exerted by an electric field.

The acceleration that an object experiences due to a force is given by Newton's second law. In vector notation we write: $\vec{F} = m\vec{a}$ (chapter 4 of *Physics in Focus Year 11*). So for a charged particle in an electric field, the force due to the electric field is $\vec{F} = q\vec{E} = m\vec{a}$, and the acceleration of the charged particle is $\vec{a} = \dfrac{\vec{E}q}{m}$.

Note that $\vec{a}$ and $\vec{E}$ are both vectors, so we need to specify direction. When the charge, $q$, is positive, $\vec{a}$ and $\vec{E}$ have the same direction. When $q$ is negative, $\vec{a}$ and $\vec{E}$ are in opposite directions.

▶ **WORKED EXAMPLE** (5.2)

What is the acceleration of an electron due to Earth's fair-weather field, $E = 100\,\text{V m}^{-1}$ down?

| ANSWER | LOGIC |
|---|---|
| $E = -100\,\text{V m}^{-1}$ | ▪ Identify the relevant data in the question. |
| $m = 9.1 \times 10^{-31}\,\text{kg},\ q = -1.6 \times 10^{-19}\,\text{C}$ | ▪ Look up other data needed. |
| $a = \dfrac{Eq}{m}$ | ▪ Relate the acceleration to the field. |
| $a = \dfrac{(-100\text{ V m}^{-1})(-1.6 \times 10^{-19}\text{ C})}{9.1 \times 10^{-31}\text{ kg}}$ | ▪ Substitute correct values with units. |
| $a = +1.76 \times 10^{13}\,\text{V C m}^{-1}\text{kg}^{-1}$ | ▪ Calculate the answer. |
| Note: $1\,\text{V} = 1\,\text{J C}^{-1} = 1\,\text{kg m}^2\,\text{s}^{-2}\,\text{C}^{-1}$ | |
| $a = +1.76 \times 10^{13}\,\text{m s}^{-2}$ pointing upwards | ▪ State the final answer with correct units, direction and appropriate significant figures. |

**TRY THIS YOURSELF**

What is the acceleration of a proton in the Earth's fair weather field? Give magnitude and direction.

## Work done on a charged particle by a uniform electric field

In chapter 12 of *Physics in Focus Year 11,* we derived the expression for the work done on a charged particle when it moves in an electric field. Starting with the force due to the field, $F = Eq$, and using the expression for work, $W = Fd$, we get $W = qEd$. Note that $E$ is the magnitude of the electric field, and $d$ is the displacement *in the direction of the field*. It is important to remember that $E$, $F$ and $d$ are really vectors. Work is only done by a force, and hence a field, when there is a displacement parallel to the direction of the field. No work is done when a charged particle moves perpendicular to the field.

For a uniform electric field, the field and potential are related by $E = -\dfrac{\Delta V}{d}$, so $Ed = -\Delta V$. Multiplying both sides by $q$ gives:

$$qEd = -q\Delta V$$

So the work done *by* the field *on the particle* when a charged particle moves through a potential difference $\Delta V$ is:

$$W = -q\Delta V$$

The negative sign tells us that when a positive charge moves through a positive potential difference, negative work is done on it by the field. This is what happens when one positively charged particle is moved closer to another positively charged particle, or towards the positively charged plate of a pair of charged plates. Either an external force must be applied to do work on the particle–field system, or, if the particle has some initial kinetic energy, it will slow down and lose this energy.

You may sometimes see this equation written as $W = qV$. This equation is not correct for the work done *by the field* on a charged particle. The lack of a negative sign implies that a positive charge will do positive work and accelerate a second positive charged particle towards it. This is not the case – positive charges do not attract each other. The $\Delta$ sign in the correct version indicates that the *change* in potential, not its value at a single point, is also very important, because a charge must move through a *change* in potential for work to be done on it. This was discussed in chapter 12 of *Physics in Focus Year 11*.

Consider a positively charged particle that starts at rest and is allowed to move freely in a uniform electric field (no other external forces act on it), as in Figure 5.5a. The particle will experience a force in the direction of the field lines, towards a position of lower potential. As the particle moves in this direction the potential energy of the field–particle system decreases and the kinetic energy of the particle increases. By conservation of energy:

$$\Delta U + \Delta K = 0, \text{ so } \Delta U = -\Delta K$$

The change in kinetic energy is also equal to the work done by the field, which is equal to the decrease in potential energy of the system.

$$W = -\Delta U = -q\Delta V$$

If the particle starts at rest then $K = \Delta K$, and $W = -\Delta U = -q\Delta V = qEd = K$.
Recall from chapter 4 of *Physics in Focus Year 11* that kinetic energy is given by $K = \dfrac{1}{2}mv^2$.

**FIGURE 5.5 a** A positively charged particle starting at rest will be accelerated in the direction of the field, to a point of lower potential. **b** A negatively charged particle starting at rest will be accelerated in the opposite direction to the field, to a point of higher potential. In both cases the particle gains kinetic energy and the particle–field system loses potential energy.

When a negatively charged particle is released from rest in an electric field it is accelerated in the opposite direction to the field lines. It moves from a position of lower to higher potential, and gains kinetic energy as it does so (Figure 5.5b).

If the particle does not start at rest then we need to remember that the work done is the *change* in kinetic energy, $\Delta K$, and $K_{\text{final}} = K_{\text{initial}} + \Delta K = K_{\text{initial}} + W$.

If external forces are acting then a positively charged particle can be pushed to a position of higher potential, increasing the potential energy of the particle–field system. In this case, work must be done by the external force.

## WORKED EXAMPLE (5.3)

An electron enters a region with uniform electric field, moving parallel with the field. The initial speed of the electron is $5.5\,\text{km s}^{-1}$ and the electric field strength is $150\,\text{V m}^{-1}$.

**a** How far does the electron travel before coming to a stop?

**b** Through what potential difference has it moved?

| ANSWER | LOGIC |
|---|---|
| $E = 150\,\text{V m}^{-1}$, $v_i = 5500\,\text{m s}^{-1}$, $v_f = 0\,\text{m s}^{-1}$ <br><br> $m = 9.1 \times 10^{-31}\,\text{kg}$, $q = -1.6 \times 10^{-19}\,\text{C}$ | ▪ Identify the relevant data in the question. <br> ▪ Look up other data needed. |
| **a**  $W = Eqd = \Delta K$ | ▪ Relate the work to the field and the change in kinetic energy. |
| $\Delta K = \dfrac{1}{2}mv_{\text{final}}^2 - \dfrac{1}{2}mv_{\text{initial}}^2 = -\dfrac{1}{2}mv_{\text{initial}}^2$ | ▪ Write the expression for change in kinetic energy, recognising that the final kinetic energy must be zero. |
| $Eqd = \Delta K = -\dfrac{1}{2}mv_{\text{initial}}^2$ | ▪ Substitute the expression for change in kinetic energy. |
| $d = -\dfrac{mv_{\text{initial}}^2}{2Eq}$ | ▪ Rearrange for distance. |
| $d = -\dfrac{(9.1 \times 10^{-31}\,\text{kg})(5500\,\text{m s}^{-1})^2}{2(150\,\text{V m}^{-1})(-1.6 \times 10^{-19}\,\text{C})}$ | ▪ Substitute correct values with units. |
| $d = 5.73 \times 10^{-7}\,\text{kg m}^3\,\text{s}^{-2}\,\text{V}^{-1}\,\text{C}^{-1}$ | ▪ Calculate the answer. |
| Note that $1\,\text{V} = 1\,\text{J C}^{-1} = 1\,\text{kg m}^2\,\text{s}^{-2}\,\text{C}^{-1}$, <br> so $\text{V}^{-1}\,\text{C}^{-1} = \text{s}^2\,\text{m}^{-2}\,\text{kg}^{-1}$ <br> $d = 5.7 \times 10^{-7}\,\text{m}$ | ▪ State the final answer with correct units and appropriate significant figures. |
| **b**  $E = -\dfrac{\Delta V}{d}$ | ▪ Relate the field to the change in potential. |
| $\Delta V = -Ed$ | ▪ Rearrange for $\Delta V$. |
| $\Delta V = -150\,\text{V m}^{-1} \times 5.73\ 10^{-7}\,\text{m}$ | ▪ Substitute correct values with units. |
| $\Delta V = -8.60 \times 10^{-5}\,\text{V}$ | ▪ Calculate the answer. |
| $\Delta V = -8.6 \times 10^{-5}\,\text{V}$ <br> The negative sign indicates the electron is moving from a point of higher potential to one of lower potential. | ▪ State the final answer with correct units and appropriate significant figures. |

### TRY THIS YOURSELF

If a proton begins at rest in this field, how far will it have to travel to reach a speed of $5.5\,\text{km s}^{-1}$?

**Electron deflection tube experiment**

If you have an electron deflection tube, you can do this experiment to see how an electric field affects an electron.

KEY CONCEPTS

- Electric field is defined as force per unit charge on a small positive test charge: $E = \dfrac{F}{q}$.
- The electric field between charged plates is uniform, and $E = -\dfrac{\Delta V}{d}$ where $\Delta V$ is the potential difference between the plates spaced distance $d$ apart. This equation is sometimes written in simplified form as $V = \dfrac{V}{d}$
- For a charged particle in an electric field $F = qE = ma$, so $a = \dfrac{Eq}{m}$.
- An electric field does work $W = qEd = -q\Delta V$ on a charged particle moving parallel to the field, changing its kinetic energy by $\Delta K = -q\Delta V = qEd$.
- When a field does positive work on a particle, the field–particle system loses potential energy and the particle gains kinetic energy.

1 Two charged parallel plates are separated by a distance of 15 mm. If the electric field between the plates is 500 V m$^{-1}$, what is the potential difference between the plates?

2 A pair of parallel plates is connected to a power supply. One plate is at a potential of +12 V and the other is at a potential of −12 V. How far apart must the plates be if the field between them is 1.2 kV m$^{-1}$?

3 A positively charged particle has initial kinetic energy $K_i$ and is moving in the direction of the field lines in a uniform field. Describe the energy transformations that occur if no other forces are acting.

4 A bubble of mass 0.01 g floats through the air. It has a charge of +1.6 × 10$^{-17}$ C. What is the acceleration of the bubble if it is subject only to the Earth's fair weather field, with a magnitude of 100 V m$^{-1}$ pointing downwards? Give the magnitude and direction of the acceleration.

5 An electron in an X-ray machine is accelerated from rest through a potential difference of 100 kV before colliding with a target and emitting X-rays.

   a What is the kinetic energy of the electron just before it hits the target?

   b Find the speed at which the electron will be moving after passing through this potential difference. Note that to get a better estimate of the speed we should really use a relativistic model, as described in chapter 12.

6 What is the speed of a particle that has been accelerated from rest through a potential difference of 1000 V when the particle is:

   a an electron?

   b a proton?

   c an **alpha particle**?

# 5.2 The trajectory of a charged particle in a uniform electric field

In a uniform electric field, a charged particle experiences a constant force parallel to the field. For a positively charged particle the force is in the direction of the field; for a negatively charged particle it is in the opposite direction (Figure 5.6).

The acceleration of the particle parallel to the field is constant and is given by:

$$a_\parallel = \frac{Eq}{m}.$$

If no other forces act, the acceleration perpendicular to the field is given by:

$$a_\perp = 0$$

The particle has some initial velocity given by $u$, which is at an angle $\theta$ to the field. Note that we define $\theta$ as the angle to the field, rather than the angle to the horizontal, as we did with projectile motion. Remember that the electric field can be in any direction.

The initial velocity can be broken into components $u_\perp$ and $u_\parallel$ (Figure 5.7).

In the direction of the field, $u_\parallel = u \cos\theta$;

perpendicular to the field, $u_\perp = u \sin\theta$.

We can apply the kinematic equations for constant acceleration (chapter 2 of *Physics in Focus Year 11*) to find the velocity at any later time.

At some time $t$ later, the velocity in the direction of the field is:

$$v_\parallel = u_\parallel + a_\parallel t = u_\parallel + \frac{Eq}{m}t$$

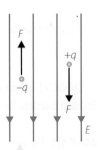

**FIGURE 5.6** A positively charged object experiences a force in the direction of the field, and a negatively charged object experiences a force opposite to the direction of the field.

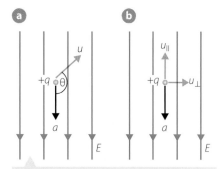

FIGURE 5.7 **a** A positively charged particle with initial velocity *u* in a uniform electric field. **b** The initial velocity can be broken into components parallel and perpendicular to the field. The acceleration is in the direction of the field.

The velocity perpendicular to the field is

$$v_\perp = u_\perp + a_\perp t = u_\perp$$

You should recognise these equations for velocity from chapter 2. They are identical to those for a projectile in a gravitational field. The only difference is that here the acceleration is due to the electric field, $E$, rather than the gravitational field, $g$. Recall that in a gravitational field the acceleration, from Newton's second law, is $a = \dfrac{F}{m} = \dfrac{mg}{m} = g$. Because the mass is always positive, the acceleration is always parallel to $g$. For a charged particle in an electric field the charge can be either positive or negative, so the acceleration, $a = \dfrac{F}{m} = \dfrac{qE}{m}$, may be parallel or anti-parallel to the field. In both cases the acceleration is non-zero parallel to the field and zero perpendicular to the field. This means that the velocity component perpendicular to the field must be constant, but the parallel component varies.

We have also defined the angle $\theta$ in the equations for the electric field case as measured from the field, rather than perpendicular to the field, as we usually do for the gravitational case. This is an arbitrary but convenient choice because the electric field can be in any direction.

To find the position of the charged particle at any time, we again apply the kinematics equations for constant acceleration. The particle begins at a position $x_0$, $y_0$. If the field is in the $y$ direction (and remember that we can define our axes to make this the case), then:

Parallel to the field:

$$y = y_0 + u_\parallel t + \frac{1}{2}a_\parallel t^2 = y_0 + u_\parallel t + \frac{1}{2}\frac{Eq}{m}t^2$$

Perpendicular to the field:

$$x = x_0 + u_\perp t$$

If we define the starting position of the projectile as the origin, so $x_0 = 0$ and $y_0 = 0$, then:

$$y = u_\parallel t + \frac{Eq}{2m}t^2 \ \text{ and } x = u_\perp t$$

Table 5.1 summarises the equations of motion for a projectile in a uniform gravitational field and a charged particle in a uniform electric field.

**TABLE 5.1** The parallel and perpendicular components of the position, velocity and acceleration of a charged particle in a uniform electric field and a projectile close to Earth's surface. In both cases the field is in the $y$ direction.

| | PARTICLE WITH MASS IN A UNIFORM GRAVITATIONAL FIELD | | CHARGED PARTICLE IN A UNIFORM ELECTRIC FIELD | |
| --- | --- | --- | --- | --- |
| | Perpendicular to gravitational field | Parallel to gravitational field | Perpendicular to electric field | Parallel to electric field |
| Position | $x = x_0 + u_x t$ | $y = y_0 + u_y t + \frac{1}{2}gt^2$ | $x = x_0 + u_\perp t$ | $y = y_0 + u_\parallel t + \frac{Eq}{2m}t^2$ |
| Velocity | $v_x = u_x$ | $v_y = u_y + gt$ | $v_\perp = u_\perp$ | $v_\parallel = u_\parallel + \frac{Eq}{m}t$ |
| Acceleration | $a_x = 0$ | $a_y = g$ | $a_\perp = 0$ | $a_\parallel = \frac{Eq}{m}$ |

9780170409131

The trajectory of a charged particle in a uniform electric field is a projectile trajectory. The particle experiences a constant force, and hence a constant acceleration, parallel to the field. Perpendicular to the field the acceleration is zero, so velocity is constant. The motion can be modelled mathematically as a combination of rectilinear constant velocity motion perpendicular to the field and rectilinear constant acceleration motion parallel to the field. This is exactly what we did for projectile motion in chapter 2.

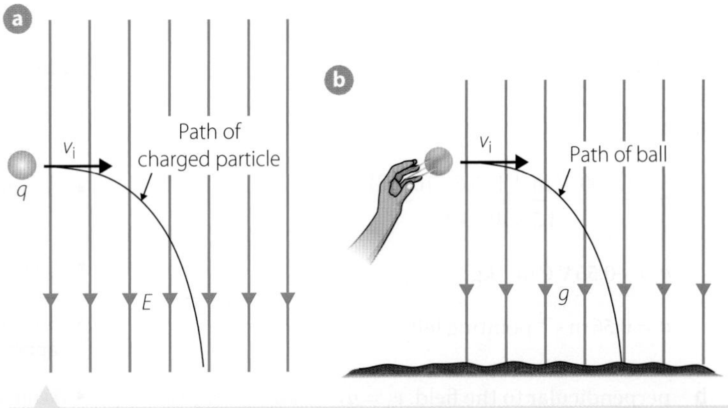

The only difference from the motion of a projectile near Earth as described in chapter 2 is that there is only one type of mass. Charge comes in two types: positive and negative. A mass always has an acceleration in the direction of the gravitational field. A charged particle may accelerate in the direction of the electric field, or in the opposite direction (Figure 5.6).

Figure 5.8 shows the trajectory of a positively charged particle in a uniform electric field and the trajectory of a ball in a uniform gravitational field. Both follow parabolic arcs and have constant velocity perpendicular to the field and constant acceleration in the direction of the field.

**FIGURE 5.8** The kinematics equations for constant acceleration describe the paths followed by both **a** a charged particle in a uniform electric field, and **b** projectiles.

The trajectory of a charged particle in a uniform electric field

▶ **WORKED EXAMPLE** (5.4)

A dust particle with mass $1.5 \times 10^{-5}$ kg and charge $-5.6 \times 10^{-9}$ C is falling vertically downwards at a speed of $0.10 \, \text{m s}^{-1}$ when it falls between two vertical parallel charged plates. The field between the plates is $1500 \, \text{V m}^{-1}$ and points to the right.

**a** Calculate the acceleration of the dust particle.

**b** Calculate the total velocity of the dust particle 1.0 s after it enters the field.

**c** Calculate the vertical and horizontal displacements of the dust particle 1.0 s after it enters the field. Assume all forces other than the electrostatic force are negligible.

| ANSWER | LOGIC |
|---|---|
| $E = 1500 \, \text{V m}^{-1}$ right; $u = 0.10 \, \text{m s}^{-1}$ down <br> $m = 1.5 \times 10^{-5}$ kg; $q = -5.6 \times 10^{-9}$ C, $t = 1.0$ s | ▪ Identify the relevant data in the question. |
| $x_0 = 0, y_0 = 0, t_0 = 0$ | ▪ Define the initial position. |
| 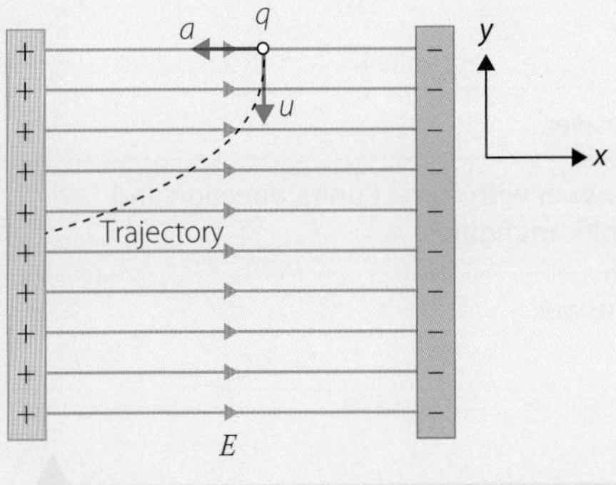 **FIGURE 5.9** The dust particle moving into the region between the plates | ▪ Draw a diagram to identify the direction of the acceleration (Figure 5.9). |

| ANSWER | LOGIC |
|---|---|
| **a** $a = \dfrac{Eq}{m}$ | ■ Relate the acceleration to the field. |
| $a = \dfrac{(1500 \text{ V m}^{-1})(-5.6 \times 10^{-9}\text{ C})}{1.5 \times 10^{-5}\text{ kg}}$ | ■ Substitute correct values with units. |
| $a = -0.56\,\text{V C m}^{-1}\text{kg}^{-1}$ | ■ Calculate the answer. |
| $a = 0.56\,\text{m s}^{-2}$ pointing left | ■ State the final answer with correct units, direction and appropriate significant figures. |
| **b** perpendicular to the field, $v_\perp = u_\perp$ | ■ Write the expression for velocity. |
| $v_\perp = 0.10\,\text{m s}^{-1}$ pointing down | ■ Substitute correct values with units. |
| parallel to the field, $v_\parallel = u_\parallel + \dfrac{Eq}{m}t$ | ■ Write the expression for velocity. |
| $v_\parallel = 0 + \dfrac{(1500 \text{ V m}^{-1})(-5.6 \times 10^{-9}\text{ C})}{1.5 \times 10^{-5}\text{ kg}}(1.0\text{ s})$ | ■ Substitute correct values with units. |
| $v_\parallel = -0.56\,\text{V C s m}^{-1}\text{kg}^{-1}$ | ■ Calculate the answer. |
| $v_\parallel = -0.56\,\text{m s}^{-1}$ pointing left | ■ State the final answer with correct units, direction and appropriate significant figures. |
| $v = \sqrt{v_\parallel{}^2 + v_\perp{}^2}$ | ■ Write the expression for total velocity. |
| $v = \sqrt{(-0.56 \text{ m s}^{-1})^2 + (-0.10 \text{ m s}^{-1})^2}$ | ■ Substitute correct values with units. |
| $v = 0.324\,\text{m s}^{-1}$ | ■ Calculate the answer. |
| $v = 0.32\,\text{m s}^{-1}$ | ■ State the final answer with correct units, direction and appropriate significant figures. |
| **c** ·vertical, perpendicular to the field $y = y_0 + u_\perp t$ | ■ Write the expression for vertical position. |
| $y = 0 + (-0.10\,\text{m s}^{-1})(1.0\,\text{s})$ | ■ Substitute correct values with units. |
| $y = -0.10\,\text{m}$ | ■ Calculate the answer and state with correct units, direction and appropriate significant figures. |
| parallel to the field $x = x_0 + u_\parallel t + \dfrac{Eq}{2m}t^2$ | ■ Write the expression for horizontal position. |
| $x = 0 + 0 + \dfrac{(1500 \text{ V m}^{-1})(-5.6 \times 10^{-9}\text{ C})}{2 \times 1.5 \times 10^{-5}\text{ kg}}(1.0\text{ s})^2$ | ■ Substitute correct values with units. |
| $x = -0.28\,\text{V C s}^2\text{m}^{-1}\text{kg}^{-1}$ | ■ Calculate the answer. |
| $x = -0.28\,\text{m}$ | ■ State the final answer with correct units, direction and appropriate significant figures. |
| The particle has moved left 0.28 m and down 0.10 m in the first second. | ■ State the final answer. |

**TRY THIS YOURSELF**

How would the answers change if the dust particle was initially moving at an angle of 45° to the field, down and to the right?

# INVESTIGATION (5.1)

## Computer simulation of the trajectory of a charged particle in a uniform electric field

This investigation can be done by modifying the spreadsheet created for Investigation 2.3 on page 49. Here we repeat the process of modelling projectile motion, but add variables for mass and charge.

**AIM**

To use spreadsheet software to model the trajectory of a charged particle in a uniform electric field

Write a hypothesis or an inquiry question for this investigation.

**MATERIALS**

- Computer with spreadsheet software

**METHOD**

Open a new file in your spreadsheet software and put a descriptive heading at the top of the sheet, then save it with a suitable filename. *Remember to save your file frequently.* It is also a good idea to save multiple pages in your file for different variations of your simulation so you can return to earlier versions.

1 Set up the initial conditions for the particle. These are the initial velocity, angle of velocity to the field, particle charge and mass, and field magnitude. In the example shown in Figure 5.10 the field points in the *y* direction. Cells are labelled above the value. Because Excel uses angles in radians, the angle, $\theta$, is entered in degrees in cell B4 and then converted to radians in cell C4 using the command **=RADIANS(B4)**.

The components of the initial velocity are then calculated below. Cell A6 contains the command **=25*COS(C4)** for the parallel component. Cell B6 contains the command **=25*SIN(C4)** for the perpendicular component. Note that all commands given in this example are for Excel.

The example shown in Figure 5.10 models a proton in Earth's fair-weather field.

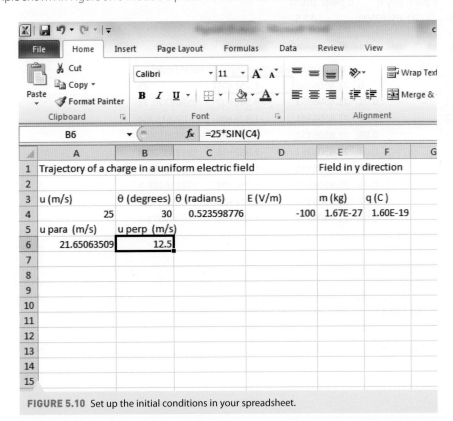

**FIGURE 5.10** Set up the initial conditions in your spreadsheet.

Critical and creative thinking

Numeracy

Information and communication technology capability

Literacy

**» 2** Implement the equations of motion to calculate the velocity as a function of time. Enter the initial time, 0, in the cell immediately under the time label. Below this cell enter a command to increase the time. In Figure 5.11 this command is **=A10+0.01**, but you can add any time increment you want to make the time steps bigger or smaller. Copy this command into as many cells below as you want.

Set up the perpendicular velocity in the next column. This is constant, and always equal to the initial perpendicular velocity. So cell B10 and all those below it contain the command **=$A$6**. Note that the $ signs tell the software that the cell reference is not to be changed.

Set up the parallel velocity in the third column. This is the constant acceleration part of the motion. The command used in cell C10 and copied to those below it is **=$B$6+(A10*$D$4*$F$4)/$E$4**. This command takes the initial velocity and adds $\frac{Eq}{m}t$ to it.

Add a fourth column to give the magnitude of the total velocity. This is found using Pythagoras' theorem, and is calculated in cell D10 using the command **=SQRT(B10*B10+C10*C10)**, which is then copied into the column of cells below.

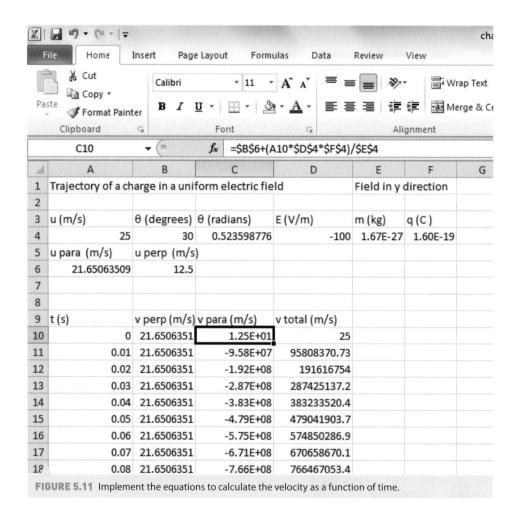

**FIGURE 5.11** Implement the equations to calculate the velocity as a function of time.

**3** Implement the equations of motion to calculate the position as a function of time, as shown in Figure 5.12. Label the next two columns x and y. The *x* position is calculated in the cell beneath the x label using the command **=$A$6*A10**, which is then copied into the cells below.

The *y* position is calculated in the cell beneath the y label using the command **=$A$6*A10+0.5*($D$4*$F$4/$E$4)*A10*A10**, which is then copied into the cells below.

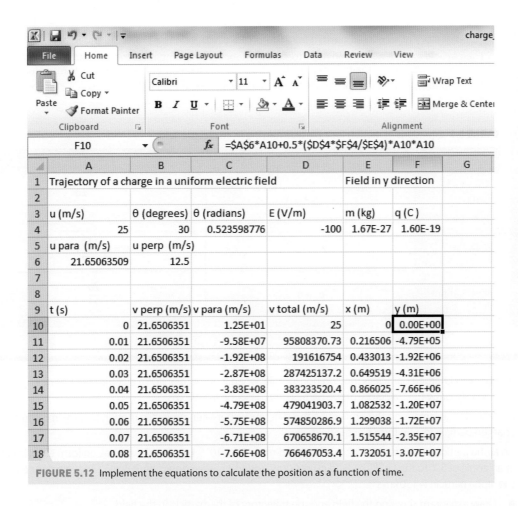

Cell reference: F10 — `=$A$6*A10+0.5*($D$4*$F$4/$E$4)*A10*A10`

| 1 | Trajectory of a charge in a uniform electric field | | | Field in y direction | | | |

| | A | B | C | D | E | F | G |
|---|---|---|---|---|---|---|---|
| 1 | Trajectory of a charge in a uniform electric field | | | | Field in y direction | | |
| 2 | | | | | | | |
| 3 | u (m/s) | θ (degrees) | θ (radians) | E (V/m) | m (kg) | q (C) | |
| 4 | 25 | 30 | 0.523598776 | -100 | 1.67E-27 | 1.60E-19 | |
| 5 | u para (m/s) | u perp (m/s) | | | | | |
| 6 | 21.65063509 | 12.5 | | | | | |
| 7 | | | | | | | |
| 8 | | | | | | | |
| 9 | t (s) | v perp (m/s) | v para (m/s) | v total (m/s) | x (m) | y (m) | |
| 10 | 0 | 21.6506351 | 1.25E+01 | 25 | 0 | 0.00E+00 | |
| 11 | 0.01 | 21.6506351 | -9.58E+07 | 95808370.73 | 0.216506 | -4.79E+05 | |
| 12 | 0.02 | 21.6506351 | -1.92E+08 | 191616754 | 0.433013 | -1.92E+06 | |
| 13 | 0.03 | 21.6506351 | -2.87E+08 | 287425137.2 | 0.649519 | -4.31E+06 | |
| 14 | 0.04 | 21.6506351 | -3.83E+08 | 383233520.4 | 0.866025 | -7.66E+06 | |
| 15 | 0.05 | 21.6506351 | -4.79E+08 | 479041903.7 | 1.082532 | -1.20E+07 | |
| 16 | 0.06 | 21.6506351 | -5.75E+08 | 574850286.9 | 1.299038 | -1.72E+07 | |
| 17 | 0.07 | 21.6506351 | -6.71E+08 | 670658670.1 | 1.515544 | -2.35E+07 | |
| 18 | 0.08 | 21.6506351 | -7.66E+08 | 766467053.4 | 1.732051 | -3.07E+07 | |

**FIGURE 5.12** Implement the equations to calculate the position as a function of time.

You now have a numerical model of the motion of a charged particle in a uniform electric field that can be easily modified to investigate the effects of changing particle charge and mass, initial velocity and field.

Experiment with changing the variables. Change the charge and mass to model an electron instead, or an alpha particle. Change the charge and mass to model a macroscopic object. Change the field to model stormy weather.

Spreadsheet

## RESULTS AND ANALYSIS

*Make sure that you have saved your file!*
Plot scatter graphs of your data. Create graphs of:

1  velocity as a function of time.

2  trajectory (*y* position as a function of *x* position).
Save and/or print these graphs.

## DISCUSSION

1  Comment on the shape of your graphs. Are they what you expect, given the kinematics equations?

2  How does the trajectory of the particle compare to a projectile in Earth's gravitational field?

3  Do a literature search on special relativity, and find out about the maximum speed at which particles can move. Are the values that you get for speed in this simulation realistic? Write about the limitations of the mathematical model that you have used in this investigation.

4  Give the answer to your inquiry question or state whether your hypothesis was supported or not.

## CONCLUSION

Write a conclusion summarising the outcomes of your investigation.

- The trajectory of a charged particle in a uniform electric field is a projectile trajectory.
- The particle experiences a constant force, and hence a constant acceleration, parallel to the field. Perpendicular to the field the particle has a zero acceleration, so it has constant velocity.
- The equations of motion for a particle moving in an electric field pointing in the $y$ direction are given in Table 5.2.

**TABLE 5.2** Equations of motion for a particle moving in an electric field

| | PERPENDICULAR TO FIELD | PARALLEL TO FIELD |
|---|---|---|
| Position | $x = x_0 + u_\perp t$ | $y = y_0 + u_\parallel t + \dfrac{Eq}{2m}t^2$ |
| Velocity | $v_\perp = u_\perp$ | $v_\parallel = u_\parallel + \dfrac{Eq}{m}t$ |
| Acceleration | $a_\perp = 0$ | $a_\parallel = \dfrac{Eq}{m}$ |

**CHECK YOUR UNDERSTANDING**

**5.2**

1 A negatively charged particle is moving horizontally when it enters a uniform electric field pointing downwards. Draw the trajectory of the particle and describe it in words.

2 Make a table summarising the similarities and differences between projectile motion and the motion of a charged particle in a uniform electric field.

3 An alpha particle, with charge $+3.2 \times 10^{-19}$ C and mass $6.6 \times 10^{-27}$ kg, enters a vertical, uniform electric field at an angle of 30° to the field. The field points upwards and has a magnitude $30\,\text{V}\,\text{m}^{-1}$. The initial speed of the alpha particle is $150\,\text{km}\,\text{s}^{-1}$.

  a Draw a diagram showing the field and the trajectory of the particle in the field.

  b Calculate the velocity of the alpha particle 5.0 s after entering the field. Give the magnitude and direction of the velocity.

4 A uniform electric field is created by charging two large parallel metal plates. A charged soap bubble drifts between the plates, with an initial velocity $v_i = 0.5\,\text{m}\,\text{s}^{-1}$ parallel to the plates, as shown in Figure 5.13. The bubble has mass $m = 0.001$ g and charge $q = 1.5$ nC, the field between the plates is $500\,\text{V}\,\text{m}^{-1}$. Assume that the gravitational force on the bubble is negligible in comparison to the electrostatic force.

  a What total distance does the bubble travel in the first second after it enters the field?

  b What is its velocity at this time?

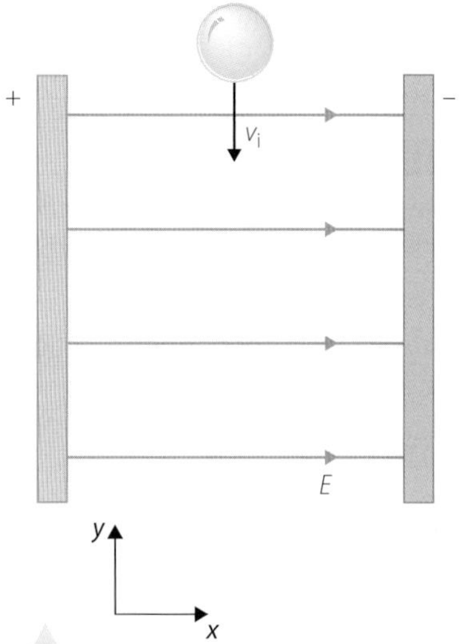

5 A proton is initially moving perpendicular to an electric field with magnitude $1.5\,\text{kV}\,\text{m}^{-1}$ at a speed of $500\,\text{m}\,\text{s}^{-1}$.

  a How long does it take before its velocity parallel to the field is equal to its velocity perpendicular to the field?

  b What is its displacement in this time?

6 An electron is moving through a horizontal uniform electric field that points left and has magnitude $1.5\,\text{kV}\,\text{m}^{-1}$. When first observed, at $t = 0$, the electron has a velocity of $3.0\,\text{km}\,\text{s}^{-1}$ at an angle of 45° to the field lines.

  a Copy the table below and complete it by calculating the velocity components, total velocity and kinetic energy at each time listed.

**FIGURE 5.13** A bubble enters the uniform electric field between a pair of charged plates.

**b** Sketch a graph of kinetic energy vs time for the electron.

**c** Calculate the work done by the field on the electron in the 5.0 s in which the electron is observed.

| TIME (s) | 0 | 1.0 | 2.0 | 3.0 | 4.0 | 5.0 |
|---|---|---|---|---|---|---|
| $v_{horizontal}$ (m s$^{-1}$) | | | | | | |
| $v_{vertical}$ (m s$^{-1}$) | | | | | | |
| $v_{total}$ (m s$^{-1}$) | | | | | | |
| $K$ (J) | | | | | | |

## 5.3 Charged particles in uniform magnetic fields

Electric fields are created by charged particles; magnetic fields are created by moving charged particles (currents). Magnetic field strength is measured in units of tesla, T, where $1\,T = 1\,kg\,s^{-2}\,A^{-1}$. Recall from chapter 14 of *Physics in Focus Year 11* that a large uniform magnetic field can be created by a current-carrying coil of wire.

A stationary charged particle experiences a force due to an electric field, but does not experience a force due to a constant magnetic field. However, when a charged particle is *moving* relative to a magnetic field it *does* experience a force. The importance of *relative* motion in electromagnetism was an observation that helped lead Einstein to his theory of relativity, as described in chapter 12.

### Force on a moving charged particle in a magnetic field

The magnitude of the force experienced by a charged particle moving in a magnetic field is proportional to its charge, its velocity and the field strength. It also depends on the direction the charged particle is moving relative to the field. The particle experiences a maximum force when it moves perpendicular to the field and a zero force if it moves parallel to the field. Mathematically, we can model this force as $F = qvB\sin\theta$, where $q$ is the charge on the particle, measured in C, $v$ is its velocity in m s$^{-1}$, $B$ is magnetic field, measured in T, and $\theta$ is the angle between the vectors $v$ and $B$. We can simplify the expression to $F = qv_{\perp}B$. The direction of the force is perpendicular to both the magnetic field and the velocity.

We need to use the right-hand rule (Figure 5.14) to find the direction of the force. Point the fingers of your *right* hand in the direction of the velocity, then curl them towards the field. Your thumb gives the direction of the force if the particle is positively charged (Figure 5.15). Note that if the charge is negative, the direction of the force is opposite to the direction for a positively charged particle.

If a positively charged particle and a negatively charged particle have the same initial velocity in a uniform magnetic field then they will experience forces in opposite directions (Figure 5.16). They will follow curved trajectories in opposite directions.

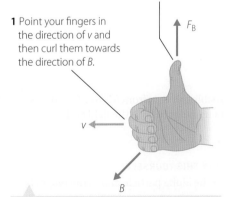

**2** Your thumb shows the direction of the magnetic force on a positively charged particle.

$F_B$

**1** Point your fingers in the direction of $v$ and then curl them towards the direction of $B$.

$v$

$B$

**FIGURE 5.14** The right-hand rule for finding the direction of the force on a moving charged particle in a magnetic field.

Charged particles
in uniform
magnetic fields

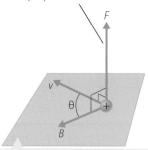

The magnetic force is
perpendicular to both *v* and *B*.

FIGURE 5.15 The force on a
charged particle in a magnetic field
is perpendicular to both the field
and the particle's velocity.

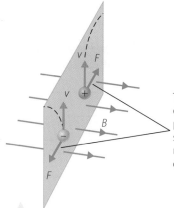

The magnetic forces
on oppositely charged
particles moving at the
same velocity in a
magnetic field are in
opposite directions.

FIGURE 5.16 The forces on positively and negatively
charged particles in a magnetic field. The dashed lines show
the trajectories of the particles.

## WORKED EXAMPLE 5.5

An alpha particle enters Earth's magnetic field at a velocity of $55\,000\,\mathrm{m\,s^{-1}}$. The local magnetic field
strength is $40\,\mathrm{mT}$. What is the range of possible forces on the alpha particle?

| ANSWER | LOGIC |
|---|---|
| $B = 40 \times 10^{-3}\,\mathrm{T}$, $v = 5.5 \times 10^4\,\mathrm{m\,s^{-1}}$ | ▪ Identify the relevant data in the question. |
| $q = +3.2 \times 10^{-19}\,\mathrm{C}$ | ▪ Look up additional data needed. |
| $F = qvB\sin\theta$ | ▪ Relate the force to the field and velocity. |
| The force has a maximum value when $\theta = 90°$.  <br><br> The force has a minimum value when $\theta = 0°$. | ▪ Recognise that the force depends on the angle. <br> ▪ Identify when it is a maximum and when it is a minimum. |
| $F_{\max} = qvB\sin 90° = qvB$ | ▪ Write the expression for the maximum force. |
| $F_{\max} = (3.2 \times 10^{-19}\,\mathrm{C})(5.5 \times 10^4\,\mathrm{m\,s^{-1}})(4.0 \times 10^{-2}\,\mathrm{T})$ | ▪ Substitute correct values with units. |
| $F_{\max} = 7.04 \times 10^{-16}\,\mathrm{C\,T\,m\,s^{-1}}$  <br><br> note that $1\,\mathrm{T} = 1\,\mathrm{kg\,s^{-2}\,A^{-1}} = 1\,\mathrm{kg\,s^{-1}\,C^{-1}}$ | ▪ Calculate the answer. |
| $F_{\max} = 7.0 \times 10^{-16}\,\mathrm{N}$ | ▪ State the answer with correct units and appropriate significant figures. |
| $F_{\min} = qvB\sin 0° = 0$ | ▪ Write the expression for the minimum force. |
| So the range of possible forces is from $0\,\mathrm{N}$ to $7.0 \times 10^{-16}\,\mathrm{N}$, in a direction perpendicular to both the field and the velocity. | ▪ State the final answer. |

### TRY THIS YOURSELF

If the alpha particle enters the magnetic field at an angle of $45°$ to the field, what magnitude force does it
experience?

**Thomson's experiment: measuring e/m**

Find out how a combination of uniform electric and magnetic fields were used to measure the charge–mass ratio of the electron.

Once we have calculated the force on the moving charged particle, we use Newton's second law to find the acceleration:

$$a = \frac{F}{m} = \frac{qvB\sin\theta}{m}$$

The acceleration is in the same direction as the force, which is perpendicular to both the field and the instantaneous velocity.

The force and acceleration depend on $\theta$, the angle between the field lines and the velocity. So the path of a charged particle in a uniform magnetic field also depends on the angle between the initial velocity and the field. The path may be a straight line, a circle or a helix.

The particle has a straight-line trajectory only when its velocity is parallel to the field. In this case the force on the particle is zero. In the previous worked example, this is when $\theta = 0°$.

If the particle is moving in a plane perpendicular to the field then the force is always perpendicular to the velocity, and the particle moves in a circle. In this case $\theta = 90°$ and the force and acceleration are a maximum. The circular motion of a charged particle in a magnetic field is mathematically the same as any other circular motion.

In chapter 3 we saw that when an object is undergoing uniform circular motion, with radius of orbit $r$, its centripetal acceleration is given by:

$$a = \frac{v^2}{r}$$

and hence by Newton's second law the net force acting on it is:

$$F = ma = \frac{mv^2}{r}$$

For a charged particle in a circular orbit in a uniform magnetic field it is the magnetic field that supplies this centripetal force. Equating the two expressions for force gives:

$$qvB = \frac{mv^2}{r}$$

where the angle between the field and velocity is $90°$, so $\sin\theta = 1$.

From this equation we can derive the radius of the orbit:

$$r = \frac{mv}{qB}$$

and orbital period:

$$T = \frac{2\pi m}{qB}$$

In chapter 4, we looked at circular motion resulting from other forces. For example, the gravitational force provides a centripetal force that keeps satellites in orbit around Earth, and Earth in orbit around the Sun. Note the similarity between Figure 5.17a and Figure 4.6 (page 85). In both cases the force and velocity are always perpendicular. This can also be seen in Figure 3.14 (page 72), which shows the general case for circular motion due to any combination of forces that creates a net centripetal force.

An object whirled on a string is kept in circular motion by a combination of the tension in the string and the gravitational force. In this case the centripetal force is the net force, which is the sum of the tension and gravitational force. For a car going around a corner the centripetal force includes the friction force. Remember that, in general, the centripetal force is the net force, not a specific single force. Any force, including the magnetic force, can produce a centripetal force. As you will see in chapter 15, the electrostatic force provides a centripetal force to keep electrons in orbit about the nucleus of an atom.

The third possible path of a charged particle in a magnetic field is a helix. This occurs if the particle has a velocity with a component in the direction of the field. This parallel component of the velocity is not altered by the field. However, the perpendicular component *is* altered by the acceleration due to the field. In this case the particle follows a helical path, with the axis of the helix in the direction of the field. Figure 5.17 shows the three types of paths of charged particles in uniform magnetic fields.

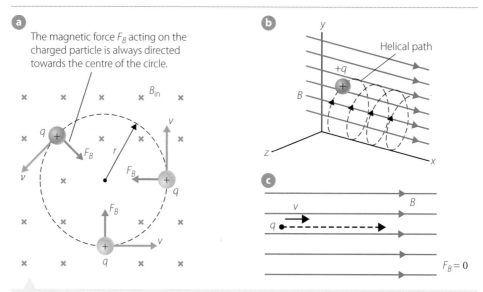

**a** The magnetic force $F_B$ acting on the charged particle is always directed towards the centre of the circle.

**b** Helical path

**c** $F_B = 0$

**FIGURE 5.17** In a uniform magnetic field, the path of a charged particle is: **a** circular when the particle's velocity is perpendicular to the field; **b** helical when the particle's velocity has components perpendicular and parallel to the field; and **c** an undeflected straight line when the velocity is parallel to the field.

▶ **WORKED EXAMPLE** (5.6)

An electron is circulating inside the ring of a synchrotron with an orbital radius of 125 m. The electron has a velocity of $1.5 \times 10^8\,\mathrm{m\,s^{-1}}$. Calculate the magnetic field required to keep the electron in orbit.

| ANSWER | LOGIC |
|---|---|
| $r = 125\,\mathrm{m}, v = 1.5 \times 10^8\,\mathrm{m\,s^{-1}}$ | ▪ Identify the relevant data in the question. |
| $q = -1.6 \times 10^{-19}\,\mathrm{C}, m = 9.1 \times 10^{-31}\,\mathrm{kg}$ | ▪ Look up additional data needed. |
| $r = \dfrac{mv}{qB}$ | ▪ Relate the radius to the field and velocity. |
| $B = \dfrac{mv}{qr}$ | ▪ Rearrange for field. |
| $B = \dfrac{(9.1 \times 10^{-31}\,\mathrm{kg})(1.5 \times 10^8\,\mathrm{m\,s^{-1}})}{(-1.6 \times 10^{-19}\,\mathrm{C})(125\,\mathrm{m})}$ | ▪ Substitute correct values with units. |
| $B = -6.83 \times 10^{-6}\,\mathrm{kg\,C^{-1}\,s^{-1}}$ | ▪ Calculate the answer. |
| $B = 6.8\,\mu\mathrm{T}$ | ▪ State the answer with correct units and appropriate significant figures. |

**TRY THIS YOURSELF**

What magnitude magnetic field would be necessary to keep a proton orbiting at this speed and radius?

This behaviour of charged particles in magnetic fields is very useful. Synchrotrons use magnetic fields to contain fast-moving charged particles. The particles are contained in giant rings, tens to hundreds of metres across. The magnetic fields keep the particles circulating in the ring. As the particles circle, they emit high-energy electromagnetic radiation. The high-energy electromagnetic radiation is used for medical, materials science and fundamental physics research. Magnetic fields are also used in mass spectrometers to determine which atoms or molecules are in a sample of material. Mass spectrometers are used in airport security and forensic analysis.

## Work done on a charged particle in a uniform magnetic field

Recall from chapter 3 that no work is done by the net (centripetal) force on an object in circular motion. This is because the force and velocity are always perpendicular to each other. Hence a magnetic field does no work on a charged particle moving in a circular orbit within it. The acceleration is always perpendicular to the path, and the magnitude of the velocity is constant, so the kinetic energy is also constant.

More generally, for any path that a charged particle takes in a uniform magnetic field, the force is always perpendicular to the velocity and no work is done. The direction of the velocity may change, but its magnitude is constant.

**Australian Synchrotron**

Find out how large the ring is, and what speeds the particles can orbit at in the Australian Synchrotron.

# INVESTIGATION (5.2)

## Computer simulation of the trajectory of a charged particle in a uniform magnetic field

This simulation models the magnetic field in a mass spectrometer.

### AIM

To use an online simulation to model the trajectory of a charged particle in a uniform magnetic field in a mass spectrometer

Write a hypothesis or an inquiry question for this investigation.

### MATERIALS

- Computer with internet access

### METHOD

1 Do a literature search online and find out how a mass spectrometer works.

2 Draw a large diagram showing the main components and write a description of what each part does. Relate this to what you have learned about charged particles in electric and magnetic fields.

3 Open the simulation at the weblink.

4 Investigate how the trajectory varies when you change the variables. Sketch your results or take screen shots, taking care to record the values of each variable for each trajectory. For each trajectory, measure the radius of the path.

   a Create a set of traces for particle trajectories with varying mass, keeping charge, velocity and field constant.

   b Create a set of traces for particle trajectories with varying charge, keeping mass, velocity and field constant. Use both positive and negative charges.

Critical and creative thinking

Numeracy

Information and communication technology capability

Literacy

**Charged particle in magnetic field**

This interactive simulation shows what happens when a charged particle is injected into a uniform magnetic field.

»    c   Create a set of traces for particle trajectories with varying velocity, keeping charge, mass and field constant.

d   Create a set of traces for particle trajectories with varying magnetic field strengths, keeping particle charge, velocity and mass constant.

## RESULTS

Record the values of the variables that are kept constant for each experiment. Make a table of data for each experiment, recording the path radius and the changing variable. For example:

| MASS ($\times 10^{-25}$ kg) | RADIUS (mm) |
| --- | --- |
| | |
| | |
| | |
| | |
| | |

## ANALYSIS OF RESULTS

Plot graphs of your data. Create graphs of:

a   radius as a function of mass.

b   radius as a function of charge.

c   radius as a function of velocity.

d   radius as a function of magnetic field strength.

## DISCUSSION

1   Comment on the shapes of your graphs. Are they what you expect, given what you know about particles in magnetic fields?

2   Explain how a mass spectrometer can be used to distinguish between different atomic species. How is that modelled by this simulation?

3   Give the answer to your inquiry question or state whether your hypothesis was supported or not.

## CONCLUSION

Write a conclusion summarising the outcomes of your investigation.

KEY CONCEPTS

- The force experienced by a charged particle moving in a magnetic field is given by $F = qvB \sin\theta = qv_\perp B$. The direction is given by the right-hand rule (Figure 5.14).
- The acceleration is in the direction of the force and is given by $a = \dfrac{F}{m} = \dfrac{qvB \sin\theta}{m}$.
- The acceleration is maximum when the field and velocity are perpendicular, $\theta = 90°$. In this case the particle follows a circular path described by $qvB = \dfrac{mv^2}{r}$.
- The acceleration is zero when the velocity and field are parallel, $\theta = 0°$, and the particle follows an undeflected straight line.
- If the velocity is at an angle to the field between 0 and 90°, the particle follows a helical path.

1  A negatively charged particle is moving horizontally when it enters a uniform magnetic field pointing downwards. Draw a diagram showing the trajectory of the particle and describe the trajectory in words.

2  Explain how the motion of a charged particle circulating in a magnetic field is similar to that of a satellite in a circular orbit or an object whirled in a circle on a string. Which force causes the acceleration in each case?

3  A mass spectrometer accelerates a particle to high speed using an electric field, then passes it through a uniform magnetic field to determine what sort of particle it is. Explain why mass spectrometers first need to ionise molecules or atoms to be able to determine what they are.

4  What is the maximum possible acceleration of a calcium ion (mass $6.7 \times 10^{-26}$ kg, charge $3.2 \times 10^{-19}$ C) moving at 500 ms$^{-1}$ in a magnetic field of magnitude 55 μT?

5  Which of the paths A, B or C shown in Figure 5.18 is not possible for a charged particle entering a region of uniform magnetic field? Explain your answer.

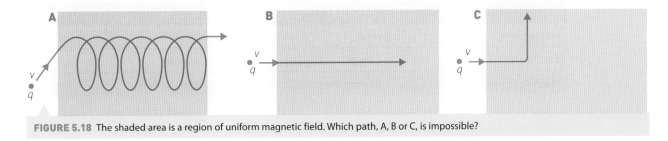

**FIGURE 5.18** The shaded area is a region of uniform magnetic field. Which path, A, B or C, is impossible?

6  A magnetic field points in the positive $z$ direction. Draw the force on an electron in the field with an initial velocity:

   a  in the positive $x$ direction.

   b  in the negative $x$ direction.

   c  in the positive $z$ direction.

   Name the direction of the force in each case.

7  What magnitude and direction of magnetic field is necessary to balance the gravitational force on an electron moving horizontally north at 100 ms$^{-1}$?

8  An electron is circulating in a synchrotron ring of radius 250 m with a period of 50 μs. Calculate:

   a  the speed of the electron.

   b  the magnetic field required to keep it in this orbit.

9  An alpha particle, with charge $+3.2 \times 10^{-19}$ C and mass $6.6 \times 10^{-27}$ kg, enters a vertical, uniform magnetic field at an angle of 30° to the field. The field points vertically upwards and has magnitude 30 mT. The initial speed of the alpha particle is 150 km s$^{-1}$.

   a  Draw a diagram showing the field and the trajectory of the particle in the field.

   b  Calculate the acceleration of the alpha particle.

   c  What is the speed of the alpha particle after 5 s in the field?

# 5.4 The gravitational, electric and magnetic fields

We use field models to explain force at a distance. You have now seen the field models associated with three different forces: gravitational, electrostatic and magnetic. The three field models are similar in some ways and different in others.

Gravitational, electric and magnetic fields are all defined as $\text{field} = \dfrac{\text{force}}{\text{property}}$, where 'property' is the thing that creates the field and on which the field acts (Table 5.3).

**TABLE 5.3** The three field models

| FIELD | CREATED BY | ACTS ON |
|---|---|---|
| Gravitational | Mass | Mass |
| Electric | Charge | Charge |
| Magnetic | Moving charge | Moving charge |

Gravitational, electric and magnetic fields have several important differences.

The electric force acts on a charged particle, and the gravitational force acts on a particle with mass, regardless of whether the particle is moving. But the magnetic force acts on a charged particle only when the particle is in motion relative to the field. So a stationary charged particle or one moving in the direction of a magnetic field experiences no force due to the magnetic field.

The gravitational and electric force vectors are parallel to the direction of the field, but the magnetic force vector is perpendicular to the magnetic field. The force due to the gravitational field is always in the direction of the field, but the force due to the electric field can be either in the direction of a field line or opposite to it. In either case, it is still parallel to the field. A charged particle moving in a magnetic field experiences a force perpendicular to the field with a direction given by the right-hand rule.

The path of a charged particle in an electric field or an object with mass in a gravitational field is that of a projectile. The acceleration is constant in the direction of the field and zero perpendicular to the field. When a charged particle moves perpendicular to a uniform magnetic field it experiences a force due to the field that is always perpendicular to both the particle's velocity and the field, resulting in a circular path. If the initial velocity has components both parallel and perpendicular to the field then it undergoes a combination of circular motion in the plane perpendicular to the field and constant velocity motion parallel to the field. This combination results in a helical path.

The gravitational and electric forces can do work in displacing a charged particle, but the magnetic force associated with a steady magnetic field does no work because the force is perpendicular to the displacement.

These field models have been extremely successful in explaining and predicting phenomena. The electric and magnetic field models are the basis for electromagnetism, which we shall study in the next two chapters. However, the field model has been unable to explain all the phenomena associated with the behaviour of charged particles such as electrons and protons. An alternative quantum model, the exchange particle model, has been developed to explain the interactions of fundamental particles. We shall look at this newer model in chapters 14 and 15. This newer model does not replace the field model any more than relativity replaces Newtonian mechanics – rather, it complements it. Scientists work with many models and representations, and choose the one that works best for a given situation. The field model has been invaluable in helping us explain electric and magnetic phenomena, and is the basis of much of the technology that we take for granted. We shall look at some of this technology in chapter 8.

The gravitational, electric and magnetic fields

9780170409131

- Field models are used to describe forces that act at a distance.
- The gravitational field is created by, and acts on, objects with mass. The gravitational force is in the direction of the field.
- The electric field is created by, and acts on, objects with charge. The electric force is in the direction of, or opposite the direction of, the field.
- The magnetic field is created by, and acts on, moving objects with charge. The magnetic force is perpendicular to the field.

CHECK YOUR UNDERSTANDING

5.4

1 Which fields will exert a force on a:

   **a** moving electron?

   **b** stationary electron?

   **c** stationary neutron?

2 Under what conditions will a charged particle not experience a force due to a uniform magnetic field?

3 A positively charged particle is moving horizontally north, in a vertical (up) electric field. The force on the particle due to the electric field is exactly balanced by a force due to a magnetic field. In what direction must the magnetic field be pointing?

4 Make a table summarising the similarities and differences between the gravitational, electric and magnetic fields.

5 Calculate the force exerted on an electron moving close to Earth's surface at $1.5 \times 10^7 \, \text{m s}^{-1}$ by:

   **a** Earth's gravitational field.

   **b** Earth's fair-weather electric field at $100 \, \text{V m}^{-1}$.

   **c** Earth's magnetic field ($50 \, \mu\text{T}$). Assume the particle is moving perpendicular to the magnetic field.

6 Explain why we do not consider the electric and magnetic forces when modelling the orbital behaviour of satellites and planets.

» The electric field between charged plates is uniform, $E = -\dfrac{\Delta V}{d}$ with magnitude $E = \dfrac{V}{d}$ where $V = \Delta V$ is the potential difference between the plates.

» For a charged particle in an electric field $\vec{F} = q\vec{E} = m\vec{a}$, so $\vec{a} = \dfrac{\vec{E}q}{m}$.

» An electric field does work $W = qEd = -q\Delta V$ on a charged particle, and its kinetic energy changes by $\Delta K = -q\Delta V = qEd$.

» A charged particle in a uniform electric field has a projectile trajectory. The particle experiences a constant force, and hence a constant acceleration, parallel to the field. Perpendicular to the field the particle has a zero acceleration, so it has constant velocity.

» The force experienced by a charged particle moving in a magnetic field is given by $F = qvB\sin\theta = qv_\perp B$. The acceleration is in the direction of the force and is given by $a = \dfrac{F}{m} = \dfrac{qvB\sin\theta}{m}$.

» The acceleration is maximum when the magnetic field and velocity are perpendicular, $\theta = 90°$. In this case the particle follows a circular path such that $qvB = \dfrac{mv^2}{r}$.

» The acceleration is zero when the velocity and magnetic field are parallel, $\theta = 0°$, and the particle follows an undeflected straight line.

» If the velocity is at an angle to the magnetic field between $0°$ and $90°$, the particle follows a helical path.

» Field models are used to describe forces that act at a distance.

» The gravitational field is created by, and acts on, objects with mass. The gravitational force is in the direction of the field.

» The electric field is created by, and acts on, objects with charge. The electric field is in the direction of, or opposite to the direction of, the field.

» The magnetic field is created by, and acts on, moving objects with charge. The magnetic force is perpendicular to the field.

# ⑤ CHAPTER REVIEW QUESTIONS

Review quiz

1 List three ways in which magnetic fields and gravitational fields are different.

2 At what angle to a magnetic field can an electron and a neutron travel to have the same path?

3 Why does Earth's magnetic field not protect us from high-energy electromagnetic radiation (photons)? How does it protect us from the other forms of cosmic radiation?

4 What is the minimum magnitude of a magnetic field necessary to apply a force of $1 \times 10^{-12}$ N to an electron moving at a speed of $500\,\text{km s}^{-1}$? What electric field is necessary to apply this magnitude force?

5 How can the motion of a moving charged particle be used to distinguish between an electric field and a magnetic field? Give a specific example.

6 Two charged particles, A and B, enter a uniform electric field, as shown in Figure 5.19. The particles have the same magnitude charge.

   a Which of the two charged plates creating the field shown is the positively charged plate?

   b What can you say about the charges and relative masses of the two particles?

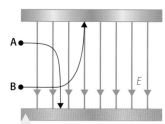

**FIGURE 5.19** Two charged particles, A and B, enter a uniform electric field.

**7** Determine the direction of each **deflection** of each charged particle as it enters the uniform electric fields shown.

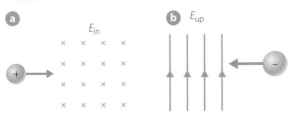

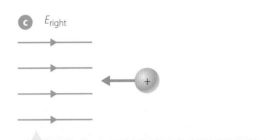

FIGURE 5.20 A charged particle enters a uniform electric field.

**8** A particle enters a uniform electric field, with initial velocity as shown in Figure 5.21. For each case, state whether the velocity and kinetic energy of the particle increase or decrease after it enters the field. Is work done by, or on, the field?

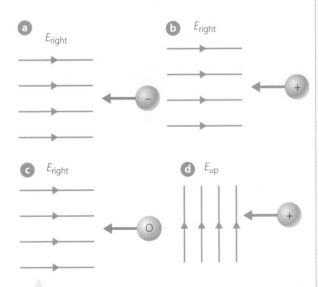

FIGURE 5.21 A charged particle enters a uniform electric field.

**9** Determine the initial direction of the deflection of each charged particle shown in Figure 5.22 as it enters the magnetic fields shown.

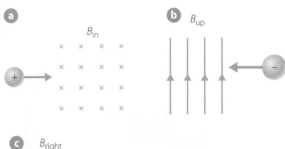

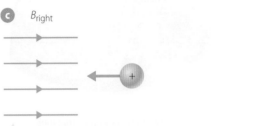

FIGURE 5.22 A charged particle entering a uniform magnetic field

**10** For the situations shown in Figure 5.22:
  **a** describe how the direction and magnitude of the particle's velocity change.
  **b** How does the kinetic energy change in each case?
  **c** Is work done on, or by, the field? Explain your answers.

**11** In Figure 5.23 a proton is moving in a magnetic field. The velocity of the proton and the direction of the magnetic force acting on it are shown. Find the direction of the magnetic field in each case.

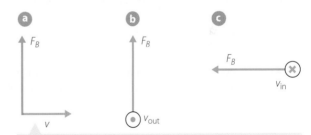

FIGURE 5.23 Force on a charged particle in a uniform magnetic field

**12** A proton and an electron enter a uniform magnetic field. The particles are travelling at the same speed perpendicular to the field. Draw a diagram showing their paths and explain the differences.

**13** A dust particle has a mass of $1.6 \times 10^{-6}$ g. What charge must this dust particle have for the electrostatic force on it due to Earth's fair weather field, $100\,V\,m^{-1}$ down, to balance the gravitational force? Give the sign and magnitude of the charge.

**14** An electron in an old television picture tube moves towards the front of the tube with a speed of $8.0 \times 10^6 \, \text{m s}^{-1}$ along the x-axis. Surrounding the neck of the tube are coils of wire that create a magnetic field of magnitude 0.035 T, directed at an angle of 60° to the x-axis and lying in the x–y plane. Calculate the acceleration of the electron.

**FIGURE 5.24** An old-style television picture tube

**15** A polystyrene bead with mass 1.0 g and charge +1.0 nC is in an electric field of $50 \, \text{V m}^{-1}$. Calculate the force on, and acceleration of, the bead due to the field.

**16** An electron moving at $25 \, \text{km s}^{-1}$ enters a uniform electric field of magnitude $550 \, \text{V m}^{-1}$ in the direction of its velocity.

  **a** What is the force on the electron?

  **b** After what distance does the electron come to a stop?

**17** An electron moving at $25 \, \text{km s}^{-1}$ to the right enters a uniform electric field of magnitude $550 \, \text{V m}^{-1}$ pointing up.

  **a** What is the acceleration of the electron? Give magnitude and direction.

  **b** What vertical distance does the electron move in the time it takes to move 25 m horizontally?

  **c** How much work does the field do on the electron in this time?

**18** A proton moving at $25 \, \text{km s}^{-1}$ enters a uniform electric field of magnitude $550 \, \text{V m}^{-1}$ moving up and to the right at an angle of 45° to the horizontal. The field points directly downwards.

  **a** What is the acceleration of the proton? Give magnitude and direction.

  **b** Draw a diagram showing the field and the trajectory of the proton.

  **c** Assuming that the proton enters the field at ground level, what is its maximum height?

  **d** What is its horizontal range?

**19** An electron moving at $20 \, \text{km s}^{-1}$ enters a uniform electric field of magnitude $500 \, \text{V m}^{-1}$ in the direction of its velocity.

  **a** What is the kinetic energy of the electron when it enters the field?

  **b** Use an energy approach to calculate the distance it takes for the electron to come to a stop.

**c** What is the potential difference between the initial point at which the electron entered the field and the point at which it comes to a stop?

**20** A proton, with mass $1.7 \times 10^{-27}$ kg, is moving at a speed of $10 \, \text{m s}^{-1}$ to the left when it enters a uniform electric field parallel to the direction of its velocity.

  **a** What is the kinetic energy of the proton when it enters the field?

After moving a distance of 5 cm in the field, the proton's velocity has doubled.

  **b** What is the kinetic energy of the proton now?

  **c** Through what potential difference has the proton moved?

  **d** What is the magnitude and direction of the electric field?

**21** An alpha particle with charge $+3.2 \times 10^{-19}$ C and mass $6.6 \times 10^{-27}$ kg is moving in a circular path in a uniform magnetic field of magnitude 55 mT. If the particle has an orbital radius of 100 m, what is its speed and orbital period?

**22** A sodium ion, with mass $3.8 \times 10^{-26}$ kg and charge $+1.6 \times 10^{-19}$ C, is injected into a uniform magnetic field, as shown in Figure 5.25, and follows a semi-circular path before leaving the field. The field has magnitude 1.5 T and the sodium ion is moving at $2.0 \times 10^6 \, \text{m s}^{-1}$.

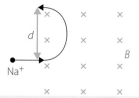

**FIGURE 5.25** A sodium ion is injected into a uniform magnetic field.

  **a** Calculate the distance d.

  **b** A chloride ion, with mass $5.9 \times 10^{-26}$ kg and charge $-1.6 \times 10^{-19}$ C, is injected into the field at the same point and with the same speed. What displacement d does this ion have before leaving the field?

  **c** Draw a scale diagram showing the paths of the two particles.

**23** A sodium ion, with mass $3.8 \times 10^{-26}$ kg and charge $+1.6 \times 10^{-19}$ C, is accelerated from rest through a potential difference of 5.0 kV. It then moves into a region of uniform magnetic field of strength 0.55 T perpendicular to the particle velocity.

  **a** What is the radius of the path of the sodium ion?

  **b** A chloride ion, mass $5.9 \times 10^{-26}$ kg and charge $-1.6 \times 10^{-19}$ C, is passed through the same combination of fields. Calculate its path radius in the magnetic field.

# The motor effect

**OUTCOMES**

**Students:**

- investigate, qualitatively and quantitatively, the interaction between a current-carrying conductor and a uniform magnetic field $F = \ell I_{\perp}B = I\ell B\sin\theta$ to establish: (ACSPH080, ACSPH081) CCT, ICT, N
  - conditions under which the maximum force is produced
  - the relationship between the directions of the force, magnetic field strength and current
  - conditions under which no force is produced on the conductor
- conduct a quantitative investigation to demonstrate the interaction between two parallel current-carrying wires
- analyse the interaction between two parallel current-carrying wires $\dfrac{F}{\ell} = \dfrac{\mu_0}{2\pi}\dfrac{I_1 I_2}{r}$ and determine the relationship between the International System of Units (SI) definition of an ampere and Newton's Third Law of Motion (ACSPH081, ACSPH106) ICT, L N

Physics Stage 6 Syllabus © 2017 NSW Education Standards Authority (NESA) for and on behalf of the Crown in right of the State of New South Wales.

FIGURE 6.1 A rail gun uses the motor effect to fire a projectile at high speeds without using explosives.

As we saw in chapter 5, a moving charged particle experiences a force in a magnetic field. If the charged particles are constrained to move within a conductor, then the charged particles will exert a force on the conductor, and pull it in the direction of the force on the particles. This is called the motor effect, because it is the force that makes the coil of wire in a motor move. The motor effect is used to convert electric potential energy into kinetic energy – to drive all sorts of machines from the tiny motor that makes your phone vibrate to gigantic industrial motors and rail guns (Figure 6.1). In chapter 8 we shall look at how different sorts of motors work.

In this chapter we develop a mathematical model for the force that a current-carrying wire experiences in a magnetic field. We shall use ideas from the previous chapter, as well as what you learned from *Physics in Focus Year 11* chapter 13 about currents and chapter 14 about magnetic fields. And, as always, we will use the key central idea of forces, and apply Newton's laws (*Physics in Focus Year 11* chapter 4).

## 6.1 Current-carrying wires in magnetic fields

A current-carrying wire experiences a force in a magnetic field because it contains lots of moving electrons. The moving electrons that make up the current experience a force due to the field. Each electron experiences a force given by $F = qvB \sin\theta$. Remember that the direction of the force is perpendicular to both the magnetic field, $B$, and the particle velocity, $v$.

Because the resistance of the wire is very low (it is a conductor) compared to any surrounding insulation or air, the electrons can only flow within the metal of the wire – they cannot move out of the wire. So the electrons are pushed in the direction of the force, and pull the wire along with them.

Imagine a section of wire, of length $\ell$, carrying a current $I$, as shown in Figure 6.2. The electrons move along the wire with some speed $v$, in the opposite direction to the current, $I$. (Note that this a simplified model. Electrons in fact follow complex paths through a wire, but they do *on average* move in the direction opposite the current.)

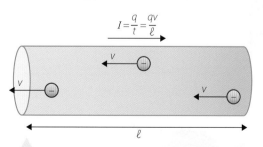

FIGURE 6.2 A section of wire carries a current, $I$. The current is carried by electrons with charge $-q$, moving with an average speed $v$.

Recall from *Physics in Focus Year 11* chapter 13 that a current is the quantity of charge per unit time flowing past a point:

$$I = \frac{q}{t}$$

where current is measured in units of amperes, A, and $1\,\text{A} = 1\,\text{C s}^{-1}$. The electrons that make up the current travel along the wire at an average speed $v$, where $v$ is distance travelled over time taken: $v = \dfrac{\ell}{t}$, where $t$ is the time it takes the electrons to travel through a displacement $\ell$ along the wire.

If we now look again at our expression for the force on these electrons due to a magnetic field we can say that:

$$F = qvB \sin\theta = q\frac{\ell}{t}B\sin\theta = \frac{q}{t}\ell B \sin\theta = \ell IB \sin\theta = \ell I_\perp B$$

Figure 6.3a shows the force that acts on a single positive charge as it moves in a magnetic field. Figure 6.3b shows the force on a section of current-carrying wire in a magnetic field. Remember that the direction of current is the direction of the flow of positive charges, so it is opposite to the direction in which the electrons move. Note that Figure 6.3 shows the case of perpendicular field and current. In general, the field and current may be at any angle, $\theta$, to each other.

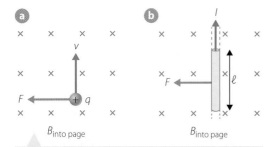

**FIGURE 6.3 a** The force on a moving charged particle in a magnetic field; **b** The force on a current-carrying conductor in a magnetic field

$F = \ell I_\perp B = \ell IB \sin\theta$ is the force that is experienced by a current of magnitude $I$ flowing through a section of wire of length $\ell$. The angle $\theta$, which is the angle between the charged particle velocity and the field, is also the angle between the section of wire and the field, as the electrons move along the direction of the wire.

Remember that we define the direction of current as the direction that positive charges would be moving to make that current. So while it is most likely to be electrons carrying a current in a wire, the direction of the current is the same as if it was positive charges flowing. We do not need to introduce a negative sign when we replace $\frac{q}{t}$ with $I$.

As we saw in the previous chapter, the magnitude of the force depends on the angle $\theta$. When the current and field are in the same direction the angle is $0°$ and the force is zero. The force is a maximum when the current and the field are perpendicular, so that $\theta = 90°$, $\sin\theta = 1$ and $F = \ell IB$.

## WORKED EXAMPLE (6.1)

An overhead power line carrying a current of $10\,000\,A$ is in a magnetic field of $30\,\mu T$ due to Earth's magnetic field. The field is perpendicular to the wire. What force per unit length does the wire experience?

| ANSWER | LOGIC |
|---|---|
| $I = 1.0 \times 10^4\,A$, $B = 3.0 \times 10^{-5}\,T$, $\theta = 90°$ | • Identify the relevant data in the question. |
| $F = \ell IB \sin\theta$ | • Relate the force to the field and current. |
| $\dfrac{F}{\ell} = IB \sin\theta$ | • Rearrange for force per unit length. |
| $\dfrac{F}{\ell} = (1.0 \times 10^4\,A)(3.0 \times 10^{-5}\,T)\sin 90°$ | • Substitute correct values with units. |
| $\dfrac{F}{\ell} = 0.30\,A\,T$ | • Calculate the answer. |
| $1\,A\,T = 1\,C\,s^{-1} \times 1\,kg\,s^{-1}\,C^{-1} = 1\,kg\,s^{-2} = 1\,N\,m^{-1}$ | • Check that the units are correct. |
| $\dfrac{F}{\ell} = 0.30\,N\,m^{-1}$ | • State the final answer with correct units and appropriate significant figures. |
| This is a large force, considering how long a span of high voltage power line is. However, power lines carry current that is alternating in direction 50 times per second. The force also alternates direction rather than constantly pulling the wire in a single direction. | |

**TRY THIS YOURSELF**

At what angle to the field would the wire need to be for the force to be half this value?

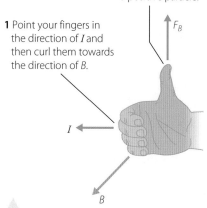

**2** Your thumb shows the direction of the magnetic force on a positive particle.

**1** Point your fingers in the direction of *I* and then curl them towards the direction of *B*.

$F_B$

*I*

*B*

**FIGURE 6.4** The right-hand rule for currents in magnetic fields

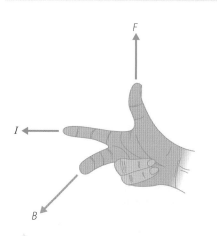

*F*

*I*

*B*

**FIGURE 6.5** A variation of the right-hand rule, in which you point along *I* with your index finger, then curl your other fingers towards *B*. Again, your thumb points in the direction of the force.

The equation $F = \ell IB \sin\theta$ gives us the magnitude of the force, but does not actually give us the direction of the force. Force is a vector – it has a direction as well as a magnitude – and we need to know what that direction is. The direction of the force is perpendicular to both the field and the direction of the current. It is found using the right-hand rule, exactly as we did for finding the direction of force on a moving charged particle in a magnetic field. Figure 6.4 shows how this is done.

Hold the fingers of your right hand together and point your thumb straight out, at a right angle to your hand. Move your hand so your fingers point in the direction of the current. Curl them towards the field. Your thumb is now pointing in the direction of the force.

You may come across other ways of applying the right-hand rule. For example, you can point your index finger towards *I*, then point your middle finger towards *B*. Your thumb will then point towards *F*, as shown in Figure 6.5. This is exactly the same as the rule shown in Figure 6.4 except that you leave your index finger behind when you curl your fingers.

It doesn't matter which variation you use, just remember to always use your *right hand*!

Figure 6.6 gives some examples of the force exerted on a current-carrying wire in a magnetic field. Note that when the field and current are parallel the force is zero. In all other cases the force is perpendicular to both the current and the field. The maximum force occurs when the current and field are perpendicular.

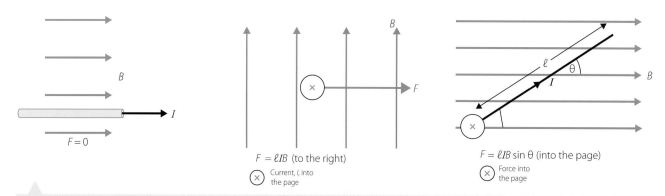

*B*

*F* = 0

*I*

*B*

×

*F*

$F = \ell IB$ (to the right)

× Current, *I*, into the page

*B*

$\ell$

$\theta$

*I*

*B*

×

$F = \ell IB \sin\theta$ (into the page)

× Force into the page

**FIGURE 6.6** The force applied to a current-carrying wire depends on the angle between the current and the magnetic field. The right-hand rule gives the direction of the force.

► **WORKED EXAMPLE** (6.2)

A magnetic field is pointing into the page and a current-carrying wire is sitting in the plane of the page carrying current to the right. What is the direction of the force on the wire due to the magnetic field?

| ANSWERS | LOGIC |
|---|---|
| 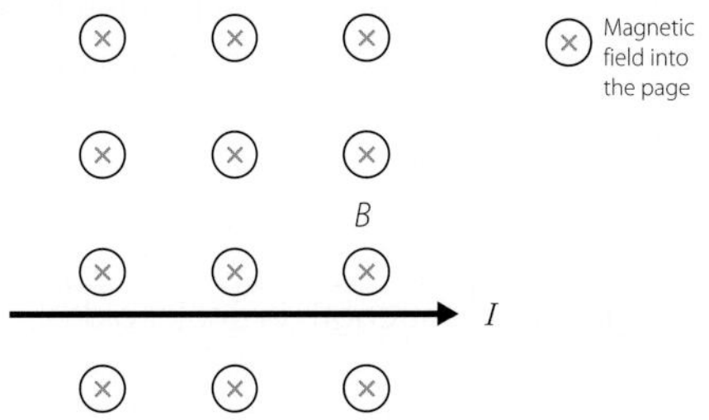  FIGURE 6.7 The current and the field | ■ Draw a diagram showing the field and the current (Figure 6.7). |
| The force must be pointing up the page.   $F = \ell IB$ (up the page) 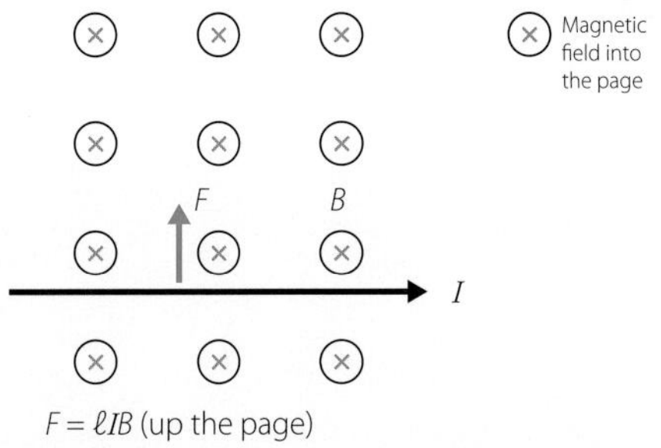  FIGURE 6.8 The force acting on the current-carrying wire | ■ Use the right-hand rule: point your fingers right, then curl them forwards (away from you). This forces your thumb to point upwards.  ■ Draw a diagram showing the force on the wire (Figure 6.8). |

**TRY THIS YOURSELF**

If the field was pointing out of the page, towards you, what would be the direction of the force?

## The definition of magnetic field

So far we have seen that magnetic fields are created by moving charged particles and currents (chapter 14 of *Physics in Focus Year 11*) and they exert forces on moving charged particles (chapter 5) and on current-carrying wires. But we have not as yet given a formal, mathematical definition of the magnetic field. It is the expression derived above for the force on a current-carrying wire that allows us to mathematically define, and to measure, a magnetic field.

The magnetic field is defined as the magnetic force per unit current, per unit length, on a current-carrying wire in a magnetic field. Mathematically we can write this as:

$$B = \frac{F}{\ell I \sin \theta}$$

where the magnetic field is measured in tesla, T, the force in newton, N, the current in ampere, A, and the length in metre, m. The angle is again the angle between the current in the wire and the field.

This definition is consistent with the definitions of the other fields as force per unit property: force per unit mass for the gravitational field, and force per unit charge for the electric field.

Critical and creative thinking

Numeracy

Literacy

# A current balance

A current balance is a current element that is placed in a magnetic field where it experiences a force. The current balance acts like a see-saw. The torque due to the magnetic force at one end is equal to the torque due to the gravitational force on the mass at the other end of the balance when the balance is in equilibrium. You may wish to review the section on torque in chapter 3, and the magnetic field due to a current-carrying coil in chapter 14 of *Physics in Focus Year 11*.

## AIM

To measure the magnetic force on a current balance to find the magnetic field strength in a solenoid
Write an inquiry question for this investigation.

## MATERIALS

### Part A

- Air core solenoid
- Materials to make a current balance:
  - Thin, stiff, lightweight insulator such as cardboard or plastic
  - Stiff conducting wire
  - Copper or zinc sheet
  - Pin
  - Fine sandpaper
  - Scissors
  - Sticky tape
- Tin snips (to cut copper or zinc sheet)
- Short pieces of wire of known mass (the weights)
- 2 DC power supplies
- 2 ammeters (0–5 A)
- 2 variable resistors or rheostats
- 2 switches
- Alligator clips and leads

### Part B

- Current balance circuit comprising power supply, rheostat, ammeter, switch
- Solenoid circuit comprising the other power supply, rheostat, ammeter, switch

**RISK ASSESSMENT**

| WHAT ARE THE RISKS IN DOING THIS INVESTIGATION? | HOW CAN YOU MANAGE THESE RISKS TO STAY SAFE? |
| --- | --- |
| Electricity can shock and cause damage to equipment. | Use low voltages and currents only. |
| Scissors can cut skin and have sharp tips. | Be very careful when using scissors. Do not run with scissors. |

What other risks are associated with your investigation, and how can you manage them?

## METHOD

### Part A

1. Cut a rectangle from the insulator so that half will fit into the solenoid and half is outside.
2. Attach a small pin to the middle of one of the short sides of the insulating rectangle so that it overhangs the end that will sit outside the solenoid (see Figure 6.9).
3. Make a rectangular half-loop of conducting wire, to sit near the edges of the insulating rectangle that goes into the solenoid. Make sure the half-loop is attached to the insulating rectangle.

4 Cut two supports out of the metal sheet, bend them and attach them to the end of the solenoid as shown. Use the alligator clips to connect them to the current balance circuit. Note that *they should not make an electrical contact with the solenoid*.

5 Bend the ends of the rectangular half-loop so that they sit on the metal supports.

6 Use the sandpaper to clean the metal to ensure a good electrical connection.

7 Measure the length of the current element that is perpendicular to the magnetic field of the solenoid. Remember that the magnetic field inside a solenoid is parallel to the long axis of the solenoid.

8 Balance the current balance by hanging pieces of wire or small weights over the pin.

## Part B

9 Connect the balance and solenoid circuits as shown in Figure 6.10.

10 Close the switch in the solenoid circuit and adjust the current to 2.5 A–3.5 A.

11 Close both switches and observe what happens to the current balance. It needs to act as a see-saw, with the inner end being pushed down by the magnetic force. Make changes to the solenoid circuit if necessary.

12 Adjust the number of weights and their positions until the current balance is balanced.

### RESULTS

Record the following in a table.

■ Current in solenoid

■ Current in current element

■ Distance from pivot to current element

■ Distance from pivot to balancing weights

Don't forget to include units on all your measurements.

### ANALYSIS OF RESULTS

Calculate the following and add them to your table:

■ Masses of weights used

■ Gravitational force on balancing masses

■ Torque by gravitational force on weights on current balance

■ Magnetic force on current element

■ Magnetic field in solenoid

The torque by the magnetic force is equal to the torque by the weight force when the current balance is balanced.

The magnetic field produced by a solenoid is given by $B = \dfrac{\mu_0 N I}{\ell}$, where $\mu_0$ is the **permeability of free space**, $\mu_0 = 4\pi \times 10^{-7}\,\mathrm{T\,m\,A^{-1}}$, $N$ is the number of turns of wire in the solenoid, $I$ is the solenoid current in A, and $\ell$ is the length of the solenoid in m.

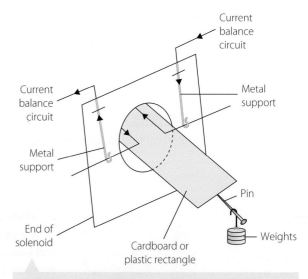

**FIGURE 6.9** Metal supports attached to the end of the solenoid allow the current balance to rotate freely when current flows in the current balance circuit.

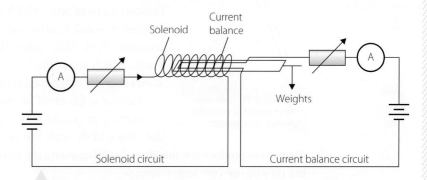

**FIGURE 6.10** Circuits for current balance and solenoid

**DISCUSSION**

1   Comment on the quality of your data and how this affects your results.

2   Write a brief summary of all the different ideas in physics that you had to use for the analysis of this experiment.

3   Give the answer to your inquiry question.

**CONCLUSION**

Write a conclusion summarising the outcomes of your investigation.

**Jumping wire**

Why does a wire 'jump' in a magnetic field?

If you have the correct equipment you can set up this experiment yourself, but be very careful because it uses large currents.

Current-carrying wires in magnetic fields

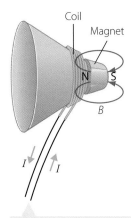

**FIGURE 6.11** The motor effect is used to make the speaker cone oscillate.

There are many applications of the motor effect, which is the force exerted by a magnetic field on a current-carrying wire. In subsequent chapters we will look at how this force can be used to produce a torque, which leads to rotation. This is how most motors work.

The motor effect is also used in speakers, which are a type of linear motor. When a sound is converted to an electrical current, the current oscillates at the same frequency as the original sound. The oscillating current flows through a coil of wire, which is wrapped around a cone in front of a large magnet. The coil is alternately attracted and repelled by the magnet's large magnetic field. This makes the speaker cone oscillate, producing sound waves, which are at the same frequency as the original sound. Figure 6.11 shows how the magnet and coil are positioned in a speaker.

Another application of the motor effect is rail guns. A rail gun uses a conducting projectile sitting across conductive rails. A current runs via the rails and through the conductive projectile. The currents in the rails themselves produce the magnetic field that exerts the force on the projectile. They require no explosive, but they do use very large currents.

**Rail guns**

Find out how a rail gun works.

**KEY CONCEPTS**

- The force experienced by a current-carrying wire in a magnetic field is given by $F = \ell IB \sin\theta = \ell I_\perp B$.
- The force is a maximum and equal to $F = \ell IB$ when the field is perpendicular to the current. The force is zero when the field and current are parallel.
- The direction of the force is given by the right-hand rule (Figure 6.4).
- The magnetic field is defined as the magnetic force per unit current per unit length, on a current-carrying wire in a magnetic field: $B = \dfrac{F}{\ell I \sin\theta}$.

**CHECK YOUR UNDERSTANDING**

**6.1**

1   What is the definition of magnetic field? Compare the definition of magnetic field to the definitions of gravitational and electric fields.

2   Explain why, in the current balance experiment shown in Figures 6.9 and 6.10, it is only the section of wire that is perpendicular to the solenoid that experiences a magnetic force.

3   Show that the units on each side of the equation $F = \ell IB \sin\theta$ are consistent.

4   A wire is carrying a current to the left in a magnetic field pointing away from you. In what direction is the force on the wire?

5   A current-carrying wire experiences a force due to a magnetic field as shown in Figure 6.12. What is the direction of the current – into or out of the page?

6   A 1 m length of wire carrying 9 A is in a perpendicular magnetic field of 0.01 T. What is the force on the wire due to the magnetic field?

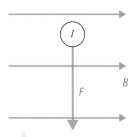

**FIGURE 6.12** What is the direction of the current?

**7** A 10 cm long wire is carrying a current of 10 A in a magnetic field of 55 μT. The wire makes an angle θ with the field.

    **a** Use a spreadsheet to calculate the force on the wire for angles from 0° to 360° in 5° increments; that is, for θ = 0°, 5°, 10° … 360°.

    **b** Plot a graph of force as a function of angle.

    **c** From your graph, find the angles at which:

      **i** the force is a maximum.

      **ii** the force is zero.

      **iii** the force is half its maximum magnitude.

Information and communication technology capability

**8** A 2.0 m length of wire carrying 5.0 A is at an angle of 30° to a magnetic field of 50 μT.

    **a** What is the force on the wire due to the magnetic field?

    **b** What is the size of the field that would give the same magnitude force if the wire was at an angle of 45° to the field?

**9** A horizontal wire with a mass per unit length of 0.05 kg m⁻¹ carries a current of 5.00 A west. What is the minimum magnetic field needed to lift this wire vertically upwards? Give magnitude and direction.

**10** A rail gun consists of two parallel conducting rails, with a length of metal wire resting across them. When the rails are connected to a power supply, a current runs up one rail, across the length of wire, and back down the other rail.

Draw a diagram showing:

    **a** the rails and length of wire and the current flowing through them.

    **b** the magnetic field due to each rail and the wire.

    **c** the force on the wire due to the magnetic field of the rails.

# 6.2 Forces on parallel current-carrying wires

**Forces on parallel current-carrying wires**

What happens when the currents are in same direction? What about when they are opposite?

A current-carrying wire both creates a magnetic field and experiences a force due to a magnetic field. Hence when two current-carrying wires are placed parallel to each other, they will each experience a force due to the field created by the other.

## INVESTIGATION 6.2

### Teacher demonstration: Forces on parallel current-carrying wires

Critical and creative thinking

Numeracy

Literacy

**CAUTION:** This investigation uses large currents. It can be dangerous and can cause damage to equipment. It should only be done as a teacher demonstration. Follow instructions carefully, *never* have the current flowing for more than a second, and always disconnect the power supply when not in use.

**AIM**

To observe the effects of the force that one current-carrying wire exerts on a second parallel current-carrying wire

Write a hypothesis for this investigation.

- High current DC power supply such as a 12V battery
- Heavy duty rheostat, 0–10Ω
- Heavy duty wires with alligator clips for making connections
- Heavy duty momentary switch
- 2 lengths of flexible cable, approximately 50cm long
- Vertical board with movable insulated pegs for holding wires
- Paper with grid marked on it
- Camera or video camera
- Tripod

**RISK ASSESSMENT**

| WHAT ARE THE RISKS IN DOING THIS INVESTIGATION? | HOW CAN YOU MANAGE THESE RISKS TO STAY SAFE? |
|---|---|
| This investigation uses large currents. Electricity can shock and cause damage to equipment. | Only perform as a teacher demonstration. Follow all instructions carefully. Keep equipment disconnected when not in use. |

What other risks are associated with your investigation, and how can you manage them?

**METHOD**

1 Attach the paper to the board so that the wires will hang in front of the grid. The grid should have markings spaced at 1 cm or 0.5 cm intervals.

2 Connect up the parallel circuit shown in Figure 6.13 but *do not* connect both terminals of the battery yet.

3 Attach the lengths of flexible wire to the pegs on the board so they hang straight down in front of the grid. They should be spaced approximately 2 cm apart.

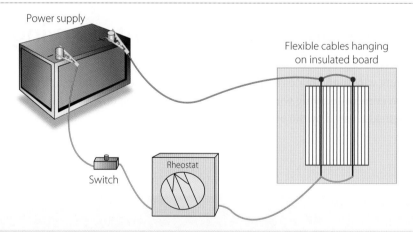

**FIGURE 6.13** The circuit for parallel current-carrying wires with current in same direction

### A Effect of varying current

4 Start with the rheostat adjusted to give the largest resistance.

5 Take a photo of the wires so that you can see how they hang against the grid. Note where you have the camera, and if possible put it on a tripod and keep it at this position.

6 *Make sure that the switch is turned off.*

Connect the alligator clips to the battery terminals.

You need two people for the next part:

7 One person turns on the switch for only *one second or less*. The other takes a photo while the current is flowing.

If any sparking occurs or if smoke is produced, immediately turn off the switch and disconnect the circuit.

9780170409131

8  Adjust the rheostat to a lower resistance.

9  Repeat steps 5 to 7.

10  Repeat steps 8 and 9 at least twice so you have a set of photos showing the behaviour of the wires with increasing current. *Do not reduce the resistance to less than 1 Ω.*

## B  Effect of varying distance

11  Repeat the experiment in part A steps 5 to 7, but keep the rheostat at a constant setting that gives a clear deflection of the wires.

12  Increase the distance between the wires and observe how the deflection changes. Repeat with at least one more distance.

## C  Effect of varying current direction

13  Disconnect the circuit and reconnect it as a series circuit as shown in Figure 6.14. This circuit gives parallel currents that flow in opposite directions.

14  Repeat the experiment in part A steps 5 to 7.

15  Disconnect the circuit.

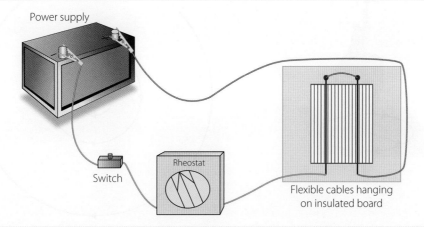

**FIGURE 6.14** The circuit for parallel current-carrying wires with current in opposite direction

### RESULTS

Record your results. You should have data showing how deflection of the wires varies with current and distance between the wires. You can measure this from the photographs taken while current was flowing, using the grid as a scale. This data can be recorded in tables.

Note also whether the wires were attracted or repelled by each other in each experiment.

### ANALYSIS OF RESULTS

If you have enough data points, your data can be graphed. Even if you do not have more than 2 or 3 data points for each experiment you should still be able to see whether deflection, and hence force, increases or decreases with current and distance. You should also be able to say whether the wires are attracted or repelled when the currents flow in the same direction or are opposite. Use this information to draw force diagrams for the wires in each circuit.

### DISCUSSION

1  Draw a diagram showing the force acting on each wire when the currents flow in the same direction and when they are in opposite directions.

2  Describe the relationship between the force on the wires and the current flowing, and the force and the distance between the wires.

3  State whether your hypothesis was supported or not.

### CONCLUSION

Write a conclusion summarising the outcomes of your investigation.

Results of experiments like Investigation 6.2 show that the force experienced by parallel current-carrying wires decreases with distance between the wires, increases with current and is attractive when the currents are in the same direction. Starting with the expression for the field due to a current-carrying wire, we can derive a mathematical expression for the force. Recall from chapter 14 of *Physics in Focus Year 11* that the magnetic field due to a long, straight current-carrying wire is:

$$B = \frac{\mu_0 I}{2\pi r}$$

where $I$ is the current in the wire, $r$ is the distance from the wire and $\mu_0$ is the permeability of free space, $\mu_0 = 4\pi \times 10^{-7}\,\mathrm{T\,m\,A^{-1}}$.

The magnetic field lines form circular loops around the wire, as shown in Figure 6.15. The direction of the field is found by pointing your right thumb in the direction of the current and curling your fingers. The direction of your fingers then gives the direction of the field lines.

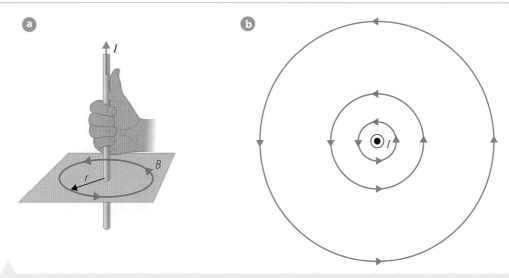

**FIGURE 6.15** Magnetic field lines form circular loops around current-carrying wires. **a** Right-hand thumb rule for finding the direction of magnetic field lines; **b** Circular field lines for a current coming out of the page

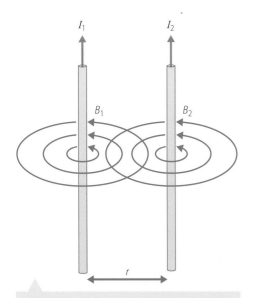

**FIGURE 6.16** Two parallel current-carrying wires. Wire 1 produces a field $B_1$, wire 2 produces a field $B_2$.

If we know the field at a point in space, then we can find the force exerted on a current-carrying wire at that position using:

$$F = \ell I_\perp B$$

Figure 6.16 shows two parallel current-carrying wires, spaced a distance $r$ apart. The wire on the left carries a current $I_1$, the wire on the right carries a current $I_2$. Wire 1 creates a magnetic field $B_1 = \frac{\mu_0 I_1}{2\pi r}$ at the position of wire 2. This field is directed into the page.

The force exerted by wire 1 on wire 2 is given by:

$$F_{\text{by 1 on 2}} = \ell I_2 B_1 \sin\theta = \ell I_2 B_1$$

because $\theta = 90°$ so $\sin\theta = 1$.

The magnetic field due to wire 1 is $B_1 = \frac{\mu_0 I_1}{2\pi r}$.

So the force by wire 1 on wire 2 is:

$$F_{\text{by 1 on 2}} = \frac{\mu_0 I_1 I_2 \ell}{2\pi r}$$

The length $\ell$ in this equation is the common length of the wires – the length for which they are parallel.

The direction of the force is found using the right-hand rule. Figure 6.17 shows the directions of the current in, the external field at, and the force acting on, wire 2. The force on wire 2 acts to the left, towards wire 1.

We can find the force acting on wire 1 due to wire 2 by following the same procedure:

The field at wire 1 due to wire 2 is $B_2 = \dfrac{\mu_0 I_2}{2\pi r}$.

$$F_{\text{by 2 on 1}} = \ell I_1 B_2 \sin\theta = \ell I_1 B_2 = \dfrac{\mu_0 \ell I_1 I_2}{2\pi r}$$

Remember that force is a vector, and the direction of the force is in the direction of the vector $r$ that points from one wire to the other.

So the magnitude of the force exerted by wire 2 on wire 1 is exactly the same as the force exerted by wire 1 on wire 2. However, the direction is opposite, as you can see from applying the right-hand rule to the current through the field at wire 1, as shown in Figure 6.18. The field at wire 1 due to the current in wire 2 points out of the page (towards you), as you can see in Figure 6.18. This results in a force acting on wire 1 to the right, towards wire 2.

So each wire experiences a force towards the other, and the two wires are attracted to each other.

## Newton's third law

We did not need to repeat the derivation of the force to find the force acting on wire 1. Once we had derived an expression for the force acting on wire 2 due to wire 1 we could simply have applied Newton's third law:

$$F_{\text{by 2 on 1}} = -F_{\text{by 1 on 2}}$$

Newton's third law tells us that both forces have the same magnitude, and act in opposite directions.

This is the third time that we have derived expressions for forces only to see that Newton's third law could have immediately given us the answer. The first two times were with Coulomb's Law for the electrostatic force between charged particles and Newton's Law of Universal Gravitation for objects with mass. Newton's third law is a powerful, unifying and ubiquitous principle in physics.

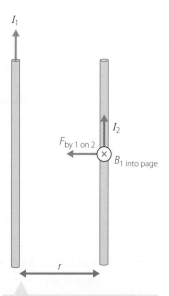

**FIGURE 6.17** Two parallel current-carrying wires. The force experienced by wire 2 due to the magnetic field created by the current in wire 1. The direction of the force is found using the right-hand rule.

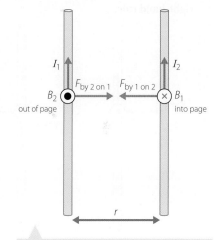

**FIGURE 6.18** Two parallel current-carrying wires. The force experienced by each wire is found using the right-hand rule. The forces are attractive when the current is in the same direction in both wires.

The circuit shown in Figure 6.14 carries a current of 15 A, and the long flexible wires are spaced 5.0 cm apart. The current is flowing up in the wire on the left and down in the wire on the right. What is the force per unit length that each wire exerts on the other? Give magnitude and direction.

| ANSWER | LOGIC |
|---|---|
| $I_1 = I_2 = 15\,\text{A}$, $r = 0.05\,\text{m}$ | ■ Identify the relevant data in the question. |
| $F = \dfrac{\mu_0 \ell I_1 I_2}{2\pi r}$ | ■ Relate the force to the current and distance. |
| $\dfrac{F}{\ell} = \dfrac{\mu_0 I_1 I_2}{2\pi r}$ | ■ Rearrange for force per unit length. |
| $\dfrac{F}{\ell} = \dfrac{(4\pi \times 10^{-7}\,\text{T m A}^{-1})(15\ \text{A})(15\ \text{A})}{2\pi(0.05\ \text{m})}$ | ■ Substitute correct values with units. |
| $\dfrac{F}{\ell} = 9.0 \times 10^{-4}\,\text{A T}$ | ■ Calculate the answer. |
| $1\,\text{A T} = 1\,\text{C s}^{-1} \times 1\,\text{kg s}^{-1}\,\text{C}^{-1} = 1\,\text{kg s}^{-2} = 1\,\text{N m}^{-1}$ | ■ Check that the units are correct. |
| $\dfrac{F}{\ell} = 9.0 \times 10^{-4}\,\text{N m}^{-1}$ | ■ State the final answer with correct units and appropriate significant figures. |
| The magnitude of the force is the same for both wires. To find the direction we draw a diagram and use the right-hand rule. 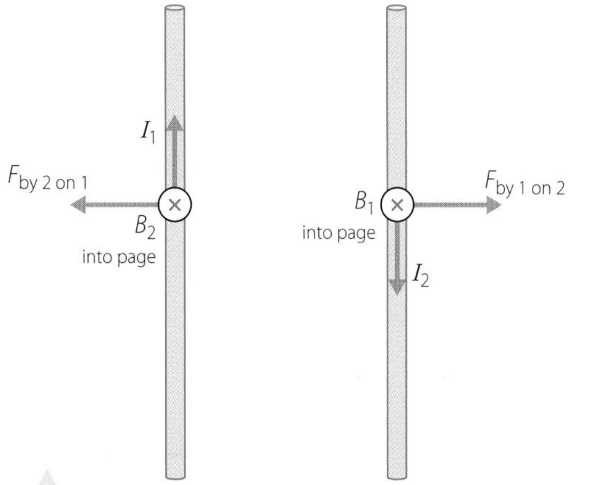<br><br><br><br>**FIGURE 6.19** Two parallel current-carrying wires. The force experienced by each wire is found using the right-hand rule. The forces are repulsive when the currents are in opposite directions.<br><br>The forces are repulsive. | ■ Draw a diagram showing the direction of the current through each wire and the field at the wire (Figure 6.19). Use the right-hand rule to find the direction of the force. |
| The force per unit length on wire 1 is $9.0 \times 10^{-4}\,\text{N m}^{-1}$ to the left.<br><br>The force per unit length on wire 2 is $9.0 \times 10^{-4}\,\text{N m}^{-1}$ to the right. | ■ State the final answer with correct units, appropriate significant figures and direction. |

**TRY THIS YOURSELF**

If the current in the circuit was reduced to half the initial value, and its direction reversed, how would the answer change?

9780170409131

We often use densities rather than absolute or total amounts in physics. For example, we might describe a string in terms of its mass per unit length when we want to know the speed of a wave on the string. When working with materials that can be modelled as continuous, such as liquids and gases, we usually use pressure (force per unit area) rather than force. The equation describing the force between two current-carrying wires is often written in terms of force per unit length as:

$$\frac{F}{\ell} = \frac{\mu_0}{2\pi} \frac{I_1 I_2}{r}$$

because this is a useful quantity that does not vary with how much wire there is.

## The definition of the ampere

The SI unit of current is the ampere, abbreviated A, named for André-Marie Ampère. The ampere is one of the seven base units in the SI system, along with the metre, second, kilogram, kelvin, mole and candela.

Current is the flow of charge, and $1\,\text{A} = 1\,\text{C s}^{-1}$. The coulomb is *not* a base unit in the SI, but rather it is derived from the ampere.

The definition of one ampere is:

*the constant current which, if maintained in two straight parallel conductors of infinite length, of negligible circular cross-section, and placed 1 metre apart in vacuum, would produce between these conductors a force equal to $2 \times 10^{-7}$ newton per metre of length.*

This definition of the ampere also fixes the value of the constant $\mu_0$, the permeability of free space. If

$$\frac{F}{\ell} = \frac{\mu_0}{2\pi} \frac{I_1 I_2}{r} = 2 \times 10^{-7}\,\text{N m}^{-1} \text{ when } I_1 = I_2 = 1\,\text{A},$$
$$\text{then } \mu_0 = 4\pi \times 10^{-7}\,\text{T m A}^{-1}.$$

This definition was agreed in 1946 by the International Committee for Weights and Measures. But, as you can imagine, measuring forces this small between current-carrying wires that are also subject to the gravitational force is not easy. In practice, the unit is measured using the power produced by a circuit in a controlled experiment.

Newton's third law is implicit in the definition of the ampere. The definition refers to the force produced *between the two conductors*, rather than the force on wire 1 or wire 2. This is because there is no need to specify which force needs to be measured, because the forces have the same magnitude. Forces are interactions between objects – whenever an interaction occurs, both objects experience the force.

**Force on parallel wires with opposite currents**

Watch this animation to see what happens when the currents flow in opposite directions.

**Force on parallel current-carrying wires**

Compare what happens when the currents flow in opposite directions to when they flow in the same direction.

**The SI unit A**

Find out how the ampere is defined and measured.

Forces on parallel current-carrying wires

**KEY CONCEPTS**

- A current-carrying wire produces a magnetic field given by $B = \dfrac{\mu_0 I}{2\pi r}$.

- Two parallel current-carrying wires each experience a force due to the other given by $F = \dfrac{\mu_0 \ell I_1 I_2}{2\pi r}$ where $I_1$ and $I_2$ are the currents in the two wires.

- The force that one current-carrying wire exerts on a parallel current-carrying wire is attractive if the currents are in the same direction. It is repulsive if the currents are in opposite directions.

- The force per unit length acting on each of a pair of current-carrying wires is given by $\dfrac{F}{\ell} = \dfrac{\mu_0}{2\pi} \dfrac{I_1 I_2}{r}$.

- The forces that parallel current-carrying wires exert on each other are a Newton's third law force pair: $F_{\text{by 2 on 1}} = -F_{\text{by 1 on 2}}$. The forces are equal and act in opposite directions.

- The SI definition of the ampere is the current through two parallel conductors 1 m apart that would produce a force of $2 \times 10^{-7}\,\text{N m}^{-1}$ between the conductors.

- The definition of the ampere fixes the value of the permeability of free space as $\mu_0 = 4\pi \times 10^{-7}\,\text{T m A}^{-1}$.

1 Explain how the ampere is defined.

2 Two long current-carrying wires are arranged so they are parallel to each other. Is it possible for the forces acting on the two wires to point in the same direction? Explain your answer.

3 Two vertical parallel current-carrying wires are observed to repel each other. The first carries a current upwards. What is the direction of current in the second wire?

4 Two parallel current-carrying wires exert a force $F$ on each other. How does the force change if:

 a the distance between the wires is doubled?

 b the current in both wires is doubled?

 c the current in one wire is halved?

 d the current in one wire is reduced to zero?

5 A hanging metal slinky (a loose coil) is attached to a battery so that a current flows through the slinky. When the current starts to flow, do the coils move closer together or further apart? Draw a diagram to help explain your answer.

6 Two long current-carrying wires are hanging 1.0 cm apart. The first wire carries a current of 1.0 A upwards and the second carries a current of 2.0 A upwards.

 a Calculate the force per unit length that each wire exerts on the other. Give magnitude and direction.

 b How would your answer change if the first wire carried a current of 2.0 A and the second carried 1.0 A?

 c How would your answer change if the currents were both flowing downwards instead?

7 A pair of 0.50 m long parallel current-carrying wires carry a current of 15 A each. At what distance should they be placed apart if the force that each exerts on the other is 1.0 mN?

8 The circuit shown in Figure 6.13 is connected to a power supply and switched on. The wires are spaced 1.0 cm apart and exert a force per unit length of 0.1 mN m$^{-1}$ on each other. What current is flowing through the circuit?

- The force experienced by a current-carrying wire in a magnetic field is given by $F = \ell IB \sin\theta = \ell I_\perp B$.

- The force is a maximum and equal to $F = \ell IB$ when the field is perpendicular to the current. The force is zero when the field and current are parallel.

- The direction of the force is given by the right-hand rule (Figure 6.4).

- The magnetic field is defined as the magnetic force per unit current, per unit length, on a current-carrying wire in a magnetic field: $B = \dfrac{F}{\ell I \sin\theta}$.

- A current-carrying wire produces a magnetic field given by $B = \dfrac{\mu_0 I}{2\pi r}$.

- Two parallel current-carrying wires each experience a force due to the other given by $F = \dfrac{\mu_0 \ell I_1 I_2}{2\pi r}$ where $I_1$ and $I_2$ are the currents in the two wires.

- The force that one current-carrying wire exerts on a parallel current-carrying wire is attractive if the currents are in the same direction. It is repulsive if the currents are in opposite directions.

- The force per unit length acting on each of a pair of current-carrying wires is given by $\dfrac{F}{\ell} = \dfrac{\mu_0}{2\pi} \dfrac{I_1 I_2}{r}$.

- The forces that a pair of parallel current-carrying wires exert on each other are a Newton's third law pair: $F_{\text{by 2 on 1}} = -F_{\text{by 1 on 2}}$. The forces are equal and act in opposite directions.

- The SI definition of the ampere is the current through two parallel conductors 1 m apart which would produce a force of $2 \times 10^{-7}\,\text{N m}^{-1}$ between the conductors.

- The definition of the ampere fixes the value of the permeability of free space as $\mu_0 = 4\pi \times 10^{-7}\,\text{T m A}^{-1}$.

# 6 CHAPTER REVIEW QUESTIONS

Review quiz

1 What is the motor effect?

2 What are the units of the constant $\mu_0$? Write these units in fundamental units.

3 Under what circumstances is a force produced on a current-carrying conductor in a magnetic field?

4 At what angle should a current-carrying wire be placed to a magnetic field to:

   a maximise the force it experiences?

   b minimise the force it experiences?

5 A current-carrying wire is placed between the poles of a magnet as shown in Figure 6.20. If the wire deflects to the left, what is the direction of the current in the wire?

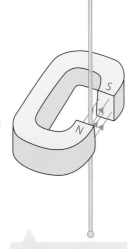

**FIGURE 6.20** A current-carrying wire between the poles of a magnet

6 A horizontal current-carrying wire experiences a force vertically upwards due to a magnetic field.

   a If the current is to the north, what must be the direction of the magnetic field?

   b If the magnetic field is to the north, what must be the direction of the current?

7 A 1.0 m length of wire carrying a current of 5.0 A lies perpendicular to a magnetic field of 50 mT. What is the force experienced by the wire due to the magnetic field?

8 A wire is carrying a large current directly upwards.

   a Draw a diagram of the magnetic field lines, as seen from above, due to this current.

   b Consider a second vertical current-carrying wire close to the first wire. If the current in the second wire is also upwards, what is the direction of the force on this second wire? Draw it on your diagram.

   c What is the direction of the force on the first wire due to the current in the second wire? Draw it on your diagram.

   d How would your answers to parts **b** and **c** change if the current in the second wire was downwards?

**9** Two long parallel wires, each carrying the same current, spaced 1 m apart, exert a force per unit length of $2 \times 10^{-7}\,\mathrm{N\,m^{-1}}$ on each other. What is the current through the wires?

**10** What is the relationship between the SI definition of the ampere and Newton's third law?

**11** If the SI definition of the ampere was the current required in two long parallel current-carrying wires for them to exert a force per unit length of $1\,\mathrm{N\,m^{-1}}$ on each other when spaced 1 m apart:

  **a** what would be the value of the constant $\mu_0$?

  **b** would a coulomb be a larger or smaller amount of charge?

**12** A wire 2.1 m long carrying a current of 0.85 A has a force of $5.0 \times 10^{-2}\,\mathrm{N}$ exerted on it by a uniform magnetic field at right angles to the wire. What is the magnitude of the magnetic field?

**13** A 5 m long current-carrying wire is at an angle of 30° to a magnetic field. It carries a current of 30 A and experiences a force of 0.02 N. How large is the magnetic field?

**14** A 2 m long wire carrying a current of 1.0 A experiences a force of $6.3 \times 10^{-6}\,\mathrm{N}$ in a magnetic field of $5.0\,\mu\mathrm{T}$. Calculate the possible angles between the current and the field.

**15** The force per unit length on a long current-carrying wire is $1.0 \times 10^{-4}\,\mathrm{N\,m^{-1}}$ when it is in a perpendicular magnetic field of 0.05 T. What is the current through the wire?

**16** Two long parallel wires carrying currents of 1.0 A and 0.5 A are spaced a distance 1.0 cm apart. What is the force per unit length experienced by the wires?

**17** In an experiment, two 50 cm long sections of wire are mounted parallel to each other. They exert a repulsive force of $4 \times 10^{-6}\,\mathrm{N}$ on each other when they each carry a current of 5 A.

  **a** How far apart are the wires spaced?

  **b** Are the currents in the wires in the same or opposite directions?

**18** A loose coil of wire of diameter $d$ has a current, $I$, running through it. The loops are separated by a vertical distance $h$.

  **a** Show that neighbouring turns in the coil attract each other. Use a diagram to help explain your answer.

  **b** Find an expression for the force that each turn of wire exerts on the turn below in terms of $I$, $d$ and $h$.

**19** Figure 6.21 shows a simple rail gun. The wire, of mass $m$, and length $d$, carrying a current $I$, experiences a force due to the magnetic field, $B$. The rails have length $L$.

  **a** Write an expression for the force acting on the wire.

  **b** Write an expression for the acceleration of the wire, assuming only the magnetic force is acting.

  **c** Write an expression for the speed of the wire as it leaves the rails.

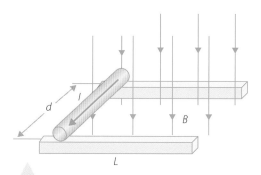

**FIGURE 6.21** A simple rail gun

**20** Figure 6.22 shows a current balance in which a loop of wire carrying a current of 3.6 A is balanced in the uniform field of a solenoid of field strength 0.20 T. The end of the loop BC has a length of 3.0 cm, while the length of side AB is 12.0 cm.

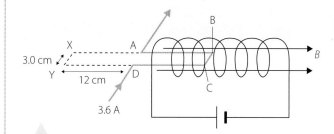

**FIGURE 6.22** A current balance

  **a** Find the magnitude and direction of the magnetic force on side:

    **i** BC of the loop.

    **ii** CD of the loop.

  **b** What length of string of mass $6.5\,\mathrm{g\,m^{-1}}$ must be placed on the end XY of the loop to restore equilibrium?

# 7 Electromagnetic induction

## INQUIRY QUESTION

How are electric and magnetic fields related?

## OUTCOMES

**Students:**

- describe how magnetic flux can change, with reference to the relationship $\Phi = B_{\parallel}A = BA\cos\theta$ (ACSPH083, ACSPH107, ACSPH109) ICT, N
- analyse qualitatively and quantitatively, with reference to energy transfers and transformations, examples of Faraday's Law and Lenz's Law $\varepsilon = -N\dfrac{\Delta\Phi}{\Delta t}$, including but not limited to: (ACSPH081, ACSPH110) ICT, N
  - the generation of an electromotive force (emf) and evidence for Lenz's Law produced by the relative movement between a magnet, straight conductors, metal plates and solenoids
  - the generation of an emf produced by the relative movement or changes in current in one solenoid in the vicinity of another solenoid
- analyse quantitatively the operation of ideal transformers through the application of: (ACSPH110) ICT, N
  - $\dfrac{V_P}{V_S} = \dfrac{N_P}{N_S}$
  - $V_P I_P = V_S I_S$
- evaluate qualitatively the limitations of the ideal transformer model and the strategies used to improve transformer efficiency, including but not limited to: CCT
  - incomplete flux linkage
  - resistive heat production and eddy currents
- analyse applications of step-up and step-down transformers, including but not limited to:
  - the distribution of energy using high-voltage transmission lines CCT

Getty Images/RainerPlendl

**FIGURE 7.1** Transformers are used to change voltage when electrical power is transmitted from a power station, via transmission lines to substations, and then via distribution lines to your home and school.

Many experiments in the 19th century showed that electricity and magnetism were intimately related. In this chapter we shall look at electromagnetic induction. Electromagnetic induction is the production of an electric field by a changing magnetic field. This is the basis for most electricity generation in Australia and elsewhere in the world. Kinetic energy is converted to electrical potential energy, which is used to drive a current. Electromagnetic induction is used in many technologies, including high-voltage electricity production and transmission (Figure 7.1), precise measuring devices, and communications technologies.

The discovery of electromagnetic induction was also important in the development of our understanding of the fundamental forces. It was the first indication that different forces, electricity and magnetism, were in fact parts of a single force. The development of electromagnetism was a crucial first step in the development of modern physics, including both relativity and quantum mechanics.

## 7.1 Magnetic flux

Electromagnetic induction is the production of an electric field by a time-varying magnetic field. An electric field in a region means that there is a potential difference between points in that region. If there is a potential difference and free charge carriers, then a current will be generated. This is the basis of electromagnetic induction, as used in electricity generation. To understand how the process works, we first need to look at the idea of magnetic flux.

Magnetic flux is a measure of the amount of magnetic field passing through an area. Recall that the density of field lines indicates the strength of a field (of any type), so flux is also a measure of how many field lines pass through an area.

Consider a uniform magnetic field, $B$, passing through an area, $A$, as shown in Figure 7.2. Both the size of the field and the size of the area will be important in determining the flux. Remember that field is a vector: it has direction as well as magnitude. So the angle between the field vector and the area is important, because it determines whether the field lines will cross the area. We define a vector, $A$, which has magnitude equal to the area and is directed normal to the area.

Mathematically, the magnetic flux, $\Phi$, is given by the product of the component of the field parallel to the vector $A$, and $A$:

$$\Phi = B_{\parallel}A = BA\cos\theta$$

The angle $\theta$ is defined to be the angle between the magnetic field vector, $B$, and the normal to the area. In Figure 7.2 the normal to the area is shown as vector $A$.

The magnetic flux, $\Phi$, has units of T m$^2$, which is given the name weber, Wb, after German physicist Wilhelm Weber who made important contributions to telecommunications.

The flux, $\Phi$, has maximum value when the field is perpendicular to the area. This occurs when the magnetic field vector, $B$, is parallel to the normal to the area, $A$, so that the angle $\theta$ is zero. The flux is zero when no field lines cross the area. This happens if the field is parallel to the surface of the area, so that $\theta = 90°$. In this case $B$ is perpendicular to the normal vector $A$.

**Wilhelm Weber**

Find out how Wilhelm Weber (1804–1891) contributed to the development of electromagnetism and communications technology.

**The vector dot product**

If you are doing Mathematics Extension II you will recognise the equation for flux as being the dot product of field and area. Find out how flux varies as the angle changes.

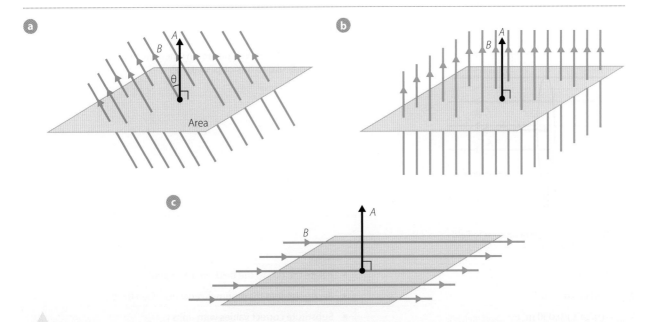

FIGURE 7.2  **a** The flux through the area depends on the angle, θ, between the normal to the area, shown as vector *A*, and the field, *B*.
**b** Flux is a maximum when θ is zero and the field is perpendicular to the area. **c** Flux is a minimum when θ = 90° and *B* is parallel to the area.

▶ **WORKED EXAMPLE** (7.1)

A loop of cross-sectional area $0.050\,\text{m}^2$ is in a uniform magnetic field of magnitude 0.24 T.

Numeracy

**a**  Draw diagrams showing the loop and field and identifying the angle, θ, between the area vector *A* and the field *B*, when the flux is a minimum and a maximum.

**b**  Find the maximum and minimum values of the flux through the loop.

| ANSWERS | LOGIC |
|---|---|
| $B = 0.24\,\text{T}, A = 0.050\,\text{m}^2$<br>The loop is circular. | ▪ Identify the relevant data in the question. |
| **a**<br><br>*(diagram of circular loop parallel to field lines, labelled B, A_out, θ = 90°)*<br><br>FIGURE 7.3  Loop is parallel to field so θ = 90° and flux is zero. | ▪ Recognise that flux is zero when the field is parallel to the loop, so θ = 90°. |

| ANSWER | LOGIC |
|---|---|
| 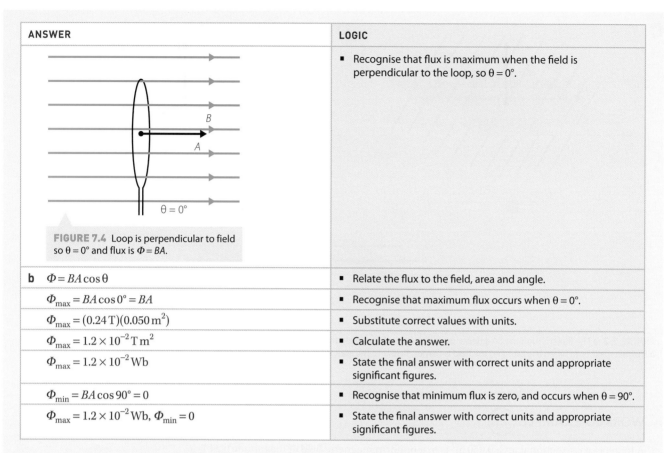 FIGURE 7.4 Loop is perpendicular to field so $\theta = 0°$ and flux is $\Phi = BA$. | <ul><li>Recognise that flux is maximum when the field is perpendicular to the loop, so $\theta = 0°$.</li></ul> |
| **b**   $\Phi = BA\cos\theta$ | <ul><li>Relate the flux to the field, area and angle.</li></ul> |
| $\Phi_{max} = BA\cos 0° = BA$ | <ul><li>Recognise that maximum flux occurs when $\theta = 0°$.</li></ul> |
| $\Phi_{max} = (0.24\,\text{T})(0.050\,\text{m}^2)$ | <ul><li>Substitute correct values with units.</li></ul> |
| $\Phi_{max} = 1.2 \times 10^{-2}\,\text{T m}^2$ | <ul><li>Calculate the answer.</li></ul> |
| $\Phi_{max} = 1.2 \times 10^{-2}\,\text{Wb}$ | <ul><li>State the final answer with correct units and appropriate significant figures.</li></ul> |
| $\Phi_{min} = BA\cos 90° = 0$ | <ul><li>Recognise that minimum flux is zero, and occurs when $\theta = 90°$.</li></ul> |
| $\Phi_{max} = 1.2 \times 10^{-2}\,\text{Wb},\ \Phi_{min} = 0$ | <ul><li>State the final answer with correct units and appropriate significant figures.</li></ul> |

**TRY THIS YOURSELF**

Find the angles for which the flux is half its maximum value.

Magnetic flux can change if any of the variables, $B$, $A$ or $\theta$, change. For example, if the magnitude of the magnetic field through the loop in Worked example 7.1 were to change, then the magnetic flux through the loop would change. If the loop itself changed, for example if it was bent into a different shape, then the flux would change. Finally, if the angle between the loop and the field changed, say by rotating the loop, the flux would change.

This third way of changing the flux, by rotating a coil, is how motors and generators work, as we shall see in the next chapter. Transformers, described in this chapter, work by changing the field.

Note that flux is a scalar: it has a magnitude but does not have a direction. Flux can be positive or negative, but for an open surface, such as the circle defined by the loop in Figure 7.4, which way is positive and which is negative is arbitrary. For a closed surface, such as a sphere, we typically define flux inwards as positive. Once we have chosen a direction for positive flux, we can define whether the change in flux is positive or negative. If a flux is positive and it increases, then the change in flux is also positive. If a positive flux decreases then the change in flux is negative.

The rate at which flux changes is constant if the flux is changing at a constant rate – in other words, if a graph of flux as a function of time is a straight line. If a graph of flux as a function of time is not a straight line, for example if it is a curve such as an exponential or sine curve, then the rate of change of flux varies with time too. This will be important when we look at induced emf in the next section.

Magnetic flux

9780170409131

The magnetic field in Worked example 7.1 decreases from 0.24 T to 0 T in 10 s. The loop is held perpendicular to the field. Assuming the rate at which the field decreases is constant, by how much does the flux change per second?

| ANSWERS | LOGIC |
|---|---|
| $B_i = 0.24\,\text{T}$, $B_f = 0\,\text{T}$, $\Delta t = 10\,\text{s}$, $A = 0.050\,\text{m}^2$ | • Identify the relevant data in the question. |
| $\Phi = BA\cos\theta$ | • Relate the flux to the field, area and angle. |
| $\Phi = BA$ | • Recognise that $\theta = 0°$ so $\cos\theta = 1$. |
| $\Delta\Phi_{1s} = (\Delta B_{1s})A$ | • Relate the change in flux to the change in field, recognising that area is constant. |
| $\dfrac{\Delta B}{\Delta t} = \dfrac{B_f - B_i}{\Delta t}$ | • Relate $\Delta B$ to the data given. |
| $\dfrac{\Delta B}{\Delta t} = \dfrac{0 - 0.24\,\text{T}}{10\,\text{s}}$ | • Substitute correct values to find the rate at which field decreases over the full 10 s. |
| $\dfrac{\Delta B}{\Delta t} = -0.024\,\text{T s}^{-1}$ | • Calculate the value. |
| $\Delta B_{1s} = \dfrac{\Delta B}{\Delta t}(1\,\text{s})$ | • Find the change in field in 1 s. |
| $\Delta B_{1s} = (-0.024\,\text{T s}^{-1})(1\,\text{s})$ | • Substitute correct values. |
| $\Delta B_{1s} = -0.024\,\text{T}$ | • Calculate change in field in 1 s. |
| $\Delta\Phi_{1s} = (-0.024\,\text{T})(0.050\,\text{m}^2)$ | • Substitute correct values into expression for change in flux. |
| $\Delta\Phi_{1s} = -1.2 \times 10^{-3}\,\text{T m}^2$ | • Calculate the answer. |
| The flux decreases by $1.2 \times 10^{-3}$ Wb in each second. | • State the final answer with correct units and appropriate significant figures. |

**TRY THIS YOURSELF**

If it took 20 s for the magnetic field to be reduced to zero, by how much would the flux change in each second?

● Magnetic flux, $\Phi$, through an area is given by $\Phi = B_{\parallel}A = BA\cos\theta$, where $A$ is the area in $\text{m}^2$, $B$ is the magnitude of the field in T, and $\theta$ is the angle between the field and the normal to the area.
● Flux is measured in units of weber, Wb. $1\,\text{Wb} = 1\,\text{T m}^2$.
● The flux through an area is a maximum when the field is perpendicular to the area, so $\theta = 0$.
● The flux is zero when the field lines are parallel to the area, so they do not cross it. This occurs when $\theta = 90°$.
● The flux through an area can be changed by changing the field, the area, or the angle between the field and the area.

**CHECK YOUR UNDERSTANDING**

**7.1**

1   What are the units of magnetic flux?

2   How is it possible for the magnetic flux through an area to be zero if neither the field nor the area is zero?

3   A rectangular loop of width 1.0 cm and length 2.0 cm is placed in a magnetic field of magnitude 55 μT.

   a   What is the maximum magnetic flux through the loop?

   b   The loop is turned through an angle of 45° from the position in which the flux is a maximum. What is the flux through the loop now?

4   A solenoid produces a magnetic field of 0.25 T in its interior. The field is approximately uniform. What is the radius of the coil, given that the flux through any loop of the solenoid is 5.0 mWb?

5   a   Use a spreadsheet to calculate the magnetic flux through an area of 0.010 m² in a magnetic field of 40 mT, for angles θ = 0°, 5°, … 360°.

    b   Plot a graph of flux as a function of angle.

    c   What is the change in flux when the angle changes from:

        i   0° to 30°?

        ii   30° to 60°?

        iii   60° to 90°?
             Read the changes from your graph.

6   A 1.0 m long length of wire is shaped into a circular loop and placed in a magnetic field of strength 0.050 T, so that the field is perpendicular to the loop. The loop is then bent into a square. By how much has the flux through the loop changed?

## 7.2 Electromagnetic induction

The ideas of potential and potential difference were introduced in *Physics in Focus Year 11* chapter 12. A potential difference exists in an electric field and is the difference in energy per unit charge between two points in an electric field. A potential difference is an example of an emf, or electromotive force. An emf is a difference in energy per unit charge, so has units of volts and is sometimes called a voltage. Even though it is called a force, it is actually an energy per unit charge, *not* a force. An emf provides the energy to make a current flow.

When the magnetic flux through an area changes with time, an electric field is induced. The induced electric field produces an induced emf. An induced emf is one that is created by a changing magnetic flux. If a loop of wire is placed in a magnetic field, and the flux through the loop changes, then an emf will be induced across the two ends of the loop.

### Faraday's Law

The magnitude of the induced emf caused by a changing magnetic flux is given by Faraday's Law:

$$\varepsilon = \frac{-(\Phi_f - \Phi_i)}{\Delta t} = \frac{-\Delta \Phi}{\Delta t}$$

where $\Phi_f$ is the final flux and $\Phi_i$ is the initial flux. $\varepsilon$ has units of $T\,m^2\,s^{-1}$, which is the same as volts, V.

Looking back at the expression for flux, we can see that:

$$\varepsilon = \frac{-\Delta(BA\cos\theta)}{\Delta t}$$

Inspecting this equation we can see that there are three ways to induce an emf:

▸   change the magnetic field, $B$

▸   change the area, $A$

▸   change the angle, $\theta$, between the area and the field.

In practice, it is usually either the magnetic field or the angle that is varied. For example, a coil connected to an alternating current (AC) produces a time-varying magnetic field. These are used in transformers and motors.

When a loop or coil of wire with area $A$ is placed in a field, the flux through the loop can be varied by spinning the loop. This changes the angle, and induces an emf in the loop. The same effect can be achieved by spinning a magnet near the loop. In both cases the flux varies in time. This is used in generators.

We can take any parameter kept constant out of the brackets. For example, if area and angle are kept constant while $B$ is varied, we can write:

$$\varepsilon = \frac{-A\cos\theta\Delta B}{\Delta t} = \frac{-A\cos\theta(B_f - B_i)}{\Delta t}$$

If the area and field are held constant but the angle is changed we can write:

$$\varepsilon = \frac{-BA\Delta(\cos\theta)}{\Delta t} = \frac{-BA(\cos\theta_f - \cos\theta_i)}{\Delta t}$$

To generate a larger emf, a coil containing multiple loops of wire is used. Each loop will have an emf induced between its ends, so connecting $N$ loops in series is like connecting $N$ batteries in series. Simply add the emf in all loops to find the total emf. Thus:

$$\varepsilon = -N\frac{\Delta\Phi}{\Delta t} = -N\frac{\Delta(BA\cos\theta)}{\Delta t}$$

Once an emf is induced, a current will flow if there are free charge carriers and a path for them to flow along. This is usually achieved by putting a metal coil in the field. This **induced current** is related to the emf by Ohm's Law (chapter 13 of *Physics in Focus Year 11*): $\varepsilon = IR$.

## Emf and potential difference

Emf is the energy per unit charge available to a charged particle. We use the term 'emf' here rather than potential difference for a reason. Potential difference and emf are often treated as if they are the same thing, but they are not. They do the same job, which is to enable a current to flow, and they have the same unit, volt (V), but they differ in an important way.

Potential difference is the *unique difference* in potential energy per unit charge between two points in an electric field. If a charged particle moves from point A to point B in an electric field then its change in energy is given by $q(V_B - V_A)$ *regardless of the path* the particle took (Figure 7.5a).

The **induced emf** between any two points in a changing magnetic field *does not* have a unique value, but *depends on the path* between the two points. This is because it depends on the flux enclosed by the path. Different closed paths between two points may contain different fluxes, and hence give different emfs (Figure 7.5b).

Potential difference is a type of emf, but not all emfs (for example induced emfs due to a changing magnetic flux) are potential differences.

Information and communication technology capability

**Faraday's Law**
Watch the simulation, then follow the video lecture and do the worked example.

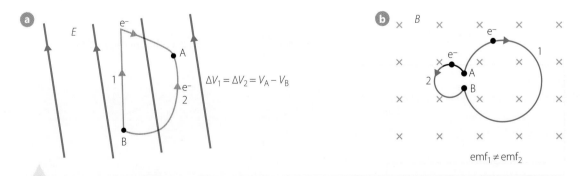

**FIGURE 7.5 a** Potential difference: an electron moves from point A to point B in an electric field. The change in potential energy of the electron is the same regardless of path taken. **b** Induced emf: an electron passes through two loops in a changing magnetic field. The emf measured across the two loops is different because they contain different magnetic fluxes, even though they have the same beginning and end points.

A wire loop of cross-sectional area $0.050\,\text{m}^2$ is in a magnetic field. The loop is perpendicular to the field. The field changes with time as shown in Figure 7.6.

**a** Sketch a graph of flux through the loop as a function of time.

**b** Sketch a graph of $\dfrac{\Delta\Phi}{\Delta t}$ as a function of time.

**c** Find the emf induced between the ends of the wire loop.

**d** Find the current induced in the loop when the loop has a resistance of $0.15\,\Omega$.

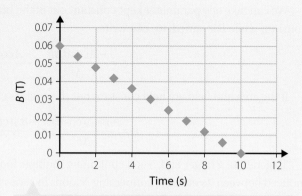

**FIGURE 7.6** Magnetic field as a function of time

| ANSWER | LOGIC |
|---|---|
| $\theta = 90°,\ A = 0.050\,\text{m}^2,\ R = 0.15\,\Omega$ | ▪ Identify the relevant data in the question. |
| $\Phi = BA\cos\theta = BA$ | ▪ Relate the flux to the field, area and angle, recognising that because $\theta = 90°$, $\cos\theta = 1$. |
| **a** <br><br>**FIGURE 7.7** Flux as a function of time | ▪ Sketch a graph of flux as a function of time using $\Phi = BA$ (Figure 7.7), and taking the values of $B$ from Figure 7.6. |
| **b** $\dfrac{\Delta\Phi}{\Delta t}$ is the gradient of the $\Phi(t)$ graph, which is constant. | |
| $\dfrac{\Delta\Phi}{\Delta t} = \dfrac{\Phi_f - \Phi_i}{t_f - t_i} = \dfrac{(0 - 0.003)\,\text{T m}^2}{(10 - 0)\,\text{s}}$ <br><br> $\dfrac{\Delta\Phi}{\Delta t} = -3.0 \times 10^{-4}\,\text{T m}^2\,\text{s}^{-1}$ | ▪ We find the gradient by taking the rise over run for a section of the $\Phi(t)$ graph. |
| **FIGURE 7.8** $\dfrac{\Delta\Phi}{\Delta t}$ as a function of time | ▪ Plot $\dfrac{\Delta\Phi}{\Delta t}$ vs time, which is a constant (Figure 7.8). |

| ANSWERS | LOGIC |
|---|---|
| c $\quad \varepsilon = -\dfrac{\Delta \Phi}{\Delta t}$ | ■ Relate emf to rate of change of flux. |
| $\varepsilon = -(-3.0 \times 10^{-4}\,\mathrm{T\,m^2\,s^{-1}})$ | ■ Substitute value with units. |
| $\varepsilon = 3.0 \times 10^{-4}\,\mathrm{V}$ | ■ State the final answer with correct units. |
| d $\quad I = \dfrac{\varepsilon}{R}$ | ■ Use Ohm's Law to relate emf to current. |
| $I = \dfrac{3.0 \times 10^{-4}\,\mathrm{V}}{0.15\,\Omega}$ | ■ Substitute values with units. |
| $I = 2.0 \times 10^{-3}\,\mathrm{V\,\Omega^{-1}}$ | ■ Calculate the answer. |
| $I = 2.0\,\mathrm{mA}$ | ■ State the final answer with correct units and appropriate significant figures. |

**TRY THIS YOURSELF**

How quickly would the field have to drop to zero to produce an induced current of 0.1 A?

# INVESTIGATION  7.1

## Electromagnetic induction

Critical and creative thinking

Numeracy

### AIM

To investigate the current produced in a coil by a changing magnetic field
Write an inquiry question or hypothesis for this investigation.

### MATERIALS

- Strong bar magnet
- Coil
- Sensitive ammeter or centre-zero galvanometer

RISK ASSESSMENT

| WHAT ARE THE RISKS IN DOING THIS INVESTIGATION? | HOW CAN YOU MANAGE THESE RISKS TO STAY SAFE? |
|---|---|
| Magnets can break or become demagnetised when dropped. | Be careful with the magnets and put them down when not in use. |

What other risks are associated with your investigation, and how can you manage them?

### METHOD

Note that it may be difficult to read the magnitude of the current in this experiment because the needle can swing quickly. You may have to simply note the current as 'large' or 'small'.

1  Connect the coil to the ammeter.
2  Slowly move the north pole of the magnet into one end of the coil and observe what happens to the ammeter or galvanometer.

»

3  Observe the ammeter or galvanometer when the north pole is held stationary within the coil.

4  Slowly pull the magnet out again. Record what happens to the ammeter or galvanometer – note the magnitude and direction of the current.

5  Quickly move the north pole of the magnet into one end of the coil. Note what happens to the ammeter or galvanometer. Pull it out again quickly and again observe the ammeter or galvanometer.

6  Predict what you will see if you put the south pole of the magnet into the coil instead. When you have written down your prediction, repeat steps 2–5 with the south pole moving into the coil.

### RESULTS

Record your results for each of steps 2–5 above. You should have noted the direction and approximate magnitude of the current. Record your results in a table such as the one shown below.

| EXPERIMENT | CURRENT DIRECTION | CURRENT MAGNITUDE |
| --- | --- | --- |
| N pole moving in slowly | | |
| N pole held still | | |
| N pole moving out slowly | | |
| N pole moving in quickly | | |
| N pole moving out quickly | | |
| S pole moving in slowly | | |
| S pole held still | | |
| S pole moving out slowly | | |
| S pole moving in quickly | | |
| S pole moving out quickly | | |

### ANALYSIS OF RESULTS

1  Sketch a graph of the current as a function of time as the magnet was moved in and out of the coil. Write down what was happening with the magnet along the time axis.

2  Draw a diagram showing the magnetic field of the magnet.

3  Relate the direction of the current to the changing magnetic flux through the coil.

### DISCUSSION

1  How is the magnitude of the current related to the speed of the magnet?

2  How does the direction of current vary with the changing magnetic flux?

3  Give the answer to your inquiry question or state whether your hypothesis was supported.

### CONCLUSION

Write a conclusion summarising the outcomes of your investigation.

## Lenz's Law

The negative sign in the equation for induced emf tells us about the direction of the induced current. Consider a loop that is in a magnetic field that is getting stronger with time, as shown in Figure 7.9. The negative sign tells us that the current must flow such that the current decreases the flux through the loop.

This is a result of conservation of energy (chapter 5 of *Physics in Focus Year 11*). The potential energy of the changing magnetic field is transformed into electric potential energy. The result is an electric field and, as we saw in chapter 6, and in chapter 12 of *Physics in Focus Year 11*, an electric field can do work by applying a force to a charged particle. Work is done on any free electrons by the induced electric field. The electrons then flow, creating the induced current. The induced current takes its energy from the changing magnetic flux (via the electric field), and so *reduces* the rate at which the flux changes.

Consider what would happen if the current flowed so that it produced a further increase in the magnetic flux through the loop. The flux would increase more, giving a bigger induced current, giving a bigger flux and so on. We would have a 'perpetual motion machine' that made more and more current without any source of energy. This would violate conservation of energy and so cannot happen. So the current must flow in the other direction and act to decrease the magnetic flux through the loop.

Lenz's Law states that an induced emf acts to produce an induced current in the direction that causes a magnetic flux that opposes the change in flux that induced the emf.

Lenz's Law is essentially a statement of *conservation of energy*.

When the magnet is moved towards the stationary conducting loop, a current is induced in the direction shown. The magnetic field lines are due to the bar magnet.

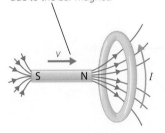

This induced current produces its own magnetic field directed to the left that counteracts the increasing external flux.

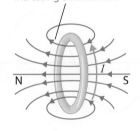

When the magnet is moved away from the stationary conducting loop, a current is induced in the direction shown.

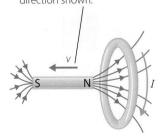

This induced current produces a magnetic field directed to the right and so counteracts the decreasing external flux.

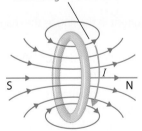

**FIGURE 7.9** A moving bar magnet induces a current in a conducting loop. The direction of the current is determined by Lenz's Law.

## Eddy currents

Induced currents are seen not only in wire loops, but also in any material in which there are free charge carriers. If a magnet is moved around over a piece of metal, the changing magnetic field will induce an emf in the metal, and the emf will induce **eddy currents** in the metal. The electrons move in circles in the region where the magnetic field is changing. They form loops and spirals of current, like eddies in a cup of tea when you stir it. These eddy currents create magnetic fields that oppose the changing flux from the moving magnet (Lenz's Law). These induced magnetic fields act to slow down or brake the magnet. This is called **magnetic braking**, and is described in the next chapter. The currents also result in heating of the material because of the resistance of the material.

Electromagnetic induction

**KEY CONCEPTS**

- A changing magnetic flux through a region induces an emf (electromotive force).
- An emf is the energy available per unit charge. An induced emf can create an induced current if there are free charge carriers and a path for them to flow along.
- The induced emf created by a changing magnetic flux is given by Faraday's Law: $\varepsilon = -N\dfrac{\Delta \Phi}{\Delta t}$.
- The flux can change if the magnetic field, area or angle between the area and the field changes.
- Lenz's Law states that the induced emf is such that it opposes the change in flux that created it.
- An induced current creates a magnetic field that acts to oppose the change in flux.

**CHECK YOUR UNDERSTANDING**

**7.2**

1  Define 'electromagnetic induction'.

2  A loop of wire is placed in a uniform magnetic field and moved in a straight line. In this case, no current is induced in the loop. Why not?

3  Describe the three ways in which an emf can be induced in a loop in a magnetic field.

4  Figure 7.10 shows the flux through a loop as a function of time.

    **a**  Sketch the rate of change of flux through the loop, $\dfrac{\Delta \Phi}{\Delta t}$. Use the same time scale.

    **b**  Sketch the induced emf across the loop as a function of time. Use the same time scale.

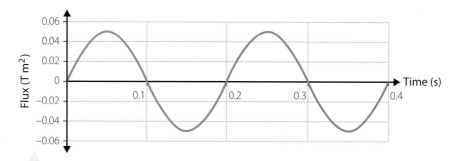

**FIGURE 7.10**  Flux through a loop as a function of time

5  Draw a flow chart that summarises Lenz's Law for a magnet being pushed into a solenoid (coil).

6  A magnet is held, south pole down, above a copper ring, as in Figure 7.11, and dropped.

    **a**  What is the direction of current in the ring as the magnet approaches the ring?

    **b**  What is the direction of current in the ring just after the magnet has passed through the ring?

7  A square loop of side length 10 cm is in a magnetic field of magnitude 0.10 T. The loop is rotated so that it takes 0.10 s to turn from parallel to the field to perpendicular to the field. Calculate the average emf induced across the loop as it is rotated.

8  A loop of cross-sectional area 0.050 m$^2$ is in a variable magnetic field. At what average rate does the field need to change to induce an emf of 1.0 V between the ends of the loop? Assume the loop stays perpendicular to the field.

**FIGURE 7.11**  A magnet is held above a copper ring and dropped.

# Transformers

We have seen how a changing magnetic field, for example caused by moving a magnet near a coil, induces an emf, and how this can be used to create an induced current.

Remember from chapter 14 of *Physics in Focus Year 11* that current-carrying wires create magnetic fields. To create a large, uniform field, a current-carrying solenoid is used. If the current varies in the solenoid, then the magnetic field also varies and an emf is induced. This emf can then be used to induce a current in a second solenoid. This is how transformers work.

Most appliances use a transformer to convert the 230 V mains power to a lower voltage. The transformer may be inside the device, or it may be a separate plug pack. There are also transformers at electricity substations. These drop the potential difference from the thousands of volts at which it is transmitted to the 230 V that is supplied to homes and businesses.

> Note that until the 1980s the standard Australian mains power supply was 240 V, but it is now 230 V to bring Australia into line with international standards.

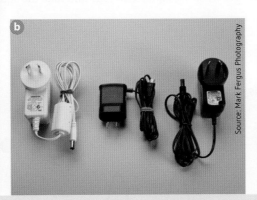

**FIGURE 7.12** Some different types of transformers: **a** high-voltage mains, **b** plug packs; **c** laboratory bench power supply

A transformer consists of two solenoids, or coils, of wire placed near each other so that an alternating current (AC) in the primary coil can induce a current in the secondary coil. An alternating current is a current that changes direction periodically. This means the magnetic field changes, and hence an emf is induced. If a direct current (DC) (constant current) was used, the field would be constant and no emf would be induced.

## AC current

AC currents vary sinusoidally with time (Figure 7.13).

We describe a sinusoidal current in terms of its amplitude and variation with time. The amplitude is the maximum current, $I_{max}$, sometimes also called the peak current $I_{peak}$. The difference between the maximum and minimum currents is the peak-to-peak current, $I_{peak-peak}$. Current is measured in A, so $I_{peak}$ and $I_{peak-peak}$ have units A. The time it takes for the current to go from its maximum value through a complete cycle and back to its maximum again is the period, $T$, measured in seconds (Figure 7.13a). The time variation can also be described in terms of the frequency, $f$, which is the number of cycles per second: $f = \dfrac{1}{T}$. Frequency has units of hertz, Hz.

Remember that currents are driven by emfs, so where an AC current is flowing, it is driven by a sinusoidal emf. For emf, the relevant quantities are peak emf, $V_{peak}$, peak-to-peak emf, $V_{peak-peak}$, period, $T$, and frequency, $f$. These are shown graphically in Figure 7.13b.

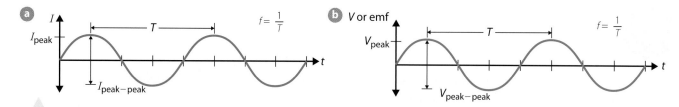

**FIGURE 7.13 a** Alternating current varies sinusoidally with time. The current, $I$, has a peak value $I_{peak}$ that is half the peak-to-peak value $I_{peak-peak}$. It describes one cycle in one period of time, $T$. **b** The voltage or emf that drives an AC current is also sinusoidal and can be described by its peak voltage, $V_{peak}$, peak-to-peak voltage, $V_{peak-peak}$, and period, $T$.

AC current and emf values vary between a peak positive value and a peak negative value, oscillating back and forth in each cycle. The average of the AC emf over one cycle is zero, yet the AC emf delivers energy during that time; that is, it delivers power to a circuit.

Power is proportional to the square of the emf (chapter 13 of *Physics in Focus Year 11*). If we square the emf and find the average, we can get a value for the average power. To convert this to a single emf that would deliver the same power as the original AC emf, we take the square root of this average. The single value of emf that we get when we square, average and take the square root is called the **root mean square** or **rms** value:

$$V_{rms} = \frac{V_{peak}}{\sqrt{2}}$$

Figure 7.14 shows this process graphically.

**AC and rms values**

More details of AC signals and how to find rms values

**FIGURE 7.14 a** A sinusoidal emf, $V(t)$; **b** We square this to get $V^2(t)$. This has a peak value of $(V_{peak})^2$. **c** The average value of $V^2(t)$, which has the value $\frac{(V_{peak})^2}{2}$. **d** Taking the square root of this value gives us the rms value: $V_{rms} = \frac{V_{peak}}{\sqrt{2}}$.

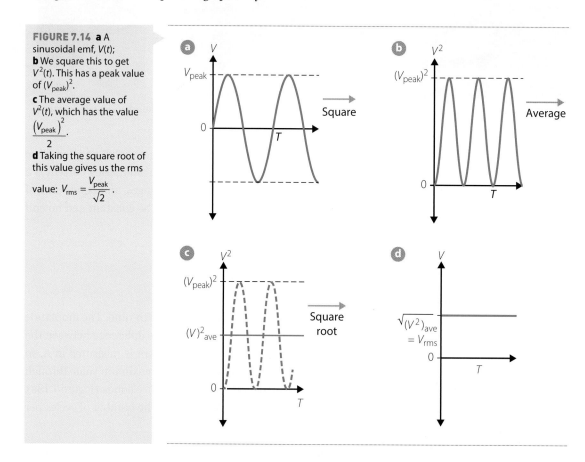

Similarly, $I_{rms} = \dfrac{I_{peak}}{\sqrt{2}}$ .

The rms emf is an average AC emf that produces the same power in a resistive component as a constant DC potential difference of the same magnitude. AC systems are usually described using rms values. For example, the 230 V mains power supply to your home provides an rms potential difference of 230 V between the active and neutral connections (Figure 7.15).

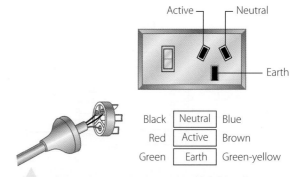

FIGURE 7.15 An Australian mains power socket and plug. The potential difference between the active and neutral is 230 $V_{rms}$. The earth is connected to the ground and is at 0 V.

### WORKED EXAMPLE 7.4

The mains supply in Australia provides an AC potential difference of 230 $V_{rms}$. To what peak value of potential difference does this correspond?

| ANSWERS | LOGIC |
|---|---|
| $V_{rms} = 230\,V$ | ▪ Identify the relevant data in the question. |
| $V_{rms} = \dfrac{V_{peak}}{\sqrt{2}}$ | ▪ Relate the rms potential difference to the peak potential difference. |
| $V_{peak} = \sqrt{2}\,V_{rms}$ | ▪ Rearrange for $V_{peak}$. |
| $V_{peak} = \sqrt{2}\,(230\,V)$ | ▪ Substitute value with units. |
| $V_{peak} = 325\,V$ | ▪ Calculate the answer. |
| $V_{peak} = 330\,V$ | ▪ State the final answer with correct units and appropriate significant figures. |

**TRY THIS YOURSELF**

The mains power supply typically fluctuates between 220 $V_{rms}$ and 250 $V_{rms}$. What peak potential differences do these values correspond to?

## Flux linkage in a transformer

The link between the input solenoid and the output solenoid in a transformer is by electromagnetic induction or flux linkage; there is no electrical connection. Flux linkage is how much flux from one coil passes through the other coil.

The coils need to be coupled so that the changing magnetic field in the primary coil causes a changing magnetic flux in the secondary coil (flux linkage). There are two ways of doing this. First, the coils can share the same space by placing one within the other. This is sometimes used in cordless appliances such as kettles. The second, and more usual, way is to link the coils using a ferromagnetic core.

Figure 7.16 shows how this works. The primary coil is wound around one side of a ferromagnetic core (see chapter 14 of *Physics in Focus Year 11*), typically iron. The current in the primary coil magnetises the whole core, not just the part within the primary coil, thus providing flux linkage from the primary coil to the secondary coil.

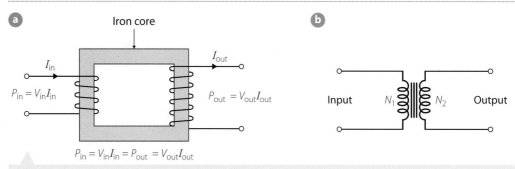

FIGURE 7.16 **a** Schematic diagram of a transformer; **b** The circuit symbol for a transformer

The alternating current in the primary coil creates a sinusoidally time-varying magnetic field, which magnetises the whole core. So the time-varying current in the primary coil creates a time-varying field in the secondary coil. The time-varying field in the secondary coil means that the flux is varying with time in the secondary coil. A time-varying flux, from Faraday's Law, creates an emf. This emf then drives a current in the secondary coil.

The overall effect is that an AC current in the primary coil induces an AC current in the secondary coil. The two currents have the same period, but not necessarily the same magnitude. The relationship between the magnitudes of the currents depends on the geometry of the coils.

For an ideal transformer, we assume the flux linkage is perfect so that the flux through any loop is the same for both coils. If the primary coil has $N_P$ turns then:

$$V_P = -N_P \frac{\Delta \Phi}{\Delta t}$$

where the subscript P refers to the primary coil. Similarly, $V_S = -N_S \dfrac{\Delta \Phi}{\Delta t}$, where the subscript S refers to the secondary coil.

Because $\dfrac{\Delta \Phi}{\Delta t}$ is the same for both coils:

$$\frac{V_S}{V_P} = \frac{N_S}{N_P}$$

Again assuming an ideal transformer so that the transformer is 100% efficient, power out = power in, $P_{out} = P_{in}$. Recall from *Physics in Focus Year 11* chapter 13 that $P = VI$, so:

$$P_{in} = I_p V_p = I_s V_s = P_{out}$$

So:

$$\frac{I_P}{I_S} = \frac{N_S}{N_P}$$

We can summarise these equations as:

$$\frac{V_S}{V_P} = \frac{I_P}{I_S} = \frac{N_S}{N_P}$$

These are known as the transformer equations.

The equation $\dfrac{I_P}{I_S} = \dfrac{N_S}{N_P}$ can be rearranged to give the current in the secondary coil as a function of the current in the primary coil:

$$I_S = \frac{I_p N_P}{N_S}$$

This tells us that if there are more turns in the primary coil than in the secondary coil ($N_P > N_S$), the current through the secondary (the output current) will be greater than the current in the

9780170409131

primary (the input current). This may seem as if it violates the law of conservation of energy, but remember that the emf also changes:

$$V_S = \frac{V_P N_S}{N_P}$$

So if the current is increased by the transformer, the emf is decreased.

**WORKED EXAMPLE** (7.5)

A transformer has a primary coil with 10 000 turns. An AC current of 5.0 A flowing in the primary coil induces a current of 1.0 A in the secondary coil.

**a** What is the number of turns in the secondary coil?

**b** What is the ratio of the potential differences across the secondary coil and the primary coil?

| ANSWERS | LOGIC |
|---|---|
| $I_P = 5.0\,A$, $I_S = 1.0\,A$, $N_P = 10\,000$ | ▪ Identify the relevant data in the question. |
| **a** $\quad I_S = \dfrac{I_P N_P}{N_S}$ | ▪ Relate the currents in the coils to the number of turns in the coils. |
| $N_S = \dfrac{N_P I_P}{I_S}$ | ▪ Rearrange for $N_S$. |
| $N_S = \dfrac{10\,000 \times 5.0\,A}{1.0\,A}$ | ▪ Substitute values with units. |
| $N_S = 50\,000$ | ▪ Calculate the answer. State the final answer with correct units and appropriate significant figures. |
| **b** $\quad \dfrac{V_S}{V_P} = \dfrac{N_S}{N_P}$ | ▪ Relate the ratio of potential differences to the ratio of number of turns in the coils. |
| $\dfrac{V_S}{V_P} = \dfrac{50\,000}{10\,000}$ | ▪ Substitute values with units. |
| $\dfrac{V_S}{V_P} = 5$ | ▪ Calculate the answer. |
| The potential difference across the secondary coil is five times that across the primary coil. | ▪ State the final answer. |

**TRY THIS YOURSELF**

A transformer has a primary coil with 10 000 turns and secondary coil with 520 turns.

**1** Find the ratio of current in the secondary coil to current in the primary coil.

**2** If the input (primary) potential difference is 230 V, what is the output (secondary) potential difference?

## Step-up and step-down transformers

The transformer in Worked example 7.5 has more turns on the secondary coil than on the primary coil. The ratio of potential differences across the primary and secondary is equal to the ratio of turns in the primary and secondary, $\dfrac{V_S}{V_P} = \dfrac{N_S}{N_P}$. So for this transformer the output voltage (potential difference or emf) is greater than the input voltage. This type of transformer is called a **step-up transformer**. Conservation of energy is not violated by this increase in potential difference. Remember that potential difference is change in energy per unit charge. As energy must be conserved, if the energy per charge is increased, the number of charges that are moved, in other words the current, must be decreased. This can be seen from the other part of the transformer equation: $\dfrac{I_S}{I_P} = \dfrac{N_P}{N_S}$.

Step-up transformers are used where a higher potential difference is required, for example to convert the 12 V supply from a car's alternator to 230 V to run an appliance. They are also used at power stations to increase the emf produced by generators to the high voltages that are then transmitted by long-distance transmission lines.

**Step-down transformers** are far more common. They are used to decrease potential difference. A step-down transformer has more turns on the primary than the secondary, so the current is greater and the potential difference lower for the secondary coil.

Most small appliances use a step-down transformer. Chargers, such as phone chargers, use a step-down transformer to decrease the supply voltage to that needed by the phone. They also use a rectifier circuit to change the current from alternating to direct (constant). Rectifiers are described in more detail in chapter 8.

Step-down transformers are also used in electricity distribution networks. At a power plant, a step-up transformer is first used to bring the potential difference up from that produced to a very high voltage, typically at around 100 kV or more. The higher voltages result in lower energy loss when electrical potential energy is transmitted through the long-distance transmission lines to substations.

At a substation, a step-down transformer reduces the voltage. The output from the substation connects to distribution lines. The distribution lines can carry high or low voltage power. The high-voltage lines, used to go longer distances or to businesses requiring high voltages, can carry a range of voltages. The type of pole and insulator used depends on the voltage. Higher voltage lines use larger insulators and higher poles.

In a city you might see power lines carried at two levels on the one pole. The higher level is the higher voltage, typically around 8 kV, and the lower connects to people's houses and is the 230 V supply.

In regional areas, outside towns, you may notice that every house has its own transformer on a pole. The distribution lines have to be much longer in regional areas, so they carry high voltages. High voltages (around 10–50 kV) from the local substation, usually close to a nearby town, are carried on distribution wires. A wire runs from the distribution line to the transformer. This is a step-down transformer, which reduces the voltage to 230 V. A wire then runs from this transformer to the house.

Information and communication technology capability

**How the grid works**
Find out how electricity gets from a power plant to your home.

**Power line types**
Use this guide to identify different high and low voltage powerlines. Similar types are used across Australia, including New South Wales.

▶ **WORKED EXAMPLE** (7.6)

A step-down transformer supplies 230 V to a house. The house is drawing 15 A to run all the various lights and appliances that are turned on. If the input voltage to the transformer is 10 kV, what current is flowing in the primary coil?

| ANSWER | LOGIC |
|---|---|
| $I_S = 15\,\text{A},\ V_P = 10\,000\,\text{V},\ V_S = 230\,\text{V}$ | ▪ Identify the relevant data in the question. |
| $\dfrac{V_S}{V_P} = \dfrac{I_P}{I_S}$ | ▪ Relate the currents in the coils to the number of turns in the coils. |
| $I_P = \dfrac{I_S V_S}{V_P}$ | ▪ Rearrange for $I_P$. |
| $I_P = \dfrac{15\,\text{A} \times 230\,\text{V}}{10\,000\,\text{V}}$ | ▪ Substitute values with units. |
| $I_P = 0.345\,\text{A}$ | ▪ Calculate the answer. |
| $I_P = 0.35\,\text{A}$ | ▪ State the final answer with correct units and appropriate significant figures |

1  If the primary coil is carrying a current of 1.0 A, how much current is the house drawing?

2  What is the ratio of turns, $\dfrac{N_S}{N_P}$, for this transformer?

## Real transformers

The transformer equations derived above were for ideal transformers and made two important assumptions. The first assumption was that the flux transfer or linkage was perfect, so that the flux through the secondary coil was exactly the same as the flux through the primary coil. The second was that the transformer was 100% efficient and no energy was lost.

The flux linkage is never completely perfect, and there is always some 'stray field'. This field, which is external to the core, can induce eddy currents in nearby materials, including the case or supports for the transformer. These eddy currents cause heating, and can even make these materials vibrate; they cause the buzzing sound that you can sometimes hear near large transformers. So energy is lost as heat, and sometimes also as sound.

If you have ever gone to unplug a transformer and noticed that it is warm, you will know that the second assumption is not true for real transformers either. Not all the energy supplied to the primary coil is available as electrical potential energy from the secondary coil. Some is lost as heat. This is why your phone charger gets warm as it charges your phone. This heating occurs both in the wires of the transformer coils and in the core. The resistance of the wires causes some energy to be lost and dissipated as heat. Ohm's Law (chapter 13 of *Physics in Focus Year 11*) tells us that the higher the resistance of the wires, the more energy is dissipated as heat. Reducing the resistance of the coils reduces this heating. The second cause of heating is eddy currents in the ferromagnetic core. Lenz's Law tells us that eddy currents will form to oppose the change in magnetic flux. The formation of these eddy currents can be reduced by making the core out of many thin layers of ferromagnetic material glued together with an insulating material (laminating). Laminated cores are more efficient than single lumps of ferromagnetic material (Figure 7.17).

Energy is also lost in the core due to magnetisation. Each time the current changes direction, so does the magnetic field. The core is magnetised to give a field in one direction, and then de- and re-magnetised to give a field the other way, over and over again. Each time a small amount of energy is lost, and again this appears as an increase in the temperature of the core. New and more efficient magnetic materials are being developed that can be magnetised and demagnetised more efficiently.

Transformers

**FIGURE 7.17** The laminated core of this transformer, visible on the outside as lots of thin layers, helps reduce eddy currents and so makes the transformer more efficient.

## Transformers

Critical and creative thinking

Numeracy

### AIMS

To observe the effect of flux changes in the secondary coil of a transformer caused by changing potential difference across the primary coil

To compare the ratio of the potential differences to the turns ratio in both step-up and step-down transformers

Write an inquiry question for your investigation.

### MATERIALS

- 2 air-core solenoids with known number of turns
- Iron rod
- Variable AC/DC power supply or separate AC and DC power supplies
- Alligator leads
- Oscilloscope (CRO)

| WHAT ARE THE RISKS IN DOING THIS INVESTIGATION? | HOW CAN YOU MANAGE THESE RISKS TO STAY SAFE? |
|---|---|
| Power supplies can cause electric shocks. | Do not use or induce potential differences above 12 V. Make sure your equipment is checked by your teacher before use. |

What other risks are associated with your investigation, and how can you manage them?

### Part A

### METHOD

1   Place the air-core solenoids end to end with the iron rod placed inside them, as shown in Figure 7.18.
2   Connect one solenoid to the DC power supply set to 2 V.
3   Connect the other solenoid to the oscilloscope.
4   Flick the switch on the DC power supply off and on.
5   Record any change in the trace on the oscilloscope. You may need to adjust the time base and the voltage scale.
6   Turn the power supply on and observe what happens.
7   Now turn it off and observe what happens. Record your results.

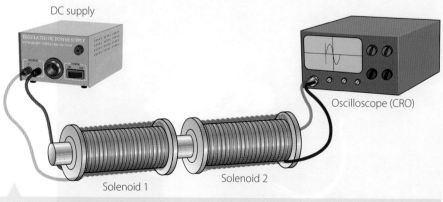

DC supply

Oscilloscope (CRO)

Solenoid 1

Solenoid 2

**FIGURE 7.18** Experimental set-up for part A

RESULTS

1 Draw a diagram showing what happened when you flicked the switch on and off.

2 Draw a diagram showing what happened when you turned the switch on and left it on.

**Part B**

METHOD

1 Change the power supply to 2V AC or swap the DC supply for a variable AC supply.

2 Turn on the power supply and, by looking at the CRO trace, observe what happens in the second solenoid.

3 Increase the supply voltage by 2V and observe how the output changes.

4 Exchange the two solenoids and repeat your measurements for steps 2 and 3.

RESULTS

1 Sketch the CRO trace for each measurement.

2 Record the maximum output voltage at the secondary coil.

ANALYSIS OF RESULTS

1 For part B, calculate the rms output voltage for each measurement.

2 Calculate the expected rms output voltage using the transformer equations, the input voltage and the number of turns in each coil.

DISCUSSION

1 Describe how the results of part A differed from those of part B. For a DC potential difference applied across the primary coil, when does the flux change? What happens in the secondary coil?

2 How well did the transformer equations predict the results of your experiment? How can you explain any discrepancies?

3 How could you reduce experimental uncertainties in this experiment?

4 State whether your hypothesis was supported or not.

CONCLUSION

Write a conclusion summarising the outcomes of your investigation.

---

**KEY CONCEPTS**

- Transformers use electromagnetic induction to change current and emf.
- Transformers consist of two coils that are magnetically linked (flux linkage).
- An alternating current in the primary coil creates a changing magnetic field in the secondary coil, which induces an emf and hence a current in the secondary coil.
- An AC current is one that varies sinusoidally with time. An AC current can be described by its maximum value $I_{peak}$ or its rms value, $I_{rms} = \dfrac{I_{peak}}{\sqrt{2}}$.
- The primary (input) and secondary (output) emfs and currents for a transformer are related to the number of turns of wire on each coil, and are given by the transformer equations: $\dfrac{V_P}{V_S} = \dfrac{N_P}{N_S}$ and $V_P I_P = V_S I_S$.
- A step-down transformer has a lower emf across its secondary coil than across its primary coil. A step-up transformer has a larger emf across its secondary coil than across its primary coil.
- An AC emf is provided as the household mains power supply in Australia with $V_{rms} = 230V$. Larger emfs are carried by transmission and distribution lines.
- Both step-down and step-up transformers are used at different stages of electricity distribution, but step-down transformers are more common.
- The transformer equations are an approximation that assumes 100% efficiency. Real transformers lose energy as heat because of eddy currents in the core, resistive heating in the coils and incomplete flux linkage.

1 What is the difference between a step-up and a step-down transformer?

2 Why is an alternating current necessary for a transformer?

3 Why must the output current in a step-up transformer be less than the input current?

4 Transformers are not 100% efficient. Their efficiency can be improved by using a laminated core; that is, a core made from thin slices of iron sandwiched together with glue. Explain how this reduces energy loss in a transformer.

5 For a transformer with a 1000-turn primary coil, what potential difference is available at the 200-turn secondary coil when the primary coil is supplied with 230V AC?

6 A transformer is connected to an AC source that can deliver 30 A. The secondary coil of the transformer can deliver a maximum current of 10 A.

   a What type of transformer is this?

   b Calculate the ratio between the number of turns in the primary coil and the number of turns in the secondary coil.

7 Draw a diagram showing the process by which electricity is transmitted from a power station to your home. Indicate whether the voltages are high or low at different stages, and show how transformers are used at different places.

8 Figure 7.19 shows the input potential difference to a transformer. Sketch a graph of the output potential difference.

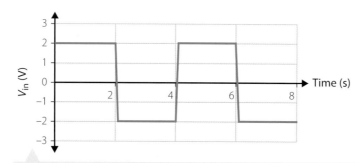

**FIGURE 7.19** Input to the primary coil of a transformer

9 A step-up transformer is connected to an AC generator that delivers 8.0 A at 120V. The ratio of the number of turns in the secondary coil to the number of turns in the primary coil is 500.

   a What is the emf across the secondary coil?

   b What is the power input?

   c What is the maximum power output?

   d What is the maximum current in the secondary coil?

10 A step-down transformer is connected to a 230V 50 Hz AC mains supply. There are 1200 turns in the primary coil. The secondary coil has three terminals, P, Q and R, as shown in Figure 7.20. When a 10 Ω resistor is connected between terminals P and Q, the current flow in the resistor is 0.60 A. When the same resistor is connected between terminals Q and R, the current flow is 1.0 A.

   a What is the potential difference between P and Q?

   b What is the potential difference between Q and R?

   c Find the number of turns in the secondary coil between terminals:

      i P and Q.

      ii Q and R.

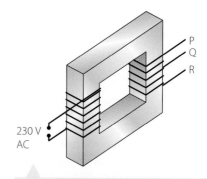

**FIGURE 7.20** A transformer with two output coils

» Magnetic flux, $\Phi$, through an area is given by $\Phi = B_{\parallel}A = BA\cos\theta$. Flux is measured in units of weber, Wb. $1\,\text{Wb} = 1\,\text{T}\,\text{m}^2$.

» The flux through an area is a maximum when the field is perpendicular to the area ($\theta = 0°$). The flux is zero when the field lines are parallel to the area ($\theta = 90°$).

» The flux through an area can be changed by changing the field, the area, or the angle between the field and the area.

» A changing magnetic flux induces an emf (electromotive force).

» The induced emf created by a changing magnetic flux is given by Faraday's Law: $\varepsilon = -\dfrac{\Delta\Phi}{\Delta t}$.

» Lenz's Law states that the induced emf is such that it opposes the change in flux that created it.

» An induced emf can create an induced current.

» Transformers use electromagnetic induction to change current and emf.

» Transformers consist of two coils that are magnetically linked (flux linkage).

» An alternating (AC) current in the primary coil creates a magnetic field in the secondary coil, which induces an emf and hence a current in the secondary coil.

» An AC current is one that varies sinusoidally with time. An AC current can be described by its maximum value $I_{\text{peak}}$ or its rms value, $I_{\text{rms}} = \dfrac{I_{\text{peak}}}{\sqrt{2}}$.

» The primary (input) and secondary (output) emfs and currents for a transformer are: $\dfrac{V_\text{P}}{V_\text{S}} = \dfrac{N_\text{P}}{N_\text{S}}$ and $V_\text{P}I_\text{P} = V_\text{S}I_\text{S}$.

» A step-down transformer has a lower emf across its secondary coil than across its primary coil. A step-up transformer has a larger emf across its secondary coil than across its primary coil.

» Both step-down and step-up transformers are used at different stages of electricity distribution.

» The transformer equations are an approximation assuming 100% efficiency. Real transformers lose energy as heat because of eddy currents in the core, resistive heating in the coils and incomplete flux linkage.

# 7 CHAPTER REVIEW QUESTIONS

Qz
Review quiz

1 What is an induced current?

2 Express the unit Wb in fundamental units.

3 How are electric and magnetic fields related? Use Faraday's and Lenz's laws to explain your answer.

4 Explain the difference between a real transformer and an ideal transformer. Give some reasons for the difference.

5 A battery is connected to the primary coil of a transformer. Why is no current observed in the secondary coil?

6 How can a loop of wire be moved within a uniform magnetic field such that the flux through the loop changes?

7 Draw a diagram showing how a loop can be oriented in a magnetic field to give:

a maximum flux.

b zero flux through the loop.

8 A transformer has a primary coil with 1000 turns and a secondary coil with 300 turns.

a Is this a step-up or step-down transformer?

b What is the ratio $\dfrac{V_\text{S}}{V_\text{P}}$ for this transformer?

9 Explain how transformers are used at different stages of power transmission and distribution. What sort of transformers are used at the different stages?

10 Explain the difference between $V_{\text{peak}}$, $V_{\text{peak–peak}}$, $V_{\text{ave}}$ and $V_{\text{rms}}$ for an AC potential difference. Draw a diagram to help explain your answer.

11 A piece of wire of length 20 cm is made into a square and held in a magnetic field of 55 µT.

a What is the maximum flux through the loop?

b At what angle between the loop and the field will the flux be $1.0 \times 10^{-7}\,\text{T}\,\text{m}^2$?

12 The magnetic flux through a wire loop of area $2.0 \times 10^{-3}\,\text{m}^2$ is $4.8 \times 10^{-4}\,\text{Wb}$. What is the magnetic field perpendicular to the plane of the loop?

13 A loop of area 10 cm$^2$ is in a uniform magnetic field of magnitude 0.10 T. Use a spreadsheet to calculate the flux through the loop for $\theta = 0°, 1°, 2° \dots 360°$ and plot a graph of $\Phi(\theta)$.

Information and communication technology capability

14 A loop of area 10 cm$^2$ is in a uniform magnetic field of magnitude 0.10 T. The loop is perpendicular to the field. If the field is increased at a steady rate of $0.01\,\text{T}\,\text{s}^{-1}$, what emf is produced between the ends of the loop?

15 A phone charger has a 230 V input and a 5 V output.

a What is the ratio of turns on the secondary coil to that in the primary coil?

**b** If the charger has an output current of 100 mA, what current is it drawing from the mains supply?

**c** What assumption have you made in your answer to part b? Would the actual current drawn be more or less than this?

**16** A potential difference is a type of emf. Explain how the emf produced by a changing magnetic flux is different from a potential difference in an electric field.

**17** A metal rod is placed across metal rails as shown in Figure 7.21. The rails are electrically connected via a resistor. The whole apparatus is in a uniform magnetic field. If the rod is moved to the right, what will be the direction of current flow in the resistor?

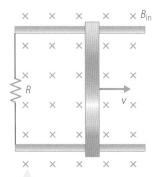

FIGURE 7.21 A metal rod is placed across two metal rails, linked by a resistor. The rod is pulled to the right.

**18** Figure 7.22 shows the north pole of a bar magnet being pushed into a solenoid. For each of the following situations, state whether a current flows, and if so whether it flows through the meter from X to Y or from Y to X.

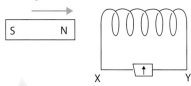

FIGURE 7.22 A magnet and a solenoid

**a** The north pole is moved towards the solenoid.

**b** The magnet is held stationary in the solenoid.

**c** The north pole of the magnet is withdrawn from the solenoid.

**d** The south pole of the magnet is moved towards the solenoid.

**e** The magnet is held stationary and the solenoid is moved to the right.

**f** Both the magnet and the solenoid are moved to the right with the same speed.

**19** A rectangular loop of area 20 cm$^2$ is placed in a magnetic field of magnitude 0.24 T. The loop is rotated so that it takes 0.10 s to go from parallel to the field

to perpendicular to the field. What is the average emf induced across the ends of the loop?

**20** Figure 7.23 illustrates a step-down transformer with a primary coil of 500 turns and a secondary coil of 250 turns. The primary coil is connected to a 240 V AC power supply.

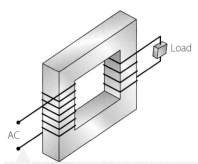

FIGURE 7.23 A step-down transformer

**a** What is the secondary potential difference?

**b** Explain how the transformer operates.

**c** If the primary current is 2.4 A, what is the current in the secondary coil?

**21** Figure 7.24 shows a rectangular coil of wire WXYZ in a uniform magnetic field of magnitude 0.06 T. The coil has 160 turns. The area of each turn is 0.025 m$^2$. A student moves the coil vertically upwards with a uniform speed. The entire coil is in the magnetic field for 0.3 s and leaves the field between 0.3 and 0.5 s.

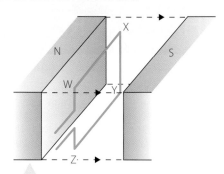

FIGURE 7.24 A loop is lifted out of a magnetic field.

**a** Draw a graph of flux through the coil as a function of time. Label your axes clearly, including units.

**b** Draw a graph of the potential difference induced across the ends of the coil with time.

**c** What is the magnitude of the emf produced across the ends of the coil for the time interval from 0.3 to 0.5 s?

**22** A metal ring of radius 5 cm and resistance 0.1 Ω is placed around a 500 turn solenoid of radius 3 cm and length of 10 cm. A power supply is attached to the solenoid and turned on so that current in the solenoid increases from 0 to 1 A in 0.01 s.

**a** What is the average current in the ring in this 0.01 s?

**b** What is the current in the ring after this time, if the power supply is left on?

9780170409131

# 8 Applications of the motor effect

INQUIRY QUESTION

How has knowledge about the Motor Effect been applied to technological advances?

OUTCOMES

**Students:**

- investigate the operation of a simple DC motor to analyse:
  - the functions of its components
  - production of a torque $\tau = nIA_{\perp}B = nIAB\sin\theta$
  - effects of back emf (ACSPH108) CCT, ICT, N
- analyse the operation of simple DC and AC generators and AC induction motors (ACSPH110) CCT
- relate Lenz's Law to the law of conservation of energy and apply the law of conservation of energy to:
  - DC motors and
  - magnetic braking CCT

Physics Stage 6 Syllabus © 2017 NSW Education Standards Authority (NESA) for
and on behalf of the Crown in right of the State of New South Wales.

Shutterstock.com/Africa Studio

**FIGURE 8.1** A household blender employs an electric motor.

At the core of nearly every modern technology is electromagnetism. Large-scale electric power generators, transformers for computers and electric motors all rely on electromagnetism. Without these technologies our lives would be very different.

Motors are used in many applications, from household appliances like blenders and hair dryers (Figure 8.1) to the electric motors that power suburban trains. How many devices have you already used today that used a motor? Electric motors are increasingly being used to run cars as hybrid and electric vehicles become more popular. Generators are used to produce electricity, turning kinetic energy into electric potential energy. Almost all our electricity supply in Australia comes from generators. In this chapter we shall look at how motors and generators work.

# 8.1 DC motors

A **DC motor** uses a current-carrying coil in a magnetic field. The coil experiences a torque due to the interaction of the field with the current. This makes the coil rotate. The net result is the conversion of *electric potential energy* into *kinetic energy*.

## Torque on a current-carrying loop in a magnetic field

We saw in chapter 6 that a current experiences a force in a magnetic field. The magnitude of the force is given by $F = \ell IB \sin \theta$, where $\theta$ is the angle between the current and the field.

Consider a loop of current-carrying wire in a magnetic field such as that shown in Figure 8.2a. The length of wire between points N and M has a force, $F = \ell IB \sin \theta$, acting upwards on it. The wire between P and Q has an equal force acting downwards on it. The net force is therefore zero, and we do not expect the loop to go either up or down. However, the loop will move because it experiences a torque.

Torque was described in chapter 3. A torque is a push or pull that acts to rotate an object. Torque is the rotational equivalent of force. The torque exerted by any force, $F$, acting at some distance, $r$, from an axis of rotation is:

$$\tau = rF \sin \theta$$

where $\theta$ is the angle between the force, $F$, and the vector $r$, which points from the axis of rotation to the point of application of the force (see Figure 3.15 on page 74).

The torque is a maximum when $r$ and $F$ are perpendicular, as in Figure 8.2. Torque is zero when $r$ and $F$ are parallel.

Like force, torque is a vector. We use the right-hand rule from chapter 3, Figure 3.16, to find its direction. Point the fingers of your right hand in the direction of $r$, then curl them towards $F$. Your thumb points in the direction of the torque. Note that the direction of the torque is parallel to the axis of rotation. The curl of your fingers gives the sense of rotation about the axis.

In Figure 8.2a the two forces acting on the loop result in a torque that acts to flip the loop over in a clockwise direction.

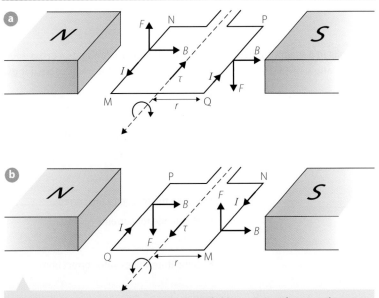

**FIGURE 8.2** A current in a loop in a magnetic field experiences forces on the arms that are perpendicular to the field. **a** The forces act in opposite directions, causing rotation. **b** After a half-circle the forces are reversed and the loop flips back the other way.

But the current has now reversed direction, and so the direction of the forces also changes. The new torque acts to push the loop back in the opposite direction, as in Figure 8.2b.

Seen from the side, the coil starts in the position shown in Figure 8.3a. It rotates through the position in Figure 8.3b and continues to rotate. At the position in Figure 8.3c the forces on each side act along the same line and the net torque is zero. The coil has sufficient rotational momentum to carry it past this balance point. Once past this balance point the net turning effect is reversed, as shown in Figure 8.3d. The coil now reverses direction and oscillates back and forth about the balance point. The oscillations get smaller due to friction. After a short time no rotation takes place and the coil comes to a stop.

The current in the coil must be stopped or reversed each half cycle of rotation to keep the coil rotating. Commutators or reversing switches do this.

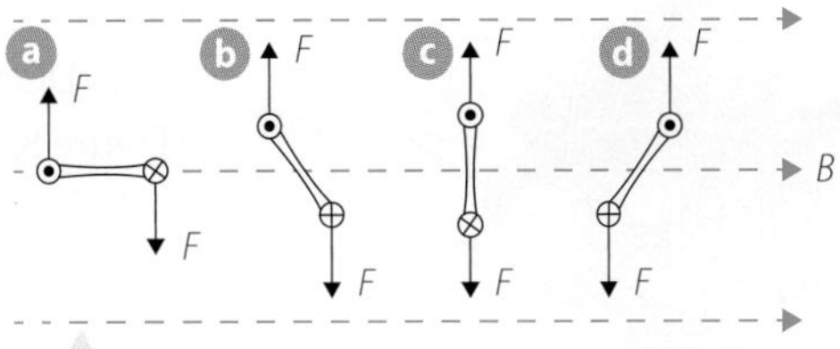

Remember that torque is greatest when the distance between the axis of rotation and the point at which the force is applied is the greatest. In Figure 8.3, the axis of rotation is through the centre of the line. The torque is greatest in 8.3a when the coil is aligned with the field and the angle between the plane of the coil and the field is zero. The torque is zero when the plane of the coil is perpendicular to the field. If we call the angle between the normal to the plane of the coil and the field θ, then the torque at any angle θ is given by:

**FIGURE 8.3** Side view of a loop rotating about an axis that is at right angles to a magnetic field. **a** Maximum net torque; **b** decreasing net torque; **c** no torque; **d** torque reversed.

$$\tau = 2Fd = 2\ell IB \times r\cos\theta = 2\ell IBr\cos\theta$$

Here $d$ is the length of sides NP and MQ (Figure 8.2), so sides MN and PQ are a distance $\dfrac{d}{2} = \dfrac{\text{NP}}{2}$ from the axis. The factor of 2 in front of the force, $F$, comes from there being two forces on the loop. While the net force is zero, both forces act to give a torque in the same direction, so they are added to give the total torque.

The product $\ell d = 2\ell r$ is the area of the loop. As in chapter 7, we use the normal to represent area $A$, so we can write the torque as:

$$\tau = IAB\sin\theta$$

▶ **WORKED EXAMPLE** (8.1)

A current-carrying loop is in a magnetic field of magnitude 0.10 T. The coil has a cross-sectional area of $20\,\text{cm}^2$. If the coil is at an angle of 45° to the field, calculate the current it must carry if it experiences a torque of 0.050 N m.

| ANSWER | LOGIC |
|---|---|
| $B = 0.10\,\text{T}, A = 20\,\text{cm}^2, \tau = 0.050\,\text{N m}, \theta = 45°$ | ▪ Identify the relevant data in the question. |
| $A = 20\,\text{cm}^2 \times \dfrac{1\,\text{m}^2}{10\,000\,\text{cm}^2} = 0.0020\,\text{m}^2$ | ▪ Convert to SI units. |
| $\tau = IAB\sin\theta$ | ▪ Relate the torque to the current. |
| $I = \dfrac{\tau}{AB\sin\theta}$ | ▪ Rearrange for current. |
| $I = \dfrac{0.050\ \text{N m}}{(0.0020\ \text{m}^2)(0.10\ \text{T})\sin 45°}$ | ▪ Substitute values with correct units. |
| $I = 35.3\,\text{N m}^{-1}\text{T}^{-1}$ | ▪ Calculate the value. |
| Note that because $1\,\text{T} = 1\,\text{kg s}^{-2}\text{A}^{-1}$ and $1\,\text{N} = 1\,\text{kg m s}^{-2}$, $1\,\text{N m}^{-1}\text{T}^{-1} = 1\,\text{A}$ | ▪ Check that units are correct. |
| $I = 35\,\text{A}$ | ▪ State the final answer with correct units and appropriate significant figures. |

**TRY THIS YOURSELF**

What maximum possible torque acts on this coil if it carries a current of 10 A in the same magnetic field?

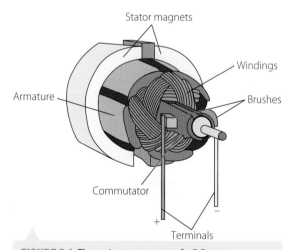

Stator magnets
Windings
Armature
Brushes
Commutator
Terminals

**FIGURE 8.4** The main components of a DC motor

**Main parts of a DC motor**

Watch this video to see how the main parts of the motor work.

## Components of a DC motor

A DC motor is one that runs on direct current. Recall from chapter 7 that direct current is current that always flows in the same direction.

A DC motor consists of two main parts; the stator and the rotor (Figure 8.4). The stator is the part of the motor that doesn't move. It includes the casing of the motor and the magnets. The input wires and brushes are also usually attached to the stator.

The rotor is the rotating part. The rotor consists of the armature, which holds the coils (windings), the coils themselves, and the commutators.

### Commutators and brushes

The commutator changes the electrical contacts on the wires as the coil's momentum carries it past its balance point. Brushes made of graphite or carbon blocks usually provide the sliding contact, as shown in Figure 8.5.

Without the commutators and brushes the motor would not spin: it would either stop or just flip back and forth. The commutators effectively act as a switch, changing the direction of the current to the coil for half of each rotation. Figure 8.6a shows the torque on a current-carrying coil in a magnetic field when the current is always in the same direction (no commutators); Figure 8.6b shows the torque when it changes direction every half rotation (with commutators).

9780170409131

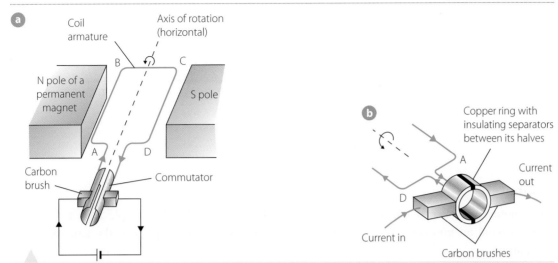

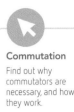

**Commutation**
Find out why commutators are necessary, and how they work.

**FIGURE 8.5** **a** The DC supply is connected to the carbon brushes. The commutator is free to move against the brushes. The insulating separators keep the two sides of the coil electrically separated. As the coil flips over the current in the coil is reversed, which allows the torque to act in the same direction over the full cycle. **b** A close-up view of the commutator.

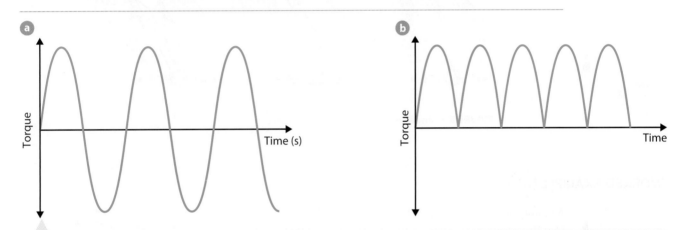

**FIGURE 8.6** The torque on the coil as a function of angle: **a** Without commutators; **b** With commutators

## The coils

Each turn or loop of the coil in a motor experiences a torque due to the magnetic field. The total torque experienced is equal to the sum of the torques experienced, so:

$$\tau_{coil} = n\tau_{loop} = \tau = nIAB\sin\theta$$

where $n$ is the number of turns in the coil.

The greater the torque on the coil, the more powerful and faster the motor will be. However, the more turns that there are on the coil, the larger and heavier the coil will be.

As the coil rotates, the torque experienced by the coil varies. If the coil is rotating at a constant speed so that $\theta$ is a function of time, then the torque varies sinusoidally, as shown in Figure 8.6b.

Most motors use more than one coil, arranged so that the plane of each coil is at an angle to the others as in Figure 8.7. This means that, as the armature rotates, while the torque on one coil decreases it is increasing on another coil (Figure 8.8). In this way the armature is always experiencing a torque. This results in both a more powerful motor and a smoother motion of the motor as the total torque is closer to constant.

**FIGURE 8.7** Most DC motors have multiple coils wound offset to each other.

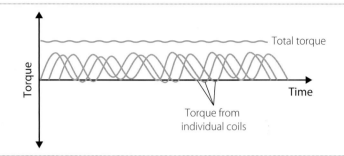

FIGURE 8.8 The total torque experienced by the armature is the sum of the torques experienced by all the coils.

## Magnets

Most motors use shaped magnets (Figure 8.9) or electromagnets with multiple coils at different angles around an iron core. These increase the magnetic field, the torque and the smoothness of the rotation.

**Build a motor**

Here is a simple design for a DC motor you can build.

**Homopolar motor**

This is an even simpler motor, with only one magnet.

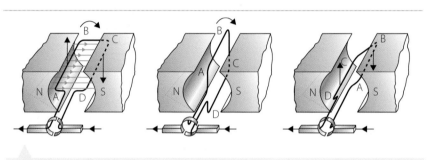

FIGURE 8.9 Shaped magnets ensure constant torque.

## ▶ WORKED EXAMPLE (8.2)

A motor uses a 50-turn coil that has dimensions 0.10 m by 0.16 m, as shown in Figure 8.10. A current of 2.0 A flows through the coil. The coil is vertical and is in a magnetic field of 0.12 T, directed upwards.

**a** What is the magnitude and direction of the force exerted on side AB?

**b** What is the magnitude and direction of the force exerted on side BC?

**c** What is the total torque acting on the coil?

**d** In which direction will the coil begin to rotate?

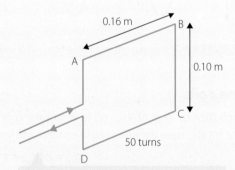

FIGURE 8.10 A 50-turn current-carrying coil

| ANSWERS | LOGIC |
|---|---|
| **a** $F = \ell IB \sin \theta$ | ▪ Relate the force per section of wire to the variables given. |
| $F = n\ell IB$ | ▪ Recognise that the angle $\theta$ is 90° and that there are $n$ lengths of wire in the section. |
| $F = 50 \times 0.16\,\text{m} \times 2.0\,\text{A} \times 0.12\,\text{T}$ | ▪ Substitute the correct values with units. |
| $F = 1.9\,\text{N}$ | ▪ Calculate the final value. |
| **b** $F = \ell IB \sin \theta$ | ▪ Relate the force per section of wire to the variables given. |
| $F = 0$ | ▪ Recognise that the angle $\theta$ is 0° and hence $\sin \theta = 0$. |
| **c** $\tau = rF \sin \theta$ | ▪ Relate torque to force. |
| $\tau_{AB} = rF_{AB}$ | ▪ Find the torque acting on side AB. |
| $\tau_{AB} = 0.05\,\text{m} \times 1.9\,\text{N}$ | ▪ Substitute the correct values with units. |
| $\tau_{AB} = 0.095\,\text{N}\,\text{m}$ | ▪ Calculate the final value. |
| $\tau_{BC} = rF_{BC} = 0$ <br> $\tau_{DC} = \tau_{AB} = 0.095\,\text{N}\,\text{m}$ <br> $\tau_{DA} = \tau_{BC} = 0$ | ▪ Find the torque acting on the other sides. |
| $\tau_{total} = \tau_{AB} + \tau_{BC} + \tau_{CD} + \tau_{DA}$ <br> $\tau_{total} = 0.095\,\text{N}\,\text{m} + 0 + 0.095\,\text{N}\,\text{m} + 0$ <br> $\quad\quad = 0.19\,\text{N}\,\text{m}$ | ▪ Calculate the total torque. |
| **d** Clockwise | ▪ Using your right hand, the torque is acting along the direction AB, so your fingers curl clockwise. |

**TRY THIS YOURSELF**

Repeat the question above but for a magnetic field directed horizontally to the right.

## Back emf

The speed of a DC motor is controlled by the **supplied emf**. The energy is being constantly supplied by the source of emf, and is being converted into kinetic energy, so it may seem that the coil should go faster and faster and faster as long as the motor is turned on. This would not violate conservation of energy, as the emf is continuing to supply energy. But this doesn't happen. Instead, the motor accelerates until it reaches a steady rotational speed. The effect of friction alone does not account for this.

We know from Lenz's Law that the changing magnetic flux through a loop, due to the movement of the loop, induces an emf in the loop. This induced emf in a motor is called a **back emf**. The back emf must be in the opposite direction to the applied potential difference that caused the movement. The faster the coil spins, the more rapidly the flux changes and the bigger the back emf. This limits the current in the coil and hence the rate at which it spins.

**Back emf**

Watch this video on how back emf arises, and make your own concept map to explain it.

DC motors

## Make your own DC motor

Critical and creative thinking

Numeracy

Information and communication technology capability

### AIM

Your first aim is to build a DC motor. Then you need to determine how well it works. Think about what measurements you can make. What will they tell you about the performance of your motor? Write a research question for your investigation.

### MATERIALS

Start by doing some research on simple motor designs. Consider the materials you will need to make your motor. You will also need some other materials and measuring devices in order to test the performance of your motor. You can follow the weblinks on page 178 to the simple DC motor and the homopolar motor, but you should also find some other designs.

### ⚠ RISK ASSESSMENT

| WHAT ARE THE RISKS IN DOING THIS INVESTIGATION? | HOW CAN YOU MANAGE THESE RISKS TO STAY SAFE? |
| --- | --- |
| Electrical equipment can cause shocks when not used correctly. | Follow the instruction manuals and instructions from your teacher for any equipment that you use. |

What other risks are associated with your investigation, and how can you manage them?

### METHOD

First you will need to collect all the equipment required to build your motor. Make a list of everything you need. If some of the equipment is not available, you will need to find a different design or modify your existing design.

Build your motor. Do not forget that it will need commutators!

Once your motor is completed you need to test it.

Think about how you can measure motor speed and power. You may need to be creative. The motor will almost certainly spin too fast for you to count revolutions by eye. What else could you do? How can you minimise uncertainties in your results?

### RESULTS AND ANALYSIS

Your results are the test data for your motor.

Record all your measurements and use them to calculate how fast the armature of your motor could spin. Do not forget to include uncertainty calculations.

### DISCUSSION

Compare your motor to that built by other students.

What design features give a high-speed motor? What features give a reliable motor?

### CONCLUSION

Write a short conclusion summarising your findings.

- A DC motor converts electric potential energy to kinetic energy.
- A DC motor uses a current-carrying coil in a magnetic field.
- A current-carrying coil in a magnetic field experiences a torque due to the field. The torque makes the coil rotate.
- The torque experienced by a current-carrying coil in a magnetic field is given by $\tau = nIAB\sin\theta$, where $n$ is the number of turns of wire in the coil, $I$ is the current through the coil, $A$ is the area of the coil, $B$ is the field strength and $\theta$ is the angle between the normal to the coil and the magnetic field.
- The main components of a DC motor are the coil, magnets, commutators and brushes.
- The coil, which is mounted on an armature, carries a current and experiences a torque due to the magnetic field that makes it rotate.
- The magnets produce the magnetic field that exerts the force that creates the torque.
- The commutators and brush system swap the direction of current every half-turn so that the direction of torque is constant.
- The speed of a motor is limited by the back emf produced by the movement of the coil in the field (Lenz's Law).

1 Name the main parts of a DC motor and describe the function of each.

2 Why does a current-carrying loop in a magnetic field experience a torque but no net force? Draw a diagram to help explain your answer.

3 Explain, using Lenz's Law, why a coil in a DC motor has a back emf. Draw a diagram to help explain your answer.

4 A single loop coil is in a magnetic field of 0.05 T. The loop has an area of $2.0 \times 10^{-4}$ m$^2$, and carries a current of 0.2 A. If the plane of the coil is at an angle of 45° to the field, what torque does it experience?

5 Figure 8.11 shows a simple electric motor. In which direction must the current flow for the coil to rotate in a clockwise direction as seen from the end of the coil AD? Copy the diagram and show the magnetic field, the current, the forces on sides AB and CD, and the torque on the coil.

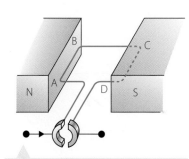

**FIGURE 8.11** A simple electric motor

6 Figure 8.12 shows a rectangular coil placed between the poles of a strong magnet as part of a DC motor. The coil has 40 turns of wire and each turn is rectangular with length 0.060 m and width 0.040 m. The magnetic field is uniform and has a magnitude of 0.050 T. The coil is free to rotate when placed between the poles of the magnet. A current of 1.5 A flows in the coil. At the instant represented in the diagram, the plane of the coil is parallel to the magnetic field.

  a What is the magnitude and direction of the force on side AB?

  b What is the magnitude and direction of the force on side BC?

  c What is the magnitude and direction of the force on side CD?

  d What is the magnitude of the torque acting on the coil?

  e In what direction – clockwise or anticlockwise – will the coil rotate?

  f How could the magnitude of the torque be increased?

7 If the coil shown in Figure 8.12 and described in Question **6** is to experience a torque of 0.015 N m, how much current must be supplied to it?

8 A motor in a blender has a single coil with 500 turns and an area of 20 cm$^2$. The coil sits in a magnetic field of 0.1 T. If the coil carries a current of 1 A, at what angle does it experience a torque of 0.05 N m?

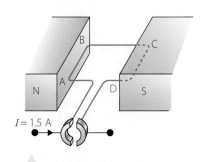

$I = 1.5$ A

**FIGURE 8.12**

Generators and AC induction motors

A motor uses a current, driven by an emf or potential difference, to produce a torque on a coil in a magnetic field. Remember that an emf or potential difference is a potential energy per unit charge. The motor transforms this potential energy into the kinetic energy of the spinning rotor.

A generator does the opposite. It converts the kinetic energy used to spin a coil in a magnetic field into electric potential energy that creates a current in the coil. A generator is effectively a motor working backwards – the armature is spun to create a current.

Almost all of the electricity that we use is produced by generators. The energy required to produce the movement may come from any source. In Australia it is mostly supplied by burning coal. A small, but increasing, fraction comes from the gravitational potential energy of water (hydroelectric power stations) and the kinetic energy of air molecules (wind turbines). Many other nations use nuclear energy to run generators. All of these energy sources require generators to produce electricity.

## AC generators

Figure 8.13 shows a very simple AC generator. Recall from chapter 7 that an alternating current is one that varies between positive and negative values. Usually AC varies sinusoidally. The coil is attached to an armature that rotates in the magnetic field between the poles of the two magnets. As the coil rotates, the flux through the loops of the coil varies, causing an emf across the ends of the coil. Each end of the coil is attached to a conducting slip ring that slides against a brush. The brushes are then connected to the external circuit that uses the emf generated.

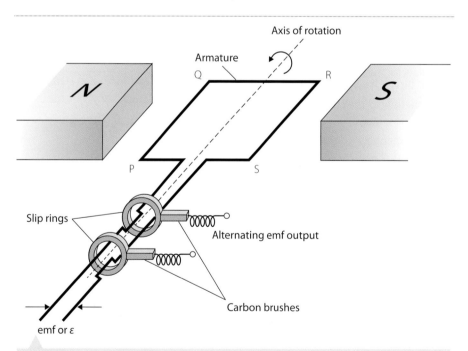

**FIGURE 8.13** A simple AC generator. The rotation is circular, so the loop's motion produces a sinusoidally varying emf.

In Figure 8.14 the flux vs time graph is a sine curve because the original flux is zero.

The flux varies with the angle $\theta$, which varies in time such that $\theta(t) = \dfrac{2\pi}{T}t = 2\pi ft$, where $T$ is the period of rotation. The frequency is $f = \dfrac{1}{T}$.

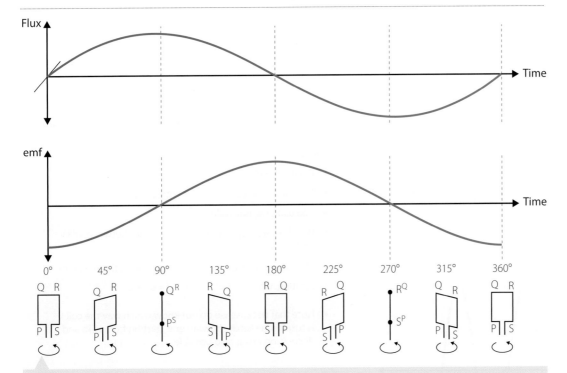

**FIGURE 8.14** A loop rotates in a circle. The flux through the loop describes a sine curve (top graph). The rate of change of flux – the gradient – also describes a sinusoidal curve, but the emf is the negative of the rate of change of flux (lower graph). The rotation of the coil is shown for each one-eighth of a turn.

Hence, the flux as a function of time is given by:

$$\Phi = nBA\sin(2\pi ft)$$

when a coil with $n$ turns of area $A$ rotates in a magnetic field of magnitude $B$.

Recall from chapter 7 that Faraday's Law tells us that the emf is the negative of the gradient of the flux as a function of time.

So the emf is given by:

$$\text{emf} = 2\pi fnBA\cos(2\pi ft)$$

This is shown in Figure 8.14. Note that when the flux is changing most rapidly the emf has its maximum values. For a sine curve, the gradient is greatest at $t = 0$, $t = \dfrac{T}{2}$, and again at the end of each cycle. This is when $2\pi ft$, or the value of the term in brackets, is 0 or $\pi$. When the flux is at a peak, at $t = \dfrac{T}{4}$ and $t = \dfrac{3T}{4}$, the gradient is momentarily zero, so the emf is zero.

The flux and emf have the same frequency.

The maximum emf is:

$$\varepsilon_{max} = 2\pi fnBA$$

The maximum emf can be changed by changing $f$, $n$, $B$ or $A$. If $f$ is changed, the period changes as well as the emf.

Usually the armature has a large coil of wire because the emf is proportional to the number of loops or turns in the coil. However, the bigger the coil the heavier it is, so sometimes it is the magnets that are rotated instead of the coil.

**Generators**
Use this simulation to find out how a generator works. How could you make a simple hand-driven generator?

## WORKED EXAMPLE 8.3

A square coil of side length 0.10 m is made up of 400 turns of wire. It is rotated at 25 Hz in a magnetic field of magnitude 0.10 T.

**a** Find the maximum emf induced.

**b** If, at $t = 0$, there is a maximum flux through the coil, sketch a graph of the emf as a function of time.

| ANSWERS | LOGIC |
|---|---|
| **a** $\varepsilon_{max} = 2\pi f n B A$ | ▪ Relate emf to parameters given. |
| $\varepsilon_{max} = 2\pi \times 25\,\text{Hz} \times 400 \times 0.10\,\text{T} \times (0.10\,\text{m} \times 0.10\,\text{m})$ | ▪ Substitute the correct values with units. |
| $\varepsilon_{max} = 63\,\text{V}$ | ▪ Calculate the final value. |
| **b** | ▪ We know from part **a** that the emf varies between −63 V and +63 V. |
| | ▪ The period of its oscillations is the same as the period of rotation, which is $T = \dfrac{1}{f} = \dfrac{1}{25}\,\text{Hz} = 0.04\,\text{s}$. |
| | ▪ Note that because we do not know which way the coil is turning, a sketch showing emf starting from zero and decreasing first is also possible. |

FIGURE 8.15 emf as a function of time

**TRY THIS YOURSELF**

What effect would doubling the frequency have on the maximum emf? Sketch emf(*t*) for this case.

## INVESTIGATION 8.2

Critical and creative thinking

Numeracy

### A simple AC generator

**AIM**

To make a simple AC generator and measure the emf it produces

Write a hypothesis for this investigation.

**MATERIALS**

- Retort stand and clamps
- Spring
- Magnet
- Weights

- Alligator leads
- Coil
- Oscilloscope or data logger connected to computer

| WHAT ARE THE RISKS IN DOING THIS INVESTIGATION? | HOW CAN YOU MANAGE THESE RISKS TO STAY SAFE? |
|---|---|
| The magnet could fly off the spring and hit someone. | Make sure the magnet is well attached and do not oscillate it too vigorously. |
| Magnets can break or become demagnetised when dropped. | Be careful with the magnets and put them down when not in use. |

What other risks are associated with your investigation, and how can you manage them?

## METHOD

1   Attach the spring to the retort stand as shown in Figure 8.16.
2   Attach the magnet and one weight to the spring.
3   Place the coil below the magnet. Adjust the height of the spring and magnet so that at equilibrium the magnet is just inside the coil.
4   Connect the coil to an oscilloscope or data logger so that you can measure the emf produced.
5   Pull the magnet down to start it oscillating. You may need to move the coil out of the way to do this, then put it back in place.
6   Record the period of oscillation and the maximum emf produced. It is more precise to measure ten complete oscillations, then divide by 10, to get the period of oscillation. Do not forget to include uncertainties in your results.
7   Repeat the measurements, adding weights to vary the frequency of oscillation. You will need to adjust the height of the spring each time.

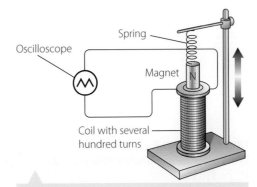

**FIGURE 8.16** A simple generator can be made by connecting a magnet to a spring and oscillating it in a solenoid.

## RESULTS

Record the maximum emf as a function of period of oscillation in a table like the one below. Do not forget to include units and uncertainties in your results.

Calculate the frequency of oscillation and add this to your table as shown.

| TIME FOR TEN OSCILLATIONS | PERIOD | FREQUENCY | MAXIMUM emf |
|---|---|---|---|
|  |  |  |  |
|  |  |  |  |

## ANALYSIS OF RESULTS

Draw a graph of maximum emf as a function of frequency of oscillation. Comment on the shape of your graph.

## DISCUSSION

1   Do your results agree with your hypothesis?
2   How could you improve this experiment to make it more accurate? What could you do to extend it?

## CONCLUSION

Write a conclusion summarising your findings.

For an AC generator, the maximum output emf is $\varepsilon_{max} = 2\pi fnBA$. The rms value of the output emf is therefore $\varepsilon_{rms} = \dfrac{\varepsilon_{max}}{\sqrt{2}} = \dfrac{2\pi fnBA}{\sqrt{2}} = \sqrt{2}\pi fnBA$.

Recall from *Physics in Focus Year 11* chapter 13 that current and emf (voltage) are related by $I = \dfrac{emf}{R}$. So the rms current produced is $I_{rms} = \dfrac{I_{max}}{\sqrt{2}} = \dfrac{\sqrt{2}\pi fnBA}{R}$, where $R$ is the resistance of the load attached to the generator output.

## WORKED EXAMPLE (8.4)

The rms potential difference produced by an AC generator is 240 V. To what peak value does this correspond?

| ANSWER | LOGIC |
|---|---|
| $V_{rms} = \dfrac{V_{peak}}{\sqrt{2}}$ | ▪ Relate $V_{rms}$ to $V_{peak}$. |
| $V_{peak} = \sqrt{2}\, V_{rms}$ | ▪ Rearrange for $V_{peak}$. |
| $V_{peak} = \sqrt{2}\,(240\,\text{V})$ | ▪ Substitute the correct values with units. |
| $V_{peak} = 340\,\text{V}$ | ▪ Calculate the final value. |

**TRY THIS YOURSELF**

What is the peak potential difference of an AC generator built for use in the United States, where $V_{rms} = 110\,\text{V}$?

## DC generators

A generator that provides an output emf that is always positive is called a DC generator. In a DC generator, a commutator is used instead of the slip rings used in an AC generator, as shown in Figure 8.17. The commutators work the same way as the commutators in a DC motor.

In the commutator, each side of the coil is connected to a conducting copper strip. The strips are separated by insulators. As the commutator rotates past the carbon brushes, the effect is the same as if the connections to the slip rings were reversed each half-cycle. The output for a single coil is shown in Figure 8.18a. The output is 'lumpy' or pulsed, rising to a maximum and dropping back to zero each half turn.

To smooth out the DC current, it is usual to use many coils that are offset relative to each other to provide a steady current, just as a DC motor typically uses many coils to produce a smooth movement. Each coil has separate pairs of connections to the commutator. The output for a two-coil DC generator is shown in Figure 8.18b. These two coils are offset by 90°. One of the problems of using multiple commutators is that there can be sparking across the insulating gap. This can be dangerous in car engines where there are flammable gases nearby.

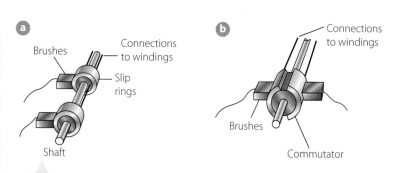

**FIGURE 8.17** **a** An AC generator uses slip rings that always have an electrical connection. **b** A DC generator uses a commutator that is connected for only part of each cycle.

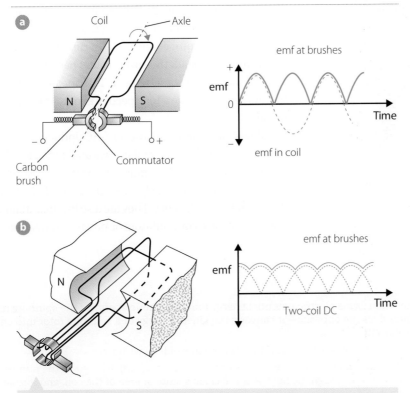

FIGURE 8.18 In a DC generator, the carbon brushes make contact with a commutator, which allows the connections to the coils to switch every half cycle. **a** A single coil gives a fluctuating output. **b** This problem is overcome by using two or more coils.

## AC induction motors

An **AC induction motor** uses the principle of electromagnetic induction, as described by Faraday's Law. No current is supplied directly to the rotating coils, but a current is induced in them by using a changing magnetic field. The changing magnetic field is produced by an AC current. The rotor (the rotating part) consists of a cylinder with metal rods embedded in it along the length of the cylinder. These are electrically connected at each end of the cylinder to form closed loops. This forms a '**squirrel cage**' (Figure 8.19).

The rotor sits between the poles of two electromagnets. The AC supplied to the coils of the electromagnets creates an oscillating magnetic field that induces a current in the squirrel cage, which acts as the rotor coils.

Transformers and induction motors are similar. The primary or stationary coils on the stator create the time-varying magnetic field. This produces a time-varying magnetic flux in the secondary coil or rotor. However, a motor is designed to rotate, so the stator coils are arranged to produce a rotating magnetic field. A force acts on the induced current in each connecting bar of the rotor due to the magnetic field from the stators. The connecting rods are arranged symmetrically so there is no net force on the rotor; however, each pair of opposite rods experiences a torque. The torques are in the same direction and add up to give a net torque on the rotor, causing it to rotate. The basic configuration of an AC induction motor is shown in Figure 8.20.

**Regenerative braking**

Electric and hybrid vehicles with regenerative braking use a motor to turn the wheels, and when braking the same motor is used as a generator to save energy. Find out how.

Generators and AC induction motors

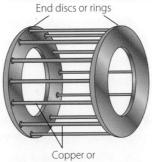

End discs or rings

Copper or aluminium conduction bars

FIGURE 8.19 A 'squirrel cage' rotor is used as the coil in an induction motor.

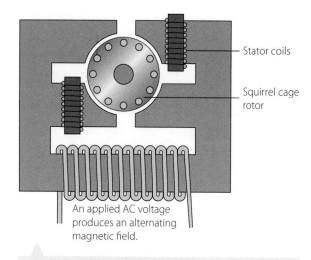

Stator coils

Squirrel cage rotor

An applied AC voltage produces an alternating magnetic field.

**FIGURE 8.20** An AC induction motor

The AC power supplied through the mains electricity has a frequency of 50 Hz. So 50 times each second the current changes direction, and a motor connected to it runs at 50 Hz without a load. When there is a load, the speed is less because the rotor cannot keep up with the magnetic field. This means that AC induction motors generally have a top speed of about 3000 rpm (50 turns per second).

Almost all motors in use are AC induction motors. Power tools such as drills and circular saws and appliances like blenders and microwaves use AC induction motors, as do large industrial machines. They are widely used because of their simplicity of design and high efficiency. They are also low maintenance because they do not have commutators or brushes to wear out.

**A very simple AC induction motor**

Follow the instructions to build your own AC induction motor.

- A motor transforms electric potential energy into the kinetic energy of the spinning rotor. A generator does the opposite – it converts the kinetic energy of a spinning rotor into an emf to drive a current.
- An AC generator produces an emf that varies sinusoidally with time.
- The emf produced by an AC generator is emf $= 2\pi fnBA \cos(2\pi ft)$, where $n$ is the number of turns on the coil, $B$ is the magnetic field, $A$ is the cross-sectional area of the coil, and it rotates with frequency $f$.
- The maximum emf produced is emf$_{max} = 2\pi fnBA$ and the rms value of the emf is $\varepsilon_{rms} = \dfrac{\varepsilon_{max}}{\sqrt{2}} = \dfrac{2\pi fnBA}{\sqrt{2}} = \sqrt{2}\pi fnBA$.
- The rms current depends on the load, $R$, and is given by $I_{rms} = \dfrac{I_{max}}{\sqrt{2}} = \dfrac{\sqrt{2}\pi fnBA}{R}$.
- A DC generator uses a commutator to provide an output emf from each coil that is always positive. Many coils, offset at different angles, are typically used to produce an output that is approximately constant.
- An AC induction motor uses electromagnets to create an AC current in the rotor via induction. No current is supplied directly to the rotating coils.
- AC induction motors run at a maximum frequency equal to that of the supplied AC current. In Australia this frequency is 50 Hz, or 3000 rpm.
- Most motors in household and industrial use are AC induction motors.

**CHECK YOUR UNDERSTANDING**

**8.2**

1  Name three sources of energy used in electricity production.

2  What is the purpose of the commutators in a DC generator?

3  Why does an AC motor not need a commutator?

4  Briefly explain the principle of operation of an AC induction motor.

5  Why do most power tools and appliances such as food processors use AC rather than DC motors?

6  The armature of an AC generator is rotating at a constant speed of 35 revolutions per second in a horizontal field of flux density 1.0 T. The diameter of the cylindrical armature is 24 cm and its length is 40 cm.

  a  What is the maximum emf induced in the armature if it has 30 turns?

  b  What is the rms emf produced by this generator?

7  A flat rectangular coil 15 cm by 25 cm has 300 turns. An alternating emf of peak value 340 V is produced when the coil rotates at 3000 revolutions per minute in a uniform magnetic field. What is the value of the magnetic field strength?

8  A rectangular coil of 30 turns and area $100\,\text{cm}^2$ rotates at 1200 revolutions per minute in a uniform magnetic field of flux density $0.50\,\text{T}$.

    a  Find the frequency of the generated emf.

    b  Find the maximum emf.

    c  What is the rms emf?

    d  Write the equation that gives the emf at any instant.

9  The armature of a $50\,\text{Hz}$ AC generator rotates in a magnetic field of strength $0.15\,\text{T}$. If the area of the coil is $2.5 \times 10^{-2}\,\text{m}^2$, how many turns must the coil contain if the maximum emf produced is $150\,\text{V}$?

# 8.3  Lenz's Law and conservation of energy

Lenz's Law states that an induced emf always acts to produce an induced current in the direction that causes a magnetic flux that opposes the change in flux that induced the emf. So if the external flux is increasing, the induced current creates a flux in the opposite direction. If the external flux is decreasing, the induced current creates a flux in the same direction. In either case, the induced current acts to reduce the *change* in flux.

Lenz's Law is essentially a statement of conservation of energy. If the induced current acted to increase an already increasing flux, then the flux would increase to infinite. Energy is stored in magnetic fields, and this energy has to come from somewhere. So the flux cannot increase to infinite.

Figure 8.21 shows the direction of induced current when magnetic flux through a loop is increased. Figure 7.9 (page 159) also shows what happens when the flux decreases.

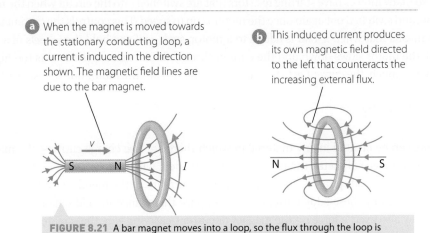

a  When the magnet is moved towards the stationary conducting loop, a current is induced in the direction shown. The magnetic field lines are due to the bar magnet.

b  This induced current produces its own magnetic field directed to the left that counteracts the increasing external flux.

**FIGURE 8.21**  A bar magnet moves into a loop, so the flux through the loop is increasing. This induces a current in the conducting loop. The current creates a magnetic field, which acts to decrease the flux through the loop.

## Lenz's Law and DC motors

When a DC motor is first turned on, the emf supplied to the coil produces a current in the coil. This current experiences a torque due to the magnetic field from the magnets. The torque makes the coil start to rotate. As the coil begins to rotate, the flux through the coil changes. This changing flux creates an induced emf across the coil.

Recall from chapter 7 that the negative sign in Faraday's Law, $\varepsilon = -\dfrac{\Delta\Phi}{\Delta t}$, is due to Lenz's Law. This negative sign tells us that the induced emf acts in the opposite direction to the emf supplied to the coil. This is called the back emf, as described earlier, and is shown in Figure 8.20.

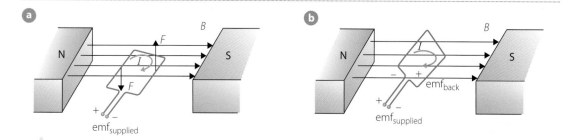

**FIGURE 8.22** **a** The coil experiences a torque that makes it turn because the supplied emf drives a current in the coil. **b** As the coil turns the flux through the loop increases, creating a back emf opposed to the supplied emf. This decreases the current through the coil.

The total emf across the coil is the sum of supplied emf (supply voltage) and the induced back emf, remembering that they have opposite signs. The current through the coil, from Ohm's Law (chapter 13 in *Physics in Focus Year 11*) is:

$$I = \frac{\text{emf}_{\text{supplied}} - \text{emf}_{\text{back}}}{R}$$

where $R$ is the resistance of the coil.

As the motor first begins to turn, it starts slowly, so the rate at which the flux changes is small, and so the induced back emf is small. However, as the motor speeds up the rate at which the flux changes increases and so the back emf also increases. This gives a maximum possible speed for the motor – the speed at which the back emf is equal to the supplied emf, at which point the current drops to zero. This is not the normal operating speed, which is usually lower than the maximum speed because of the load attached to the motor.

One important result of this is that the current flowing through a motor is much larger when it is first turned on, and so many motors have starting resistors that are switched into the circuit when the motor is first switched on, and switched out again once the motor is up to speed. This controls the current through the coil. It also means that when a load is added to a motor, such as if you hold the wheels of a remote control car still, the motor slows down and the current through it increases. If the load is too high, and slows the motor too much, the current gets too large and can burn out the motor due to resistive heating.

## Magnetic braking

Induced currents can be created in any material in which there are free charge carriers. If a magnet is moved around over a piece of metal, the changing magnetic field will induce an emf in the metal, and the emf will induce loops and spirals of current, called eddy currents, in the metal. From Lenz's Law, the magnetic fields created by these eddy currents act to oppose the change in field (change in flux) that created them. They also exert a force on the magnet or current-carrying wire that created the changing field in the first place.

Look again at Figure 8.21, and now imagine dropping a magnet, north pole first, down a copper pipe. Copper is not ferromagnetic so the magnet is not attracted to it. But as the magnet falls into the pipe, there is a current induced in the pipe. You can think of the pipe as made of lots and lots of rings of metal, stacked on top of each other. The current induced in each ring creates a magnetic field with field lines that point upwards, just as if they were coming from a north pole below the falling magnet (Figure 8.23).

The falling magnet experiences a repulsive force, which pushes upwards, and hence slows it down. This is called magnetic braking.

Note that it doesn't matter whether it is the magnet that is moving or the pipe that is moving. The interaction between the eddy currents and the magnet acts to reduce *relative* movement, so whichever one is moving will be slowed down.

This means that whenever a moving conductor is in a magnetic field, it will be slowed down, or braked. This applies to both straight-line motion and rotation. Magnetic braking has many uses in

**Back emf**

Watch this video, which explains back emf in motors.

Lenz's Law and conservation of energy

transport and in industry. Some high-tech trains use magnetic braking, and industrial machines may use magnetic braking to slow motors. This is an important safety feature, for example on large blades in saw mills where it is desirable that blades stop spinning quickly once the saw is turned off. Magnetic braking is also used to control the flow of molten metals, for example in steel casting to control the flow into moulds. Usually for these applications large electromagnets are used rather than permanent magnets because they can be quickly switched on and off to control the braking.

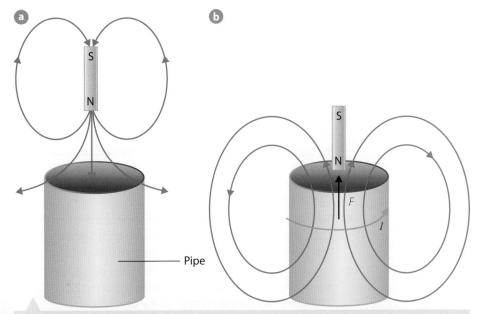

FIGURE 8.23 **a** As the magnet approaches the pipe the flux through the pipe increases. **b** The increasing flux through the pipe creates loops of current around the pipe. These current loops create magnetic fields that exert a force on the magnet, opposing its motion.

# INVESTIGATION (8.3)

## Magnetic braking

### AIM

To investigate magnetic braking of a magnet falling through different pipes
Write a hypothesis for this investigation.

Critical and creative thinking

Numeracy

### MATERIALS

- Retort stand and clamps
- Copper pipe, at least 50 cm long, approximately 5 cm in diameter
- Copper pipe of same dimensions with lengthwise slit cut into it
- Plastic pipe of same dimensions
- Magnet
- Metre ruler
- Stopwatch
- Small cushion
- Plumb-bob (weight on a string)

RISK ASSESSMENT

| WHAT ARE THE RISKS IN DOING THIS INVESTIGATION? | HOW CAN YOU MANAGE THESE RISKS TO STAY SAFE? |
|---|---|
| Magnets can break or become demagnetised when dropped. | Be careful with the magnets and put them down when not in use. Always place the cushion below the pipe before the magnet is dropped. |
| A magnet can hurt you if it falls on you. | Keep heads and hands clear of the bottom of the pipe when the magnet is dropped. |

>> What other risks are associated with your investigation, and how can you manage them?

**METHOD**

1 Measure the length of the pipes, and ensure they are all the same. Record the length.

2 Attach the metre ruler to the retort stand. Use the plumb-bob to adjust it so it is vertical.

3 Place the cushion below the bottom of the ruler. Hold the magnet close to, but not touching, the top of the ruler. Use the stopwatch to time how long it takes the magnet to fall from rest through a height equal to the length of the pipes. Repeat your measurement at least three times so you can estimate the uncertainty in the time.

4 Remove the ruler and attach the plastic pipe to the retort stand so it stands vertically (use the plumb-bob). Make sure the cushion is below the bottom of the pipe (Figure 8.24).

5 Drop the magnet, north pole down, down the pipe. Time how it long it takes to fall from rest at the top of the pipe until it emerges from the bottom of the pipe. Make repeat measurements.

6 Replace the plastic pipe with the copper pipe with the slit and repeat step 5.

7 Replace the copper pipe with the slit with the copper pipe without the slit and repeat step 5.

8 Repeat step 5 with the magnet falling south pole down through the complete copper pipe.

FIGURE 8.24 Experimental setup

**RESULTS**

Record your results for time taken to fall through the height (length) of the pipes in a table as shown:

| PIPE | $t_1$ (s) | $t_2$ (s) | $t_3$ (s) | $t_{ave}$ (s) | $v_{ave}$ (m s$^{-1}$) | $v_{final}$ (m s$^{-1}$) | $K$ gained (J) | $U_g$ lost (J) | Mechanical energy lost (J) |
|---|---|---|---|---|---|---|---|---|---|
| none | | | | | | | | | |
| plastic | | | | | | | | | |
| slit Cu | | | | | | | | | |
| solid Cu | | | | | | | | | |

Also record the length of the pipes, and note whether the times were different when the magnet was dropped south pole down compared to north pole down.

**ANALYSIS OF RESULTS**

1 Copy and complete the table above.

2 Calculate the average time from your three (or more) repeat measurements. Calculate the uncertainty in the average time from the range of the measurements.

3 Calculate the average speed of the magnet for each pipe and no pipe. Use the kinematics equation $v_{ave} = \dfrac{\text{distance}}{\text{time}}$.

4 Calculate the final speed of the magnet as it emerged from each pipe (or no pipe) by assuming constant acceleration so that $v_{final} = 2v_{ave}$. *This is an approximation, particularly in the case where resistive forces such as magnetic braking are acting.*

5 Calculate the kinetic energy gained by the magnet as it fell: $K = \dfrac{1}{2}mv_{final}^2$.

6 Calculate the gravitational potential energy lost as the magnet fell: $U_g = mg\Delta h$, where $\Delta h$ is the length of the pipes.

7 Calculate the mechanical energy lost due to resistive forces: mechanical energy lost $= U_g - K$.

>>

1 How did the energy lost to resistive forces for the magnet falling through the pipes, in particular magnetic braking, compare to when the magnet was allowed to fall freely? Where does this energy go?

2 How can you explain the difference in results for the copper pipe with the slit and the copper pipe without a slit?

3 Does it matter whether the magnet is dropped north or south pole down?

4 Was your hypothesis supported?

## CONCLUSION

Write a conclusion summarising the outcomes of your investigation.

In the investigation above you can observe how magnetic braking slows the fall of a magnet dropping through a metal pipe. The falling magnet induces eddy currents in the pipe, which create magnetic fields, which act to slow the magnet. This is a demonstration of Lenz's Law. The induced emf acts to decrease the change in flux that created it. If the induced emf acted to increase the flux, this would act to make the magnet go even faster, and we would gain more kinetic energy than the loss in gravitational potential energy – a violation of conservation of energy. There are many other examples of magnetic braking that you can experiment with. For example, you can make a pendulum where the bob is a small sheet of copper or aluminium, as in Figure 8.25. If you set the pendulum going so the bob moves between the poles of a horseshoe magnet, it will stop oscillating very quickly. You can experiment with using different pendulum bobs, such as a thin strip, a plate, or a plate with slits cut in it. Figure 8.25 illustrates one possible variation.

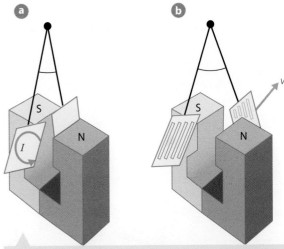

**FIGURE 8.25** Magnetically braked pendulum. **a** A metal plate is braked because of the eddy currents formed in the plate. **b** A metal plate with many slits is not effectively braked because current cannot flow across the slits.

KEY CONCEPTS

- Lenz's Law states that an induced emf always acts to produce an induced current in the direction that causes a magnetic flux that opposes the change in flux that induced the emf.
- Lenz's Law is a statement of conservation of energy.
- The change in flux due to the rotation of a coil in a DC motor creates an induced back emf across the coil.
- The back emf acts against the supplied emf and reduces the current in the coil:
  $I = \dfrac{emf_{supplied} - emf_{back}}{R}$. The faster the motor turns, the more the current is reduced.
- The back emf limits the speed at which a DC motor can run.
- Induced currents can cause magnetic braking.
- Magnetic braking is an application of Lenz's Law.
- Magnetic braking occurs when the relative movement of a magnet and a conductor is reduced because of induced currents in the conductor. These induced currents create a magnetic field that opposes the relative motion of the conductor and magnet.
- Magnetic braking is used in industry to slow moving parts and to control the flow of molten metals.

1 Describe how the motion of the coil in a DC motor creates a back emf.

2 A magnet is dropped through a copper pipe north pole downwards, and observed to accelerate at less than the acceleration due to the gravity. When the same magnet is dropped through the same pipe south pole down, will it accelerate at more or less than the acceleration due to gravity? Explain your answer.

3 Why are electromagnets rather than permanent magnets used to provide magnetic braking for large machinery?

4 Jose is making vegetable soup using his blender-stick. He puts in a carrot, and the blender blades jam. Why is it a good idea for him to turn it off right away?

5 Imagine you work for an engineering company and someone tries to sell you a design for a DC motor that runs faster and faster as long as it is connected to a battery, with no limit to its top speed. Would you invest in the patent? What advice would you give to the company director on this?

6 In Figure 8.23 (page 191), as the magnet falls north pole down through the pipe, the direction of the induced current *below* the magnet is always anticlockwise if viewed from above. What is the direction of the induced current in the section of pipe *above* the magnet? Use a diagram to help explain your answer.

» A motor transforms electric potential energy into the kinetic energy of the spinning rotor. A generator does the opposite – it converts the kinetic energy of a spinning rotor into an emf to drive a current.

» A DC motor uses a current-carrying coil in a magnetic field. The coil experiences a torque due to the field and the torque makes the coil rotate.

» The torque experienced by a current-carrying coil in a magnetic field is given by $\tau = nIAB \sin \theta$, where $n$ is the number of turns of wire in the coil, $I$ is the current through the coil, $A$ is the area of the coil, $B$ is the magnetic field strength and $\theta$ is the angle between the normal to the plane of the coil and the magnetic field.

» The main components of a DC motor are the coil, the magnets, the commutators and the brushes.

» The coil, which is mounted on an armature, carries a current and experiences a torque and rotates.

» The magnets produce the field that creates the torque.

» The commutators and brush system swap the direction of current every half-turn so that the direction of torque is constant.

» The change in flux due to the rotation of a coil in a DC motor creates an induced back emf across the coil (Lenz's Law). The back emf acts against the supplied emf and reduces the current in the coil: $I = \dfrac{\text{emf}_{\text{supplied}} - \text{emf}_{\text{back}}}{R}$. The faster the motor turns, the more the current is reduced.

» The back emf limits the speed at which a DC motor can run.

» An AC induction motor uses electromagnets to create an AC current in the rotor via induction. No current is supplied directly to the rotating coils (squirrel cage).

» AC induction motors run at a maximum frequency equal to that of the supplied AC current. In Australia this frequency is 50 Hz, or 3000 rpm, for mains power.

» Most motors in household and industrial use are AC induction motors.

» An AC generator produces an emf that varies sinusoidally with time.

» The emf produced by an AC generator is emf $= 2\pi f n BA \cos (2\pi f t)$, where $n$ is the number of turns on the coil, $B$ is the magnetic field, $A$ is the cross-sectional area of the coil which rotates with frequency $f$.

» The maximum emf produced is emf$_{\text{max}} = 2\pi f n BA$, the rms value of the emf is $\varepsilon_{\text{rms}} = \dfrac{\varepsilon_{\text{max}}}{\sqrt{2}} = \dfrac{2\pi f n BA}{\sqrt{2}} = \sqrt{2}\pi f n BA$.

» A DC generator uses a commutator to provide an output emf from each coil that is always positive. Many coils, offset at different angles, are used to produce an output that is approximately constant.

» Induced currents can cause magnetic braking.

» Magnetic braking is an application of Lenz's Law.

» Magnetic braking occurs when the relative movement of a magnet and a conductor is reduced because of induced currents in the conductor. These induced currents create a magnetic field that opposes the relative motion of the conductor and magnet.

» Magnetic braking is used in industry and transport.

# ⑧ CHAPTER REVIEW QUESTIONS

**1** What is Lenz's Law, and how is it related to energy?

**2** Why do DC motors usually have multiple coils? What is the advantage of this?

**3** Name the main components of a generator.

**4** Explain the difference between AC and DC generators. Sketch the voltage output as a function of time for each type.

**5** Why is it necessary for the direction of current in the coil of a DC motor to be reversed every half turn?

**6** How does a motor differ from a generator and in what ways are they similar?

**7** What is back emf, and how does it limit the speed of a DC motor?

**8** A generator produces an output current of $I(t) = 3.0A \sin(6.2\,s^{-1}\,t)$. For this generator, what is:

  **a** $I_{peak}$?

  **b** $I_{peak-peak}$?

  **c** $I_{ave}$?

  **d** $I_{rms}$?

**9** Figure 8.26 shows the basic features of a small DC electric motor. WXYZ is the rotating coil, connected to a DC battery. The direction of the magnetic field is parallel to the plane of the coil. The magnetic field has magnitude of 0.48T and the coil has side lengths all of 0.080 m. A current of 2.0 A flows in the coil.

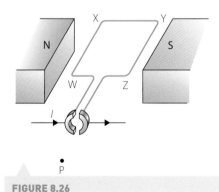

**FIGURE 8.26**

  **a** With the coil in the position shown in Figure 8.26, and looking from point P, will the coil rotate clockwise or anticlockwise?

  **b** What is the force on the side ZY?

  **c** What is the torque on the coil?

**10** A rectangular coil of length 15 cm and width 12 cm, having 60 turns, rotates in air about its axis at 1200 revolutions per minute in a uniform magnetic field of 0.40T.

  **a** What is the emf when the plane of the coil is:

    **i** perpendicular to the magnetic field?

    **ii** parallel to the magnetic field?

  **b** Draw a graph to show the time variation of the induced emf over an interval of 0.05 s.

  **c** What is the magnitude of the peak emf?

  **d** What is the rms emf?

**11** By how much does the output emf of a generator change if:

  **a** the frequency at which the coil rotates is doubled?

  **b** the magnetic field in which the coil rotates is doubled?

  **c** the number of turns on the coil is doubled?

**12** A generator is manufactured to produce an alternating emf of peak value 340V when the coil rotates at 3000 revolutions per minute in a uniform magnetic field.

  **a** Why would this be a desirable output for a generator?

  The magnetic field strength in which the coil rotates is 0.25T and the coil has a cross-sectional area of 0.040 m$^2$.

  **b** How many turns does the coil have?

**13** An AC generator has a square coil of side length 12 cm, 500 turns and a total resistance of 10 Ω. If the coil is rotated at 50 Hz in a 0.5T magnetic field, what current flows in the loop?

**14** Sketch the current through the loop in Question 13 as a function of time.

**15** **a** Explain the difference between AC and DC motors.

  **b** If you were designing a remote-control car, what sort of motor would you use and why?

  **c** If you were designing a hair dryer, what sort of motor would you use and why?

**16** Explain how magnetic braking arises and give an example of an application of magnetic braking.

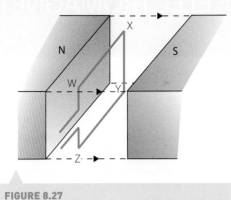

**FIGURE 8.27**

**17** Figure 8.27 shows a rectangular coil of wire WXYZ in a uniform magnetic field of magnitude 0.06 T. The coil has 160 turns. The area of each turn is 0.025 m².

If the coil has a current of 0.20 A running through it in the direction from W to X, and is in the position shown in Figure 8.27, what is:

**a** the net force acting on the coil?

**b** the net torque acting in the coil?

**c** the direction of rotation of the coil?

If the coil is turned through 90° so that it is parallel with the field and side WX is on the left, what is:

**d** the net force acting on the coil?

**e** the net torque acting in the coil?

**f** the direction of rotation of the coil?

**18** A windscreen wiper motor in normal operation is connected to the 12 V car battery, and carries a current of 1.2 A. If the resistance of the coil is 5.0 Ω, what back emf is being induced in the coil?

**19** A magnet is falling towards a loop of wire as shown in Figure 8.28.

**a** Copy Figure 8.28 and add the magnetic field lines for the magnet, the direction of induced current in the loop, the magnetic field lines due to the current, and the direction of force on the magnet.

**b** Draw a diagram showing the magnetic field lines for the magnet, the direction of induced current in the loop, the magnetic field lines due to the current, and the direction of force on the magnet just *after* it has fallen through the loop.

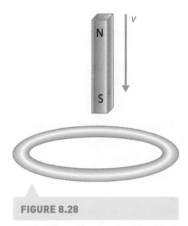

**FIGURE 8.28**

**20** Draw a flow chart showing how back emf arises.

# MODULE ⑥: ELECTROMAGNETISM

**Answer the following questions.**

1  A pair of large charged parallel plates are spaced 2.0 cm apart, one above the other. They have an approximately uniform field of $1200\,\text{N C}^{-1}$ between them, pointing upwards.

   a  If the lower plate is at a potential of 12V, what potential is the upper plate at?

   b  Draw a diagram of the plates and the field, labelling the potential difference between them and showing the electric field lines.

   c  Draw lines of equipotential on your diagram. Show the lines for 6V, 0V, and −6V.

2  An electron is initially moving upwards at $100\,\text{m s}^{-1}$ when it enters a uniform electric field of magnitude $100\,\text{V m}^{-1}$ pointing left.

   a  Sketch the path of the electron.

   b  Calculate the acceleration of the electron. Give magnitude and direction.

   c  How long after the electron enters the field does it have a velocity with a horizontal component equal to $100\,\text{m s}^{-1}$?

   d  Through what potential difference has the electron moved at this time?

3  An electron is initially moving upwards at $100\,\text{m s}^{-1}$ when it enters a uniform magnetic field of magnitude 0.10T pointing left.

   a  Sketch the path of the electron.

   b  Calculate the acceleration of the electron. Give magnitude and direction.

   c  How long after the electron enters the field does it have a velocity with a horizontal component equal to $100\,\text{m s}^{-1}$?

   d  How much has the kinetic energy of the electron changed in this time?

4  A pair of long 1 m long current-carrying wires are arranged vertically and parallel to each other, wire A on the left and wire B on the right. Wire A carries a current upwards, and experiences a force to the right.

   a  What is the direction of the current in wire B?

   b  What is the direction of the force acting on wire B?

   c  If the wires are spaced 10 cm apart and each carries a current of 2 A, what is the magnitude and direction of the force exerted on wire B?

   d  If the current in wire A is unchanged, but the current in wire B is halved, what is the force exerted on wire B now?

5  A current carrying wire is placed in a magnetic field.

   a  Under what conditions does the wire experience a force?

   b  Write the expression for the force on a current-carrying wire in a magnetic field.

   c  Calculate the minimum magnitude magnetic field required to exert a force of 1 N on a long straight 1- wire carrying a current of 1 A.

   d  At what angle between the wire and the field is the force:

      i  half the maximum value?

      ii  one quarter the maximum value?

      iii  zero?

   e  Draw a diagram showing the force acting on a current-carrying wire if the current is into the page and the magnetic field points left.

6  A loop of wire is placed in a magnetic field as shown in Figure EOM6.1, so that the loop is perpendicular to the field.

   a  Which of the following will induce an emf across the loop?

      i  rotating the loop

      ii  moving the loop to the right

      iii  changing the shape of the loop

      Explain your answer.

   b  Imagine the magnetic field shown is confined to the region shown in the figure. If the loop starts above the area shown and is gradually pulled down into the field, what is the direction of current in the loop?

   c  What will be the direction of the current as the loop continues to moves down and out of the field? Explain your answer, with reference to Lenz's Law.

   d  Sketch a graph of current in the loop as a function of time as it moves, starting above the field, until it has completely passed out of the field.

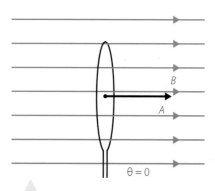

**FIGURE EOM6.1**

**7** A transformer has 300 turns on its primary coil and 600 turns on its secondary coil.

    **a** Is this a step-up or step-down transformer?

    Both coils are wound around a laminated soft iron core.

    **b** Explain the function of the iron core, and the reason why it is laminated instead of solid.

    **c** The primary coil is connected to a DC generator that supplies 12 V. What is the emf across the secondary coil?

    **d** The primary coil is connected to an AC generator that outputs $240 V_{rms}$. What is the emf across the secondary coil? Give both rms and peak value.

**8** Figure EOM6.2 shows a current carrying coil between two magnets.

    **a** Draw a diagram of the magnetic field lines between the magnets.

    **b** Draw the forces acting on sides WX and YZ.

    **c** Identify the direction of the torque on the coil.

    **d** Describe the function of the commutator, where the current enters the coil.

    **e** Calculate the torque acting on the coil if the magnetic field has magnitude 0.050 T, the input current is 1.0 A and the coil has 120 turns and side lengths of 5 cm for WX and YZ and 3 cm for XY and ZW.

    **f** Explain why the current in the coil decreases as the coil starts to rotate even though the supplied emf is constant.

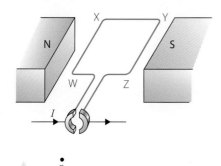

**FIGURE EOM6.2**

**9** Figure EOM6.3 shows a graph of output emf as a function of time for a generator.

    **a** Is this generator an AC or DC generator?

    **b** Identify the times at which the flux is a maximum, minimum and zero.

    **c** If the magnetic field in the generator that produces this output is 0.5 T, and the coil has an area of $0.010 m^2$, how many turns are on the coil?

    **d** Draw a graph of output emf as a function of time for this generator if the magnetic field in the generator is doubled.

    **e** Draw a graph of output emf as a function of time for this generator if the frequency at which the coil rotates in the generator is doubled.

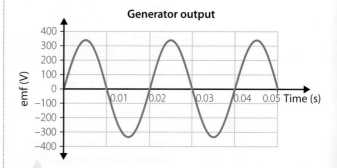

**FIGURE EOM6.3**

**10** Write a paragraph comparing AC and DC motors. Discuss the advantages and disadvantages of each, and describe where each type of motor is used.

→ Investigate the history of electromagnetism. Make a timeline of the major discoveries.

→ Perform a secondary-sourced investigation to find out how a mass spectrometer works. Make a webpage or poster summarising your findings.

→ Perform a first-hand investigation of the force exerted on current-carrying wires in magnetic fields.

→ Create a computer simulation of the forces exerted on parallel current-carrying wires using a numerical model based on the equations in chapter 6.

→ Investigate how electricity is transmitted from the power station to your home. Create a map or flow chart showing the various generators and transformers involved.

→ Build a transformer for a particular purpose, or investigate the transformers in your house.

→ Research where your household power comes from. How much comes from sources that use generators, and what energy transformations are involved?

→ Design and build a small hand-powered generator.

→ Extend Investigation 8.1 on building a motor. Design, build and test your own motor.

→ Perform a secondary-sourced investigation into industrial uses of magnetic braking. Use the internet and first-hand sources such as local manufacturers.

# THE NATURE OF LIGHT

Shutterstock.com/frankie's

# 9 Electromagnetic spectrum

OUTCOMES

**Students:**

- investigate Maxwell's contribution to the classical theory of electromagnetism, including:
  – unification of electricity and magnetism
  – prediction of electromagnetic waves
  – prediction of velocity (ACSPH113) CCT ICT
- describe the production and propagation of electromagnetic waves and relate these processes qualitatively to the predictions made by Maxwell's electromagnetic theory (ACSPH112, ACSPH113)
- conduct investigations of historical and contemporary methods used to determine the speed of light and its current relationship to the measurement of time and distance (ASCPH082) CCT ICT
- conduct an investigation to examine a variety of spectra produced by discharge tubes, reflected sunlight or incandescent filaments
- investigate how spectroscopy can be used to provide information about: ICT
  – the identification of elements
- investigate how the spectra of stars can provide information on: CCT ICT
  – surface temperature
  – rotational and translational velocity
  – density
  – chemical composition

Physics Stage 6 Syllabus © 2017 NSW Education Standards Authority (NESA) for and on behalf of the Crown in right of the State of New South Wales.

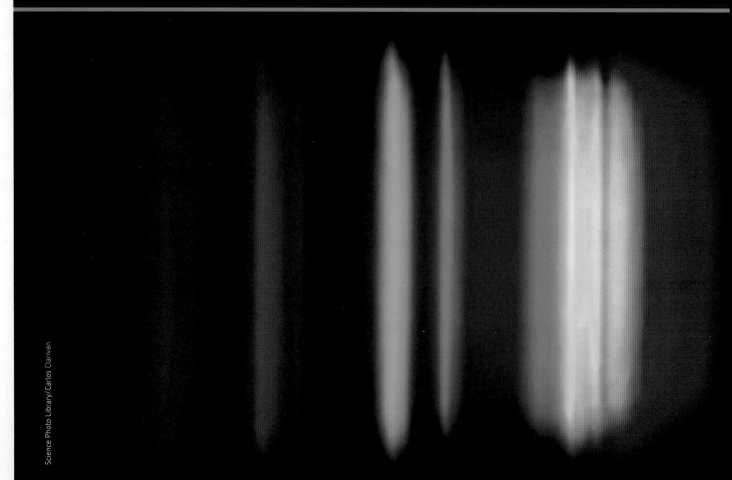

Science Photo Library/Carlos Clarivan

The attempt to understand the nature of light has attracted inquisitive minds for thousands of years. The Greek philosopher Empedocles, born around 490 BCE, proposed that we see objects because light emanates instantaneously from our eyes and touches them. French mathematician and philosopher René Descartes, foreshadowing the idea of a luminiferous aether, thought space was filled with tiny, inflexible invisible spheres he called *plenum*. Light from a distant source pushed these spheres. The pressure this put on our eyeballs caused our perception of light. Sir Isaac Newton believed light itself consisted of particles. In 1678 Christiaan Huygens put forward the wave theory of light. The particle/wave debate would inspire scientists and drive research for hundreds of years.

But nobody saw an intrinsic connection between light, electricity and magnetism, until James Clerk Maxwell developed his famous equations (Figure 9.1).

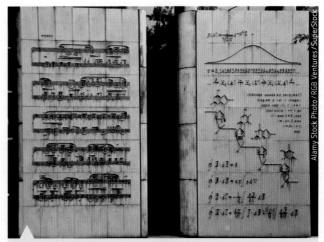

**FIGURE 9.1** Maxwell's equations on the walls outside Warsaw University (lower right)

## 9.1 Maxwell's contributions to the theory of electromagnetism

That there was a relationship between electricity and magnetism was well known by the early 19th century. One of the first people to suggest such a link was Henry Elles, who claimed in 1757 that he had written to the Royal Society in 1755 stating that 'there are some things in the power of magnetism very similar to those of electricity', although he thought them to be actually different phenomena. By the early 19th century, scientists and mathematicians were putting these ideas into a theoretical framework, backed up by practical investigations.

### The velocity of electricity

One influential experiment that was conducted in 1834 was Charles Wheatstone's demonstration, and calculation, of the velocity of electricity. He used a vessel called a Leyden jar that holds a charge (essentially a large capacitor) as a source of electricity. He arranged two pieces of wire so that one had a small gap between its end and the jar's positive terminal, the other had a small gap between its end and the negative terminal, and there was a third gap between the two wires themselves. When in operation, sparks would jump across each of the gaps.

Wheatstone mounted a tiny mirror on the works of a watch so that it would rotate rapidly, and used this to observe the three sparks. The gaps were arranged so that if the sparks were simultaneous, they would appear in a straight line in the rotating mirror. What was actually observed was that the middle one lagged behind the others, showing that it was an instant behind them. This meant that the electricity had to take a minuscule but real amount of time to travel from the ends of the wire near the jar to the gap in the middle. Knowing the speed of rotation of the mirror and the length of the half-wire, he was able to calculate that time, and arrived at the figure of $463\,491\,\text{km s}^{-1}$. We know today that this estimated value is far too high because it is well in excess of the speed of light.

Critical and creative thinking

In 1856, Wilhelm Eduard Weber and Rudolf Kohlrausch discharged a Leyden jar to measure the ratio between the magnetic and electric constants, the permeability of free space ($\mu_0$) and permittivity of free space ($\varepsilon_0$). They discovered that $\dfrac{1}{\sqrt{\mu_0\varepsilon_0}}$ was very close to the speed of light as measured by Fizeau (page 209). The next year, Gustav Kirchhoff built on the work of Wheatstone to show that the speed of electricity in a resistanceless wire was also the same. None of this was lost on Maxwell.

Maxwell was a keen experimental scientist, dedicated natural philosopher and outstanding mathematician. He occupied the position of Cavendish Professor of Mathematics at Cambridge University and was fascinated with electricity, magnetism and light. He combined work by Gauss, Faraday and Ampère, with suitable modifications, to make a series of four equations that set out the theory of electromagnetism. These he published in 1865.

His theory allowed for electromagnetic radiation to propagate through space at a rate of $3.1 \times 10^8 \, \mathrm{m\,s^{-1}}$, given by the ratio $\dfrac{1}{\sqrt{\mu_0\varepsilon_0}}$ measured by Weber and Kohlrausch. Since this was indeed close to Fizeau's speed of light, and the ratio was known to be a velocity vector by dimensional analysis, he surmised that light was in fact electromagnetic radiation. Maxwell also predicted that a large range of frequencies was possible for electromagnetic waves, well beyond the visible spectrum.

## Maxwell's equations

Maxwell found the following four laws of particular importance.

### Gauss's laws for electricity and magnetism

Gauss's Law for Electricity was formulated in 1813, based on mathematical work by Joseph-Louis Lagrange (1773). This relates the electric flux through a closed or Gaussian surface to the amount of electric charge enclosed by the surface. If there is a net charge inside a Gaussian surface, there will always be some amount of electric flux leaving it (Figure 9.2).

Gauss also stated a Law for Magnetism based on the same mathematical principles. In this case he formalised an idea, first propounded in 1269 by Petrus Peregrinus de Maricourt, that there were no magnetic monopoles. This means that there are no independent north or south magnetic poles. It implies that, if any magnetic field lines leave a Gaussian surface, they must also enter it. Therefore, the net magnetic flux through such a surface is zero.

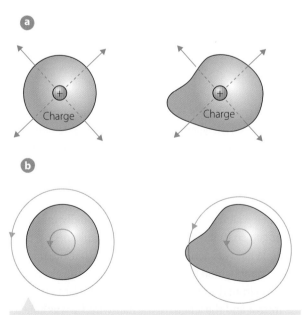

**FIGURE 9.2** Gaussian surfaces for: **a** electric; **b** magnetic fields. The shape of the surface is not important, as long as it has no sharp points and doesn't double back on itself.

### Faraday's Law of Induction

Michael Faraday discovered electromagnetic induction in 1831. As we saw in chapter 6, this law reveals a very close relationship between magnetism and electricity, which has led to many of the basic applications upon which the world now depends. Faraday was the originator of the concept of lines of force. Maxwell formulated Faraday's ideas mathematically and they appear as the third of his equations.

### Ampère's Circuital Law

In 1826 André-Marie Ampère discovered the relationship of the net magnetic field around a closed loop to the strength of the

9780170409131

electric current flowing through the loop. This relationship is illustrated in Figure 9.3. It was formalised by Maxwell in 1855, based on work he had done in hydrodynamics. In this form it was only applicable to situations where the current and the magnetic field enclosing it were steady.

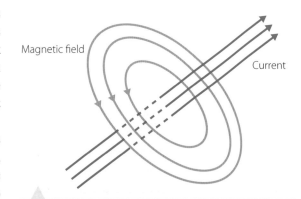

**FIGURE 9.3** Ampère's Circuital Law

To make the equation work for situations where the currents and charges were variable, such as in a capacitor (similar to the electric plates studied in Year 11), Maxwell had to make a correction. He added a new element called the 'displacement current', which represents the slight separation of electrons from their atoms in a material when a varying electric field passes through it. This ultimately required the use of the permeability and permittivity of free space. It allowed the equation to describe both magnetic and electric fields changing in space as sine waves, and so permitted electromagnetic radiation.

In fact, the wave equations show that $c^2 = \dfrac{1}{\mu_0\varepsilon_0}$, which meant that the speed of propagation of these magnetic and electric fields could be predicted if $\mu_0$ and $\varepsilon_0$ could be measured, as they had been by Weber and Kohlrausch. Figure 9.4 shows the equations in their final form.

$$\nabla \cdot \mathbf{E} = \frac{\rho}{\varepsilon_0} \qquad \oint \mathbf{E} \cdot d\mathbf{A} = \frac{q_{enc}}{\varepsilon_0}$$

$$\nabla \cdot \mathbf{B} = 0 \qquad \oint \mathbf{B} \cdot d\mathbf{A} = 0$$

$$\nabla \times \mathbf{E} = -\frac{\partial \mathbf{B}}{\partial t} \qquad \oint \mathbf{E} \cdot d\mathbf{s} = -\frac{d\Phi_B}{dt}$$

$$\nabla \times \mathbf{B} = \mu_0\mathbf{J} + \mu_0\varepsilon_0\frac{\partial \mathbf{E}}{\partial t} \qquad \oint \mathbf{B} \cdot d\mathbf{s} = \mu_0\varepsilon_0\frac{d\Phi_E}{dt} + \mu_0 i_{enc}$$

**FIGURE 9.4** Maxwell's equations

WS

Maxwell's contribution to the theory of electromagnetism

**KEY CONCEPTS**

- Wheatstone first measured the speed of electricity in a wire in 1834. This was refined by Kirchhoff in 1857.
- Weber and Kohlrausch experimentally measured the ratio of $\mu_0$ and $\varepsilon_0$. They discovered that $\dfrac{1}{\sqrt{\mu_0\varepsilon_0}} = 3.1 \times 10^8\,\mathrm{m\,s^{-1}}$.
- Maxwell set out the theory of electromagnetism in four equations.
  He did this by extending and mathematically formalising separate observations that others had made.
- Maxwell's equations state:
  1  Point charges radiate an electric field outwards.
  2  Magnetic field lines are always loops – there are no magnetic monopoles.
  3  A changing magnetic field creates a changing electric field.
  4  A changing electric field creates a changing magnetic field.
- Maxwell's fourth equation involves the electric constant and the magnetic permeability of free space and permittivity of free space ($\mu_0$ and $\varepsilon_0$), which can be measured in a laboratory.
- Since the speed of propagation of electromagnetic waves is related to $\mu_0$ and $\varepsilon_0$, it could be predicted theoretically.
- The predicted speed of these waves closely matched the measured value of the speed of light, indicating that light was electromagnetic radiation.
- Maxwell also predicted that a large range of frequencies was possible for electromagnetic waves, well beyond the visible spectrum.

1 Gauss's Law for Electricity is the basis of Maxwell's first equation.

   **a** Provide a detailed explanation of what this law means.

   **b** Why does it imply that electric fields originate on charged particles?

2 Gauss's Law for Magnetism is the basis of Maxwell's second equation.

   **a** Provide a detailed explanation what this law means.

   **b** Why does it imply that there are no magnetic monopoles?

3 What was important about the way Maxwell modified Ampère's circuital law?

4 What does the speed of light have to do with electricity in a resistanceless wire?

# 9.2 The production and propagation of electromagnetic waves

Critical and creative thinking

Information and communication technology capability

## Electric force

As Maxwell's first equation (Gauss's Law for Electricity) implies, every charge gives off an electric field. In *Physics in Focus Year 11* chapter 12, we noted that the strength of the fields drops off as the square of distance. Other charges 'feel' this force and are either attracted or repelled according to Coulomb's Law of force:

$$|F| = k_e \frac{|q_1 q_2|}{r^2}$$

Here, $|F|$ is the absolute value of the force, $q_1$ and $q_2$ are the charges, $r$ is the separation and $k_e$, Coulomb's constant, is given by

$$k_e = \frac{1}{4\pi\varepsilon_0}$$

Two things to notice here are how similar this is to Newton's Universal Law of Gravitation, and that it involves the permittivity of free space.

Because there is a force between the charges, they are accelerated ($F = ma$) and so, in the absence of any restriction, they will move.

Now consider a point in space past which one of these charges is moving, as shown in Figure 9.5. The electric field, emanating from the moving charge as it approaches, will be increasing – not around the charge itself, but at that point in space. It will decrease as the charge recedes. So, at this point, we have a changing electric field.

Maxwell's fourth equation, the modified version of Ampère's circuital law, says that a changing electric field creates a changing magnetic field. His third equation, Faraday's Law of Induction, says that a changing magnetic field creates a changing electric field. This then creates a changing magnetic field, and so on. These fields propagate away at the speed of light, again according to Maxwell's fourth equation.

Remember that an electric field on its own, without an originating charged particle nearby, is sufficient to make a charge move because it provides a force $F$ on a charge $q$, with magnitude $F = Eq$. In this way, energy that is propagated by the action of mutually interacting electric and magnetic fields can be transformed into the kinetic energy of a charge at an unlimited distance (Figure 9.6). Notice also that no medium is required: the electric and magnetic fields only need each other, nothing more. This also implies that moving charges generate electromagnetic radiation.

The production and propagation of electromagnetic waves

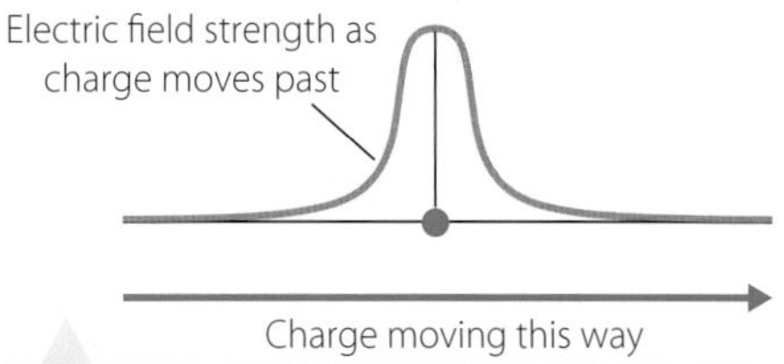

Electric field strength as charge moves past

Charge moving this way

**FIGURE 9.5** The electric field at a point in space varies as a charge moves past it.

## Magnetic force

As we saw in chapter 5, magnetic fields also apply a force on moving charged particles, given by:

$$F = qvB\sin\theta$$

where $F$ is the force, $q$ is the charge, $v$ is the velocity of the charged particle, $B$ is the strength of the magnetic field and $\theta$ is the angle between the particle's trajectory and the direction of the magnetic field. Note that the force is greatest when the magnetic field direction is perpendicular to the charged particle's path ($\sin\theta = 1$).

Maxwell's equations, plus the two equations governing the force produced on a charged particle by electric and magnetic fields, give us a complete theory of electromagnetic interactions.

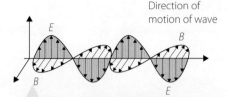

**FIGURE 9.6** The electric and magnetic fields oscillate at right angles to each other, while the wave travels in the third dimension at right angles to both the electric and magnetic fields.

**KEY CONCEPTS**

- Moving charges generate electromagnetic radiation.
- When a charge moves past any point in space, that point experiences a changing electric field.
- A changing electric field creates a changing magnetic field.
- A changing magnetic field creates a changing electric field.
- These mutually creating fields do not need a medium to propagate through; they only need each other.
- This radiation can travel an unlimited distance at the speed of light.
- Any charge at any distance will respond to the electric field in the radiation and be accelerated according to $F = qE$.
- Similarly, any charge will experience a force $F$ from a magnetic field according to $F = qvB\sin\theta$.
- Thus electromagnetic radiation can be transformed to kinetic energy at any distance.

**CHECK YOUR UNDERSTANDING**

**9.2**

1. An electron and a proton are separated by a distance of $2.5 \times 10^{-11}$ m.
   a. What is the force experienced by each particle?
   b. What is the acceleration of the electron?
   c. What is the magnitude of the electric field that each particle experiences?

2. An object travelling in a circular path of radius $r$ has an acceleration $a = \dfrac{v^2}{r}$. Assuming that the electron in Question **1** is in a circular orbit:
   a. what is the velocity of the electron?
   b. how long does it take the electron to travel one complete orbit?
   c. what is the frequency of the radiation this electron should generate?
   d. what is the wavelength of the radiation the electron should generate?

# 9.3 Determining the speed of light

Until the 17th century, people thought that light travelled any distance instantaneously. The first known attempt to measure the speed of light was by Galileo. In 1638, when he was in Florence, he and his assistant each had covered lanterns and stood on hilltops with a known distance between them. Galileo uncovered his light and his assistant, upon seeing Galileo's light, immediately opened the shutter on his own. When Galileo saw the return light he noted the time, probably using a water clock. His conclusion was that the speed of light was at least ten times the speed of sound, although he realised that his results were likely to be inaccurate due to the small time interval, human reaction time, and the limits of the equipment he was using. He correctly concluded that light speed was too fast to be measured by this method.

Critical and creative thinking

Information and communication technology capability

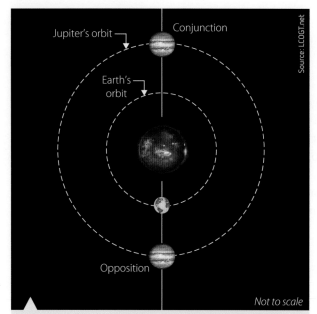

FIGURE 9.7 Rømer proposed that variations in the times of eclipses of Jupiter's moon Io were due to light having to travel different distances across Earth's orbit.

In the figure: Jupiter's orbit, Conjunction, Earth's orbit, Opposition, Not to scale, Source: LCOGT.net

## Measuring time

Knowing your longitude is vital for oceanic navigation. Longitude can be worked out if the time the Sun rises can be measured, but in the 17th century, before the invention of accurate mechanical timepieces, there was no good way of telling the time aboard ship.

Galileo proposed a method based on observing Jupiter's moon Io going into and coming out of eclipses caused by Jupiter's shadow. These were generally predictable but turned out to be not as accurate as expected.

Danish astronomer Ole Rømer noticed that the variations in times of the eclipses – a maximum of 16.6 minutes – coincided with times that Jupiter was closer to, or further away from, Earth due to their different orbital speeds around the Sun. He explained the time variations as being mostly caused by light having to cross the diameter of Earth's orbit to a greater or lesser degree (Figure 9.7). In 1676, he made a rough estimate of the time it would take light to travel the distance from the Sun to Earth of about 11 minutes (about $214\,000\,\mathrm{km\,s^{-1}}$).

The actual diameter of Earth's orbit was not known accurately at the time and so his results were well off the value we understand today (8 minutes 19 seconds at $300\,000\,\mathrm{km\,s^{-1}}$). However, the significance of Rømer's work is that he was the first person to suggest that light travelled at a finite speed. As Christiaan Huygens wrote to Jean-Baptiste Colbert, the French Minister of Finance at the time, 'I have seen recently, with much pleasure, the beautiful discovery of Mr. Rømer, to demonstrate that light takes time in propagating, and even to measure this time'.

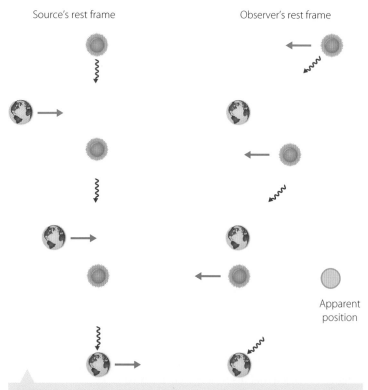

Source's rest frame          Observer's rest frame

Apparent position

FIGURE 9.8 Aberration of light (not to scale). The apparent position of the source of light is ahead of its actual position when the observer is moving with respect to it.

## A different angle

Englishman James Bradley was both an astronomer and priest. He worked with Samuel Molyneux to try to determine the parallax of the star Gamma Draconis. This would have allowed them to work out the distance to the star. They did not succeed, thereby showing that the stars were much further away than was thought at the time. However, they did discover an unexplained circular motion of the star's apparent position. Shortly after Molyneux's death in 1728, Bradley realised how the phenomenon – the aberration of light – came about.

From Earth's point of view, the stars move. The fact that the Earth's journey around the Sun is the cause of this apparent motion makes no difference except to change the direction of a star's apparent motion, yielding a circular pattern. If we move towards a star (not head-on), the light we see was emitted some time previously. Since then, the star has moved towards us from our point of view. The light we see shows the star's position as it was then, not as it is now, so it appears to be ahead of where it should be (Figure 9.8).

The angles involved are very small and depend only on $\frac{v}{c}$, where $v$ is Earth's velocity around the Sun and $c$ is the speed of light. Knowing Earth's orbital velocity, and using careful measurements of the angle, Bradley was able to determine the speed of light. He calculated a value of $301\,000\,\mathrm{km\,s^{-1}}$, a considerable improvement on Rømer's result.

## Down to Earth

In 1848–49, Armand Hippolyte Louis Fizeau used a method similar to that of Galileo, but with better technology. He generated a light that was reflected from a distant mirror, giving a total path length of about 19 km.

The light was focused by lenses coming and going, and shone through a wheel with 720 teeth evenly placed around its rim, as shown schematically in Figure 9.9. The gaps between the teeth were the same width as the teeth themselves, and the wheel was capable of rotating at hundreds of times per second. The light would shine out through a gap and the rotational speed of the wheel was adjusted until no light shone back through the gap. This allowed the time for the light's round trip to be measured based on the distance the wheel had rotated at the set speed. Using this method, Fizeau determined the speed of light to be about $315\,000\,\mathrm{km\,s^{-1}}$.

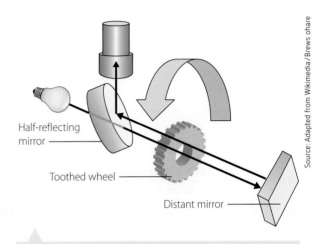

Half-reflecting mirror

Toothed wheel

Distant mirror

**FIGURE 9.9** Fizeau's apparatus

## Reflecting on light

Léon Foucault had been a friend and collaborator with Fizeau. They evidently had a falling out and Fizeau pursued his method for determining the speed of light on his own.

Foucault used an improved version of Fizeau's equipment. It had been recommended to him that he consider using Wheatstone's idea of a rapidly revolving mirror in his investigations of the speed of light, and he replaced the rotating cogwheel with a rotating mirror (Figure 9.10). A light shone onto the rotating mirror, which reflected it onto a fixed mirror. This fixed mirror reflected it back onto the rotating mirror, which in turn reflected it back to the source, having first rotated through a slight angle. Measuring this angle let Foucault determine

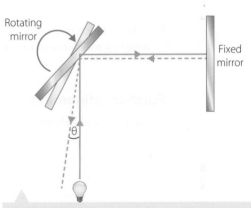

Rotating mirror

Fixed mirror

$\theta$

**FIGURE 9.10** Foucault's apparatus

how far the mirror had rotated, and this permitted the calculation of the speed of light.

Let the distance between the rotating and fixed mirrors be $d$. The time it takes for light to travel from the rotating mirror to the fixed one and back again is:

$$t = \frac{2d}{c}$$

where $c$ is the speed of light.

Let the rotating mirror rotate at a constant $\omega$ radians per second. The angle it moves through during the light's round trip, $\theta$, is given by:

$$\theta = \frac{2d\omega}{c} = \omega t$$

and so it follows that:

$$c = \frac{2d\omega}{\theta}$$

In 1862 Foucault arrived at a value of $298\,000\,\mathrm{km\,s^{-1}}$. His equipment had the added advantage that the space that the light passed through could be filled with various substances, so that the speed of light through these substances could also be measured.

A student trying to duplicate Foucault's method of determining the speed of light used an experimental set-up where the distance between the rotating and fixed mirrors was $10\,\text{m}$. The mirror rotated at $1000\,\text{rps}$ ($=2\pi \times 1000$ radians per second). In order to achieve the same result as Foucault, what angle (degrees) must the mirror have moved through? Note that one complete revolution is $2\pi$ radians $= 360°$.

| ANSWER | LOGIC |
|---|---|
| $d = 10\,\text{m}$, $\omega = 360 \times 1000$ degrees s$^{-1}$ | ▪ Identify the relevant data in the question. |
| $c = \dfrac{2d\omega}{\theta}$ | ▪ Use the formula relating angle, distance and angular speed. |
| $\theta = \dfrac{2d\omega}{c}$ | ▪ Rearrange to find the angle. |
| $\theta = \dfrac{2 \times (10\ \text{m}) \times (360 \times 1000\ \text{degrees s}^{-1})}{2.98 \times 10^{8}\ \text{m s}^{-1}}$ | ▪ Substitute the correct values with units. |
| $\theta = 0.024°$ | ▪ State the final answer with correct units and appropriate significant figures. |

**TRY THESE YOURSELF**

1 What separation between the mirrors, $d$, would be needed to measure the speed of light at $3.00 \times 10^{8}\,\text{m s}^{-1}$ with an angle of $0.030°$ if the rotation rate were $900\,\text{rps}$?

2 What rotation rate would be needed to measure the speed of light as $3.00 \times 10^{8}\,\text{m s}^{-1}$ given a mirror separation distance of $15\,\text{m}$ and a measured angle of $0.05°$?

## Further refinements

At the urging of Urbain le Verrier, Director of the Paris Observatory, Marie Alfred Cornu conducted a number of experiments between 1872 and 1876, using refinements of Fizeau's method. His last experiment used a light path almost three times as long as Fizeau's, and yielded a value of $300\,400\,\text{km s}^{-1}$, which is within 0.2% of the value we know today.

Albert Michelson also ran a series of experiments between 1877 and 1931. He used Foucault's method, with improvements. In 1879 he achieved a value of $299\,944 \pm 51\,\text{km s}^{-1}$, within about 0.05% of the modern value. His 1926 experiment, incorporating even more refinements, achieved a value of $299\,796 \pm 4\,\text{km s}^{-1}$, just $4\,\text{km s}^{-1}$ above today's accepted value.

### Modern measurements

The definition of a metre was changed in 1983 to be exactly $\dfrac{1}{299\,792\,458}$ times the distance that light travels in a second, in a vacuum. Therefore, today we have an exact value for the speed of light: $299\,792\,458\,\text{m s}^{-1}$. Measurement techniques that produce even higher accuracy will have implications for the length of the metre, not the speed of light.

Determining the speed of light

### Resonant cavity

Further improvements have been made on methods already discussed. However, a series of measurements made by Louis Essen and A C Gordon-Smith in the period from 1946 to 1950 used a different technique known as cavity resonance. This involved independently measuring the frequency $f$, and the wavelength $\lambda$, of light, and then determining the speed $c$ through the relationship $c = f\lambda$.

A **resonant cavity** is a physical container that will sustain standing waves with particular frequencies. The condition is that an integral number of half-wavelengths must fit exactly between the walls of the cavity, as shown in Figure 9.11. If such transverse waves meet a solid wall, they are reflected at 180°. Points where the wave amplitude is always zero are called nodes, and the waves must have a node at each cavity wall, much like the bridge and the nut on a guitar.

For the Essen–Gordon–Smith experiment, the wavelength could be found exactly by knowing the distance between the opposite walls of the resonant cavity. They determined their distance using gauges calibrated by interferometry and achieved an accuracy of ±0.8 μm. They set the frequency using an electric circuit that oscillates at a known frequency.

## Interferometry

An interferometer is a device that splits a beam of electromagnetic radiation of known frequency $f$ into two beams that are directed to follow different paths. When the beams are recombined they form an interference pattern, as shown in Figure 9.12. The pattern depends on the difference in length between each path and the wavelength of the radiation. When the recombined beams are in phase there is constructive interference and a bright spot will be detected. A phase difference of 180° will cause a dark spot.

By carefully adjusting the path length and measuring the **path difference** while observing the interference pattern, the radiation's wavelength can be determined. The speed of light can then be calculated using $c = f\lambda$.

Originally, coherent radio sources were used for such measurements. However, radio wavelengths are long, the shortest being about half a centimetre, so precision was limited. Precision was considerably improved with the advent of lasers using coherent beams of light and having wavelengths in the hundreds of nanometres. This made it more difficult to directly measure the frequency but, using advanced techniques, the speed of light was measured by the United States National Bureau of Standards (now NIST) with a fractional uncertainty of $3.5 \times 10^{-9}$.

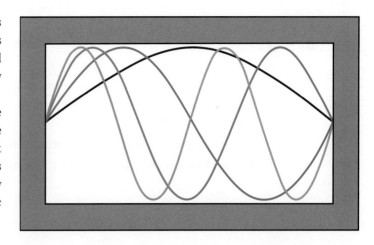

| Number of wavelengths | Nodes |
|---|---|
| $\frac{1}{2}$ | 2 |
| 1 | 3 |
| $1\frac{1}{2}$ | 4 |
| 2 | 5 |

**FIGURE 9.11** Resonant cavity with electromagnetic standing waves

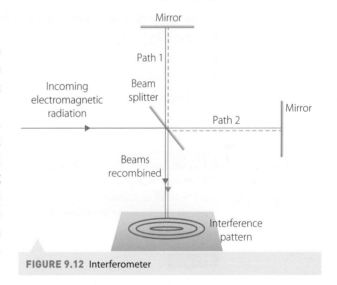

**FIGURE 9.12** Interferometer

---

▶ **WORKED EXAMPLE** (9.2)

An antenna for a radio station transmits a signal with frequency of 106.3 MHz. What is the wavelength of this electromagnetic wave?

| ANSWER | LOGIC |
|---|---|
| $f = 106.3\,\text{MHz}$ | ▪ Identify the relevant data in the question. |
| $c = f\lambda$ | ▪ Use wave equation for velocity of light to relate frequency and wavelength. |
| $\lambda = \dfrac{c}{f}$ | ▪ Rearrange for wavelength. |

| $\lambda = \dfrac{3.00 \times 10^8 \text{ m s}^{-1}}{106.3 \text{ MHz}}$ | ▪ Substitute values with correct units. |
|---|---|
| $\lambda = 2.82 \text{ m}$ | ▪ State the final answer with correct units and appropriate significant figures. |

**TRY THESE YOURSELF**

1 A microwave oven has a frequency of 2.45 GHz. What is its wavelength?

2 A red laser has a frequency of $4.76 \times 10^{14}$ Hz. What is its wavelength?

## INVESTIGATION  9.1

# Determining the speed of light with a resonant cavity

Critical and creative thinking

### AIM

To determine the speed of light

### MATERIALS

- Microwave oven of known frequency (should be written on the back)
- Either 1 kg block of chocolate or equivalent **or** 500 g margarine
- Scissors
- Knife or spatula
- Baking paper
- Ruler
- Calculator

**RISK ASSESSMENT**

| WHAT ARE THE RISKS IN DOING THIS INVESTIGATION? | HOW CAN YOU MANAGE THESE RISKS TO STAY SAFE? |
|---|---|
| Microwaved substances can be hot. | Take care when removing chocolate or margarine from the microwave oven. |
| Possible slight leakage of microwaves | Stand away from the microwave door when in operation. |

What other risks are associated with your investigation, and how can you manage them?

### METHOD

1 Remove the turntable from the microwave oven.

2 Cut enough baking paper from its roll to completely cover the floor of the microwave oven.

3 Make a small hole in the baking paper to enable it to go over the spindle that normally drives the microwave turntable.

4 a If using margarine, cover the baking paper to a depth of about 4 mm with the margarine and then insert it into the microwave.

b If using chocolate, put baking paper into the microwave. Put chocolate face down (smooth side up) to cover the baking paper.

5 Close the door and run a 1000 W microwave at high power for about 35 seconds. Your time may vary depending on the power of your microwave.

6 Examine the margarine or chocolate for hot spots. They should be separated by a number of centimetres. Do not wait too long because the hot spots will spread out via thermal conduction, making it more difficult to determine the centre of the spots.

7 Measure the distance between the centres of adjacent hot spots. 》

### ANALYSIS OF RESULTS

1   Note the frequency $f$ of the microwave oven.
2   Tabulate the distances between hotspots you discovered $\left(\dfrac{\lambda}{2}\right)$.
3   For each measured distance, calculate the speed of light using the formula $c = f\lambda$. Remember to use the correct units.

### DISCUSSION

Write a discussion that includes an analysis of your results.

1   Reliability and validity.
2   How does your result vary from the published value of the speed of light ($299\,792\,458\,\text{m s}^{-1}$)?
3   How might you improve the results in further trials?

### CONCLUSION

With reference to the data obtained and its analysis, write a conclusion based on the aim of this investigation.

## INVESTIGATION 9.2

## Teacher demonstration: Determining the speed of sound with an analogue of Rømer's method

While determining the speed of light using Rømer's method is beyond the facilities of most high schools, the basic method can be applied to determining the speed of another kind of wave – sound.

Critical and creative thinking

### AIM

To determine the speed of sound

### MATERIALS

- Starter's pistol (borrow from PDHPE department)
- Ear muffs
- Stopwatch

| WHAT ARE THE RISKS IN DOING THIS INVESTIGATION? | HOW CAN YOU MANAGE THESE RISKS TO STAY SAFE? |
| --- | --- |
| Firing a starter's pistol can cause temporary impairment to hearing. | The person firing the starter's pistol should wear ear muffs.<br>Other people should stand well away when the pistol is fired. |

RISK ASSESSMENT

What other risks are associated with your investigation, and how can you manage them?

### METHOD

1   The person firing the pistol should go to one end of the school's oval, or any large open space available. Observer(s), with stopwatch, should stand approximately half-way across the oval, watching the person with the pistol.
2   The pistol-bearer fires the pistol upwards. The observers start the stopwatch when they see the flash, and stop it when they hear the report.
3   The observers note the time taken between flash and report.

4 The observers move to the opposite end of the oval, pacing (not measuring) to get an estimate of the distance (it should be about 50 metres).

5 The pistol-bearer again fires the pistol and the observers record the time, using the stopwatch, between the flash and the report.

## ANALYSIS OF RESULTS

1 Determine the difference in arrival times between the first and second observations.

2 Using the approximate distance between the first and second positions, calculate the speed of sound.

## DISCUSSION

1 In what ways is this investigation similar to the method Rømer used to determine the speed of light?

2 In what ways do the two investigations differ significantly?

3 Comment on the accuracy and reliability of this investigation.

4 Compare these comments with the accuracy and reliability inherent in Rømer's method.

## CONCLUSION

With reference to the data obtained and its analysis, write a conclusion based on the aim of this investigation.

KEY CONCEPTS

- Galileo had a reasonable method for determining the speed of light but lacked an accurate timing device.
- Danish astronomer Ole Rømer, following a suggestion made by Galileo, used variations in the times for Io's eclipses to estimate the time it took light to travel across the diameter of Earth's orbit around the Sun.
- James Bradley discovered the aberration of light and used it to obtain a better estimate of the speed of light than Rømer.
- Armand Hippolyte Louis Fizeau used a method similar to that of Galileo, but with better technology. He shone a light through the gaps in a wheel with 720 cogs, which was reflected back by a mirror for a total path length of about 19 km. Knowing the angular speed of the wheel, Fizeau could work out the speed of light when the return beam was blocked by a tooth on the wheel.
- Léon Foucault used a rapidly revolving mirror and a shorter path to work out a better value for the speed of light than Fizeau.
- Marie Alfred Cornu refined and improved upon Fizeau's design and achieved a result just $4 \, km \, s^{-1}$ above today's accepted value.
- A resonant cavity is a container for waves that requires a standing wave inside it to have stationary points (nodes) on its walls.
- Louis Essen and A C Gordon-Smith generated a beam of known frequency and measured its wavelength, achieving an accuracy of $\pm 0.8 \, \mu m$, using a resonant cavity. They then used $c = f\lambda$ to work out $c$.
- Interferometry splits a beam into two different beams that travel different paths, are reflected back and recombine. They form an interference pattern that can be used to determine the wavelength.
- The United States National Bureau of Standards used interferometry to measure $c$ with a fractional uncertainty of $3.5 \times 10^{-9}$.
- The definition of a metre in was changed in 1983 to be exactly $\dfrac{1}{299\,792\,458}$ times the distance that light travels in a second, in a vacuum, so today we have an exact value for the speed of light: $299\,792\,458 \, m \, s^{-1}$.

1   With the aid of a diagram, describe the aberration of light as discovered by James Bradley.

2   If the distance between the rotating and fixed mirrors in a Foucault apparatus has a total path length of 25 m, and the rotating mirror spins at 1500 rps, what is the angle that the mirror travels through?

3   Describe how a resonant cavity can be used to determine the frequency of an electromagnetic wave that it contains.

4   A radio station has a frequency of 92.9 MHz. What is its wavelength?

5   How does an interferometer work?

6   What limits the precision of radio-based interferometers?

# 9.4  The electromagnetic spectrum

## Refractive index

The permeability and permittivity of free space appear in Maxwell's equations because they relate to the rate at which magnetic and electric fields propagate in a vacuum. Magnetic and electric fields can still propagate in substances, but usually at a slower rate. This is because of atomic interactions, all of which take time. At a macro level, this all adds up to the beam slowing down while it is in the substance.

This means that, in a refractive medium, the value of the *permittivity* ($\varepsilon_r$) in particular becomes greater than it is in a vacuum ($\varepsilon_0$). The value of a material's permeability, $\mu_r$, is very close to $\mu_0$ for most normal materials so we can ignore the difference. So, $\dfrac{1}{\sqrt{\mu_r \varepsilon_r}}$ is smaller than $\dfrac{1}{\sqrt{\mu_0 \varepsilon_0}}$, and therefore the speed of light in the material, $v$, is less than in a vacuum. We call the ratio of $\dfrac{c}{v}$ the refractive index of the material and give it the symbol $n$.

$$n = \frac{c}{v}$$

The refractive index actually changes according to wavelength of the radiation in a material. The change in refractive index also causes a change in the angle of refraction. This causes the different wavelengths and frequencies of electromagnetic radiation to spread out when leaving a material if the exit surface is not parallel to the entry surface. We call this phenomenon dispersion. It is responsible for separating out the colours that compose white light when a light beam passes through a prism or water droplet. The range of electromagnetic wavelengths laid out in this manner is called the electromagnetic spectrum (Figure 9.13).

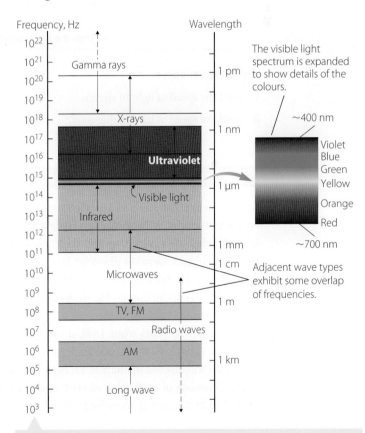

**FIGURE 9.13** The electromagnetic spectrum

**TABLE 9.1** Refractive index of some common materials

| MATERIAL | REFRACTIVE INDEX |
|---|---|
| Air | 1.0003 |
| Water | 1.33 |
| Ethanol | 1.36 |
| Zircon | 1.92 |
| Diamond | 2.42 |

Consider Table 9.1 above. What is the speed of light in ethanol?

| ANSWER | LOGIC |
|---|---|
| $n = 1.36$ | • Identify the relevant data in the question. |
| $n = \dfrac{c}{v}$ | • Use the formula relating relative refractive index to speed of light in the material. |
| $v = \dfrac{c}{n}$ | • Rearrange for $v$. |
| $v = \dfrac{3.00 \times 10^8 \text{ m s}^{-1}}{1.36}$ | • Substitute values with correct units. Note that refractive index has no units because it is a ratio. |
| $v = 2.21 \times 10^8 \text{ m s}^{-1}$ | • State the final answer with correct units and appropriate significant figures. |

**TRY THESE YOURSELF**

1 Find the speed of light in zircon.

2 Find the speed of light in diamond.

## Spectral lines and identifying elements

Spectroscopy is the study of how electromagnetic radiation interacts with matter. One of its earliest uses was in astronomy. William Wallace Hyde was an English physician. He was examining the Sun's spectrum in 1802 when he noticed several dark gaps. He thought that these might be the natural boundaries between colours. German physicist Joseph von Fraunhofer saw the lines in 1814 and catalogued about 500 of them. He noticed that the 'D line' in the yellow part of the spectrum corresponded to a known bright line in candle flames. He showed that the spectrum of Venus matched that of the Sun, and also took spectra of a number of stars.

For about 20 years from 1848, a number of scientists, including Foucault, Bunsen and Kirchhoff, identified various solar spectral lines with the known emission lines of various elements.

As we will discuss in chapter 11, light can be considered as a stream of particles. A light particle, or photon, consists of oscillating electric and magnetic fields, and has an exact amount of energy that is related to its frequency, according to the formula:

$$E = hf$$

Here, $E$ represents the energy of the photon, $h$ is Planck's constant, and $f$ is the frequency. The particle model is useful because it explains why atoms absorb or emit specific wavelengths of electromagnetic radiation.

Every electron around each atom has an exact energy level. The lowest possible energy level an electron can occupy is called its ground state, although there can be only one electron in any particular state. Now, Maxwell's equations say that all moving charges generate electromagnetic radiation. This means they should lose energy. However, ground state electrons do not radiate energy and these states are stable (see chapter 10). They can, however, absorb energy if, in so doing, they are able to transition to a higher energy level within the atom, or to leave it altogether.

For instance, the single electron in a hydrogen atom is in its ground state in the lowest shell, or lowest energy state possible. If a photon interacts with it with exactly the right amount of energy, it can be promoted to the second shell. At this point it is unstable and will in due course decay back to the first shell, emitting a photon of exactly that amount of energy in the process.

Different photons with different energies could promote such an electron to the third, fourth, fifth or higher shells, or remove it altogether. Higher-shell electrons will fall back to lower energy levels, or shells, giving out photons of the appropriate energies. Figure 9.14 illustrates a transition between the second and third energy levels.

For single atoms, there is therefore a set of energy transitions that its electrons may undergo. The set is like a signature, and every different type of atom – every element – has its own unique set. This is because each element has a different number of protons so the electrostatic attraction of each nucleus is different, leading to different energy levels of the shells, subshells and orbitals.

Transitions from the first atomic shell require more energy than those from shells further out. The energy range for first shell electrons is generally in the ultraviolet so we can't detect them with our eyes, although we can with suitable instruments.

From thermodynamics (*Physics in Focus Year 11* chapter 11), you will recall that an incandescent light source is one that is heated, such as an electric globe with a tungsten filament, or a star. The radiation from an incandescent source contains all frequencies, with its peak wavelength related to its temperature. When the peak wavelength of a continuous spectrum corresponds to green, the object appears white, not green. This is because at such a temperature there is a significant amount of blue being emitted. Our brain interprets this colour mixture as white.

If the gaseous atoms of an element are viewed against the background of an incandescent source, a spectrum will be observed, but there will be dark lines where certain frequencies are missing. These frequencies will have been absorbed by the atoms in the sample. We call these black lines absorption lines.

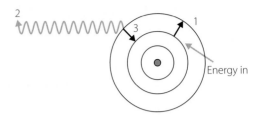

An electron jumps up an energy level (1) when it absorbs energy. It releases the energy as a photon of light (2) with a set frequency when it returns to its original energy level (3).

**FIGURE 9.14** The source of emission spectra

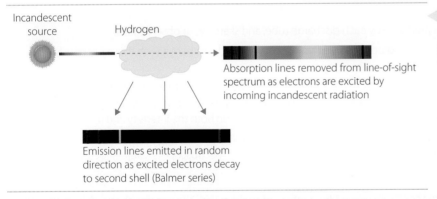

Incandescent source
Hydrogen

Absorption lines removed from line-of-sight spectrum as electrons are excited by incoming incandescent radiation

Emission lines emitted in random direction as excited electrons decay to second shell (Balmer series)

**FIGURE 9.15**
Absorption and emission spectra of hydrogen

As electrons that have absorbed photons fall back to their original energy levels, only a small fraction of the photons of light they emit continue in the same direction as the photons that were absorbed. If you are viewing the sample from an angle, you do not have the incandescent background. What you see are emission lines emitted by the previously excited electrons as they decay to lower energy levels.

Whether the sample is in laboratory equipment, a star or a nebula many light years away, the principle is the same. The emission and/or absorption lines can identify the elements involved. This technique also applies to electrons in molecules. Each different molecule also has its own unique signature. Figure 9.15 depicts the absorption and emission spectra of hydrogen.

The electromagnetic spectrum

## Teacher demonstration: Investigating visible spectral lines

### AIM

To examine a variety of spectra produced by discharge tubes, reflected sunlight, or incandescent filaments
Write an inquiry question for this investigation.

### MATERIALS

- Hand-held spectroscope
- Induction coil
- 12V incandescent globe
- 12V power supply
- Leads and alligator clips
- Discharge tubes containing a variety of different gases (sodium and mercury are two that are suitable; see Figure 9.16)

| WHAT ARE THE RISKS IN DOING THIS INVESTIGATION? | HOW CAN YOU MANAGE THESE RISKS TO STAY SAFE? |
| --- | --- |
| Sunlight can be blinding. | If observing outside, use reflection from a white wall. Do not look directly at the Sun or a reflection in a mirrored surface. |
| Irradiation with X-rays | Students should stand at least 3 m from operating discharge tubes. |

What other risks are associated with your investigation, and how can you manage them?

**FIGURE 9.16** The emission spectrum produced by a mercury discharge tube

### METHOD

1 Set up the induction coil and discharge tubes in a darkened room.

2 Observe the spectra produced by each discharge tube, and sketch your observations.

3 Next, observe the spectrum produced by a fluorescent light and compare it with those previously observed from discharge tubes.

4 Observe the spectrum produced from the incandescent globe (without other sources of light present) and contrast this spectrum with those observed in steps 2 and 3. Repeat using different voltage settings on the power supply, noting the effects of the change in temperature on both the intensity and the range of colour.

5 Finally, go outside and observe the spectrum from *reflected* sunlight. *Never point the spectroscope towards the Sun – damage to your retina may result!* (You must ensure that the spectroscope is carefully focused for this part.)

### DISCUSSION

Compare and contrast the spectrum viewed from reflected sunlight with the spectra observed from the other sources.

### RESULTS

1 Record all your observations.

2 Give an answer to your inquiry question.

### CONCLUSION

Write a conclusion summarising your findings.

## WORKED EXAMPLE (9.4)

**TABLE 9.2** Emission line wavelengths for the hydrogen Balmer series

| TRANSITION OF $n$ | COLOUR | WAVELENGTH (nm) |
|---|---|---|
| $3 \rightarrow 2$ | Red | 656.3 |
| $4 \rightarrow 2$ | Aqua | 486.1 |
| $5 \rightarrow 2$ | Blue | 434.1 |
| $6 \rightarrow 2$ | Violet | 410.2 |

Table 9.2 shows the wavelengths of light emitted by excited electrons as they drop to the second shell of a hydrogen atom from higher energy shells. The $n$ in the heading of column 1 refers to the shell numbers. The set of transitions into shell 2 is called the Balmer series.

What is the transition energy from shell 3 to shell 2? Planck's constant $h = 6.63 \times 10^{-34}$ J s. Energies involving electrons at atomic levels are often stated in electron volts (eV) rather than joules, where $1\,\text{eV} = 1.602 \times 10^{-19}$ J. Express your answer in eV.

| ANSWER | LOGIC |
|---|---|
| $h = 6.63 \times 10^{-34}$ J s, $\lambda = 656.3$ nm, $1\,\text{eV} = 1.602 \times 10^{-19}$ J | • Identify the relevant data in the question. |
| $E = hf$ and $f = \dfrac{c}{\lambda}$ | • Use the formulae relating energy to frequency, and frequency to wavelength. |
| $E = \dfrac{hc}{\lambda}$ | • Rearrange to find the energy. |
| $E = \dfrac{6.63 \times 10^{-34}\ \text{J s} \times 3.00 \times 10^{8}\ \text{m s}^{-1}}{656.3 \times 10^{-9}\ \text{m}}$ | • Substitute the correct values with units. |
| $= 3.03 \times 10^{-19}$ J | • Calculate the final value. |
| $= 1.89$ eV | • State the final answer with correct units and appropriate significant figures. |

**TRY THESE YOURSELF**

1   What is the energy transition from shell 4 to shell 2?

2   What is the energy transition from shell 6 to shell 2?

- The speed at which electric and magnetic fields are propagated depends on the permittivity of free space and permeability of free space ($\varepsilon_0$ and $\mu_0$) or of the refractive medium they are travelling through ($\varepsilon_r$ and $\mu_r$).
- The refractive index, $n$, of a material is given by $n = \dfrac{c}{v}$, where $v$ is the speed of light in the material.
- Refractive index changes according to the wavelength of radiation in a material.
- This causes dispersion, leading to the spreading out of spectra.
- Electromagnetic radiation can be thought of as a stream of particles called photons, each having a particular frequency and wavelength.
- An atomic electron can absorb packets with the right energy to promote them to a higher energy level in the atom.
- Bulk material that absorbs particular frequencies gives rise to absorption lines in spectra.
- Electrons that decay back to lower energy states radiate photons with a frequency that represents the energy difference between higher and lower states.
- Such radiation gives rise to emission lines.
- Each element has a unique signature of emission/absorption lines.

1 Define 'spectroscopy'.

2 What physical property changes inside a material and gives rise to a change in the speed of light in that material?

3 Why does white light separate into a spectrum when passed through a prism?

4 What is the speed of light in water?

5 How much energy is required for an electron to transition from shell 2 to shell 5 in a hydrogen atom?

# 9.5 Astronomical applications of spectroscopy

The classification of stars by their spectra began before astronomers fully understood the link between the patterns within the spectra and the surface temperature of the star. Angelo Secci, an Italian astrophysicist, was one of the first scientists to authoritatively declare that the Sun is a star. In the 1860s Secci divided the spectra of about 4000 stars into five basic groups, with subtypes. He used the stars' colours and the number of spectral lines as part of his classification scheme.

The classification system used the letters A through to O based on the strength of the visible hydrogen absorption lines present (the Balmer series). It happens that a surface temperature of about 10 000 K produces the strongest Balmer series absorption lines. These stars were assigned the spectral type A based upon this. Cooler red stars exhibit very weak Balmer series lines in their spectra, and were assigned the letter M for their spectral type. Very hot stars have no discernible Balmer series lines. These stars were assigned the letter O. The reason for the lack of hydrogen lines in such stars is that at temperatures above 20 000 K hydrogen is completely ionised. That is, the electron in hydrogen responsible for producing the hydrogen lines in the spectrum is no longer associated with the nucleus of the hydrogen atom (a single proton). It exists as a free electron.

Subsequent observation of **black body radiation** experiments and laboratory observations of hot gas spectra enabled astronomers to match the spectral types to surface temperatures of stars. (Black body radiation is the characteristic pattern of electromagnetic radiation given off by objects due to their temperature.) The previous alphabetical order was found to need a complete overhaul. Rather than re-assign all the letters, the spectral types were simply placed in the order hottest to coolest and simplified to eliminate overlapping and confusing spectra. This work was mainly done at Harvard University from 1918 to about 1924. The re-organised order, O B A F G K M, is still used today in the **Hertzsprung–Russell diagram**, a useful tool used by astronomers to assist in the classification of stars.

> Black body radiation is discussed further in chapter 11.

**Information from astronomical spectra**

Explore this survey of information that can be obtained from stellar spectra and the methods used to obtain it.

Key features of a star's spectrum are now used by astronomers when classifying the star. These include the appearance and intensity of spectral lines, the relative thickness of certain absorption lines, and the wavelength at which peak intensity occurs.

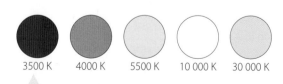

| 3500 K | 4000 K | 5500 K | 10 000 K | 30 000 K |

**FIGURE 9.17** How the colour of a star varies with surface temperature

## Stellar surface temperature

A star is a **black body**, emitting all frequencies of electromagnetic radiation from their cores. As with any incandescent body, the peak wavelength indicates the temperature. Therefore, we can determine the surface temperature of a star by its apparent colour, as illustrated in Figure 9.17.

Table 9.3 lists some spectral features of stars that are related to their surface temperature.

**TABLE 9.3** The characteristics of the different spectral classes of stars

| SPECTRAL CLASS | EFFECTIVE TEMPERATURE (K) | COLOUR | H BALMER FEATURES | OTHER FEATURES |
|---|---|---|---|---|
| O | 28 000–50 000 | Blue | Weak | Ionised He$^+$ lines, strong UV continuum |
| B | 10 000–28 000 | Blue–white | Medium | Neutral He lines |
| A | 7500–10 000 | White | Strong | Strong H lines, ionised metal lines |
| F | 6000–7500 | White–yellow | Medium | Weak ionised Ca$^+$ |
| G | 4900–6000 | Yellow | Weak | Ionised Ca$^+$, metal lines |
| K | 3500–4900 | Orange | Very weak | Ca$^+$, Fe, strong molecules, CH, CN |
| M | 2000–3500 | Red | Very weak | Molecular lines, e.g. TiO, neutral metals |
| L? | <2000 | Tentative new (2000) classification for very low mass stars | | |

**Black body radiation**

Use the animations to set the temperature and note the wavelength at which peak intensity occurs.

Information and communication technology capability

A Doppler effect simulation for the effect on sound waves

## Rotational and translational velocity

The relative velocity of a star either approaching or moving away from an observer can be measured by the blue shift or red shift exhibited in the star's spectrum. The **Doppler effect** is the shortening of the wavelength of the light from a source that is approaching an observer and the lengthening of wavelengths from sources moving away (Figure 9.18). The Doppler effect with sound may be noticed when an emergency vehicle passes with the siren on. The relative speed of the vehicle causes an increase in the pitch of the siren and then a decrease after it passes. This effect is also very noticeable for racing cars, as their high speed is a significant fraction (almost one-third) of the speed of sound.

Star not moving

Star moving away from us

**FIGURE 9.18** An example of red shift evident in a star's spectrum

If a star is rotating, one side is moving towards us while the other side is moving away, as shown in Figure 9.19. This results in the absorption lines within the spectrum being both red and blue shifted simultaneously, so that they appear broader than expected. Careful measurement of the amount of broadening, along with an estimate of the size of the star, can lead to the calculation of the rotational velocity of the star.

Many astrophysical objects emit spectra. Nebulae are concentrations of gas and dust that are illuminated from within, and we see their emission spectra. Edwin Hubble was able to determine the **red shift** of galaxies by examining their spectral lines.

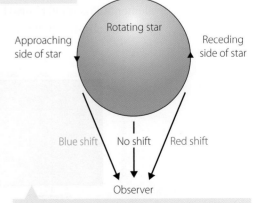

**FIGURE 9.19** How a rotating star has its spectral lines broadened due to the Doppler effect

## Density

It is very useful to know the density of a star's atmosphere. The largest supergiant stars have the lowest densities, while Main Sequence stars have higher densities. Finding the luminosity class of a star and its spectral type allows for a very good estimate of the star's absolute magnitude, or total amount of light the star is emitting, from which its distance can be calculated. Lower density stellar atmospheres produce sharper, narrower spectral lines. This is due to the motion of the atoms and ions that are absorbing the radiation and producing the lines. The particles in lower density gases travel further before each collision with other particles. The absorption lines they produce are sharper. As giant stars have less gravity near their surface, the pressure of the gases in the star's atmosphere, where the absorption spectrum is being produced, is also lower. A gas under lower pressure produces finer spectral lines, so giant stars have finer spectral lines.

## Chemical composition

Photons are absorbed by gases in a star's outer layers, and then re-emitted in all directions, as shown in Figure 9.20. This enables us to determine which elements are present, and hence their chemical composition.

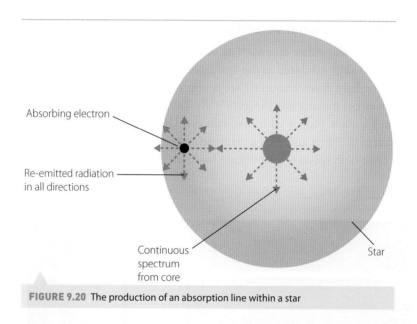

**FIGURE 9.20** The production of an absorption line within a star

Stars like the Sun have small quantities of many elements in their atmospheres. Each of these elements produces its own characteristic spectral lines. Calcium, potassium and iron are three such elements. Fraunhofer lines (Figure 9.21) are named after Joseph von Fraunhofer, who carefully observed the Sun's absorption lines in the 1800s (see page 216). They are due to the many elements absorbing

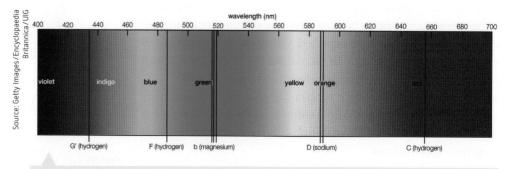

Source: Getty Images / Encyclopaedia Britannica / UIG

**FIGURE 9.21** Fraunhofer lines visible in the Sun's spectrum

9780170409131

radiation at particular wavelengths. Figure 9.21 shows the Fraunhofer lines in the spectrum of the Sun. Matching the absorption lines found in a star's spectrum with absorption lines produced by an element under laboratory conditions verifies the existence of that element in the star's atmosphere. The relative intensity of the absorption lines indicates the abundance of that element.

Astronomical applications of spectroscopy

**CHECK YOUR UNDERSTANDING**

**9.5**

1  What was the basis of Secci's classification of stars?
2  Why do very hot (O) stars have no discernible Balmer series lines?
3  What key features of a star's spectrum are now used by astronomers when classifying the star?
4  How is a star's temperature determined?
5  What feature of a star's spectrum can distinguish between rotational and translational velocity?
6  Why is it important to know a star's absolute magnitude?
7  How is it possible to determine the relative abundances of elements present in a star?

- Wheatstone first measured the speed of electricity in a wire in 1834. This study was refined by Kirchhoff in 1857.

- Weber and Kohlrausch experimentally measured the ratio of $\mu_0$ and $\varepsilon_0$. They discovered that $\dfrac{1}{\sqrt{\mu_0 \varepsilon_0}} = 3.1 \times 10^8\,\text{m s}^{-1}$.

- Maxwell set out the theory of electromagnetism in four equations.

- Maxwell's equations state:

  **1** Point charges radiate an electric field outwards.

  **2** Magnetic field lines are always loops – there are no magnetic monopoles.

  **3** A changing magnetic field creates a changing electric field.

  **4** A changing electric field creates a changing magnetic field.

- Maxwell's fourth equation involves the permeability of free space and the permittivity of free space ($\mu_0$ and $\varepsilon_0$), which can be measured in a laboratory.

- Since the speed of propagation of electromagnetic waves is related to $\mu_0$ and $\varepsilon_0$, it could be theoretically predicted.

- The predicted speed of these waves closely matched the measured value of the speed of light, indicating that light was electromagnetic radiation.

- Maxwell also predicted that a large range of frequencies was possible for electromagnetic waves, well beyond the visible spectrum.

- A changing electric field creates a changing magnetic field.

- A changing magnetic field creates a changing electric field.

- These mutually-creating fields do not need a medium to propagate through; they only need each other.

- This radiation can travel an unlimited distance at the speed of light.

- Galileo had a reasonable method for determining the speed of light but lacked an accurate timing device.

- Cornu refined and improved upon Fizeau's design and achieved a result just $4\,\text{km s}^{-1}$ above today's accepted value.

- The definition of a metre was changed in 1983 to be exactly $\dfrac{1}{299\,792\,458}$ times the distance that light travels in a second, in a vacuum, so today we have an exact value for the speed of light: $299\,792\,458\,\text{m s}^{-1}$.

- The speed at which electric and magnetic fields are propagated depends on the permeability and permittivity of free space ($\mu_0$ and $\varepsilon_0$) or the refractive medium they are travelling through ($\mu_r$ and $\varepsilon_r$).

- The refractive index, $n$, of a material is given by $n = \dfrac{c}{v}$, where $v$ is the speed of light in the material.

- Refractive index changes according to the wavelength of radiation in a material.

- This causes dispersion, leading to the spreading out of spectra.

- Electromagnetic radiation can be thought of as a stream of particles called photons, each having a particular frequency and wavelength.

- An atomic electron can absorb packets with the right energy to promote them to a higher energy level in the atom.

- Bulk material that absorbs particular frequencies gives rise to absorption lines in spectra.

- Electrons that decay back to lower energy states radiate photons with a frequency that represents the energy difference between higher and lower states.

- Such radiation gives rise to emission lines.

- Each element has a unique signature of emission/ absorption lines.

- The spectral types of stars are placed in order from hottest to coolest and simplified to eliminate overlapping and confusing spectra. The order is O B A F G K M. An even cooler, lower-mass star, type L, has been suggested.

- The key features of a star's spectrum include the appearance and intensity of spectral lines, the relative thickness of certain absorption lines, and the wavelength at which peak intensity occurs.

- Stars are black bodies, so the peak wavelength emitted indicates the temperature. Therefore, we can determine the surface temperature of a star by its apparent colour.

- The relative velocity of a star either approaching or moving away from an observer can be measured by the blue shift or red shift in the star's spectrum (Doppler effect).

- If a star is rotating, one side is moving towards us while the other side is moving away. This results in the absorption lines being both red- and blue-shifted simultaneously.

- Careful measurement of the amount of broadening, along with an estimate of the size of the star, can lead to the calculation of the rotational velocity of the star.

- Finding the luminosity class of a star and its spectral type allows for a very good estimate of the star's absolute magnitude, from which its distance can be calculated.

- Lower density stellar atmospheres produce sharper, narrower spectral lines.

- Giant stars have less gravity, and hence lower pressure near their surface, so their spectral lines are finer.

- Matching the absorption lines found in a star's spectrum with absorption lines produced by an element under laboratory conditions shows which elements are in the star's atmosphere.

- The relative intensity of the absorption lines indicates elemental abundances.

# ⑨ CHAPTER REVIEW QUESTIONS

Review quiz

1 How did Newton's idea of light differ from that of Maxwell?

2 The charge on a proton is $1.602 \times 10^{-19}$ C. Protons are typically separated by about $1 \times 10^{-15}$ m within an atomic nucleus. What is the magnitude of the electrostatic force, in Newtons, between two such protons?

3 Rømer noticed that the variations in times of the eclipses of Io had a maximum of 16.6 minutes.

  a Based on this information, and given that $c = 3.00 \times 10^8\,\text{m}\,\text{s}^{-1}$, calculate the radius of Earth's orbit.

  b What astronomical factors might create errors in this estimate?

4 Crown glass is often used in scientific and medical instruments. Its refractive index is 1.52. Calculate the speed of light in crown glass.

5 The wavelength of the photon needed for an electron in the principal energy shell or 'ground state' ($n = 1$) to escape completely from a hydrogen atom is 91.18 nm. What is the energy of the photon?

6 Compare and contrast an absorption spectrum, an emission spectrum and a continuous spectrum.

7 How is a continuous spectrum produced?

8 The spectrum from a distant galaxy appears to be a continuous spectrum. Why?

9 The individual spectral lines within a star's spectrum all appear slightly shifted towards the red end. What does this tell us about the relative motion of this star?

10 Outline the information that can be found from the analysis of a star's spectrum.

11 A star that has been observed for many years does not seem to be moving relative to nearby stars; however, its spectral lines are shifted towards the blue end of the spectrum. What does this tell us about the motion of the star relative to us?

12 Use a diagram to explain how the spectral lines of a rotating star are red- and blue-shifted simultaneously.

13 How would the spectrum of a star with an atmosphere rich in metallic elements differ from that of a star that has a very low or non-existent abundance of metallic elements?

14 How would the peak wavelength of radiation emitted from a red star be different from the peak wavelength emitted by a white star?

15 Outline how the density of a star and hence its luminosity class can be found from its spectrum.

# 10 Light: wave model

## INQUIRY QUESTION

What evidence supports the classical wave model of light and what predictions can be made using this model?

## OUTCOMES

**Students:**

- conduct investigations to analyse qualitatively the diffraction of light (ACSPH048, ACSPH076)  ICT
- conduct investigations to analyse quantitatively the interference of light using double slit apparatus and diffraction gratings $d\sin\theta = m\lambda$ (ACSPH116, ACSPH117, ACSPH140) ICT, N
- analyse the experimental evidence that supported the models of light that were proposed by Newton and Huygens (ACSPH050, ACSPH118, ACSPH123) CCT
- conduct investigations quantitatively using the relationship of Malus' Law $I = I_{max}\cos^2\theta$ for plane polarisation of light, to evaluate the significance of polarisation in developing a model for light (ACSPH050, ACSPH076, ACSPH120) ICT, N

Physics Stage 6 Syllabus © 2017 NSW Education Standards Authority (NESA) for and on behalf of the Crown in right of the State of New South Wales.

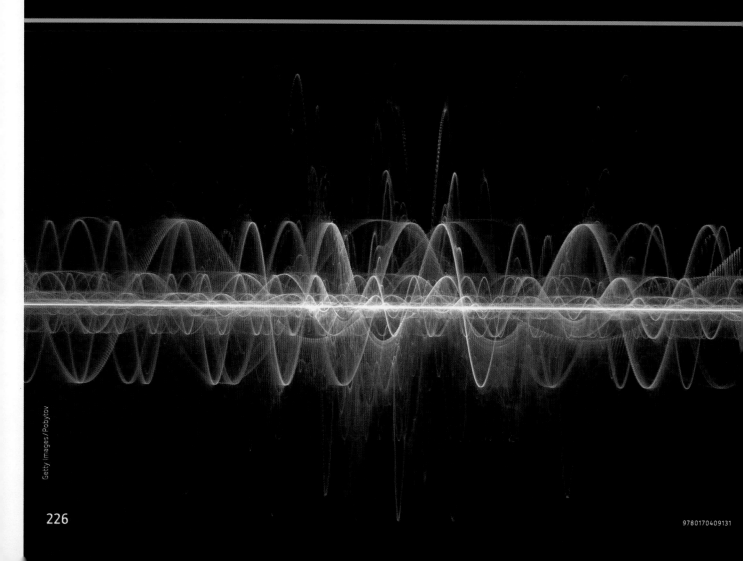

9780170409131

The nature of light has long intrigued philosophers. In Greece in the 5th century BCE, Democritus, who along with Leucippus proposed the original idea of atoms, argued that everything is composed of indivisible components, including particles of light.

Around the year 1000 CE, the Arab scholar Ibn al-Haytham (also known as Alhazen) wrote a seven-volume treatise on the behaviour of light called the *Book of Optics*. In it he debunked the Greek philospher Ptolemy's theory that light emanated from the eyes, allowing us to see what it touched, by the evidence that eyes suffer damage from bright light. Ibn al-Haytham also discussed the behaviour of light, such as reflection and refraction, but always from the point of view of light as particles.

Source: GIPhotoStock/SCIENCE PHOTO LIBRARY

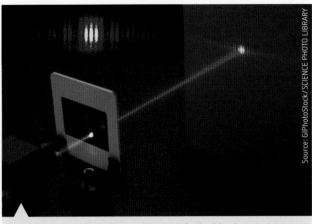

**FIGURE 10.1** The interference pattern of a double-slit experiment

What was missing from these early ideas was evidence, other than that light travelled in a straight line.

# 10.1 Competing models

By the 17th century, there were two competing models for light – the 'corpuscular' or particle model favoured by Isaac Newton, and the 'undulatory' or wave model proposed by Christiaan Huygens.

## Newton's corpuscular theory

Newton published his second major book on physical science, *Opticks,* in 1704 (Figure 10.2). It included the topics of refraction, diffraction and colour mixtures using lights and pigments. He held the view that light consisted of tiny particles or **corpuscles** that had mass. The basic principles of his theory were:

- Light consists of small particles.
- These particles have mass and obey the laws of physics in the same way as any other particles, such as billiard balls or planets.
- The particles are so small that, when two beams cross, they do not scatter each other.

OPTICKS:
OR, A
TREATISE
OF THE
REFLEXIONS, REFRACTIONS, INFLEXIONS and COLOURS
OF
LIGHT.
ALSO
Two TREATISES
OF THE
SPECIES and MAGNITUDE
OF
Curvilinear Figures.

LONDON,
Printed for SAM. SMITH, and BENJ. WALFORD, Printers to the Royal Society, at the *Prince's Arms* in St. *Paul's* Church-yard. MDCCIV.

Alamy Stock Photo/Granger Historical Picture Archive

**FIGURE 10.2** Isaac Newton's *Opticks*, published 1704

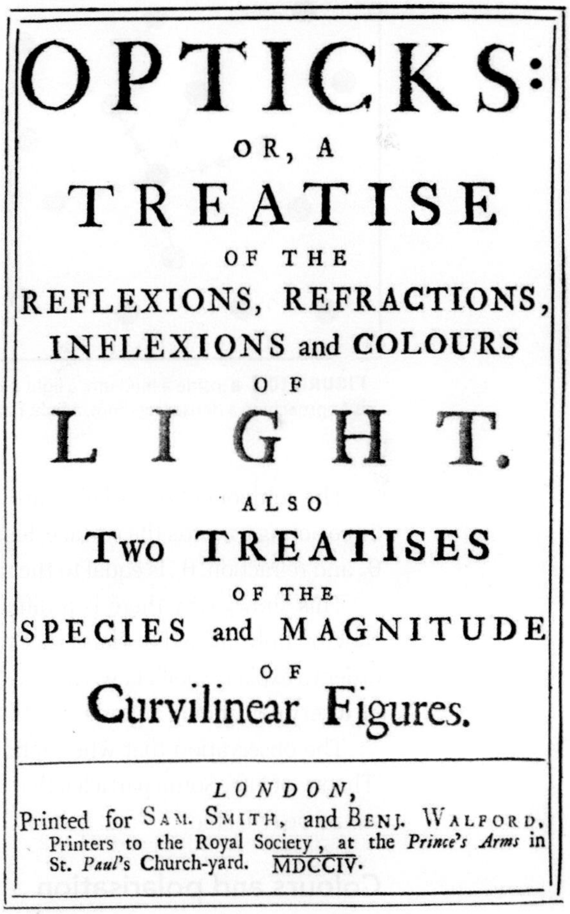

### Shadows, diffraction and reflection

Newton viewed the idea of **rectilinear propagation** as evidence for the theory. If light particles travelled in straight lines, they would make sharp shadows, which he observed. One of the implications of this theory is that the particles would be subject to gravity and would therefore undergo parabolic motion in a gravity field such as that of Earth.

You studied Snell's Law and refraction in chapter 10 of *Physics in Focus Year 11*.

This also led to Newton's explanation of **diffraction**. If light travels in straight lines, it would not travel around corners. He explained observed diffraction effects as the interaction of particles running into each other at the edges of objects.

The theory could explain the law of reflection of light. The angle of incidence equals the angle of reflection. This is the same way a moving particle, such as a ball bearing, undergoes an elastic, frictionless collision on a smooth surface.

## Newton and Snell's Law

Refraction could also be explained, with the further assumption that particles of light were attracted to particles of matter. According to this model, a light particle deep within a refractive medium would experience no net force because it is attracted equally in all directions (Figure 10.3a). However, as it approaches the surface of a denser medium from an oblique angle, the short-range attractive force from all the particles in that medium would yield a net force in the direction normal to the surface as shown in Figure 10.3b. The particle would accelerate and its path (the light ray) would therefore bend in a parabolic arc towards the normal as it entered the denser medium, which matches observations.

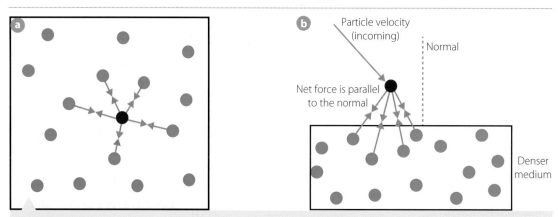

**FIGURE 10.3** **a** Inside a medium, a light particle would be attracted in all directions so there would be no net force. **b** Approaching a denser medium, a light particle would be attracted in a direction normal to the surface.

The component of the light particle's velocity parallel to the boundary would be unaffected by the net force acting towards the surface. Figure 10.4 shows how the ratio of the sines of the angles of incidence, $\theta_i$, and refraction, $\theta_r$, is equal to the ratio of the velocities of the light particle in water and air.

This shows why there is a different value for the index of refraction between different materials. Newton only needed to claim that the speed of light was different in different transparent materials in order to explain Snell's Law. He believed this to be a real triumph, and wrote in *Opticks* that 'I take this to be a very convincing argument of the full truth of this proposition.'

The observation that when refraction occurs, some light will be also reflected he explained by the 'Theory of Fits'. Some particles that could fit between atoms would be refracted. Those that couldn't fit would be reflected.

## Colours and polarisation

Noting that the refraction of light depends on its colour, with red refracting least and violet refracting most, Newton's solution was that the mass of a light particle varied with its colour, giving them more, or less, inertia. The force on these particles would produce less, or more, acceleration as they approached the optical medium interface, causing them to bend different amounts – red less than violet.

Look at a light source through two sheets of polarised plastic material and rotate one against the other. We observe that the intensity of the light varies from a maximum to a minimum with a rotation of 90°, with further rotation bringing it back to maximum again. This shows polarisation. If light particles

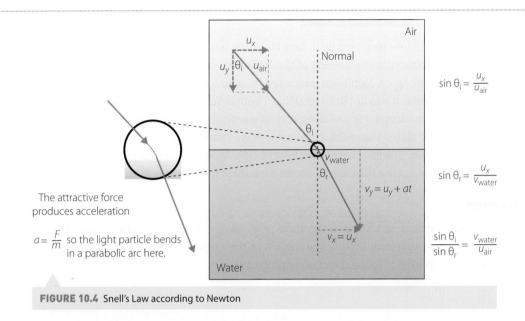

$$\sin \theta_i = \frac{u_x}{u_{air}}$$

$$\sin \theta_r = \frac{u_x}{v_{water}}$$

$$v_y = u_y + at$$

$$v_x = u_x$$

$$\frac{\sin \theta_i}{\sin \theta_r} = \frac{v_{water}}{u_{air}}$$

The attractive force produces acceleration

$a = \frac{F}{m}$ so the light particle bends in a parabolic arc here.

**FIGURE 10.4** Snell's Law according to Newton

were spherical, this could not happen. Newton wrote that light particles must therefore have sides although the exact shape – from rectangular to plate-like – was difficult to visualise.

## The decisive test

We saw in chapter 9 that Foucault, in the 19th century, used a rotating mirror to measure the speed of light. He applied this technique to testing Newton's particle theory of light, which depended on the speed of light being faster in water than in air.

Foucault reflected a light beam off a mirror that rotated at 800 revolutions per second, driven by a small steam turbine, to a fixed mirror 9 m away. The light was reflected back, taking 60 ns to return, by which time the mirror had rotated slightly. The beam was therefore deflected slightly below the source. When the rotating mirror was not at the appropriate angle to send the beam to the fixed mirror, the beam was reflected elsewhere and lost.

A 3 m tube filled with water was then introduced into the light path between the mirrors, as shown in Figure 10.5. The light therefore had to pass through it twice. If, as Newton had suggested, light travelled faster through the water, then the mirror should rotate less during its journey and the return beam should be deflected closer to the light source. In fact, the reverse was true. Light that travelled through the water deflected to a point further away from the source than the reference beam, proving that light travelled more slowly in water than in air.

The experiment was decisive and, since Newton's theory depended entirely on his ideas about light speed being faster in denser optical media, the theory had to be abandoned. This is a classic example of how quickly a major theory can collapse. As Albert Einstein observed in 1919, 'A theory can thus be recognized as erroneous if there is a logical error in its deductions, or as inadequate if a fact is not in agreement with its consequences. But the truth of a theory can never be proven'. This is often paraphrased as 'No amount of experimentation can ever prove me right; a single experiment can prove me wrong.'

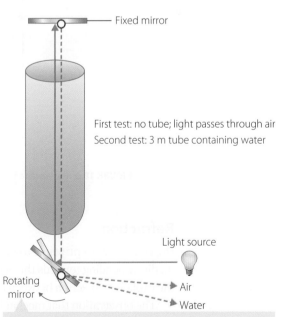

Fixed mirror

First test: no tube; light passes through air
Second test: 3 m tube containing water

Light source

Rotating mirror

Air

Water

**FIGURE 10.5** Foucault's mechanism for testing Newton's particle theory of light

## Huygens and the wave model

The wave model of light was proposed by Robert Hooke in 1665 and improved by Huygens. In particular, Huygens' principle explained some aspects of the behaviour of light where Newton's theory was weak, although it was not strong on rectilinear propagation, which is why Newton rejected it. Huygens was largely ignored at the time. It was not until 1802 that Thomas Young performed his double-slit experiment and determined the wavelengths of visible light, thus vindicating Huygens' ideas.

Huygens' principle states that each point on a wave behaves as a point source for waves in the direction of propagation. The line tangent to these circular waves is the new position of the wave front a short time later. We do not see the circular waves because of the principle of superposition.

You met many of the properties of waves in chapter 8 of *Physics in Focus Year 11.*

### Reflection

Wavefronts meet a reflecting surface at the angle of incidence. At this point, new wavelets are created, with their associated wavefront leaving the surface at the angle of reflection (Figure 10.6). Remember that Huygens' principle refers to the direction of propagation.

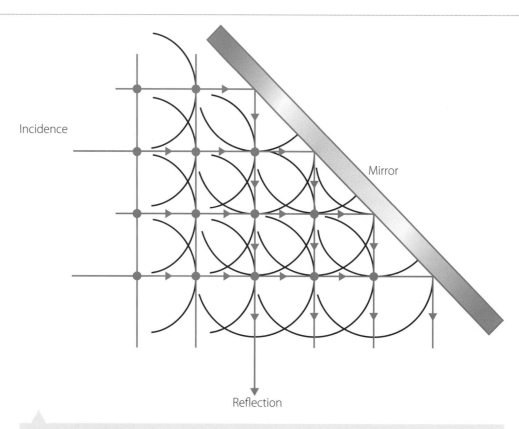

**FIGURE 10.6** The wavefront occurs where the tangents to the circular wavelets line up.

### Refraction

Huygens used his principle to propose that waves slow down when they encounter an optically denser medium, bending towards the normal and obeying Snell's Law. Note that Huygens did not need to make the explicit connection between optical density and material density as did Newton.

Competing models

The observation that mechanical waves show partial refraction and partial reflection when there is a change of velocity gave strong support to Huygens. This effect depends on the angle of incidence and is most pronounced when wave velocity increases.

9780170409131

## Diffraction and rectilinear propagation

Diffraction is the bending of light as it passes the edge of an object or goes through a gap. The amount of diffraction depends on the relationship between the size of the opening and the wavelength (Figure 10.7). Light rays going through an open doorway into a room leave sharp shadows. On the other hand, a sound coming through a doorway can easily be heard throughout the room. This is because the wavelengths of sounds are very approximately of the same order of magnitude as doors. Once the wavelength of visible light was determined to be between 400 and 700 nm, it was clear that

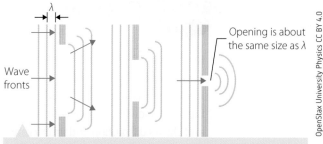

OpenStax University Physics CC BY 4.0

**FIGURE 10.7** Diffraction, and therefore the sharpness of shadows, depends on the relative sizes of the wavelength and the opening.

the wave model was better than the particle model at explaining diffraction. This also solved Huygens' issue with rectilinear propagation. With such small wavelengths, very little diffraction would occur.

**KEY CONCEPTS**

- In the 17th century, Newton proposed that light consisted of particles. At about the same time, Christiaan Huygens was working on his wave model.
- Newton's corpuscular theory of light proposed that:
  - Light consists of small particles.
  - These particles have mass and obey the laws of physics in the same way as any other particles, such as billiard balls or planets.
  - The particles are so small that, when two beams cross, they do not scatter each other.
  - This theory could explain refraction and Snell's Law.
  - Light travelled faster in optically denser media.
- Huygens' principle states that each point on a wave behaves as a point source for waves in the direction of propagation. The line tangent to these circular waves is the new position of the wave front a short time later.
  - Huygens used his principle to propose that waves slow down when they encounter an optically denser medium.
  - This theory could explain refraction and Snell's Law.
  - The amount of diffraction depends on the relationship between the size of the opening and the wavelength.
  - This solved Huygens' issue with rectilinear propagation. With such small wavelengths, very little diffraction would occur.
- Foucault's experiment found that the speed of light through water is slower than the speed of light in air.
- Foucault therefore disproved Newton's theory of light, since Newton required the speed of light to be faster in water.

**CHECK YOUR UNDERSTANDING**

10.1

1  What were the basic principles of Newton's theory of light?

2  From chapter 2 we know that the change in vertical height of a particle with no initial vertical velocity moving in a gravitational field is given by $\Delta y = \frac{1}{2} gt^2$. If Newton knew that light travels 300 m in 1 microsecond:

  **a**  how far would a particle of light fall in that time?

  **b**  what would have been the implications of this result for Newton's corpuscular theory?

3  How did Newton explain refraction?

4  How did Newton explain the fact that some reflection occurs when light is refracted?

5  What is Huygens' principle?

6  How did Huygens' theory of light differ from Newton's theory concerning refraction?

7  Huygens' theory of light had an issue concerning diffraction.

  **a**  What was that issue?

  **b**  How did the discovery that light had wavelengths in the hundreds of nanometres solve Huygens' problem?

# 10.2 Diffraction patterns and interference

In the early 19th century experiments such as Young's double-slit experiment provided convincing evidence that light acts like a wave. The classical wave model, coupled with the electromagnetic field model as expounded by Faraday and refined by Maxwell, can explain many of the behaviours of light. Light is modelled as an electromagnetic wave. Light behaves just like other waves – it reflects, refracts and shows diffraction and interference patterns, just as do mechanical waves such as sound.

## INVESTIGATION (10.1)

### Interference patterns

**AIM**

To observe interference patterns from an incandescent light source using a diffraction grating

Write an inquiry question for this investigation.

**MATERIALS**

- Incandescent light source such as a desk lamp
- Magnifying lens
- Diffraction grating such as a CD or DVD disc.
- Mount for disc
- Screen such as a wall, whiteboard or large sheet of paper

**RISK ASSESSMENT**

| WHAT ARE THE RISKS IN DOING THIS INVESTIGATION? | HOW CAN YOU MANAGE THESE RISKS TO STAY SAFE? |
| --- | --- |
| Temporary disturbance to vision from magnified desk lamp | Do not look directly at the magnified light beam. |
| | Take care when reflecting the light beam off the diffraction grating that you do not stare directly at the undiffracted light. |

What other risks are associated with your investigation, and how can you manage them?

**METHOD**

1 Shine the incandescent light source through the magnifying glass so that the beam is focused on the diffraction grating. Use the outer edge where the grooves closest to parallel.

2 Observe the light that is diffracted by the grating.

**RESULTS**

Sketch your observations, noting carefully any colour sequences.

**DISCUSSION**

1 How can you explain pattern of colours you see in terms of light as a wave and noting what happens to different wavelengths?

2 Give an answer to your inquiry question.

**CONCLUSION**

With reference to the data obtained and its analysis, write a conclusion based on the aim of this investigation.

## Young's double-slit experiment

Thomas Young was a polymath. Born in 1773, he taught himself to read by age two, and had learned Latin by age six. His interests included Egyptology, pure science and medicine. He was elected to the Royal Society at age twenty-one.

Young had read Newton's *Opticks* when he was seventeen and was impressed with Newton's corpuscular theory – light as particles – but by 1800 he saw some problems with it.

▶ At some boundaries, such as between water and air, some light was refracted, some was reflected. Particle behaviour could not easily account for this.

▶ Corpuscular theory could not explain why different colours of light were refracted by different amounts.

Young was aware that sound in air was a compression wave, and that interference between two waves of different frequencies caused beats (*Physics in Focus Year 11* chapter 9).

In the double-slit experiment, light is shone through a pair of narrow, closely spaced slits. The resulting interference pattern is observed on a distant screen. The interference pattern results from the path difference between the light waves coming from the two different slits. The path difference results in areas of constructive and destructive interference, giving a pattern of bright and dark fringes. The path difference is calculated as shown in Figure 10.8.

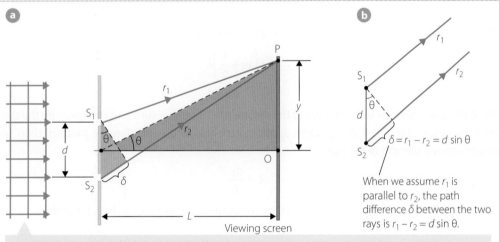

**FIGURE 10.8** The double-slit experiment. Bright fringes occur when $d\sin\theta = m\lambda$, where $m$ is an integer.

When the screen is a long way from the slits, the two rays, $r_1$ and $r_2$, are approximately parallel. Figure 10.8b shows a closeup of the approximately parallel rays at the slits. The interference pattern produced at the screen is the result of the two rays travelling different distances to reach a given point on the screen.

The difference in distance travelled is called the **path difference**. The path difference is shown as $\delta$ in Figure 10.8, and is given by:

$$\delta = r_2 - r_1$$

We define an angle, $\theta$, as the angle between the normal to the line joining the two slits (shown as length $L$) and the point of interest, P, on the distant screen. We assume that the screen is also perpendicular to this normal line. This allows us to write the path difference in terms of the angle $\theta$:

$$\delta = r_2 - r_1 = d\sin\theta$$

where $d$ is the distance between the slits. Recall from *Physics in Focus Year 11* chapter 8 that constructive interference occurs when two waves have the same phase, in other words when peaks line up with peaks. At these points there is an **antinode** or bright spot in the pattern. This occurs whenever the path difference is equal to a whole number of wavelengths, or when:

$$d\sin\theta = m\lambda, \ m = 0, 1, 2, \ldots$$

At the positions where peaks meet troughs, the waves are always half a cycle out of phase and a node (dark spot) appears. This occurs when:

$$d\sin\theta = \left(m + \frac{1}{2}\right)\lambda, \, m = 0, 1, 2, \ldots$$

The angle, $\theta$, can be related to the height, $y$, above the point where the normal line reaches the screen (point O in Figure 10.8a) and the distance $L$ to the screen. For small angles ($L \gg d$),

$$\tan\theta \approx \sin\theta = \frac{y}{L}$$

Therefore, the path difference can also be written as:

$$\delta = d\frac{y}{L} = \frac{dy}{L}$$

Relating this to the conditions for constructive and destructive interference tells us where the bright and dark spots in the pattern will occur.

The points of constructive interference (bright spots) in a double-slit interference pattern occur at:

$$\frac{dy}{L} = m\lambda$$

Rearranging,

$$y = \frac{Lm\lambda}{d}$$

The points of destructive interference (dark spots) occur at:

$$y = \frac{L\left(m + \frac{1}{2}\right)\lambda}{d}$$

What this also means is that, if we know the distance between the slits, and can measure the distance between successive ($m = 1, 2, 3, \ldots$) bright spots (antinodes) in the interference pattern, then we can determine the wavelength of the light:

$$\lambda = \frac{dy}{L}$$

Conversely, if we know the wavelength, we can determine the distance between the slits.

## Single-slit diffraction

In our analysis so far, we have assumed that each slit has been so narrow that it acts effectively as a point source of light. In fact, such slits have a finite width, and light passing through them will give an interference pattern based on the width of the slit itself.

The phenomenon is called Fraunhofer diffraction and depends on the fact that each side of the single slit acts as a point source of waves that can interfere with the other. The pattern is different from that produced by a double slit, with the central peak large and the other peaks small, as shown in Figure 10.9. The screen needs to be at a large distance compared to the slit width.

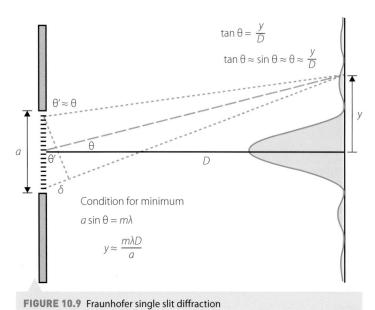

$$\tan\theta = \frac{y}{D}$$

$$\tan\theta \approx \sin\theta \approx \theta \approx \frac{y}{D}$$

Condition for minimum

$$a\sin\theta = m\lambda$$

$$y \approx \frac{m\lambda D}{a}$$

**FIGURE 10.9** Fraunhofer single slit diffraction

According to the Huygens principle, each position along the slit acts as a point source of waves. To simplify things, let us divide the slit into halves. The first minimum will occur when each of the rays emanating from the first half cancels out each of the rays in the second half. Suppose there are 40 positions, with 20 in the first half ($a_1 \dots a_{20}$) and 20 in the other half ($a_{21} \dots a_{40}$). At the position of the first *minimum*, $a_1$ will cancel $a_{21}$, $a_2$ will cancel $a_{22}$ and so on. Therefore the condition for the minimum is:

$$\frac{a}{2}\sin\theta = \frac{m\lambda}{2}$$

So $a\sin\theta = m\lambda$, $m = \pm 1, 2, 3 \dots$ where $a$ is the slit width.

This is also shown in Figure 10.9.

# INVESTIGATION (10.2)

## Single-slit diffraction

### AIM

To observe the behaviour of light incident on a single slit
Write an inquiry question for this investigation.

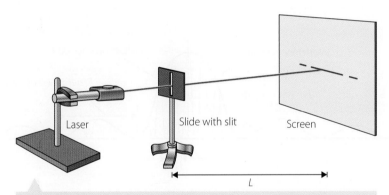

**FIGURE 10.10** Experimental setup for Fraunhofer single-slit diffraction

### MATERIALS

- Red laser pointer
- Laser mount
- Prepared slides with a single slit of different widths
- Slide mount, such as boss head and clamp
- Projection screen
- Metre ruler

RISK ASSESSMENT

| WHAT ARE THE RISKS IN DOING THIS INVESTIGATION? | HOW CAN YOU MANAGE THESE RISKS TO STAY SAFE? |
| --- | --- |
| Temporary disturbance to vision from laser | Take strict precautions when using the laser. Do not point it at anybody. Do not look directly at the beam. Take particular care that all eyes are well out of the path of any reflections from the back of the slide. |

What other risks are associated with your investigation, and how can you manage them?

»

1 Arrange the laser, a single-slit slide and the screen as shown in Figure 10.10.

2 Measure the distance from the slide to the screen.

3 Turn on the laser and observe the pattern formed on the screen.

4 Repeat for each of the other slides.

**RESULTS**

Sketch each of the patterns you observed. Note the relative brightness of the various spots.

**DISCUSSION**

1 Refer back to what you have learned about waves. How can the patterns you observed be explained?

2 Give an answer to your inquiry question.

**CONCLUSION**

With reference to the data obtained and its analysis, write a conclusion based on the aim of this investigation.

Double-slit experiments will generally yield patterns that have contributions from both the slit width (Fraunhofer diffraction pattern) and interference substructure, caused by separation between the slits. This is shown in Figure 10.11.

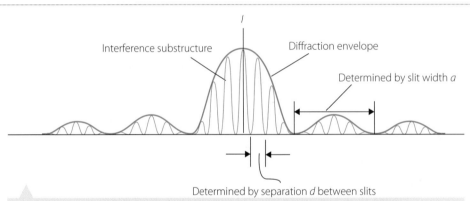

**FIGURE 10.11** Double-slit interference with diffraction

We can now work out the number of interference fringes within the single-slit diffraction maximum. Remember that $m$ is a counting factor representing the number of wavelengths difference in the pathlengths (and also the number of a particular fringe from the central maximum), and applies, separately, to both the single-slit and double-slit conditions.

We have seen that, for double-slit interference, interference maxima occur at $d\sin\theta = m\lambda$, $m = 0, 1, 2, ...$

We also saw that the condition for a single slit minimum is $a\sin\theta = m\lambda$, $m = \pm0, 1, 2, 3 ...$

The first single-slit minimum ($m = 1$) is therefore given by $a\sin\theta = \lambda$. To have the first single-slit minimum coincide with a double-slit maximum, we require $\dfrac{d\sin\theta}{a\sin\theta} = \dfrac{m\lambda}{\lambda}$.

Therefore $m = \dfrac{d}{a}$.

This particular $m$ is the number of the interference fringe that coincides with the first diffraction minimum. We cannot see this particular fringe since it is 'swallowed up' by the diffraction minimum, so the total number of fringes on each side of the central maximum is $(m - 1)$. Including the central fringe itself, we therefore arrive at:

$$N = 2(m-1) + 1 = 2m - 1$$

## Double-slit diffraction

Shining a light beam past a thin wire acts very much like shining it through a double slit. Each side of the wire acts like a point source of light waves from an incoming parallel wave front. The waves from each side of the wire will then spread out and interfere with each other, forming a diffraction pattern on the screen. The thickness of the wire is the equivalent of the distance between two slits.

The leads that connect power packs with various electrical components used in junior high school experiments are often composed of fine woven strands of copper wire. One single strand works very well for this investigation (Figure 10.12). Depending on the length it might require several such strands to complete the winding on the dowel. The ends do not need to be physically joined for this exercise.

### AIM

To determine the distance between two slits on a microscope slide using double-slit diffraction

Write an inquiry question for this investigation.

### MATERIALS

- 2 razor blades and clip
- Sticky tape
- Red laser pointer
- Laser mount
- Microscope slide
- Retort stand
- Tongs (to hold microscope slide in flame)
- Slide mount, such as boss head and clamp
- Candle and matches
- Putty
- 1 m thin, flexible wire
- Thin dowel rod
- Projection screen
- Metre ruler
- Wire cutters

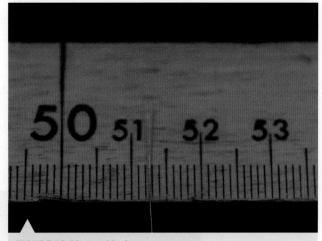

**FIGURE 10.12** Suitable thin copper wire

RISK ASSESSMENT

| WHAT ARE THE RISKS IN DOING THIS INVESTIGATION? | HOW CAN YOU MANAGE THESE RISKS TO STAY SAFE? |
|---|---|
| Temporary disturbance to vision from laser | Take strict precautions when using the laser. Do not point it at anybody. Do not look directly at the beam. Take particular care when shining the laser through the double slit that all eyes are well out of the path of any reflections from the back of the slide. |
| Cut from razor blades | Carefully tape the blades together. When marking the slide, handle the blades via the clip. |
| Burn from candle flame | Waft the slide carefully through the yellow flame. |

What other risks are associated with your investigation, and how can you manage them?

»

## Part 1 Determine the wavelength of the laser

### METHOD

1 Put a small lump of putty on the bench about 2 m from the screen.

2 Cut about 2 cm off the thin flexible wire. Mount this vertically, protruding upwards from the putty.

3 Mount the laser so it points directly towards the mounted 2 cm vertical wire and the screen beyond, as shown in Figure 10.13.

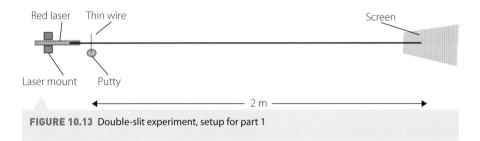

**FIGURE 10.13** Double-slit experiment, setup for part 1

4 Turn on the laser. This will generate a diffraction pattern on the screen (a series of bright red dots).

5 Measure from the leftmost visible dot to the rightmost visible dot with the metre ruler (Figure 10.14).

6 Determine and record the average separation between dots. The dots will be smeared out but we are interested in the centre points of the dots.

7 Wind the remaining thin flexible wire onto the dowel rod, keeping a careful count of the number of windings, until you have covered about 2 cm of the dowel. Make sure the windings are tight, with no spacings or overlap.

8 Measure the distance from the first winding to the last along the dowel as precisely as possible.

9 Divide the measured distance by the number of windings to determine the thickness of the wire as accurately as possible.

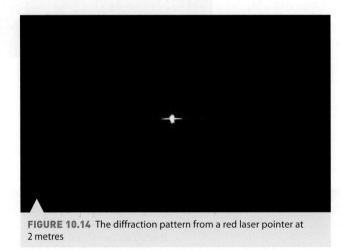

**FIGURE 10.14** The diffraction pattern from a red laser pointer at 2 metres

### RESULTS

Using the antinode separation distance and the thickness of the wire, calculate the wavelength of the laser.

### DISCUSSION

1 Red laser pointers typically produce a beam with a wavelength between 630 and 650 nm. How do your results compare?

2 Comment on the accuracy of your results.

3 Comment on the reliability of your results.

4 Why are the dots in the diffraction pattern smeared out?

## » Part 2 Determine the distance between two slits on a microscope slide using double-slit diffraction

1 Bind the two razor blades together with sticky tape. Put the two bound blades in a clip. You might find it useful to put a thin scrap of paper between the two blades.

2 Light the candle (or a Bunsen burner set to yellow safety flame).

3 Carefully and gently waft the microscope slide through the flame. Incomplete combustion in the yellow flame produces carbon particles, which should form a thin coating on the slide.

4 Using the bound razor blades, gently make a double slit on the carbon-covered microscope slide. This requires only minimal action.

5 Mount the slide so that the slits are vertical. Your setup should be as shown in Figure 10.15.

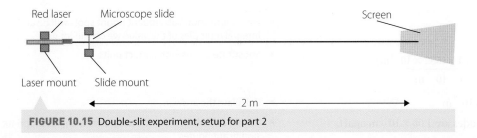

**FIGURE 10.15** Double-slit experiment, setup for part 2

6 Shine the laser through the double slit onto the screen 2 m distant.

7 Determine and record the spacing between adjacent antinodes on the screen.

### RESULTS

Using the antinode separation distance and the calculated wavelength of the light, calculate the distance between the slits on the microscope slide.

Did you notice the interference substructure? This is the smaller scale fringe pattern and depends on the separation between the slits rather than the slit width.

### ANALYSIS OF RESULTS

1 Determine the ratio of the separation of the slits ($d$) to the slit width ($a$).

2 Determine the slit width.

3 How many interference fringes are there within the first diffraction maximum?

### DISCUSSION

1 Comment on the accuracy of your results. Compare this to physical measurement of the slit separation, including any practical advantages/impediments.

2 Comment on the reliability of your results.

3 What could you do to improve accuracy and reliability?

4 Give an answer to your inquiry question.

### CONCLUSION

With reference to the data obtained and its analysis, write a conclusion based on the aim of this investigation.

## WORKED EXAMPLE (10.1)

Information and communication technology capability

Numeracy

A green laser has a wavelength of 532 nm. When light from the laser passes through a slit of width 0.1 mm and travels a distance $L = 2$ m to the screen, what would be the separation of the antinodes on the screen?

| ANSWERS | LOGIC |
|---|---|
| $\lambda = 532$ nm, $L = 2$ m, $d = 0.1$ mm | ▪ Identify relevant data in the question. |
| $y = \dfrac{Lm\lambda}{d}$ | ▪ Choose the appropriate equation. |
| Let $m = 1$. | ▪ First bright spot. Subsequent bright spots will be at integral multiples of the value we obtain for $y$. |
| $y = \dfrac{(2 \text{ m}) \times 1 \times (532 \times 10^{-9} \text{ m})}{1 \times 10^{-4} \text{ m}}$ | ▪ Substitute values with correct units. |
| $y = 1.06 \times 10^{-2}$ m | ▪ Calculate the answer. |
| The antinodes are $1.06 \times 10^{-2}$ m apart. | ▪ State the final answer with correct units and appropriate significant figures. |

### TRY THESE YOURSELF

A blue laser has a wavelength of 458 nm. It passes through two parallel slits, each 0.1 mm wide, and separated by a distance of 0.4 mm. The distance to the screen is 2 m.

**a** What would be the antinode separation caused by Fraunhofer diffraction?

**b** What would be the antinode separation caused by the double-slit interference?

## Diffraction gratings

WS

Diffraction patterns and interference

Diffraction gratings work in the same way as double-slit experiments do, but with many parallel slits very close together. We assume the incoming light is planar, and that diffraction will spread the light over a large number of grating lines that will each generate interference. The difference in the light path length between adjacent lines is, again, $\delta = d\sin\theta$ and the condition for maxima is $d\sin\theta = m\lambda, m = 0, 1, 2, \ldots$

If the path difference at a particular point on the screen obeys this condition for all pairs of slits, then there will be a principal maximum (bright spot). The location of the maxima does not depend on the number of grating lines, $N$. However, it can be shown that the thickness of the maxima is inversely proportional to the number of lines. That is, the greater the value of $N$, the thinner the maxima, as shown in Figure 10.16.

Sources with mixed wavelengths, such as incandescent light, will generate separate peaks for each wavelength. They will therefore generate a spectrum, as you would have found in Investigation 10.1.

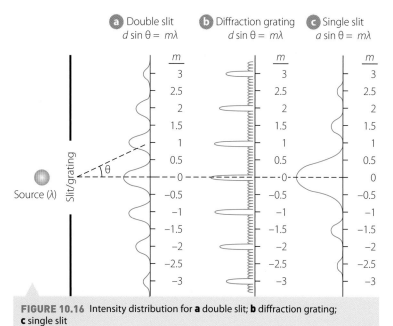

**a** Double slit $d\sin\theta = m\lambda$  **b** Diffraction grating $d\sin\theta = m\lambda$  **c** Single slit $a\sin\theta = m\lambda$

**FIGURE 10.16** Intensity distribution for **a** double slit; **b** diffraction grating; **c** single slit

# Interference of monochromatic light

## AIM

To observe interference of monochromatic light using a diffraction grating

Write an inquiry question for this investigation.

## MATERIALS

- Red laser pointer
- Laser mount
- Diffraction grating such as a CD or DVD disc
- Two-dimensional diffraction grating such as nylon stocking
- Diffraction grating mount
- Projection screen
- Metre ruler

RISK
ASSESSMENT

| WHAT ARE THE RISKS IN DOING THIS INVESTIGATION? | HOW CAN YOU MANAGE THESE RISKS TO STAY SAFE? |
| --- | --- |
| Temporary disturbance to vision from laser | Take strict precautions when using the laser. Do not point it at anybody. Do not look directly at the beam. Take particular care that when the laser is diffracted off the grating, all eyes are well out of the path of any reflections. |

What other risks are associated with your investigation, and how can you manage them?

## METHOD

1 Set the diffraction grating parallel to the floor at about 1 m away from the projection screen.

2 Shine the laser onto the CD diffraction grating so that the beam is forward diffracted onto the projection screen. Use the outer edge where the grooves are closest to parallel.

3 Observe the light pattern that is diffracted off the grating.

4 Measure the distances between the spots on the screen.

5 Replace the CD grating with the two-dimensional grating and repeat steps 2–4.

## RESULTS

Sketch the patterns you observe in both cases, indicating distances between the spots you have selected.

## ANALYSIS AND DISCUSSION

Investigation 10.3 yielded the wavelength of the laser, so use this information to calculate the spacing of the gratings in each case.

1 Comment on the accuracy of your results. Compare this to physical measurement of the slit separation, including any practical advantages/impediments.

2 Comment on the reliability of your results.

3 What could you do to improve accuracy and reliability?

4 Give an answer to your inquiry question.

## CONCLUSION

With reference to the data obtained and its analysis, write a conclusion based on the aim of this investigation.

- In the early 19th century experiments such as Young's double-slit experiment provided convincing evidence that light acts like a wave.
- The electromagnetic field model, as expounded by Faraday and refined by Maxwell, can explain many of the behaviours of light as an electromagnetic wave.
- Diffraction patterns of incandescent light are dependent on the wavelengths and so form peaks and troughs that reveal the spectrum.
- Young noted that:
  - At some boundaries, such as between water and air, some light was refracted and some was reflected. Particle behaviour could not account for this easily.
  - Particle theory could not explain why different colours of light were refracted by different amounts.
- Fraunhofer diffraction depends on the fact that each side of a single slit acts as a point source of waves that can interfere with each other. It produces a large central peak.
- Young's double-slit experiment is based on light from one slit forming interference patterns with light from the other slit.
- The interference pattern results from the path difference between the light waves coming from the two different slits.
- For double-slit interference, interference maxima occur at $d \sin \theta = m\lambda, m = 0, 1, 2, \ldots$
- The distance between peaks in a double-slit experiment is given by $y = \dfrac{Lm\lambda}{d}$.
- The points of destructive interference (dark spots) occur at:
$$y = \frac{L\left(m + \dfrac{1}{2}\right)\lambda}{d}$$
- Wavelength can be determined from $\lambda = \dfrac{dy}{L}$.
- If slits have a width $a$ and are separated by a distance $d$, then the counting factor $m$ for the interference fringe that coincides with the first diffraction minimum is given by $m = \dfrac{d}{a}$.
- Diffraction gratings work in the same way as double-slit experiments, but with many parallel slits very close together.

**CHECK YOUR UNDERSTANDING**

**10.2**

1 Name two models of light that you have studied.

2 Name three phenomena that the wave model of light can explain.

3 Why is it only possible for an interference pattern to form in the double-slit experiment if light travels as a wave and hence is delocalised? Draw a diagram to help explain your answer.

4 In the double-slit experiment, state if the spacing of the light and dark fringes increases, decreases or stays the same if:

   a a shorter wavelength of light is used.

   b the screen is moved closer to the slits.

   c there is less space between the slits.

5 In a double-slit experiment light of wavelength 630 nm is incident on a pair of slits spaced a distance 0.015 mm apart. If the screen is a distance of 2.0 m from the slits, at what positions do the first three bright spots appear?

6 In a measurement to find the wavelength of a light source, a viewing screen is placed a distance of 4.8 m from a pair of slits with separation 0.030 mm. The first dark fringe is a distance of 4.5 cm from the centre line on the screen.

   a Find the wavelength of the light.

   b Find the distance between any two adjacent bright spots.

7 A pair of slits spaced 0.015 mm apart is illuminated with light of two wavelengths at the same time: $\lambda_1 = 630$ nm and $\lambda_2 = 420$ nm. The viewing screen is a distance of 3.0 m from the slits. At what position on the screen, other than at $y = 0$, do the maxima from the two interference patterns first line up? Give the $m$ values for the bright spot for each wavelength at this position.

9780170409131

# 10.3 Polarisation

The wave model of light says that electromagnetic radiation consists of oscillating electric and magnetic fields, at right angles to each other, propagating in the direction at right angles to both as a wave, at the speed of light. Explanations of observable phenomena such as reflection, refraction, diffraction and interference, combined with successful quantitative predictions, give good support to this model. We have seen some of these results from the double-slit investigations carried out above.

Another such phenomenon is polarisation. French physicist Etienne Louis Malus initiated the study of polarisation through his studies of the reflection of light from polished surfaces. According to the wave model, polarisation distinguishes waves from 'non-wave' material objects, such as a stream of particles. Polarisation refers to the particular direction of oscillation of the electric field in the transverse plane.

The direction of the oscillating electric field depends on the direction of motion of the originating charged particle. Electrons in a vertical wire, for instance, will oscillate vertically, giving rise to vertical polarisation of the emitted electromagnetic radiation. Incandescent sources tend to consist of particles vibrating in all directions at different speeds, so the light they give off is said to be unpolarised.

## Plane polarisation

Plane polarisation means the electric fields in the electromagnetic radiation are all oscillating in the same two-dimensional plane. Unpolarised light can be plane-polarised in several ways, including by reflection or scattering, and passing through a polarising filter (Figure 10.17).

When light from an incandescent source such as the Sun reflects off a plane surface such as a lake, the oscillating charged particles at the surface respond by oscillating if they are able to do so. Even water at pH 7 has some $H^+$ and $OH^-$ ions, while seawater contains other ions as well. These ions are free to oscillate in a direction parallel to the surface but not perpendicular to it. Incoming light that does not have the correct orientation will be scattered or absorbed, while light oriented parallel to the surface will cause the ions to oscillate. These ions in turn will emit electromagnetic radiation, all of which is polarised in a direction parallel to the surface. Hence, reflected sunlight will be seen to be plane-polarised.

This polarisation should be detectable by an analyser that only allows light of a particular polarisation to pass.

**Light passing through crossed polarisers**

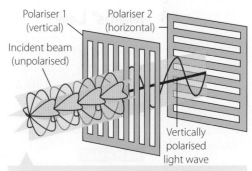

FIGURE 10.17 Polarising a beam of unpolarised light with a polarising filter (polariser 1). The polarised light cannot pass the crossed second polariser (polariser 2).

## Malus' Law

Based on the wave model, it can be shown that the intensity, $I$, of the electric field in electromagnetic radiation is proportional to the square of its amplitude, $A$:

$$I \propto A^2$$

Let the electric field vector of the plane-polarised light be $A_0$. This vector can be decomposed into a component parallel to the analyser orientation ($A_0 \cos\theta$) and a component perpendicular to the analyser orientation ($A_0 \sin\theta$), as shown in Figure 10.18.

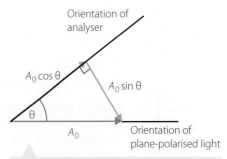

FIGURE 10.18 The amplitude of the plane-polarised electric field passed by an analyser at angle $\theta$

Let the intensity of the plane-polarised light be $I_0$. The intensity $I$ of the electromagnetic radiation passed by the analyser should therefore be:

$$I \propto (A_0 \cos\theta)^2$$

$$\frac{I}{I_0} = \frac{(A_0 \cos\theta)^2}{A_0^2} = \cos^2\theta$$

$$I = I_0 \cos^2\theta$$

Polarisation

The minimum amount of light, 0, will be passed when the analyser and the plane of polarisation are perpendicular and $\cos\theta = 0$. The maximum amount of light will be transmitted when the polarisation plane and the analyser are parallel and $\cos^2\theta = 1$. If we call the maximum amount of light transmitted $I_{max}$, then we can restate our equation as:

$$I = I_{max} \cos^2\theta$$

This is Malus' Law. (Note that $\cos^2\theta = (\cos\theta)^2$.)

## INVESTIGATION (10.5)

## Malus' Law

The SI unit for light intensity is the **lux**. Light *sources* are often rated in **lumen**, which is very approximately the output of a candle. To allow for distance and the inverse square law we need the idea of flux – light output in a specified area. 1 lux = 1 lumen m$^{-2}$.

### AIM

To test the accuracy of Malus' Law

### MATERIALS

- Dissecting tray or other large, shallow and flat container
- Water
- Incandescent light source, such as a desk lamp, light kit or direct sunlight
- Large sheet of cardboard to use as a shield
- Scissors
- Plane polarising filter (analyser), such as polarising plastic or sunglasses
- Stand/holder for analyser
- Diffraction grating mount
- Protractor
- Light meter or appropriate phone app

A horizontal polariser may be substituted for the flat tray of water.

RISK ASSESSMENT

| WHAT ARE THE RISKS IN DOING THIS INVESTIGATION? | HOW CAN YOU MANAGE THESE RISKS TO STAY SAFE? |
| --- | --- |
| Sunlight can be blinding. | Do not look directly at the Sun or its reflection. |
| Desk lamps can cause temporary vision disturbance. | Do not look directly at a desk lamp. |
| Scissors can cause injury. | Be careful when carrying or cutting with scissors. |

What other risks are associated with your investigation, and how can you manage them?

»

## METHOD

1   Set up the tray and almost fill it with water.

2   Cut a slit in the cardboard shield and set the shield in position, as shown in Figure 10.19.

3   Set the light meter in place. Take and record a light reading.

4   Turn on the light source and record the reading on the light meter.

5   Interpose the analyser between the shield and light meter as shown in Figure 10.19.

6   Record the reading on the light meter in a table with suitable headings. (See 'Results' below.)

7   Rotate the analyser 10°.

8   Repeat steps 6 and 7 until you have recorded 18 readings in the table you constructed in step 6.

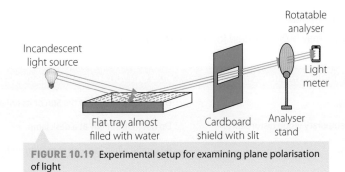

FIGURE 10.19 Experimental setup for examining plane polarisation of light

## RESULTS

1   Construct a table for your results. The first column is for the relative orientation of the analyser. Column two is for your measured results. The value recorded in step 3 is unavoidable ambient light. It should be subtracted from all other readings. Write the adjusted values in the third column.

The value recorded in step 3 represents $I_{max}$. This should help you determine which of your readings had the analyser most nearly parallel to the plane polarisation direction. This sets the reading where $\theta = 0$. Use column 4 to record $\theta$ with respect to this orientation.

Now that we know $I_{max}$ and $\theta$, calculate the values expected from Malus' Law and put them in column 5.

2   Plot your results in column 3 against angle of rotation of the analyser. On the same graph plot the calculated values in column 5.

### ANALYSIS OF RESULTS

1   Comment on the accuracy of your measured results. Estimate the error.

2   Discuss the degree of similarity or otherwise of the plot of measured results vs expected results.

3   Discuss the reliability and validity of your results.

4   Discuss any deficiencies in your investigation. How might this investigation be improved?

### CONCLUSION

Discuss the validity of the investigation. Did your investigation meet its aim? How well do your results agree with the predictions of Malus' Law? Given that Malus' Law is based on the wave model of light, does your experimental evidence support this model?

## Does monochromatic light obey Malus' Law?

**AIM**

To determine whether monochromatic light obeys Malus' Law
  Write an inquiry question for your investigation.

**MATERIALS**

- As for Investigation 10.5
- Filters for red, green and blue
- A horizontal polariser may be substituted for the flat tray of water.

| WHAT ARE THE RISKS IN DOING THIS INVESTIGATION? | HOW CAN YOU MANAGE THESE RISKS TO STAY SAFE? |
| --- | --- |
| Sunlight can be blinding. | Do not look directly at the Sun or its reflection. |
| Desk lamps can cause temporary vision disturbance. | Do not look directly at a desk lamp. |
| Scissors can cause injury. | Be careful when carrying or cutting with scissors. |

What other risks are associated with your investigation, and how can you manage them?

**METHOD**

1  Place a red filter between the light source and the polariser (tray of water).
2  Repeat steps 3–7 in Investigation 10.5.
3  Repeat this method for green and blue filters.

**RESULTS**

Tabulate and calculate your results as you did in Investigation 10.5.

**ANALYSIS OF RESULTS**

1  Comment on the accuracy of your measured results. Estimate the error.
2  Is there any difference in $I_{max}$ between the results for any of the individual colours and the result achieved in Investigation 10.5?
3  Discuss the similarity, or otherwise, of your results for the different filters, and compare them to that observed in Investigation 10.5.
4  Discuss the reliability and validity of your results.
5  Discuss any deficiencies in your investigation. How might this investigation be improved?

**CONCLUSION**

Discuss the validity of the investigation. Did your investigation meet its aim? Do coloured filters have much effect on the results you achieved in Investigation 10.5? How well do your results agree with the predictions of Malus' Law?

- Polarisation refers to the particular direction of oscillation of the electric field in the transverse plane.
- The direction of the oscillating electric field depends on the direction of motion of the originating charged particle.
- Incandescent sources consist of particles vibrating in all directions at different speeds, so the light they give off is unpolarised.
- Plane polarisation means the electric fields in the electromagnetic radiation are all oscillating in the same plane.
- The amount of plane-polarised light that can pass through a polariser oriented at $\theta°$ to the original plane is given by Malus' Law.
- Malus' Law: $I = I_{max} \cos^2\theta$.
- The SI unit for the intensity of a light source is a lumen.
- $1\,\text{lux} = 1\,\text{lumen}\,\text{m}^{-2}$

1   Name three ways of plane polarising light.

2   Describe how sunlight can be plane polarised from the surface of a body of water.

3   How can we determine that light is polarised?

4   A beam of incandescent light is passed through a polarising filter. How would the intensity of the original beam compare with the filtered beam if the filter were at:

   **a**   0°?

   **b**   45°?

   **c**   90°?

5   A beam of plane-polarised light is passed through a filter oriented at 45° to the direction of the light's plane of polarisation. What is the intensity of the filtered beam compared to the original beam?

6   A beam of light is passed through a polarising filter. The intensity of the light is determined to be 0.866 times the intensity of the beam before it passed through the filter. What was the angle of the filter with respect to the angle of polarisation of the original light beam?

- In the 17th century, Newton proposed that light consisted of particles. At about the same time, Christiaan Huygens was working on his wave model.

- Newton's corpuscular theory of light proposed that:
  - Light consists of small particles.
  - These particles have mass and obey the laws of physics in the same way as any other particles, such as billiard balls or planets.
  - The particles are so small that, when two beams cross, they do not scatter off each other.
  - This theory could explain refraction and Snell's Law.
  - Light travelled faster in optically denser media.

- Huygens' principle states that each point on a wave behaves as a point source for waves in the direction of propagation. The line tangent to these circular waves is the new position of the wave front a short time later.
  - Huygens used his principle to propose that waves slow down when they encounter an optically denser medium.
  - This theory could explain refraction and Snell's Law.
  - The amount of diffraction depends on the relationship between the size of the opening and the wavelength.
  - Because light has very small wavelengths, very little diffraction would occur, solving Huygens' issue with rectilinear propagation.

- Foucault's experiment found that the speed of light through water is less than the speed of light in air.

- Foucault therefore disproved Newton's theory of light, since Newton required the speed of light to be faster in water.

- In the early 19th century experiments such as Young's double-slit experiment provided convincing evidence that light acts like a wave.

- The electromagnetic field model, as expounded by Faraday and refined by Maxwell, can explain many of the behaviours of light as an electromagnetic wave.

- Diffraction patterns of incandescent light are dependent on the wavelengths and so form peaks and troughs that reveal the spectrum.

- Young noted that:
  - At some boundaries, such as between water and air, some light was refracted, some was reflected. Particle behaviour could not easily account for this.
  - Particle theory could not explain why different colours of light were refracted by different amounts.

- Fraunhofer diffraction depends on the fact that each side of a single slit acts as a point source of waves that can interfere with each other. It produces a large central peak.

- Young's double-slit experiment is based on light from one slit forming interference patterns with light from the other slit.

- The interference pattern results from the path difference between the light waves coming from the two different slits.

- For double-slit interference, interference maxima occur at $d \sin \theta = m\lambda$, $m = 0, 1, 2, \ldots$

- The distance between peaks in a double-slit experiment is given by $y = \dfrac{Lm\lambda}{d}$

- The points of destructive interference (dark spots) occur at
$$y = \frac{L\left(m + \dfrac{1}{2}\right)\lambda}{d}.$$

- Wavelength can be determined from $\lambda = \dfrac{dy}{L}$.

- If slits have a width $a$ and are separated by a distance $d$, then the counting factor $m$ for the interference fringe that coincides with the first diffraction minimum is given by $m = \dfrac{d}{a}$.

- Diffraction gratings work in the same way as double-slit experiments, but with many parallel slits very close together.

- Polarisation refers to the particular direction of oscillation of the electric field in the transverse plane.

- The direction of the oscillating electric field depends on the direction of motion of the originating charged particle.

- Incandescent sources consist of particles vibrating in all directions at different speeds, so the light they give off is unpolarised.

- Plane polarisation means the electric fields in the electromagnetic radiation are all oscillating in the same plane.

- The amount of plane-polarised light that can pass through a polariser oriented at $\theta°$ to the original plane is given by Malus' Law: $I = I_{max} \cos^2 \theta$.

- The SI unit for the intensity of a light source is a lumen.

- 1 lux = 1 lumen m$^{-2}$

**1** What is the key difference between the theories of Newton and Huygens with respect to light?

**2** Why was Newton successful in explaining reflection?

**3** Why did Newton require that the speed of light be faster in an optically denser medium?

**4** Explain how Foucault's experiment definitively disproved Newton's theory of light.

**5** Why can we hear around corners but not see around corners?

**6** Why does a single-slit investigation show a pattern of maxima and minima?

**7** Why does a double-slit investigation show an overall pattern of maxima and minima with a further pattern of narrower bands?

**8** What similarities does a diffraction grating have to a double slit?

**9** If an incandescent light were used in a double-slit experiment, would you expect to see any spectral separation? Why, or why not?

**10** What is the relationship between the oscillating electric field in electromagnetic radiation and the polarisation of the radiation?

**11** How do polarising sunglasses work?

**12** Describe how the surface of a lake can polarise sunlight.

**13** In a single-slit investigation with a slit width of 0.4 mm and a monochromatic light source with a wavelength of 460 nm, at what angle does the first ($m = 1$) minimum occur?

**14** At what distance from the central maximum will the first minimum in Question **13** occur, assuming a slit-to-screen distance of 2.5 m?

**15** The angle, θ, between the normal to the line joining two slits in a double-slit experiment and a point P on a distant screen is 3°. If the slits are separated by 0.5 mm, what is the wavelength of the light?

**16** In a double-slit investigation, a monochromatic light source of wavelength 530 nm produces a second diffraction maximum ($m = 2$) 18 cm from the central peak. If the distance to the screen is 1.5 m, what is the separation between the two slits?

**17** In a double-slit investigation, the slits have a width $a = 0.3$ mm and a separation $d = 1.5$ mm. How many interference fringes will there be between the central bright spot and the first diffraction minimum?

**18** Explain how diffraction gratings produce spectra.

**19** A plane-polarised light source has an intensity of 60 lumen. What is the intensity of the light after it passes through a polarising filter at an angle of:

  **a** 30°?

  **b** 60°?

  **c** 90°?

**20** After passing through a polarising filter, the intensity of the plane-polarised light is 50% of the unfiltered value. What is the angle of the polariser?

**21** Explain why a wave model of light is needed to understand the interference pattern produced in the double-slit experiment. Give two other examples of the wave-like behaviour of light.

What evidence supports the particle model of light and what are the implications of this evidence for the development of the quantum model of light?

**OUTCOMES**

**Students:**

- analyse the experimental evidence gathered about black body radiation, including Wien's Law related to Planck's contribution to a changed model of light (ACSPH137) CCT, ICT, N

  - $\lambda_{max} = \dfrac{b}{T}$

- investigate the evidence from photoelectric effect investigations that demonstrated inconsistency with the wave model for light (ACSPH087, ACSPH123, ACSPH137) CCT, ICT

- analyse the photoelectric effect $K_{max} = hf - \phi$ as it occurs in metallic elements by applying the law of conservation of energy and the photon model of light (ACSPH119) ICT, N

Physics Stage 6 Syllabus © 2017 NSW Education Standards Authority (NESA) for and on behalf of the Crown in right of the State of New South Wales.

The nature of light is a question that has occupied scientists for hundreds of years. In the 17th century, Englishman Robert Hooke and Dutch scientist Christiaan Huygens independently suggested light could be thought of as a wave. Huygens developed the idea in great detail. At about the same time, Isaac Newton put forward a 'corpuscular' theory – that is, that light was made up of a stream of small particles. Due to Newton's great reputation, the corpuscular theory dominated, especially in England. In the middle of the 19th century, French physicist Léon Foucault used careful measurement of the speed of light in air and in water to show that Newton's theory failed. Newton's ideas said light should be faster in water than in air, and Foucault showed that it was slower.

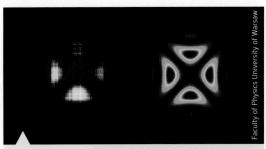

**FIGURE 11.1** Hologram of a single photon, reconstructed from raw measurements (left) and theoretically predicted (right)

In the early 20th century, the idea of the particle nature of light returned with the notion of light having a dual nature. Sometimes it acted like a wave and sometimes it acted like a particle (Figure 11.1). We talked about the wave picture – electromagnetic radiation – in chapter 10. This chapter discusses the particle picture. Light particles have been known as photons since 1926. They have no mass, but they do have energy.

Some phenomena suggest light is a wave. When we perform a diffraction or refraction experiment, we explain the behaviour best if we consider light as a wave. Other evidence, such as the photoelectric effect, suggests that light behaves like a stream of particles.

Late in the 19th century there were a number of problems that suggested that classical physics was not the whole story. Explaining the spectrum (colours and their intensities) of light emitted by a heated body (like a coal in a fire) was one. Another was the way shining light onto a material could cause electrons to be emitted. These two experiments were crucial in the development of quantum physics. And it is quantum physics that gives us computers, smart phones and so much of our modern technology.

## 11.1 Black body radiation

All objects continuously radiate energy in the form of electromagnetic waves. If the body is at a temperature above 0 K, as it must be, then its atoms and molecules have some energy. For a solid, this means the atoms and molecules are vibrating back and forth. Since this implies they are accelerating back and forth, and accelerating charges emit electromagnetic radiation, it follows that all objects are always emitting. In fact, when we warm up by standing in front of a radiating fire, we are radiating too. We warm up because we are absorbing more than we are emitting. All objects are emitting and absorbing all the time.

It is important to note that this way of producing radiation is different from what occurs in, for example, a light-emitting diode (LED). Here, we are talking about the emission due to the object's temperature. Light sources like LEDs and fluorescent lights use different methods to generate light.

So, at any non-zero temperature, a body emits electromagnetic radiation of all wavelengths, but the distribution of intensities in the spectrum of wavelengths depends on the temperature. If an object is very hot you can see the light that is being emitted; for example, you can see the glowing coals in a fire or the filament of an old-fashioned light globe. At low temperatures, the wavelengths of the emitted radiation are mainly in the low-frequency infrared region and cannot be seen, although you may still be able to feel the radiation as heat on your skin. Measurements show that the hotter an object is, the more electromagnetic radiation is emitted, and more of that radiation is at shorter

**FIGURE 11.2** Lava emits a wide range of wavelengths of radiation.

wavelengths. Hence, if you take a piece of metal and heat it up slowly, it will first glow a dim red, then bright yellow and eventually very bright white – though it may be liquid by then. Figure 11.2 shows lava pouring from a volcano. The apparent colour depends on which wavelengths are being emitted most strongly. The hotter lava looks white, the cooler red, the coolest black – it is too cool to emit visible light. It may still be too hot to touch, and may be emitting strongly in the infrared range.

Measurement of the intensity of emitted radiation as a function of wavelength shows that it is a continuous distribution of wavelengths from the infrared, through the visible and ultraviolet. This distribution is called a continuous spectrum. A continuous spectrum is shown in Figure 11.3a. Figure 11.3b shows a graph of the intensity of the light as a function of the wavelength for a black body. The shape of this curve depends only on the temperature of the object and not on any of its other properties. It is this kind of curve that could not be modelled using classical physics.

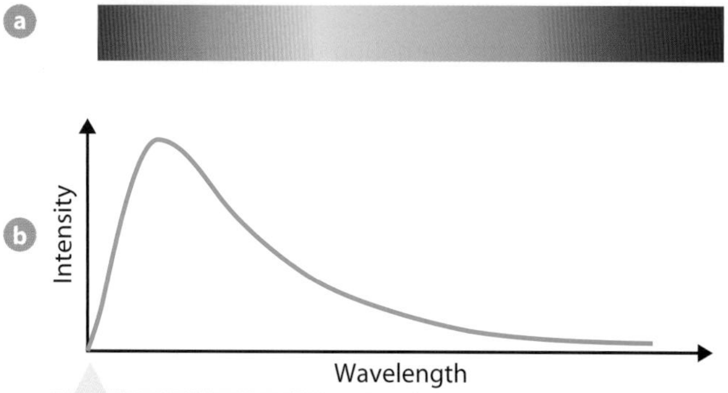

**a**

**b**

**FIGURE 11.3 a** A continuous spectrum **b** Intensity is a function of wavelength for a continuous black body spectrum.

## What is a black body?

A black body is an ideal surface that completely absorbs all wavelengths of electromagnetic radiation falling on it. Hence it is a *black* body. Such a surface will also be a perfect emitter of electromagnetic radiation at all wavelengths. The radiation is characteristic of the temperature of the black body but not of its composition (that is, what atoms it is made of). When the body is at room temperature, the strongest emissions are in the infrared part of the electromagnetic spectrum.

Although a true black body is a theoretical concept, it can be closely simulated in a laboratory. Consider a cavity (hollow space) that has the interior walls blackened and which is kept at a constant temperature (Figure 11.4). If a small hole is made in the wall of the cavity, the hole will act like a black body radiator. Any radiation that falls on the hole from the outside will pass through it. After multiple reflections, the radiation will be absorbed by the interior surfaces. As the cavity is in thermal equilibrium with its surroundings, the interior surfaces will emit radiation at the same rate at which it is absorbed. The radiation that escapes depends only on the temperature of the cavity. It is not affected by the size of the cavity or the material of which it is made.

**FIGURE 11.4** The opening to a cavity is a good approximation of an ideal black body. Note that it is the *opening* to the cavity that is the black body, not the entire hollow object. The hole acts as a perfect absorber.

Remember that this is an idealised object, not a real one. In practice, materials that absorb most of the light incident on them are good approximations of a black body. This is where the term 'black body' comes from – black objects absorb most of the light incident on them, regardless of wavelength.

You met the use of the black body spectrum to determine the temperature of stars in chapter 9.

## The black body spectrum

The black body model is useful because it allows us to determine the temperature of distant objects. For example, we can estimate the surface temperature of the Sun by measuring its electromagnetic spectrum.

Figure 11.5 shows the spectra of emitted radiation for a black body at various temperatures. Note how the peak in the radiation curve gets higher and shifts to shorter wavelengths as the temperature increases. Also, the area under the curve is greater for higher temperatures. So as $T$ increases, more radiation is emitted and on average it has a shorter wavelength.

## Wien's Law

In 1893 Wilhelm Wien derived a relationship between the position of the peak (the wavelength at which the radiation is most intense) and the temperature of a black body. He used the idealised black body cavity model to derive the relationship, now known as Wien's Displacement Law or simply Wien's Law.

The position of the peak wavelength is given by Wien's Law:

$$\lambda_{max} = \frac{b}{T}$$

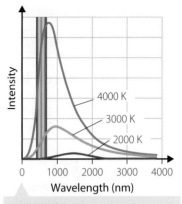

**FIGURE 11.5** Intensity distribution of wavelengths of radiation from a black body at different temperatures. The rainbow lines show where visible light sits on the horizontal scale.

where $\lambda_{max}$ is the wavelength where the peak of the curve, like those shown in Figure 11.5, appears. $T$ is the absolute temperature in K, and $b$ is Wien's constant, $b = 2.898 \times 10^{-3}$ m K (metre kelvin, not millikelvin – we can see this by using dimensional analysis, since it relates a length to a temperature).

We can use Wien's Law to estimate the temperature of a body based on the colour (frequency or wavelength) of the brightest light it emits.

▶ **WORKED EXAMPLE** (11.1)

The surface of the Sun is at a temperature of approximately 5800 K. If we treat the Sun as a black body, what is the peak wavelength of the emitted radiation?

| ANSWER | LOGIC |
|---|---|
| $T = 5800$ K | ▪ Identify the relevant data in the question. |
| $\lambda_{max} = \dfrac{b}{T}$ | ▪ Relate wavelength to temperature. |
| $\lambda_{max} = \dfrac{2.898 \times 10^{-3} \text{ m K}}{5800 \text{ K}}$ | ▪ Substitute values with correct units. |
| $\lambda_{max} = 5.0 \times 10^{-7}$ m $= 500$ nm | ▪ Calculate the final value. Note that we have rounded the answer to two significant figures because the temperature was given to two significant figures. |

**TRY THIS YOURSELF**

Figure 11.6 shows the black body spectrum for the star Antares. What is the surface temperature of the star Antares?

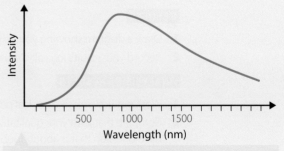

**FIGURE 11.6** Black body spectrum for Antares

## Black body radiation from a light globe filament

It is possible to qualitatively explore some of the properties of black body radiation using an incandescent light bulb. These are becoming rare for household lights. However, many car light bulbs such as tail and interior lights still use incandescent bulbs, and so do many light box kits you might use in physics experiments.

### AIM

To observe black body radiation from a light globe filament as the temperature of the filament changes
   Write an inquiry question for this investigation.

### MATERIALS

- 12 V incandescent light globe
- Continuously variable 12 V DC power supply, or a steady voltage supply with a variable resistor
- Voltmeter (or multimeter)
- Alligator leads
- Light globe holder
- Ammeter (or multimeter)

**RISK ASSESSMENT**

| WHAT ARE THE RISKS IN DOING THIS INVESTIGATION? | HOW CAN YOU MANAGE THESE RISKS TO STAY SAFE? |
| --- | --- |
| Electrical shock | Ensure the transformer has been electrically tested. Make sure it is unplugged from the wall while setting up the experiment. When adjusting the voltage, do not touch the wires or bulb. |
| Burns from a hot bulb | Do not touch the bulb once the experiment starts. Wait until it has cooled once the experiment is over before dismantling. |

What other risks are associated with your investigation, and how can you manage them?

### METHOD

1 Make sure the power supply is switched off and that the voltage is turned to zero. Using a proper light bulb holder, connect the light globe across the terminals of the power supply.

2 Turn on the power supply and *very slowly* turn up the voltage from zero until the globe just starts to glow.

3 Note the voltage and your observations. Don't touch the bulb, but hold your hand a few centimetres away and note what, if anything, you feel. Use the voltmeter and ammeter to measure the voltage across the globe (the power supply dial may not be very accurate) and the current through the globe.

4 Increase the voltage *very slowly*, observing what happens to the globe. Make observations at five or so different voltages between the first one and the maximum. Note the current and voltage at each point, and make some qualitative observations about the brightness of the light. You could also hold your hand a safe distance from the bulb and comment on the temperature as well.

### RESULTS

1 Draw a diagram showing your experimental set-up. Label all the parts clearly.

2 Record your results logically. You could use a table for the voltages and currents.

### ANALYSIS OF RESULTS

1 Work out how much electrical power the globe is using at each measurement.

2 Compare that with your qualitative comments on temperature and brightness. For example, if you have one step that is about double the power of another, does the light look twice as bright? Recalling Figure 11.5, is this what you would expect?

### DISCUSSION

In your discussion, there are some things worth thinking about:

1 Was the filament emitting radiation even before it was glowing?

2 What does that suggest about the fraction of emitted radiation that is visible and how it changes with voltage?

In other words, as the filament gets hotter, is a greater or lesser fraction of the radiation coming off it likely to fall in the visible spectrum? At what temperature do you think the visible fraction would be maximum? Think about Wien's Law and the fact that the human eye is not equally sensitive to all colours. It is most sensitive to greeny-yellow light.

4   Give an answer to your inquiry question.

> When the light is reasonably bright, the human eye is most sensitive to light around green to yellow.

### CONCLUSION

1   How does the apparent brightness of the filament change with the voltage applied?

2   How does it change with the actual power the bulb is consuming?

3   Can you qualitatively explain your results with reference to:

   **a**   the sensitivity of the human eye to different colours?

   **b**   the fraction of the black body spectrum that is likely to be visible for different filament temperatures?

4   Can you suggest how the experiment might be made more quantitative?

---

Wien's Law was a successful model in that it accurately predicted the position of the peak wavelength. However, there were still two problems. First, there was no theory that explained the shape of the wavelength–intensity curve. Second, Wien's Law was based on an idealised system – a cavity with a small hole. It is difficult to see how this theoretical model could represent the surface of a solid piece of material or a star such as the Sun.

Classically, it was thought that the thermal radiation originated from oscillating charged particles near the surface of an object. In chapter 10 we saw that oscillating charges are a source of electromagnetic waves. This is how antennas work. The oscillating charges in the antenna produce an electromagnetic wave of the same frequency as the oscillations. Recall also from chapter 11 of *Physics in Focus Year 11* that the temperature of a material is a measure of the average kinetic energy of the atoms of that material. In a gas or a liquid the particles are free to move. The higher the temperature, the more kinetic energy the particles have and the faster they move. In a solid material, the atoms are not free to move, so this kinetic energy is observed as vibrations, and the higher the temperature the higher the frequency of vibration. And as you know, atoms are made up of smaller particles including protons and electrons, which are charged. Hence, this theory provided the oscillating charges needed to produce the electromagnetic radiation.

> You learnt about standing waves in chapter 9 of *Physics in Focus Year 11*.

Now consider again the ideal model of the black body cavity. If the atoms on the inside surface are acting as little antennas, we would see standing waves set up between the walls of the cavity. The waves produced by the vibrating atoms in the inside surface would reflect from the opposite surface. If the waves have the right wavelength, a standing wave is set up, just like standing waves in a string. We call these standing waves **modes of vibration**.

Classically, all the possible modes of vibration would be equally probable, and the total energy would be divided equally between them all. However, more short wavelength modes would be able to fit in the cavity. This means more short wavelength radiation should be emitted through the hole. As the temperature of the cavity increases, so should the total energy. As the energy increases, the energy associated with the short wavelengths (ultraviolet, X-rays and gamma rays) would approach infinity. According to this theory, even a regular heater should be emitting dangerous amounts of X-rays and gamma rays! Figure 11.7 shows a comparison of a theoretical spectrum based on this model and a measured spectrum. This mismatch between theory and experiment was called the 'ultraviolet catastrophe'. Of course, it was only a catastrophe for the theory that predicted it.

A new theory was needed to solve these problems.

> The classical theory (yellow curve) shows intensity growing without bound for short wavelengths, unlike the experimental data (blue curve).

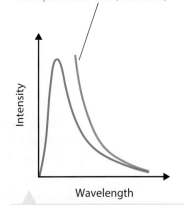

**FIGURE 11.7** Comparison of the classically predicted and experimental black body spectra

## Planck's quanta of energy

In 1900 a German physicist named Max Planck used 'lucky guesswork' (as he called it) to derive a formula that correctly matched the experimentally observed spectrum. Planck proposed that the atoms could only oscillate with discrete energies, given by:

$$E_n = nhf$$

where $n$ is an integer, $f$ is the frequency of oscillation and $h$ is a constant. The constant $h$, now known as the Planck constant, is $h = 6.626 \times 10^{-34}$ Js $= 4.14 \times 10^{-15}$ eVs. These two values are related by the electronic charge, $e$, because $h$ in Js divided by $e$ gives $h$ in eVs.

This was a radical proposition. It means that the energy of the oscillators is quantised. It may only take discrete values, given by the equation above, rather than any possible value in a continuous range.

From this Planck deduced that the oscillators could only emit and absorb electromagnetic radiation (light) in packets of energy of specific sizes (Figure 11.8). He called these packets of energy 'quanta'. The amount of energy emitted is equal to the amount of energy lost by an oscillator when it goes to a lower energy state. For example, if an oscillator goes from an energy of $E_3 = 3hf$ to $E_2 = 2hf$, the energy lost is $E_3 - E_2 = 3hf - 2hf = hf$. One quantum of light has energy $E = hf$.

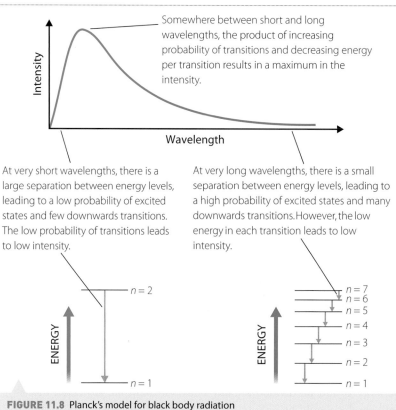

FIGURE 11.8 Planck's model for black body radiation

Planck combined this idea of quantisation with two ideas from classical statistical mechanics. First, the probability of an oscillator having a particular energy decreases as the energy increases. Hence, the probability of an atom being in a higher energy state (called an excited state) is lower. This means that the intensity of radiation at high frequencies (short wavelengths) is small. Second, the probability of a *change* in energy decreases with the relative gap between energy levels. The relative gap is larger for lower energies, or long wavelengths, so intensity is again low at these wavelengths. In between, at intermediate wavelengths, we see the peak observed in the experimental spectra, as illustrated in Figure 11.7.

The quantisation of energy was such a revolutionary departure from classical physics that Planck was reluctant to accept his own idea. Although Planck had discovered a mathematical way of explaining the shape of the black body spectrum, he was concerned that there was no physical model for how the energy could be in these discrete packets. Planck proposed that the oscillators were quantised. That is, they gained or lost energy in these discrete amounts. He did not propose that light itself was quantised. It was Albert Einstein who put physical meaning to Planck's quantum hypothesis.

Black body radiaton

▶ **WORKED EXAMPLE** (11.2)

A quantum of energy has wavelength $5.8 \times 10^{-5}$ m.

**a**  What is the frequency of this quantum?

**b**  What is the energy of this quantum?

| ANSWERS | LOGIC |
|---|---|
| $\lambda = 5.8 \times 10^{-5}$ m | ■ Identify the relevant data in the question. |
| **a**  $c = f\lambda$ | ■ Use the dispersion relation for light to relate frequency to wavelength. |
| $f = \dfrac{c}{\lambda}$ | ■ Rearrange for frequency. |
| $f = \dfrac{3.00 \times 10^8 \text{ m s}^{-1}}{5.8 \times 10^{-5} \text{ m}}$ | ■ Substitute values with correct units. |
| $f = 5.2 \times 10^{12}$ Hz | ■ Calculate the final value. |
| **b**  $E = hf$ | ■ Relate energy to frequency. |
| $E = (6.626 \times 10^{-34} \text{ J s})(5.2 \times 10^{12} \text{ Hz})$ | ■ Substitute values with correct units. |
| $E = 3.3 \times 10^{-21}$ J | ■ State the final answer with correct units and appropriate significant figures. |

**TRY THESE YOURSELF**

**1**  Using the equations used in this worked example, find an expression for $E$ as:

    **a**  a function of $f$.

    **b**  a function of $\lambda$.

**2**  What happens to $E$ when:

    **a**  $f$ increases?

    **b**  $\lambda$ increases?

## Earth as a black body

We can think about Earth as a black body. It is being hit by the radiation of everything around it (mainly the Sun), and is also emitting radiation itself. If the energy coming in is equal to that going out, its temperature will be roughly stable.

Now, at our distance from the Sun, the power per unit area coming from the Sun is about $1360 \text{ W m}^{-2}$; that is, more than a kilowatt per square metre. This is called the solar constant, $S$. If we call Earth's radius $R$, then the area Earth presents to the Sun is $\pi R^2$. Therefore, we intercept a total of $S\pi R^2$ W of the Sun's power. That's just the area times the power per area.

Now, the surface area of Earth is $4\pi R^2$. So, if Earth is to radiate that energy away again and not heat up, we need to radiate $\dfrac{S\pi R^2}{4\pi R^2} = \dfrac{S}{4} = 340 \text{ W m}^{-2}$ from each square metre of the Earth's surface. Except Earth is not a black body. About a third of this energy is reflected back into space, so Earth needs to radiate more like $240 \text{ W m}^{-2}$.

9780170409131

**CHAPTER 11** » LIGHT: QUANTUM MODEL **257**

The area under the black body curve relates to the rate at which the black body is emitting energy – that is, the power. Black body radiation calculations can show that for Earth to radiate $240\,\mathrm{W\,m^{-2}}$, it would have to be at a temperature of about 255 K, or about $-18°\mathrm{C}$!

Earth is warmer than that because it has a blanket – the atmosphere. The effectiveness of that blanket depends on what it is made of. Carbon dioxide lets visible light past but not infrared, so it traps heat. This is good – it stops us freezing! But, because humans have added a lot of carbon dioxide (and other gases) to the air, Earth's blanket is working too well, and Earth is heating up.

So understanding black body radiation is central to understanding what is probably the biggest issue of our time – climate change.

**CHECK YOUR UNDERSTANDING**

**11.1**

1 Name two classical models in physics.

2 Define 'black body'.

3 Show, based on $E = hf$, that the units for Planck's constant must be J s.

4 Vega is a blue star and Antares is a red star. Which is hotter? Explain your answer.

5 Imagine an oven being used to bake a cake. What is the most likely wavelength of electromagnetic radiation in the oven?

6 An atomic oscillator has frequency $f = 6.1 \times 10^{12}\,\mathrm{Hz}$, and is in the $n = 3$ state.

   a What is the energy of this oscillator?

   b What frequency light will be emitted if it transitions to the $n = 2$ state?

7 Briefly outline the nature of the 'ultraviolet catastrophe' and its implications for the development of modern physics.

8 Figure 11.9 shows the black body radiation spectrum for the star Vega.

   a What is the peak wavelength?

   b Calculate the surface temperature of Vega. Give your answer to an appropriate number of significant figures.

9 The filament of an incandescent light globe can be modelled as a black body. A tungsten filament reaches a temperature of 2900 K.

   a What wavelength does it emit most strongly?

   b Explain why such light globes emit more radiation in the infrared than in the visible part of the electromagnetic spectrum.

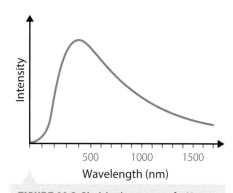

**FIGURE 11.9** Black body spectrum for Vega

10 It has been suggested that one way to stop climate change would be to spray reflective material high into Earth's atmosphere. Discuss how this might help prevent the planet from warming up.

In chapter 17 you will look at the discovery of many more particles, and how they are related, when you explore the Standard Model.

# 11.2 Investigating the photoelectric effect

As we have seen, Planck introduced the idea of quantisation of energy as a mathematical trick to explain the black body spectrum.

When the quantum model was being developed, the idea of quantisation of matter was already well established and accepted, but the idea of quantisation of energy was new. It also contradicted the accepted model of light as a wave.

The photoelectric effect provided the evidence needed for quantisation of energy to be accepted as more than just a mathematical trick.

## The photoelectric effect

The photoelectric effect was first observed by Heinrich Hertz in 1887. He observed that when light is shone on a highly polished metal surface, electrons can be emitted from the surface. One of Hertz's assistants, Philipp Lenard, performed experiments to investigate the photoelectric effect in detail. Lenard developed much of the equipment needed to make quantitative measurements of the intensity and energy of the emitted 'cathode rays', as the beams of electrons are known. Other physicists, including Robert Millikan, also investigated the effect.

Their data showed that:

» no electrons were emitted unless the frequency of the light was above some minimum or critical frequency, regardless of the intensity of the light

» the number of electrons (the current) was proportional to the intensity. It did not vary with the frequency of the light, as long as the frequency was high enough.

A photoelectric effect apparatus is shown in Figure 11.10.

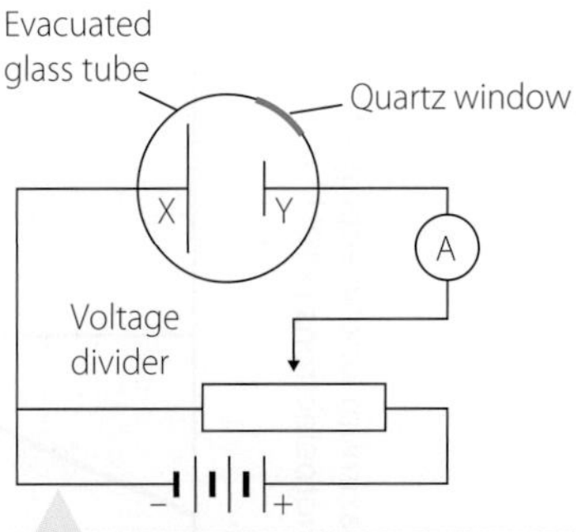

When light is shone through the quartz window on to the polished metal plate, X, **photoelectrons** are emitted. The photoelectrons are attracted to the positively charged metal plate, Y. The ammeter, A, measures the current of photoelectrons produced – the **photocurrent**.

FIGURE 11.10 A typical setup for a photoelectric experiment

Using this apparatus, experiments show that:

» a photocurrent is only produced when the frequency of the light is above some minimum value, called the **cut-off frequency**, $f_0$. This implies a cut-off wavelength, $\lambda_0 = \dfrac{c}{f_0}$, above which no photocurrent is produced. Even if the light is very bright, if the frequency is not high enough no electrons are emitted. This goes against classical theory

» the size of the current (the number of photoelectrons produced) depends on the intensity of the light but not the frequency, as long as the frequency is above $f_0$

» there is no time delay between light being incident on the metal and photoelectrons being emitted, regardless of intensity

» different metals have different characteristic cut-off frequencies.

The voltage divider is used to vary the potential difference between X and Y. When the potential difference is reversed, the maximum kinetic energy of the emitted photoelectrons can be measured. This is called a reverse bias voltage. In this case, plate Y is negative and repels the photoelectrons. The electrons lose energy as they move from a point of higher potential to one of lower potential in the direction of an electric field. Hence, the electrons lose kinetic energy as they move in the direction of X to Y.

Potential is potential energy per unit charge. As a negatively charged electron moves from a point of higher to lower potential, its potential energy increases so its kinetic energy must decrease.

The reverse bias voltage between X and Y is slowly increased and the current observed. When the current just drops to zero, the potential difference is equal to the maximum energy per unit charge of the electrons. It is just big enough to stop the most energetic (fastest) emitted electrons. So, it measures the maximum energy a photoelectron can have (for this experiment). This potential difference is called the **stopping voltage**, $V_s$. Hence, the potential difference multiplied by the electron charge is equal to the maximum kinetic energy of the photoelectrons.

$$K_{max} = V_s q = V_s e$$

This experiment has the following results.

The maximum kinetic energy (measured via the stopping voltage) depends on the frequency of light but *not* on the intensity, as shown in Figure 11.11. The y-axis on Figure 11.11 gives the maximum energy of an emitted electron at each frequency of the light. For an electron to have zero energy when emitted (that is, it 'gets out' of the metal but has no velocity when it does), the light must have a frequency *greater* than zero. It is about $9 \times 10^{14}$ Hz for magnesium, for example. That suggests that it takes some 'work' to get out of the metal. The energy an electron needs to just get out of the metal is the metal's **work function**.

Different metals have different characteristic stopping voltages, which depend on frequency.

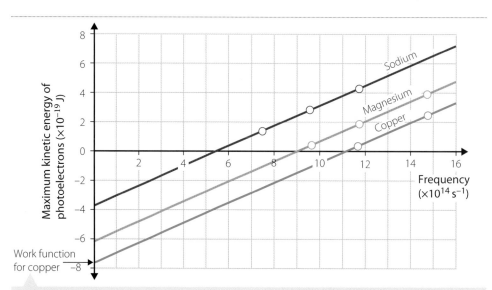

**FIGURE 11.11** Maximum kinetic energy (in units of $10^{-19}$ J) as a function of frequency for three different metals. The data are shown as circles and the lines are extrapolated to the energy axis to find the minimum energy required to eject an electron from each metal. This energy is known as the work function of the metal concerned.

The electromagnetic wave model of light cannot explain all of these observations. Table 11.1 compares the results of these experiments with the predictions of classical electromagnetic wave theory.

It was for his explanation of the photoelectric effect that Einstein won his Nobel Prize, and not for the development of his theories of relativity (chapter 12). Although relativity theory is fascinating, in terms of impact on our daily lives it is **quantum theory** that has made the most difference.

**TABLE 11.1** Comparison of experimental results with predictions of classical electromagnetic wave theory

| EXPERIMENTAL OBSERVATION | PREDICTION FROM CLASSICAL ELECTROMAGNETIC WAVE MODEL |
|---|---|
| Photocurrent only occurs for frequencies above $f_0$, and $f_0$ is characteristic of the material. | Electrons should be emitted at any frequency, as long as the intensity is high enough. |
| The size of the current depends on intensity but not frequency. | The current should depend on both intensity and frequency. |

| EXPERIMENTAL OBSERVATION | PREDICTION FROM CLASSICAL ELECTROMAGNETIC WAVE MODEL |
|---|---|
| There is no time delay between the absorption of light and emission of a photoelectron at any intensity. | At low intensities it takes time for enough energy to be absorbed by the atoms. Hence, there should be a delay between the light being turned on and electrons being emitted. The delay should be longer for lower intensities. |
| The maximum kinetic energy of the electrons depends on the frequency of light but *not* on the intensity. | The kinetic energy should only be related to the intensity, and *not* to the frequency. |

## WORKED EXAMPLE

In a photoelectric effect experiment, the stopping voltage, $V_s$, has a magnitude of 3.75 V.

**a** What is the kinetic energy of the fastest emitted electron?

**b** How fast is it going?

| ANSWERS | LOGIC |
|---|---|
| $V_s = 3.75$ V | ▪ Identify the relevant data from the question. |
| **a** $K_{max} = V_s q = V_s e$ | ▪ Use the relationship between kinetic energy and stopping voltage |
| $K_{max} = 3.75$ V $\times 1.6 \times 10^{-19}$ C | ▪ Substitute values with correct units. |
| $K_{max} = 6.0 \times 10^{-19}$ J | ▪ State the final value with correct units and appropriate significant figures. |
| **b** $K = \dfrac{1}{2}mv^2 = V_s e$ | ▪ Use the equation for kinetic energy |
| $v = \sqrt{\dfrac{2V_s e}{m}}$ | ▪ Rearrange to make $v$ the subject. |
| $v = \sqrt{\dfrac{2(3.75 \text{ V})(1.6 \times 10^{-19} \text{ C})}{9.1 \times 10^{-31} \text{ kg}}}$ | ▪ Substitute values with correct units. |
| $v = 1.1 \times 10^6 \text{ m s}^{-1}$ | ▪ State the final answer with correct units and appropriate significant figures . |

**TRY THIS YOURSELF**

If an electron was ejected from the surface with velocity $8.6 \times 10^5 \text{ m s}^{-1}$, what would the stopping voltage be?

**KEY CONCEPTS**

- When a light is shone on a metallic surface, no electrons are emitted unless the frequency of the light is above some minimum or critical frequency, the cut-off frequency $f_0$.
- When the frequency is above $f_0$, the number of electrons (the current) is proportional to the intensity.
- A cut-off frequency, $f_0$, implies a cut-off wavelength, $\lambda_0 = \dfrac{c}{f_0}$, above which no photocurrent is produced.
- There is no time delay between light being incident on the metal and photoelectrons being emitted, regardless of intensity. That means the metal does not need to absorb energy over time before it can emit an electron.
- Different metals have different characteristic cut-off frequencies.

1 Classical theories predicted that electrons should be emitted once the metal had absorbed enough energy, and that this should be possible at all frequencies of light and depended on the intensity of the light. How do the results of the photoelectric effect experiment show this to be wrong?

2 Define:
   a cut-off frequency.
   b cut-off wavelength.
   c work function.

3 What do we measure when we apply a reverse bias voltage to the apparatus illustrated in Figure 11.10?

4 Explain how you can find Planck's constant from a graph of frequency against stopping voltage from a photoelectric experiment.

5 It requires more energy to remove an electron from the surface of a polished piece of copper than from a polished piece of lithium.
   a Which metal has the larger work function?
   b Which metal has the greater cut-off wavelength?

6 Sodium has a cut-off frequency of $5.94 \times 10^{14}$ Hz. What is the cut-off wavelength?

7 An electron is ejected from sodium with an energy of 2.1 eV.
   a What voltage is required to stop it?
   b How fast is it going?

# 11.3 Analysing the photoelectric effect

Just as with black body radiation, a new theory was needed to explain the results of photoelectric effect experiments.

It was Einstein who came up with a new model in 1905. His explanation combined two ideas – the very familiar one of conservation of energy, and Planck's more recently introduced idea of quantisation.

## Conservation of energy

An ion is a charged particle, usually an atom that has either lost or gained electrons. The ionisation energy is the energy needed to remove an electron from an atom. In the case of the photoelectric effect, electrons are lost from the whole metal lattice rather than from individual atoms.

You are already familiar with the idea of conservation of energy. Einstein explained the photoelectric effect by saying that electromagnetic radiation, or light, is quantised, or at least behaves as if it is quantised. When it interacts with matter, such as the metal plate X in Figure 11.10, it can only give up its energy in discrete amounts. Each quantum of light has energy $E = hf$, where $h$ is the Planck constant and $f$ is the frequency of the light. This is the relationship between energy and frequency first introduced by Planck to explain black body radiation.

When an electron in the metal plate X (Figure 11.10) absorbs a photon, it gains this energy. However, to leave the metal plate costs it an amount of energy; effectively an ionisation energy. Hence, the cut-off frequency, which is characteristic of the metal, is a measure of this ionisation energy. This energy, as discussed in the previous section, is called the work function of the metal, and is given by:

$$\phi = hf_0$$

where $\phi$ is the work function, $h$ is Planck's constant and $f_0$ is the cut-off frequency.

Putting this together with conservation of energy, Einstein said that:

$$K_{max} = hf - hf_0 = hf - \phi$$

where $K_{max}$ is the maximum kinetic energy an electron can have. This is the photoelectric equation. It says that if an electron absorbs light energy $hf$, and is emitted from the metal, it can have a maximum kinetic energy of $hf$, minus the energy needed to leave the plate, which is $\phi = hf_0$.

Looking again at Figure 11.11, we can now see that the gradient of each line must be equal to Planck's constant. The extrapolated straight lines of best fit in Figure 11.11 meet the *y*-axis at the value of the work function. This graphical representation of the experimental data allows us to quickly find values for both the Planck constant and the work function of the metal used.

**TABLE 11.2** Work functions of some metals

| METAL | $\phi$ (eV) | $\phi$ (J) |
|-------|-------------|------------|
| Na | 2.46 | $3.94 \times 10^{-19}$ |
| Al | 4.08 | $6.53 \times 10^{-19}$ |
| Fe | 4.50 | $7.20 \times 10^{-19}$ |
| Cu | 4.70 | $7.52 \times 10^{-19}$ |
| Zn | 4.31 | $6.90 \times 10^{-19}$ |
| Ag | 4.73 | $7.57 \times 10^{-19}$ |
| Pt | 6.35 | $1.02 \times 10^{-19}$ |
| Pb | 4.14 | $6.62 \times 10^{-19}$ |

Note that these are typical values for these metals. Measured values vary depending on whether the metal is a single crystal or polycrystalline. For single crystals the value also depends on which face of the crystal is illuminated.

## Millikan's experiments

Einstein's hypothesis that energy was quantised was confirmed by Robert Millikan. Millikan did not initially believe Einstein's explanation for the photoelectric effect and spent 10 years performing experiments to test it. He was a brilliant experimentalist. He improved the photoelectric effect experiment to make it precise enough to really test Einstein's model. He developed ways of cleaning the components and assembling the experiment in vacuum, so that surfaces would be clean and pure enough for the experiment to work. Using his equipment, he produced graphs similar to that shown in Figure 11.11. He found that the gradient of his graphs was consistently within **uncertainty** of Planck's constant. He eventually came to the conclusion that Einstein's model explained the experimental data, and that no other existing model did. He wrote in his autobiography that his data 'scarcely permits of any other interpretation than that which Einstein had originally suggested'.

▶ **WORKED EXAMPLE**

Using the graph in Figure 11.12, find the value of the work function for caesium. Give your answer in electron-volts and joules.

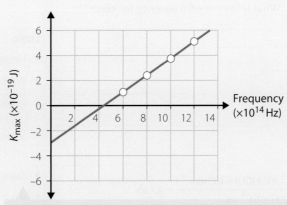

**FIGURE 11.12** A plot of photoelectron maximum kinetic energy as a function of frequency of incident light for caesium

| ANSWER | LOGIC |
|---|---|
| $f_0 = 4.4 \times 10^{14}\,\text{Hz}$ | ▪ Identify the relevant data from the question. Read $f_0$ from the graph. |
| $\phi = hf_0$ | ▪ Relate $f_0$ to the work function. |
| $\phi = 6.63 \times 10^{-34}\,\text{J s} \times 4.4 \times 10^{14}\,\text{Hz}$ | ▪ Substitute values including correct units. |
| $\phi = 2.9 \times 10^{-19}\,\text{J}$ | ▪ State the final value with correct units and appropriate significant figures. |
| $\phi = \dfrac{2.9 \times 10^{-19}\,\text{J}}{1.6 \times 10^{-19}\,\text{J eV}^{-1}} = 1.8\,\text{eV}$ | ▪ Convert from J to eV. |

**TRY THIS YOURSELF**

Use the graph in Figure 11.12 to find a value for Planck's constant, $h$.

You may have noticed that we have been talking about the *maximum* kinetic energy of the photoelectrons. The photoelectrons have all values of energy *up to* this maximum value. After absorbing the energy $hf$ from the light, the electrons have energy $hf$. Recall from chapter 13 of *Physics in Focus Year 11* that in a metal there are lots of conduction electrons. These are electrons that are free to move through the metal and are not bound to any particular atom. It is these conduction electrons that can be ejected as photoelectrons. The valence electrons, which are bound to the atoms, cannot gain enough energy to be ejected in this way.

Electrons that have absorbed $E = hf$ lose a minimum energy of $\phi$ to escape the material, if they are at the very surface and have no interactions with other electrons or nuclei as they escape. However, most of the electrons will lose some of the absorbed energy in collisions with other electrons and with interactions with the atomic nuclei in the metal. Many do not leave the metal at all because they lose all of this energy, which shows up as the increased temperature of the metal. Hence, there is a continuous spectrum of electron energies from zero to the maximum value of $hf - \phi$.

▶ **WORKED EXAMPLE** (11.5)

Ultraviolet light of wavelength 200 nm is incident on a polished silver plate. The work function for silver is 4.7 eV.

**a** What is the kinetic energy of the fastest moving electrons?

**b** What is the kinetic energy of the slowest moving electrons?

**c** What is the cut-off frequency for silver?

| ANSWERS | LOGIC |
|---|---|
| $\lambda = 200\,\text{nm}, \phi = 4.7\,\text{eV}$ | ▪ Identify the relevant data from the question. |
| **a** $K_{max} = hf - \phi$ | ▪ Relate maximum $K$ to other values given. |
| $f = \dfrac{c}{\lambda}$ | ▪ Relate frequency to wavelength. |
| $K_{max} = \dfrac{h}{\lambda} - \phi$ | ▪ Substitute the expression for $f$. |
| $K_{max} = \dfrac{(4.14 \times 10^{-15}\,\text{eV s})(3.0 \times 10^8\,\text{m s}^{-1})}{200 \times 10^{-9}\,\text{m}} - 4.7\,\text{eV}$ | ▪ Substitute values including correct units.<br>▪ $\phi$ is in eV, so convert $h$ to eV s. |
| $K_{max} = 1.5\,\text{eV}$ | ▪ Calculate the final value with correct units and appropriate significant figures. |

9780170409131

| | |
|---|---|
| **b**  $0\,\text{eV}$ | ■ The slowest moving electrons only just make it out of the metal. |
| **c**  $\phi = hf_0$ | ■ Relate the cut-off frequency to known values. |
| $f_0 = \dfrac{\phi}{h}$ | ■ Rearrange for $f_0$. |
| $f_0 = \dfrac{4.7\ \text{eV}}{4.14 \times 10^{-15}\ \text{eV s}}$ | ■ Substitute values with correct units. |
| $f_0 = 1.1 \times 10^{15}\,\text{Hz}$ | ■ Calculate the final value with correct units and appropriate significant figures. |

**TRY THIS YOURSELF**

Draw a quantitative graph of maximum photoelectron kinetic energy as a function of frequency of incident light for silver.

# INVESTIGATION (11.2)

## The photoelectric effect

It is possible to observe the photoelectric effect using some very simple, equipment, but for this Investigation it is probably better to use a commercial electroscope rather than making your own as described in the Weblink.

### AIM

To observe the photoelectric effect

### MATERIALS

- Electroscope
- Polished zinc plate
- Steel wool or fine sandpaper
- Glass rod
- Polythene rod
- Fur or wool fabric
- Ultraviolet light source
- Other light sources, for example lasers and torches
- Stopwatch

**Soft drink can electroscope**

Build your own electroscope using a soft drink can. If you are health-conscious, pour out the soft drink – do not drink it.

⚠ **RISK ASSESSMENT**

| WHAT ARE THE RISKS IN DOING THIS INVESTIGATION? | HOW CAN YOU MANAGE THESE RISKS TO STAY SAFE? |
|---|---|
| Ultraviolet light can damage your eyes. | Do not look directly at the ultraviolet light source. Turn it off when not in use. |

What other risks are associated with your investigation, and how can you manage them?

### METHOD

1  Clean the zinc plate *thoroughly* with the steel wool or sandpaper, then place it on the electroscope. Try not to leave any fingerprints on it as you do so.

2  Charge up the glass rod by rubbing it vigorously with the fur or wool.

3  Touch the rod to the zinc plate. You should observe the leaves of the electroscope separate.

»

**4** Time how long it takes for the leaves to fall back together. If it takes more than two minutes, just record your result as 'more than two minutes' and discharge the electroscope by touching it with your hand. Remember to avoid touching the zinc plate.

**5** Repeat steps 2 to 4, but this time shine one of your light sources on the plate.

**6** Repeat steps 2 to 5 with each of your different light sources.

**7** Charge up the polythene rod by rubbing it vigorously with the fur or wool fabric.

**8** Use the rod to charge the electroscope.

**9** Time how long it takes for the leaves to fall back together.

**10** Repeat steps 7 to 9, but this time shine a light source on the zinc plate. Repeat this for each light source.

**11** Charge the zinc plate with the polythene rod and then shine the ultraviolet light on it from various distances away. This varies the intensity of the light on the plate.

RESULTS

**1** Draw a diagram showing your experimental setup. Label all the parts clearly.

**2** Record your results in a table, such as that shown below. The glass rod becomes positively charged. The plate is also positively charged when charged with the glass rod. The polythene rod becomes negatively charged. The plate is also negatively charged when charged with the polythene rod.

**3** Don't forget to include units and uncertainties on your data.

| CHARGE ON PLATE | LIGHT USED | TIME TO DISCHARGE (s) |
|---|---|---|
|  |  |  |
|  |  |  |

**4** If you measured the discharge time using the ultraviolet light at various distances (step 11 above), record your results in a table like the one below.

| DISTANCE TO LIGHT (m) | TIME TO DISCHARGE (s) |
|---|---|
|  |  |
|  |  |

ANALYSIS OF RESULTS

**1** Explain what is happening in each case you investigated.

**2** Plot a graph of time to discharge as a function of distance to light, using the data in your second table. Do you expect this graph to be linear? If not, what shape do you expect it to be? Can you check whether it is?

DISCUSSION

**1** Discuss the shape(s) of your graph(s). Are they what you expected? If not, why might this be the case? Think about any assumptions you have made.

**2** What could you do to improve the accuracy and precision of this experiment?

**3** Extend the method to investigate another aspect of your electroscope.

CONCLUSION

Write a conclusion to your investigation based on your results and analysis.

## Photons: quantisation of energy

Analysing the photoelectric effect

Einstein's explanation of the photoelectric effect gave a physical meaning to the idea of quantisation of energy of electromagnetic radiation. It meant that, in some circumstances, light behaved like particles. The term 'photon' was introduced by the chemist Gilbert Lewis in 1926 to describe these particles. Further experiments by Arthur Holly Compton gave evidence for the existence of photons. Compton scattered single photons from electrons and found that only particular energies were absorbed. Photons are now accepted as particles with zero rest mass, and with energy given by $E = hf$. They also have momentum, which we can calculate from $E = pc$, where $p$ is momentum.

▶ **WORKED EXAMPLE** (11.6)

Find the range of energies of photons in the visible spectrum. The visible spectrum ranges from blue light with wavelength approximately 400 nm to red light with wavelength approximately 700 nm.

| ANSWER | LOGIC |
|---|---|
| $E = hf$ | ▪ Relate energy to other values. |
| $f = \dfrac{c}{\lambda}$ | ▪ Relate frequency to wavelength. |
| $E = \dfrac{hc}{\lambda}$ | ▪ Substitute expression for $f$. |
| $E_{max} = \dfrac{hc}{\lambda_{min}}$ | ▪ Recognise that the highest energy corresponds to the smallest wavelength. |
| $E_{max} = \dfrac{(4.14 \times 10^{-15}\,\text{eV s})(3.0 \times 10^{8}\,\text{m s}^{-1})}{400 \times 10^{-9}\,\text{m}}$ | ▪ Substitute values including correct units. |
| $E_{max} = 3.1\,\text{eV}$ | ▪ Calculate the final value with correct units and appropriate significant figures. |
| $E_{min} = \dfrac{hc}{\lambda_{max}}$ | ▪ Recognise that the lowest energy corresponds to the longest wavelength. |
| $E_{max} = \dfrac{(4.14 \times 10^{-15}\,\text{eV s})(3.0 \times 10^{8}\,\text{m s}^{-1})}{700 \times 10^{-9}\,\text{m}}$ | ▪ Substitute values including correct units. |
| $E_{min} = 1.8\,\text{eV}$ | ▪ Calculate the final value with correct units and appropriate significant figures. |

**TRY THIS YOURSELF**

Photons in the visible spectrum have energies ranging from 1.8 eV to 3.1 eV. Convert these energies to J.

**CHECK YOUR UNDERSTANDING**

**11.3**

1  What principle did Einstein use to explain the energies of photoelectrons?

2  What is a photon?

3  How is a photoelectron different from any other electron?

4  If a metal has work function $\phi$ and is irradiated with light of frequency $f$, what are the possible energies of any emitted photoelectrons?

**5** The graph in Figure 11.13 shows the results of a photoelectric experiment using a magnesium metal plate.

    **a** Find a value for Planck's constant from this graph. Estimate the uncertainty in your value.

    **b** Find a value for the work function of magnesium. Estimate the uncertainty in your value.

    **c** Imagine that silver had been used in this experiment instead of magnesium. Silver has a work function of 4.7 eV. Draw a line on the graph showing where the data points for silver would be located.

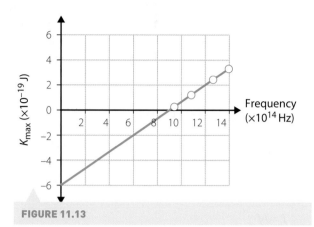

**FIGURE 11.13**

**6** Figure 11.14 shows a photoelectric tube with light of frequency $f$ and intensity $I$ incident on a metal cathode. Electrons emitted from the cathode are collected at the anode. The potential difference between the anode and cathode is varied, and the resulting photocurrent is measured. Figure 11.15 shows the results of this experiment.

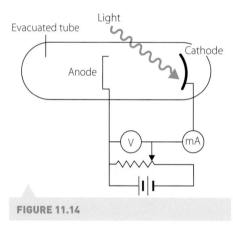

**FIGURE 11.14**

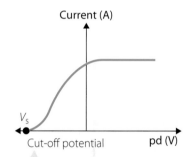

**FIGURE 11.15** Photocurrent as a function of applied potential difference (pd)

    **a** Why is the photocurrent constant at positive values of pd?

    **b** If the frequency of the light is varied, which of the curves A–F in Figure 11.16 represents the relationship between the stopping voltage, $V_s$, and $f$?

**7** A photon has wavelength $1 \times 10^{-10}$ m. What is its momentum?

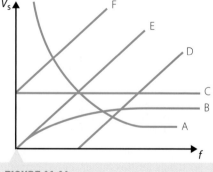

**FIGURE 11.16**

- Black body radiation is emitted by all objects at non-zero temperature.

- Black body radiation has a spectrum characteristic of the temperature of the body. The black body spectrum is a continuous spectrum.

- The spectrum cannot be explained using classical physics, and so was an important early piece of evidence that 'new' physics remained to be discovered. This was quantum physics.

- The peak wavelength in a black body spectrum is given by Wien's Displacement Law: $\lambda_{max} = \dfrac{b}{T}$

- Black body radiation can be explained by the quantisation of electromagnetic energy. We call these quanta of energy 'photons'. Using this idea, Planck was able to correctly model the dependence of intensity on wavelength as it varied with temperature of the body.

- Photons are 'particles' of electromagnetic energy. Each photon has energy $E = hf$.

- Earth behaves somewhat like a black body, and this is important in maintaining its temperature balance and climate.

- The photoelectric effect is the ejection of electrons from a polished metal surface by incident light. The light must have a minimum frequency for this to occur.

- The photoelectric effect was another experiment that showed that classical physics could not explain some observations. Einstein explained the effect using the idea of photons.

- The minimum frequency corresponds to a minimum energy, $E = hf_0$.

- The minimum energy corresponds to the work function, $\phi$, of the metal. It depends on the metal, not on the light.

- The maximum kinetic energy of the photoelectrons is $K_{max} = hf - \phi$. This is a statement of conservation of energy.

- The photocurrent, which is proportional to the number of electrons, depends on the intensity of the light. The intensity is a measure of the number of incident photons.

- The ideas of quantisation of energy and light behaving like particles (photons) were very successful at explaining these experiments. This contributed to the acceptance of the quantum theory, which was one of the most important developments in physics in the 20th century.

# (11) CHAPTER REVIEW QUESTIONS

Review quiz

1 Name three physicists who contributed to the development of the quantum theory and briefly describe their contributions.

2 When was the photoelectric effect first observed, and by whom?

3 An incandescent light globe is connected to a variable power supply and the voltage gradually increased. Describe the sequence of colours produced by the filament of the globe.

4 Figure 11.17 shows the black body spectra for two stars, A (red curve) and B (blue curve). Which star is hotter? Explain your answer.

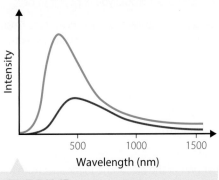

**FIGURE 11.17**

**5** Why are fluorescent globes and LEDs so much more energy-efficient than incandescent globes?

**6** Which metal in Table 11.2 (page 263) would require the highest frequency light for photoelectrons to be emitted?

**7** A polished lead surface is illuminated with light of wavelength 200 nm. What is the effect on the photocurrent if:

  **a** the wavelength is halved?

  **b** the intensity of the light is doubled?

**8** Figure 11.18 shows a graph of maximum kinetic energy as a function of frequency for a particular metal. Copy the graph and, on the same set of axes, draw a line showing maximum kinetic energy for:

  **a** the same metal but with higher intensity light.

  **b** light of the same intensity but for a metal with a larger work function.

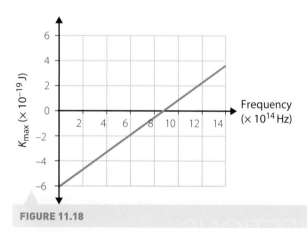

**FIGURE 11.18**

**9** What is the peak wavelength emitted by a toaster element at 700°C? What colour would you expect it to be?

**10** Space itself is not at absolute zero temperature. Cosmic background radiation (see chapter 13) has a spectrum similar to that produced by a black body at 2.7 K. What is the peak wavelength of this radiation?

**11 a** A photon has energy 5.5 eV. What is its:

   **i** energy in J?

   **ii** wavelength?

   **iii** frequency?

  **b** To what part of the electromagnetic spectrum (radio waves, microwaves, all the way through to X-rays and gamma rays) does the photon belong?

**12** A polished sodium surface is illuminated by light.

  **a** Find the cut-off wavelength for sodium. What colour does this correspond to?

  **b** Find the maximum kinetic energy of ejected photoelectrons when light of wavelength 300 nm is used.

**13** The graph in Figure 11.19 shows the results of a photoelectric experiment using an unknown metal. Calculate the work function for this metal. What is this metal?

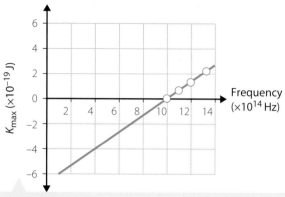

**FIGURE 11.19**

**14** A microwave oven produces electromagnetic radiation with a wavelength of 12.2 cm.

  **a** What energy does this correspond to? Give your answer in J and eV.

  **b** Is this enough energy to produce a photocurrent from a metal surface? Why do you think it might be a bad idea to put metal objects in a microwave? *Caution: Don't try it!*

**15** The following data was collected by students doing an experiment on black body radiation.

  **a** Plot a graph showing power as a function of wavelength.

  **b** Calculate the temperature of the black body that produced this radiation. Explain your answer.

| WAVELENGTH (nm) | POWER AT DETECTOR (W) |
|---|---|
| 100 | 0 |
| 200 | 0.01 |
| 300 | 0.18 |
| 400 | 0.30 |
| 500 | 0.34 |
| 600 | 0.32 |
| 700 | 0.27 |
| 800 | 0.21 |
| 900 | 0.17 |
| 1000 | 0.14 |
| 1100 | 0.10 |
| 1200 | 0.09 |

16 The table below shows data collected in a photoelectric experiment. Plot an appropriate graph and find:

  a Planck's constant.

  b the work function for this metal.

| WAVELENGTH (nm) | $K_{max}$ OF PHOTOELECTRONS (eV) |
| --- | --- |
| 588 | 0.67 |
| 505 | 0.98 |
| 445 | 1.35 |
| 399 | 1.63 |

17 Research Robert Millikan. Why did Millikan initially object to Einstein's explanation of the photoelectric effect? Why did he eventually accept the photon explanation? What other important experiments did Millikan perform?

18 Why does the decrease in Earth's **albedo** (basically, how reflective it is) result in more extreme weather, such as storms, floods and droughts? You will need to think about what you learned when you studied thermodynamics in Year 11, as well as what you have learned in this chapter.

19 Research on the internet the views of climate sceptics and climate scientists. What evidence does each group use to support their opinions? Evaluate the arguments put forward by both sides and summarise and justify your own opinion.

# Light and special relativity

INQUIRY
QUESTION

How does the behaviour
of light affect concepts of
time, space and matter?

OUTCOMES

**Students:**

- analyse and evaluate the evidence confirming or denying Einstein's two postulates:
  - the speed of light in a vacuum is an absolute constant
  - all inertial frames of reference are equivalent (ACSPH131)
- investigate the evidence, from Einstein's thought experiments and the subsequent experimental validation, for time dilation $t = \dfrac{t_0}{\sqrt{\left(1 - \dfrac{v^2}{c^2}\right)}}$ and length contraction $l = l_0\sqrt{\left(1 - \dfrac{v^2}{c^2}\right)}$, and analyse quantitatively situations in which these are observed, for example:
  - observations of cosmic-origin muons at the Earth's surface ICT, N
  - atomic clocks (Hafele–Keating experiment) CCT, ICT, N
  - evidence from particle accelerators CCT, ICT, N
  - evidence from cosmological studies ICT
- describe the consequences and applications of relativistic momentum with reference to:
  - $p_v = \dfrac{m_0 v}{\sqrt{\left(1 - \dfrac{v^2}{c^2}\right)}}$ ICT, N
  - the limitation on the maximum velocity of a particle imposed by special relativity (ACSPH133) CCT
- Use Einstein's mass–energy equivalence relationship $E = mc^2$ to calculate the energy released by processes in which mass is converted to energy, for example: (ACSPH134) ICT, N
  - production of energy by the sun
  - particle–antiparticle interactions, e.g. positron–electron annihilation
  - combustion of conventional fuel

Physics Stage 6 Syllabus © 2017 NSW Education Standards Authority (NESA) for
and on behalf of the Crown in right of the State of New South Wales.

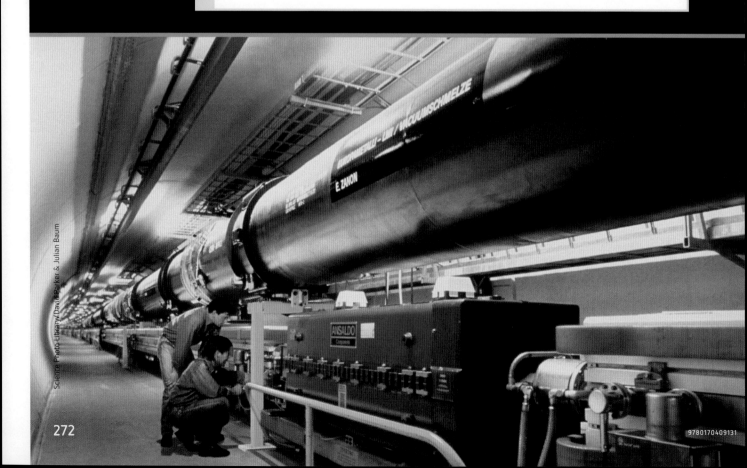

9780170409131

Towards the close of the 19th century, Newton's work on motion had been thoroughly tested and found to be valid in all circumstances. Scientists felt that they understood interactions between matter and energy. However, within 30 years the most fundamental laws of physics would be challenged and replaced. Small but significant discrepancies were becoming apparent between predictions based on Newton's equations of motion and experimental results. Refinement of the theory of electromagnetism by James Clerk Maxwell, which Hertz later successfully tested by experiment, was one trigger for this change. The man who had the most impact was Albert Einstein, a German-speaking Swiss patent clerk. He developed a different way of looking at time, space, matter and energy. A fundamental question was: what is the speed of light? Does light travel instantly from one place to another, or is its speed large but finite?

Many early scientists conjectured that light had a finite speed. This was first demonstrated by Rømer at the end of the 17th century. It was confirmed by the astronomer James Bradley in 1728. Bradley measured the difference between where a star was expected to be seen and where it was actually observed. This **stellar aberration** is due to the finite speed of light and the speed of Earth combining to affect the position from which the starlight appears to come.

In Figure 12.1, if Earth were stationary relative to the star (or if the speed of light were infinite), then the star would appear at some angle $\theta$. If Earth moves in the direction of the arrow, the apparent angle is decreased to $\phi$. Since there is no way to know the 'correct' angle $\theta$, this is not much help. However, six months later Earth is on the other side of the Sun and moving in the opposite direction. If the same star is observed, the angle is found to *increase*. This is evidence of the finite nature of the speed of light.

Bradley determined the speed to be about $298\,000\,\mathrm{km\,s^{-1}}$ (in modern units). This was later refined using other methods by Fizeau (1849), Foucault (1852) and Michelson (1879, 1883 and 1926). The standard speed of light in a vacuum is now $299\,792\,458\,\mathrm{m\,s^{-1}}$.

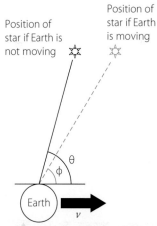

**FIGURE 12.1** Stellar aberration. A star appears at a different angle from that expected. This is caused by the movement of Earth and the finite speed of light.

> The speed of light in a vacuum is $299\,792\,458\,\mathrm{m\,s^{-1}}$. In this chapter, we will use $c = 2.998 \times 10^8\,\mathrm{m\,s^{-1}}$ and sometimes $3.0 \times 10^8\,\mathrm{m\,s^{-1}}$.

# 12.1 Frames of reference and relativity

Galileo described a thought experiment in which a sailor drops an object from the mast of a sailing ship moving at steady velocity (Figure 12.2). He asked the question: 'Where would the object land relative to the deck of the ship?' In his frame of reference, the sailor would see the object fall straight down, parallel to the mast; however, a nearby observer who is on land would see from his frame of reference that the object would follow a parabolic path.

Newton agreed with Galileo. He described frames of reference that were stationary or moving at constant velocity as **inertial frames of reference**. They are not accelerating or changing direction. Inertial reference frames could include a spaceship, a table and a cruising aeroplane. Non-inertial frames of reference include merry-go-rounds and aeroplanes taking off or landing – in both cases, there is an acceleration. If the frame is accelerating in any way, someone in that frame can tell that they are moving and it is not inertial.

An inertial frame is one in which Newton's first law applies to a very good approximation. Any departures from the law are negligible.

## Galilean transformations

Classical physics, the physics of Galileo and Newton, relies on the 'sensible' idea that all inertial coordinate systems are equivalent.

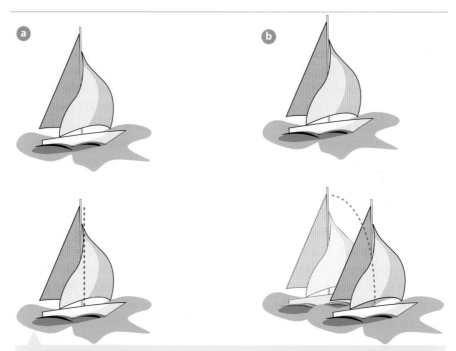

**FIGURE 12.2** Path of a falling object in the reference frame of **a** a sailing ship and **b** Earth (dotted lines). The first (top) image in each figure gives the position of the boat relative to the observer when the object is dropped. The second (lower) gives the position relative to the observer when it hits the deck. In the lower image in **b** the original position is also shown.

Let us analyse this assumption for the case of two inertial frames of reference in two dimensions (Figure 12.3). The 'privileged' frame of reference, P, uses coordinates $(x, y)$. It is stationary with respect to a moving frame of reference, P′, which uses coordinates $(x', y')$. The two sets of coordinates can be made to coincide for a moment, but then P′ moves further and further away from P. The distance between them depends on the relative speed, $v$, of P′ with respect to P and the time elapsed, $\Delta t$, after the two frames coincided.

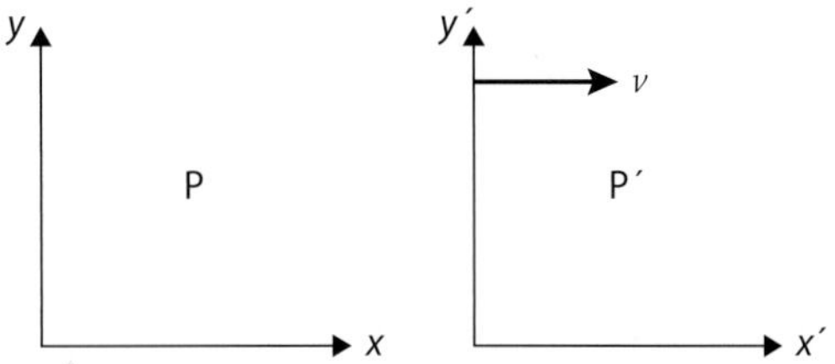

The time interval in each frame of reference is the same. That is, for classical physics, clocks in all frames of reference are identical in their timekeeping. Time is invariant. That is, observers in all frames of reference experience time moving at the same rate. Thus, $\Delta t$ in P and $\Delta t'$ in P′ are equal. We shall denote the time interval as $\Delta t$.

**FIGURE 12.3** Two inertial frames that are moving relative to each other. One frame (P) is regarded as stationary.

Suppose an observer located at (0, 0) in frame P sees a rocket some distance $y$ away. It is travelling parallel to the $x$-axis at speed $v$. Initially (at $t = 0$) the rocket's coordinates in P′ coincide with those in P. But as the rocket travels in frame P′ its $x$ coordinates increase in the time interval $\Delta t$ by $v\Delta t$. This is just the distance between the two coordinate systems. Thus, the $x$ coordinates in P relate to the $x'$ coordinates in P′ by the Galilean transformation:

$$x = x' + v\Delta t$$

Similar arguments can be used to show that if the rocket is travelling at speed $v$ parallel to the $y$- and $y'$-axes:

$$y = y' + v\Delta t$$

The transformation equations rely on the relative speeds of the two frames of reference. If P′ had been chosen as the stationary (or privileged) frame, then P would be travelling at a negative speed with respect to P′. This needs to be taken into account in the Galilean transformation equations. Finally, if the rocket were travelling with vector velocity $v$, the transformation equations in the $x$ and $y$ directions require the use of the $x$ and $y$ components of the velocity respectively.

9780170409131

A car travels at $12\,\text{m s}^{-1}$ through an intersection. After $10\,\text{s}$, what is the position of the car with respect to coordinates based on the:

**a** intersection?

**b** car?

The car is the primed $(x', y')$ coordinate frame.

| ANSWERS | LOGIC |
|---|---|
| $v = 12\,\text{m s}^{-1},\ t = 10\,\text{s}$ | ▪ Identify the relevant data from the question. |
| **a** $x = x' + v\Delta t$ | ▪ Write the equation connecting $x$ and $x'$. |
| $x = 0\,\text{m} + 12\,\text{m s}^{-1} \times 10\,\text{s}$ | ▪ Substitute the values with correct units. |
| $x = 120\,\text{m}$ | ▪ State the final answer with correct units and appropriate significant figures. |
| **b** $x' = x - v\Delta t$ | ▪ Rearrange the equation to make $x'$ the subject. |
| $x' = 120\,\text{m} - 12\,\text{m s}^{-1} \times 10\,\text{s}$ | ▪ Substitute the values with correct units. |
| $x' = 0\,\text{m}$ | ▪ State the final answer with correct units and appropriate significant figures. Note that the result makes sense if we just ask 'how far away from the car is the car' – of course the answer is zero! |

**TRY THESE YOURSELF**

A train passes through a station at $30\,\text{m s}^{-1}$. A person on the train walks towards the front of the train at $2\,\text{m s}^{-1}$. After $5.0\,\text{s}$, the person has moved a distance from the common origin of train and station. What are the coordinates with respect to:

**a** the station?

**b** the train?

## Principles of classical relativity

Galileo and Newton argued that the laws of motion are the same in all inertial frames of reference (relativity principle). Observers in different inertial frames would record different velocities, and therefore determine different values of energy and momentum. Nevertheless, they would agree on *conservation* of energy and *conservation* of momentum.

This suggests that there is no privileged inertial frame. No inertial reference frame is more fundamental than any other. If you are on a train travelling at a steady velocity of $80\,\text{km h}^{-1}$ west across the Nullarbor Plain, it is valid for you to argue that, from your reference frame, you are stationary and Earth is moving at $80\,\text{km h}^{-1}$ east. Providing your ride is smooth and at steady velocity, there is no experiment you can perform to test whether you are moving or stationary.

In other words:

▸ the laws of motion are the same in all inertial frames of reference

▸ the laws of conservation of energy and conservation of momentum apply in all inertial frames of reference

▸ all inertial frames are equivalent. All are equally valid.

### Relative velocities

It is straightforward to calculate the velocity of an object in one inertial frame of reference compared to another inertial frame. Take the example of an aeroplane flying from Sydney to Perth. At the cruising altitude of passenger aircraft, about $10\,000\,\text{m}$, there is nearly always a strong westerly wind. It varies from

$50 \, \text{km} \, \text{h}^{-1}$ to more than $300 \, \text{km} \, \text{h}^{-1}$. The direction of the aircraft's movement is almost exactly parallel to the direction of the wind. The westbound flight is scheduled at 4 hours 25 minutes (Figure 12.4a), while the eastbound flight is 3 hours 50 minutes (Figure 12.4b). The aeroplane's velocity, relative to the ground, is higher on the way to Sydney than on the way to Perth.

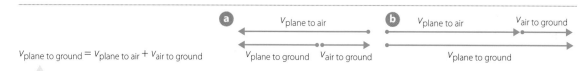

**FIGURE 12.4** Vector addition of the relative velocities of an aeroplane flying from **a** Sydney to Perth and **b** Perth to Sydney

In general, the velocity of A relative to B is the velocity of A relative to C plus (vector addition) the velocity of C relative to B:

$$v_{AB} = v_{AC} + v_{CB}$$

This agrees with what intuition tells us.

## WORKED EXAMPLE 12.2

An aeroplane is aiming due north at a speed of $400 \, \text{km} \, \text{h}^{-1}$. There is a westerly wind (i.e. coming from the west) of $80 \, \text{km} \, \text{h}^{-1}$.

**a** Draw a vector diagram to show the velocity of the plane relative to the ground.

**b** What is the speed of the plane relative to the ground?

**c** What is the resultant velocity of the plane relative to the ground?

| ANSWERS | LOGIC |
|---|---|
| **a** $v_{\text{air to ground}}$ (80 km h$^{-1}$), $v_{\text{plane to air}}$ (400 km h$^{-1}$), $v_{\text{plane to ground}}$, $\theta$ <br><br> **FIGURE 12.5** | ▪ Construct a correct vector diagram using the data given in the question (Figure 12.5). |
| **b** From Figure 12.5 we can use Pythagoras' theorem to find: <br> $v^2 = (400 \, \text{km} \, \text{h}^{-1})^2 + (80 \, \text{km} \, \text{h}^{-1})^2$ <br> $v = 408 \, \text{km} \, \text{h}^{-1}$ | ▪ State the final answer with correct units and appropriate significant figures. |
| **c** $\tan\theta = \dfrac{v_{\text{air to ground}}}{v_{\text{plane to air}}} = \dfrac{80 \, \text{km} \, \text{h}^{-1}}{400 \, \text{km} \, \text{h}^{-1}}$ <br> $\theta = 11.3°$ | ▪ Find the correct angle. |
| The velocity of the plane relative to the ground is $408 \, \text{km} \, \text{h}^{-1}$, N11°E. | ▪ State the final answer of velocity with direction. |

A person rows at $1.0\,\mathrm{m\,s^{-1}}$ through the water of a river that is flowing at $0.5\,\mathrm{m\,s^{-1}}$. The rower rows perpendicular to the bank.

**a** Draw a vector diagram to show the velocity of the rower from the reference frame of a person on the bank.

**b** Using your diagram, specify completely the velocity of the boat relative to the bank.

# INVESTIGATION

## Relative motion and frames of reference

There are two types of reference frame: inertial and non-inertial (accelerating). The path travelled by an object will be different in these two frames of reference.

### AIM

To observe motion in different frames of reference
  Write an inquiry question for this investigation.

### MATERIALS

- Several sheets of white paper
- Carbon paper
- Large metal ball bearing
- Ramp
- Potter's wheel, lazy susan or record turntable
- 2 dissecting boards or similar-sized flat surfaces
- Graph paper
- Scissors
- Adhesive tape
- Video recording camera or smartphone
- Frame to support the camera; a ripple tank may make a good frame

**RISK ASSESSMENT**

| WHAT ARE THE RISKS IN DOING THIS INVESTIGATION? | HOW CAN YOU MANAGE THESE RISKS TO STAY SAFE |
| --- | --- |
| Large metal ball bearings can cause injuries. | Keep the ball bearing safely on the table and use only in the experiment. |

What other risks are associated with your investigation, and how can you manage them?

### METHOD

**Part A: Inertial frame of reference**

1 Refer to Figure 12.6 for the set-up. Place the carbon paper face down on one sheet of white paper.

2 Place a sheet of graph paper (facing up) on top to form a sandwich. The graph paper establishes the grid system.

3 Tape the sandwich to one dissecting board (P), with the carbonised side down.

4 Abut a second dissecting board, Q, next to P.

5 Secure the ramp on Q.

6 Place the frame (or ripple tank) over P and Q and secure the camera in position to view the arrangement.

7 Move P at constant speed relative to Q. (This may take a little practice.)

8 While P is moving at constant speed, release the ball bearing from the ramp. Use several different release heights. Do some tests to make sure the ball bearing will leave enough of a trace on the paper.

9 Video record the motion in each case.

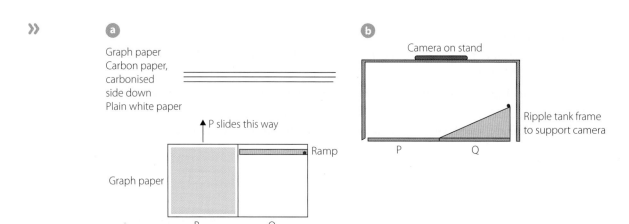

FIGURE 12.6 Set-up for Part A: **a** View from above, camera omitted; **b** Front view

Labels in figure a: Graph paper / Carbon paper, carbonised side down / Plain white paper; P slides this way; Ramp; Graph paper; P; Q

Labels in figure b: Camera on stand; Ripple tank frame to support camera; P; Q

### Part B: Accelerating frame of reference

1 Refer to Figure 12.7 for the set-up. Place the carbon paper face down on one sheet of white paper.

2 Place a sheet of graph paper (facing up) on top to form a sandwich.

3 Cut the sandwich to fit the circular turntable.

4 Tape the sandwich to the turntable, with the carbonised side down.

5 Arrange the turntable and ramp so that the ball bearing rolls seamlessly onto the turntable.

6 Rotate the turntable at a constant rate.

7 Release the ball bearing from the ramp. Use several different release heights.

8 Video record the motion in each case.

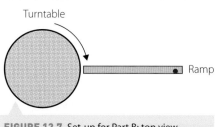

FIGURE 12.7 Set-up for Part B: top view, camera omitted

Labels: Turntable; Ramp

### RESULTS

Identify each path and relate it to the height of release of the ball bearing.

### ANALYSIS OF RESULTS

1 From the video recording find, for the camera frame of reference, the speed of:

   **a** the ball bearing.    **b** board Q.

2 Qualitatively compare the tracks for:

   **a** Part A.    **b** Part B.    **c** Part A relative to Part B.

3 For Part A:

   **a** establish coordinate systems for the camera frame $(x, t)$ and the frame moving at constant speed $(x', t')$.

   **b** produce data tables for both coordinate systems.

   **c** use the Galilean transformations to quantitatively compare the data from the two frames of reference.

4 Give an answer to your inquiry question.

### DISCUSSION

1 How well did your results for Part A demonstrate the Galilean transformations?

2 The carbon paper records what is happening in the moving frame. The camera records what happens in the non-moving frame. What differences were there between the observations in the inertial frame and those in the non-inertial frame?

3 How can the observations in Part B be reconciled with Newton's laws?

- The speed of light is finite.
- Frames of reference that are stationary or moving at constant velocity are inertial frames of reference.
- Galilean, or classical, relativity says that the laws of physics are the same in all inertial frames.
- In Galilean relativity, time is invariant.
- When relating the velocity of an object (A) measured in one frame (B) to that measured in another (C), the velocity of A relative to B is the velocity of A relative to C plus (vector addition) the velocity of C relative to B: $v_{AB} = v_{AC} + v_{CB}$.

1   Name one contribution to the understanding of light by these scientists.

   **a**   Rømer

   **b**   Bradley

2   Define 'inertial frame of reference'.

3   Write the Galilean relativity transformations.

4   The path followed by an object dropped from the mast of a moving ship can appear to be both a straight line and a parabola. Explain why.

5   Can you tell if you are in an inertial frame of reference? Give a reason.

6   A bus is travelling at $12\,\mathrm{m\,s^{-1}}$ when a passenger at the very back starts to walk at $2.0\,\mathrm{m\,s^{-1}}$ towards the front. A stationary person outside the bus also observes the situation.

   **a**   What is the speed of the person relative to the:

     **i**   bus?

     **ii**   person outside the bus?

   **b**   After $5.0\,\mathrm{s}$, what is the position of the person relative to the:

     **i**   bus?

     **ii**   person outside the bus?

7   The tide is running south at $3.0\,\mathrm{m\,s^{-1}}$. At the same time, a yacht is steering at $4.0\,\mathrm{m\,s^{-1}}$ directly towards a buoy in the east. What is the velocity of the yacht relative to the water? Show your answer in vector form.

## 12.2 The aether and problems with classical relativity

The Galilean principle of relativity states that there is no absolute frame of reference and that all velocities are relative. This principle became known as classical relativity. This was the predominant view of physicists through the 18th and 19th centuries.

Maxwell's equations of electromagnetism presented a problem to this view of the laws of physics. His equations indicated that light acts like an electromagnetic wave, and that its speed is a constant in a particular medium, irrespective of the frame of reference of the observer or the source of light.

Physicists in the mid-19th century thought that light required a medium, just like sound waves require air molecules as a vibrating material to move through. They called this medium the aether. It was thought to be a transparent weightless substance that enabled light (electromagnetic waves) to vibrate through space. This aether was also thought to be fixed in space. The Sun and the planets moved relative to this medium, or 'through it' if you prefer. That would make it an absolute frame of reference – one against which all others could be measured. If $c$ was $3.00 \times 10^8\,\mathrm{m\,s^{-1}}$ in this aether, it should be faster or slower in a reference frame moving through the aether – and Earth was such a frame.

## The Michelson–Morley experiment

If the speed of light was $c$ relative to the aether and an observer was moving at velocity $v$ in the same direction as the light, then Galilean relativity said the light should seem to be moving at $c - v$.

In 1887, Albert Michelson teamed up with Edward Morley to test this idea. Their apparatus consisted of an interferometer (see chapter 9) sitting on a large stone block floating on a bath of mercury so that it could be rotated smoothly. The apparatus had a number of mirrors designed to create two light paths at right angles to each other (Figure 12.8).

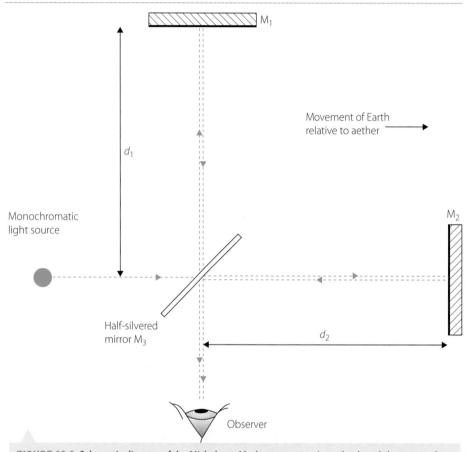

**FIGURE 12.8** Schematic diagram of the Michelson–Morley apparatus. Lengths $d_1$ and $d_2$ were made equal to very high precision.

We talked about interference patterns and waves combining after reflecting or diffracting in chapter 10.

The aether and problems with classical relativity

Part of a beam of light would be transmitted through the half-silvered mirror $M_3$ and travel to $M_2$, where it would be reflected. Some of that beam would then reflect off $M_3$ to the observer. The other part of the beam from the source would be reflected off $M_3$ and travel to $M_1$, where it would be reflected. Some of that beam would then be transmitted back through $M_3$ to the observer. The two beams of light travel the same distance, arrive together, superimpose and produce an interference pattern.

Because the beams are perpendicular, they cannot both be parallel to Earth's motion through the aether. Therefore, if the speed of light was relative to the aether (and not the observer) then it would be different along the two paths. As the instrument is rotated, one path and then the other would be more parallel to the aether. That would cause the interference fringes to change position as the instrument was rotated. Their measurements failed to detect any evidence of the presence of the aether. Over the next 50 years, Michelson, and other scientists, used a variety of techniques to increase the sensitivity of the test, but no evidence for the aether was found. Some argued that perhaps Earth dragged the aether along with it and this would explain Michelson and Morley's results. Later work showed this to be incorrect.

- Classical relativity says there is no absolute frame of reference.
- Maxwell's laws of electromagnetism say that the speed of light is a constant regardless of the frame of reference of the observer or the source of light.
- Light was thought to require a medium, which was dubbed the 'aether', which was considered to be fixed. Stars and planets moved through it.
- If light travelled at a fixed speed in the aether, and Earth moved through the aether, then the speed of light should be seen to vary with Earth's motion.
- This was not found to be the case. Therefore light did not appear to obey conventional relativity.

CHECK YOUR UNDERSTANDING

12.2

1   What is the aether? Why did people go looking for the aether?

2   What was the point of the Michelson–Morley experiment? What did they find?

3   Why did Michelson and Morley use light interference in their eponymous experiment?

4   In the Michelson–Morley experiment, how would the shift of the interference pattern confirm the existence of the aether?

# 12.3   Einstein's Theory of Special Relativity

Albert Einstein tackled issues around the behaviour of light by challenging the most fundamental assumptions. He investigated the nature of space and time. The result was the theory of special relativity. This was a bold step but a well-trodden path to discovery: questioning taken-for-granted assumptions leads to fresh, more powerful ways of understanding the world.

In 1905, Einstein published four papers that would change the way we view the universe. One of these papers introduced relativity. It did away with the aether, and argued that there was no absolute frame of reference.

His theory was based on two propositions.

» *First postulate of special relativity* – The laws of physics are the same in all inertial frames of reference.

» *Second postulate of special relativity* – The speed of light has the same value, $c$, in all inertial frames. It does not depend on the speed of either the source or the observer.

The first postulate did away with the aether, since the aether would be one privileged reference frame. Thus, the aether either did not exist, or its existence could not be demonstrated by experiments conducted completely within an observer's inertial reference frame. How could such an observer demonstrate the motion of their own reference frame?

Consider the second postulate. Let's say we have two rockets, one moving away from a distant pulsar and one towards it (Figure 12.9). The pulsar emits a short pulse of light. Observers on each rocket attempt to measure the speed of this pulse of light by using light-sensitive cells set on the outsides of the rockets ($A_1$ and $B_1$, and $A_2$ and $B_2$, respectively).

Each rocket uses a timer and the distance between the sensors to measure the speed of light relative to its own reference frame. Rocket 1 measures the speed of light to be $3.0 \times 10^8 \, \text{m s}^{-1}$. The second rocket also determines it to be $3.0 \times 10^8 \, \text{m s}^{-1}$.

Newtonian or Galilean relativity says that the rocket heading towards the pulsar would have registered a light speed of $1.25c$ and the rocket moving away from the pulsar would have registered a light speed of $0.75c$.

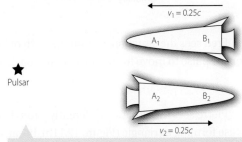

FIGURE 12.9 Passing rockets both view light from a pulsar.

The speed of light in a vacuum is $c$ regardless of the speed of either the observer or the source. This is consistent with the results of the Michelson–Morley experiment. When later asked, Einstein was not sure whether he knew of the Michelson–Morley experiment when he came up with relativity. He was focused on Maxwell's equations when he gradually came to realise that the aether was not necessary for the equations to work. The Michelson–Morley results were, however, highly influential in the acceptance of the theory of special relativity. The other early evidence in support of Einstein's theory relied on his conclusions from the more complex theory of general relativity.

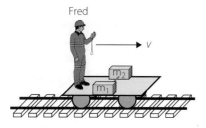

## Time dilation

The theory of special relativity asks us to give up our Newtonian view of space and time, and accept some very strange ideas. To illustrate this, we will use a technique that Einstein used himself: simple thought experiments (*Gedanken* experiments, in German) that are based on the two postulates of special relativity.

Let us imagine another experiment using a train, Einstein's favourite thought experiment apparatus. The train is running on a smooth track at a high velocity of $v$ relative to the ground. On one carriage is a set of two mirrors ($m_1$ and $m_2$), set up to allow a series of light pulses to bounce forwards and back between them. Pierre, standing on the ground nearby, watches the train pass (Figure 12.10). His timer is capable of measuring very small increments.

Fred, an observer on the train (also using a very accurate watch), measures the time ($t_0$) it takes for a pulse of light to travel from $m_1$ to $m_2$ and back again. Pierre and Fred agree that the distance between the mirrors is $w$ ($w$ is perpendicular to the direction of movement).

**FIGURE 12.10** A passing high-speed rail cart

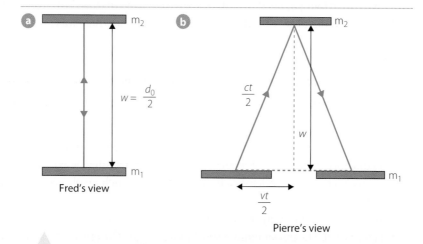

**FIGURE 12.11** The reflection of the light as viewed from **a** Fred's frame and **b** Pierre's frame of reference. The time measured by Fred is the proper time – the mirrors are stationary relative to him.

Fred sees the situation as a simple path of light between two mirrors, and therefore the time it takes for the pulse to move from $m_1$ to $m_2$ and back again to $m_1$ (Figure 12.11a) will be:

$$t_0 = \frac{2w}{c}$$

Pierre sees the situation differently. From his viewpoint, the velocity of the mirrors results in the light pulse forming a triangle (Figure 12.11b), because both mirrors are moving to the right with a velocity of $v$ as the light pulse moves from one mirror to the other and back again. Pierre measures the time for one pulse to move from $m_1$ to $m_2$ to be $t$. One half of the journey forms a right-angled triangle with sides of length $w$, $\frac{vt}{2}$ and $\frac{ct}{2}$.

Proper time is the time interval between two events occurring at the same place in an inertial frame, as measured by an observer in that inertial frame. Thus, from Fred's viewpoint, proper time is:

$$t_0 = \frac{2w}{c}$$

The time measured by Pierre ($t$) is not the proper time, because Pierre is not travelling with the mirrors. It is possible to relate $t$ and $t_0$.

From Pythagoras' theorem we see that:

$$\left(\frac{ct}{2}\right)^2 = \left(\frac{vt}{2}\right)^2 + w^2$$

$$\frac{t^2}{4}\left(1 - \frac{v^2}{c^2}\right) = \frac{w^2}{c^2}$$

$$\frac{t}{2}\sqrt{1 - \frac{v^2}{c^2}} = \frac{w}{c}$$

$$t\sqrt{1 - \frac{v^2}{c^2}} = \frac{2w}{c} = t_0$$

$$t = \frac{t_0}{\sqrt{1 - \frac{v^2}{c^2}}}$$

So the time for a single reflected pulse, as recorded by Pierre's watch (which is identical in every way to Fred's), is longer than that recorded by Fred's watch. Pierre's time measurement has been dilated. Einstein argued that the postulates of special relativity led to the understanding that observers in different inertial frames would not agree on time measurements. Recall that in Galilean relativity time was invariant. Here, it is not.

Time dilation will only be large enough to be observable when inertial reference frames are moving relative to each other at speeds close to the speed of light. If $v$ is small, such as the speeds we normally encounter every day, then $t = t_0$ because the ratio $\frac{v^2}{c^2}$ approaches zero.

The factor $\frac{1}{\sqrt{1 - \frac{v^2}{c^2}}}$ appears repeatedly in relativity. It is given the symbol $\gamma$. It is also common to denote the ratio $\frac{v}{c}$ as $\beta$, so that $\gamma = \frac{1}{\sqrt{1 - \beta^2}}$. Figure 12.12 shows how $\gamma$ varies with $v$. We can see from the expression for $\gamma$ that $v$ must be less than $c$ if we are to avoid a negative number under the square root sign. Since the square root of a negative number does not make physical sense, so we know that $v < c$.

TABLE 12.1 Approximate values of $\gamma$ for various values of $\beta$

| $\beta = \dfrac{v}{c}$ | $\gamma$ |
|---|---|
| 0 | 1 |
| 0.0010 | 1.000 000 5 |
| 0.010 | 1.000 05 |
| 0.10 | 1.005 |
| 0.20 | 1.021 |
| 0.50 | 1.155 |
| 0.80 | 1.667 |
| 0.90 | 2.294 |
| 0.94 | 2.931 |
| 0.99 | 7.089 |
| 0.999 | 22.37 |

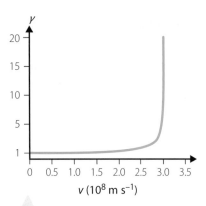

When β gets very close to 1 – say 0.999 999 999 – computer and calculator calculations can become prone to rounding errors and may be unreliable.

**FIGURE 12.12** Graph of γ vs v. As speed approaches c, γ increases rapidly.

Let us refer back to Pierre and Fred. What would happen if the mirrors were on the ground with Pierre and the experiment was repeated? Fred could argue that his train was stationary and that Pierre was moving at –v with respect to him. This means he would find that Pierre's clock is slow compared to his. How do we reconcile these two observations?

Time dilation is about measurements in *different* inertial frames. It is a result of relative movement between the frames. The clocks do not physically change. Time dilation is about what an observer in one frame observes about an event in another, and it must be reciprocal because there is no absolute frame of reference.

▶ **WORKED EXAMPLE**

A pilot in a rocket travelling with a velocity of 0.250c presses a button to flash a 'Hello' sign for 5.00 s at a space station as the rocket passes.

**a** For how long is the flash seen by an observer on the space station?

**b** Explain your reasoning.

| ANSWERS | LOGIC |
|---|---|
| **a** $t = \dfrac{t_0}{\sqrt{1 - \dfrac{v^2}{c^2}}}$ | ▪ Identify the relevant data from the question.<br>▪ Write the equation connecting time in the two frames of reference. |
| $= \dfrac{5.00 \text{ s}}{\sqrt{1 - 0.25^2}}$ | ▪ Substitute the values with correct units. |
| $= \dfrac{5.00 \text{ s}}{\sqrt{0.9375}}$<br>$= 5.16 \text{ s}$ | ▪ Calculate the final answer with correct units and appropriate significant figures. |
| **b** The observer in the space station views the rocket travelling towards it at a velocity of 0.250c. From this viewpoint, the clock on the rocket will be slow compared to the one in the space station. Any observer regards a clock that is moving relative to their frame of reference as running slow. | ▪ Recognise that time dilation is important at this relative speed. |

**TRY THIS YOURSELF**

Looking at the graph of γ against v in Figure 12.12, if the rocket's speed was doubled, would you expect the amount of time dilation to double, more than double, or less than double? What does this say about what happens to time dilation as v gets close to c?

## Length contraction

If time measurements are different in different inertial frames, what happens to measurements of length? Let us look at another thought experiment.

A train travels at a velocity of $v_0$. On the train, a source of light emits a short pulse that travels to a mirror and back (Figure 12.13). An observer in the same frame of reference as the mirrors measures the

9780170409131

time ($t_0$) for the pulse to travel up and back. If $w_0$ is the distance from the source to the mirror, we can write:

$$t_0 = \frac{2w_0}{c}$$

where $w_0$ is the **proper length** because it is measured in the mirror reference frame.

$$w_0 = \frac{ct_0}{2}$$

Light pulse emitted and received

Source

Mirror

**FIGURE 12.13** An event as viewed within an inertial reference frame

On the ground, an observer who is stationary relative to the moving frame of the train measures the time of transit of the light pulse to the mirror to be $t_1$. She sees the initial distance between the source and mirror to be $w$ and the distance between the original location of the source and the mirror to be $L_1$, as shown in Figure 12.14a.

In this stationary reference frame, the observer notes that, during the transit of the light pulse to the mirror, the mirror has moved forwards a distance of $v_0t_1$, as shown in Figure 12.14b. Then:

$$L_1 = w + v_0t_1$$

However, $L_1$ is also equal to $ct_1$, because the speed of light is the same in all reference frames. Therefore:

$$ct_1 = w + v_0t_1$$

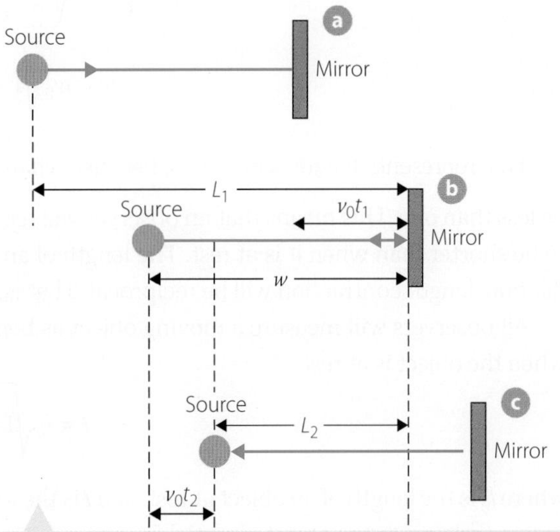

So, $t_1 = \dfrac{w}{c - v_0}$.

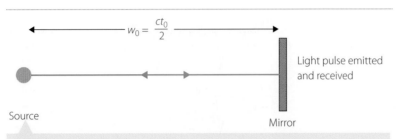

**FIGURE 12.14** The event as viewed in the stationary frame of reference

In Figure 12.14c, the light pulse bounces off the mirror and travels back to the source in a time measured in the stationary reference frame to be $t_2$. So:

$$L_2 = w - v_0t_2$$

However, $L_2$ is also equal to $ct_2$, so:

$$ct_2 = w - v_0t_2$$

and $t_2 = \dfrac{w}{c + v_0}$.

From the perspective of the observer in the stationary reference frame, the total time for the journey is $t_1 + t_2$, and:

$$t_1 + t_2 = \frac{w}{c - v_0} + \frac{w}{c + v_0} = \frac{w(c + v_0) + w(c - v_0)}{c^2 - v_0^2} = \frac{2wc}{c^2 - v_0^2}$$

From the time dilation equation we know that:

$$t_1 + t_2 = \frac{t_0}{\sqrt{1 - \dfrac{v_0{}^2}{c^2}}}$$

Therefore, we have:

$$\frac{2wc}{c^2 - v_0{}^2} = \frac{t_0}{\sqrt{1 - \dfrac{v_0{}^2}{c^2}}} = \frac{2\dfrac{w_0}{c}}{\sqrt{1 - \dfrac{v_0{}^2}{c^2}}}$$

$$\frac{wc^2}{c^2 - v_0{}^2} = \frac{w_0}{\sqrt{1 - \dfrac{v_0{}^2}{c^2}}}$$

$$\frac{w}{1 - \dfrac{v_0{}^2}{c^2}} = \frac{w_0}{\sqrt{1 - \dfrac{v_0{}^2}{c^2}}}$$

$$w = w_0\sqrt{1 - \frac{v_0{}^2}{c^2}}$$

This represents length contraction, because whatever the value of $v_0$, the factor $\sqrt{1 - \dfrac{v_0{}^2}{c^2}}$ will always be less than one. This means that an observer will regard the length of an object moving relative to them to be shorter than when it is at rest. The length of an object at rest is called its proper length. Like time dilation, length contraction will be reciprocal. That is, to each frame the other will seem shortened.

All observers will measure a moving object as being shorter in the direction of relative motion than when the object is at rest:

$$l = l_0\sqrt{1 - \frac{v_0{}^2}{c^2}}$$

Einstein's Theory of Special Relativity

where $l_0$ is the length of an object at rest, and $l$ is the length of the object measured by an observer who is moving relative to the object's inertial frame.

It is important to note that the contraction is only in the direction of relative motion. There is no contraction in length at right angles to the motion. Just like time dilation, length contraction is only observable at very high speeds.

## Lorentz factor

Before Einstein, two scientists, George Fitzgerald from Ireland and Hendrik Lorentz from Holland, independently suggested an alternative explanation for the null result for the aether from Michelson and Morley's experiment. They wanted an explanation that allowed Maxwell's equations to be applied to moving charges. Their explanation was that all bodies shrink in the direction of motion relative to the stationary aether by a modifying factor of $\sqrt{1 - \dfrac{v^2}{c^2}}$.

Lorentz saw the consequence of the equation as a physical contraction of objects in space. However, Einstein argued that there was no physical change in the length of high-speed objects, but a change in the properties of time and space itself. The modern version of this factor, $\gamma = \dfrac{1}{\sqrt{1 - \dfrac{v^2}{c^2}}}$, is sometimes called the Lorentz factor. The effect is sometimes called Lorentz–Fitzgerald contraction.

9780170409131

> ## WORKED EXAMPLE (12.4)

1. An observer on the Moon notices a spacecraft travelling past at a speed of $2.08 \times 10^8\,\mathrm{m\,s^{-1}}$. The spacecraft has a proper length of 120 m.

   What length will the observer on the Moon measure?

2. A crewed mission is to be sent to a newly discovered exoplanet 8 light years away. It will travel at a velocity of $0.5c$ to get there.

   a According to the mission crew:

      i how far away from Earth is the exoplanet?

      ii how long will the journey take?

   b According to the mission command on Earth, how long will the journey take?

| ANSWERS | LOGIC |
|---|---|
| 1 $l_0 = 120\,\mathrm{m}$, $v_0 = 2.08 \times 10^8\,\mathrm{ms^{-1}}$ | ▪ Identify the relevant data from the question. |
| $l = l_0 \sqrt{1 - \dfrac{v_0^{\,2}}{c^2}}$ | ▪ Write the equation connecting length in the two frames of reference. |
| $l = 120\,\mathrm{m} \times \sqrt{\left(1 - \dfrac{(2.08 \times 10^8\ \mathrm{m\ s^{-1}})^2}{(3.00 \times 10^8\ \mathrm{m\ s^{-1}})^2}\right)}$ | ▪ Substitute the values with correct units. |
| $l = 120\,\mathrm{m} \times \sqrt{1 - 0.481}$ <br> $= 120\,\mathrm{m} \times 0.72$ <br> $= 86.5\,\mathrm{m}$ | ▪ Calculate the answer with correct units and appropriate significant figures. |
| 2 $l_0 = 8$ light years, $v_0 = 0.5\,c$ | ▪ Identify the relevant data from the question. |
| a i $l = l_0 \sqrt{1 - \dfrac{v_0^{\,2}}{c^2}}$ | ▪ Write the equation connecting length in the two frames of reference. |
| $l = 8 \text{ light years} \times \sqrt{1 - \dfrac{(0.5c)^2}{c^2}}$ <br><br> $= 8 \text{ light years} \times \sqrt{1 - 0.25}$ | ▪ Substitute the values with correct units. |
| $= 8 \times 0.866 \text{ light years}$ <br> $= 6.9 \text{ light years}$ | ▪ Calculate the answer with correct units and appropriate significant figures. |
| ii $t_{\mathrm{crew}} = \dfrac{s}{v} = \dfrac{6.9 \text{ light years}}{0.5c} = 14 \text{ years}$ | ▪ Substitute the values with correct units and calculate the answer with correct units and appropriate significant figures. |
| b $t_{\mathrm{Earth}} = \dfrac{s}{v} = \dfrac{8 \text{ light years}}{0.5c} = 16 \text{ years}$ | ▪ Substitute the values with correct units and calculate the answer with correct units and appropriate significant figures. |

### TRY THESE YOURSELF

A spaceship travels at an average velocity of $0.4c$ to an exoplanet 4.4 light years away.

a What is the distance from Earth to the exoplanet in the frame of reference of the spacecraft?

b What is the difference between the time taken from Earth's perspective and the time from the perspective of the spacecraft?

- *First postulate of special relativity* – The laws of physics are the same in all inertial frames of reference.
- *Second postulate of special relativity* – The speed of light has the same value, $c$, in all inertial frames. It does not depend on the speed of either the source or the observer.
- The factor $\dfrac{1}{\sqrt{1 - \dfrac{v^2}{c^2}}}$ is given the symbol $\gamma$. The ratio $\dfrac{v}{c}$ is often known as $\beta$, so that $\gamma = \dfrac{1}{\sqrt{1 - \beta^2}}$.
- Proper time is the time interval between two events occurring at the same place in an inertial frame, as measured by an observer in that inertial frame.
- Proper length is measured in a frame of reference in which the thing being measured is stationary.
- If $t_0$ is the proper time between two events, then the observed time allowing that the frame of reference of the observer may be moving is $t = \gamma t_0$.
- If $l_0$ is a proper length, the observed length allowing that the frame of reference of the observer may be moving is $l = l_0 \sqrt{1 - \dfrac{v_0^2}{c^2}}$.
- Thus times appear longer and lengths appear shorter when measured from a frame that is moving relative to the objects in question.

**CHECK YOUR UNDERSTANDING**

**12.3**

1  Define the following terms.

   **a**  Proper time

   **b**  Proper length

2  Write down Einstein's two postulates of relativity.

3  Define the following terms.

   **a**  Time dilation

   **b**  Length contraction

4  A spacecraft has a proper length of 50 m. It travels past an observer at 0.6c. What was its length in the observer's frame of reference?

5  A train carriage travels at 0.75c. To an observer outside the carriage, it fits exactly between two markers, P and Q. To an observer on the train, does the carriage appear to fit exactly between P and Q? If not, what can you conclude about events in the different frames of reference?

6  A pion has a mean lifetime of 26 ns in Earth's inertial frame of reference. What is its mean lifetime in the pion's frame of reference if it has velocity $v = 0.75c$?

7  An astronaut travels to a star 20 light years away at a constant speed of 0.9c as measured by an astronomer on Earth.

   **a**  How long does the journey take according to the astronomer?

   **b**  How long does it take according to the astronaut?

# 12.4 Experimental evidence for relativity

A range of experiments and observations support relativity. In section 12.2 we discussed the Michelson–Morley experiment. Section 12.5 talks about relativistic mass and how we have to allow for it in particle accelerators. Some cosmological observations agree with the predictions of special relativity. When massive stars explode it is called a supernova. The fragments of a supernova can fly out at speeds of over 3% of the speed of light. If the speed of light depended on the speed of the source, the light coming from fragments moving *away* from us would be slower than light coming from fragments moving towards us. Such an effect would show up in astronomers' observations, and it does not. Also, no particle, whether produced on Earth or from space, has ever been reliably observed to travel faster than light.

## Muon decay

Muons are a class of particles with a mass about 200 times greater than an electron. They are produced naturally by cosmic ray bombardment of the upper atmosphere. They travel at close to the speed of light and have a mean lifetime of $2.2 \times 10^{-6}$ s (2.2 µs) as measured in the laboratory. The mean lifetime, $t_{mean}$, is a different measure from the half-life you will meet in chapter 13.

Clearly, muons do not exist for very long but, because they have relativistic speeds, they allow demonstration of relativistic effects. D H Frisch and J H Smith published their work on this in 1963. Muons travelling down through the atmosphere were collected on a mountaintop at an altitude of 1907 m above sea level (Figure 12.15). They were travelling at 0.995$c$ relative to Earth.

At that speed, an observer travelling with the muon, that is, in the rest frame of the muon, would expect to reach a second detector at sea level in a proper time given by $v = \dfrac{s}{t}$ where $s$ = height, $H$:

$$t = \frac{H}{v}$$
$$= \frac{1907 \text{ m}}{0.995 \times 2.998 \times 10^8 \text{ m s}^{-1}}$$
$$= 6.39 \text{ µs}$$

The mean lifetime for particle decay is related to half-life by $t_{mean} = 1.44 t_{1/2} = 2.2$ µs. This is in the rest frame of the decaying particle, which is a muon (µ), so call it $t_{mean}^{\mu}$. To travel from the mountaintop to sea level takes $\dfrac{t}{t_{mean}^{\mu}} = \dfrac{6.39 \text{ µs}}{2.2 \text{ µs}} = 2.9$ mean lifetimes. The rate of decay of muons is governed by an exponential law. If we measure $N_{MT}$ muons in an experiment at the mountaintop (MT), we would expect

to measure $N_{SL} = N_{MT} e^{-t/t_{mean}^{\mu}}$ muons in an experiment of the same duration time $t$ later at sea level (SL). As a fraction:

$$\frac{N_{SL}}{N_{MT}} = e^{-t/t_{mean}^{\mu}} = e^{-2.9} = 0.055$$

So we might expect to measure about $0.055 \times 100 = 5.5\%$ as many muons at sea level as on the mountain top. However, the detectors are in Earth's rest frame while the muon is decaying in its own rest frame. So as seen in Earth frame, time dilation is occurring for the muon. If $t_{mean}^{E}$ is the mean lifetime *as measured in Earth frame*, then:

$$t_{mean}^{E} = \frac{t_{mean}^{\mu}}{\sqrt{1 - \dfrac{v_0^2}{c^2}}} = \gamma t_{mean}^{\mu}$$

And at $v = 0.995c$ we find that $\dfrac{1}{\sqrt{1 - \dfrac{v_0^2}{c^2}}} = \gamma = 10$ so $t_{mean}^{E} = 22$ µs, which means the trip is now

$\dfrac{t}{t_{mean}^{E}} = \dfrac{6.39 \text{ µs}}{22 \text{ µs}} = 0.29$ mean lifetimes.

So the fraction we measure at sea level would be:

$$\frac{N_{SL}}{N_{MT}} = e^{-t/t_{mean}^{E}} = e^{-0.29} = 0.748$$

In their experiment, Frisch and Smith measured an average of 563 muons per hour in their upper detector. Assuming no muons are created between detectors, the number of muons arriving per hour at the second detector would be about $563 \times 0.055 \approx 30$ if the muon experienced no time dilation, but $563 \times 0.748 \approx 420$ if it did. In fact, they measured 412 muons per hour – close to the relativistic estimate.

For an observer riding with the muon, Earth would seem to be moving at 0.995c. For them, the apparent distance between the detectors would seem shorter by a factor of $\gamma = \dfrac{1}{\sqrt{1 - \dfrac{v_0^2}{c^2}}}$ (due to relativistic length contraction). That would then explain *to them* why so many muons reach the second detector.

Thus, an observer on a moving muon (muon frame of reference) and an observer on Earth (Earth frame of reference) will agree about the:

» relative speed, 0.995c

» number of physical decays of muons

» number of elapsed mean lifetimes between detectors.

They will disagree about the:

» mean lifetime

» distance between the detectors.

 Without relativistic considerations, according to an observer on Earth, muons created in the atmosphere and travelling downwards with a speed close to c travel only about 656 m before decaying with a mean lifetime of 2.2 μs. Therefore, very few muons would reach the surface of Earth.

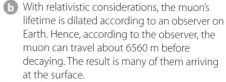 With relativistic considerations, the muon's lifetime is dilated according to an observer on Earth. Hence, according to the observer, the muon can travel about 6560 m before decaying. The result is many of them arriving at the surface.

Muon is created

≈ 656 m

Muon decays

Muon is created

≈ 6560 m

Muon decays

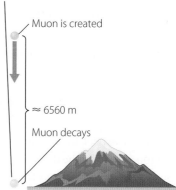

FIGURE 12.15 Travel of muons according to an observer **a** on the muon, and **b** on Earth

## WORKED EXAMPLE (12.5)

Muons with a mean lifetime of $2.2 \times 10^{-6}$ s and speed 0.999c were observed at a height of 10 000 m above Earth.

**a** From the viewpoint of a stationary observer on Earth:

  **i** how long would it take for a muon to travel to the ground (assuming it does not decay)?

  **ii** ignoring relativity how many mean lifetimes will elapse before this muon arrives at the ground?

  **iii** qualitatively, does it seem likely this muon will reach the ground?

**b** From the viewpoint of an observer travelling with the muon (muon reference frame), how many mean lifetimes will elapse before the muon arrives at the ground?

| ANSWERS | LOGIC |
|---|---|
| **a** **i** $t = \dfrac{s}{v} = \dfrac{10\ 000\ \text{m}}{0.999 \times 3.0 \times 10^8\ \text{m s}^{-1}}$ | • Substitute the values with correct units. |
| $t = 3.3 \times 10^{-5}\ \text{s}$ | • Calculate the answer with correct units and appropriate significant figures. |
| **ii** Let $n$ = Number of mean lifetimes $n = \dfrac{3.3 \times 10^{-5}\ \text{s}}{2.2 \times 10^{-6}\ \text{s}} = 15.2$ | • Calculate the answer with correct units and appropriate significant figures. |
| **iii** The probability is almost zero. Use relativity! | • Correct use of probability ideas. |
| **b** $l = l_0 \sqrt{1 - \dfrac{v^2}{c^2}}$ | • Relate the length in the moving frame to that in the Earth frame. |
| $l = (10\ 000\ \text{m}) \times \sqrt{1 - 0.999^2}$ $= 447.10\ \text{m}$ | • Substitute the values with correct units and calculate the intermediate answer. |
| $n = \dfrac{t}{t_{\text{mean}}} = \dfrac{1}{t_{\text{mean}}} \times \dfrac{l}{v}$ $n = \dfrac{1}{2.2 \times 10^{-6}\ \text{s}} \times \dfrac{447.10\ \text{m}}{0.996 \times 3.00 \times 10^8\ \text{m s}^{-1}} = 0.68$ | • Use the definition of velocity $\left(v = \dfrac{l}{t}\right)$ to find the time taken and compare to the mean lifetime. |

**TRY THESE YOURSELF**

Muons are formed 3.0 km above Earth. They travel at $0.996c$ and have a mean lifetime, in the laboratory, of 2.2 μs.

**a** For an observer on a muon:

   **i** how long does it take for the muon to travel to the ground?

   **ii** how many mean lifetimes will elapse before the muon arrives at the ground?

   **iii** explain why it is possible for the muon to arrive at the ground.

**b** For an observer on the ground, how many lifetimes will elapse before the muon arrives at the ground?

## Atomic clocks

Atomic clocks use the frequency of radiation emitted during electronic transitions in atoms to measure time. They can achieve precisions of around one second every hundred million years. This means they are able to directly measure time dilation even when speeds are not very large compared with $c$. In October 1971, Joseph Hafele and Richard Keating sent atomic clocks on flights around the world on commercial passenger jets. Figure 12.16 shows one such clock. Clocks flew twice around the world, some east and some west. They were then compared with a clock that had not moved. The clocks that had flown were found to disagree with the fixed clock, and the disagreements were as predicted by relativity.

The results depend on special relativity (as discussed here) and also on general relativity, a more complex theory involving gravity. However, we can think about the special relativity aspect of the experiment. Compared to a clock on the ground, a

**FIGURE 12.16** One of the atomic clocks used in the Hafele–Keating experiment

moving clock will experience time dilation. If the clock on the ground measures a time interval as $t_0$, the moving clock will measure the same interval as $t$ where:

$$t = \gamma t_0$$

$$= \frac{t_0}{\sqrt{1 - \dfrac{v^2}{c^2}}}$$

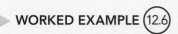

**WS**

Experimental evidence for relativity

There is a problem. If the clock is on an aeroplane, $v$ is so small compared with $c$ that the calculation will not work using your calculator. But, when $v < 0.2c$ we can use an approximation:

$$\gamma \approx 1 + \frac{1}{2}\left(\frac{v}{c}\right)^2$$

So now we can write:

$$t \approx t_0\left(1 + \frac{1}{2}\left(\frac{v}{c}\right)^2\right)$$

The time difference on the clocks would be $t - t_0$, so:

$$t - t_0 \approx \frac{t_0}{2}\left(\frac{v}{c}\right)^2$$

---

▶ **WORKED EXAMPLE** (12.6)

Find the time difference, $t - t_0$, between a clock fixed on the ground and one flown around Earth twice at a speed of about $1000\,\text{km h}^{-1}$. The circumference of the Earth is about $40\,000\,\text{km}$. Two circuits of Earth would therefore be expected to take 80 hours. What time dilation do we expect for 80 hours at $1000\,\text{km h}^{-1}$ (about $300\,\text{m s}^{-1}$)? Round numbers will do!

| ANSWER | LOGIC |
|---|---|
| $t - t_0 \approx \dfrac{t_0}{2}\left(\dfrac{v}{c}\right)^2$ | ▪ Find the expression for $t - t_0$. We use the approximation because $v$ is so small compared with $c$. |
| $t_0 \approx 2.5 \times 10^{12}\,\mu\text{s}$ | ▪ Convert $t_0$ to a more useful unit. Since the difference will be small, we can convert 80 h to microseconds. This helps make sure the top line in the equation is not too small compared with the bottom. |
| $t - t_0 \approx \dfrac{(1.25 \times 10^{12}\,\mu\text{s})(300\,\text{m s}^{-1})^2}{(3 \times 10^8\,\text{m s}^{-1})^2}$ | ▪ Substitute values with correct units. |
| $\approx 1\,\mu\text{s}$ <br> This is measurable with an atomic clock. We are really getting an order of magnitude estimate, so one significant figure is reasonable. | ▪ Evaluate, and state to sensible precision. |

**TRY THIS YOURSELF**

Find $t - t_0$ if the moving clock spent 80 hours on the supersonic airliner Concorde (speed $2000\,\text{km h}^{-1}$).

---

**Atomic clocks**

Atomic clocks are so precise they would gain or lose less than a second in 100 million years.

**KEY CONCEPTS**

- The predictions of the theory of special relativity can be seen in experiments performed on Earth and in astronomical observations.
- The behaviour of light from distant, powerful sources like supernovas is what we would expect if Einstein's theory were correct.
- The time subatomic particles like muons take to decay depends on the relative speeds of the muon and the observer, as Einsteinian relativity predicts.
- Atomic clocks, which are extraordinarily precise, show the effects of relativity even when moving at the speed of a commercial jetliner. These speeds are slow compared to $c$.

1 Describe three experiments or observations that support Einstein's relativity.

2 Carbon-15 has a half-life of 2.449 s and so $t_{mean} = 1.44t_{1/2} = 3.53$. If such atoms were travelling at $0.9c$ relative to an observer, what would the mean lifetime, $t_{mean}$, be as measured by that observer?

3 If the detector used to measure the speed of these $^{15}C$ atoms was 8 m long as measured by the observer, how long would it seem to the atoms?

4 While orbiting the Earth, the now-retired Space Shuttle travelled at around 28 000 km h$^{-1}$. Find $t - t_0$ for an 80 hour journey made at this speed, and compare with Worked example 12.6. Before doing any calculation, state whether you expect the result to be bigger or smaller. Why?

# 12.5 Relativistic mass and momentum

Length and time have been shown to have values that depend on relativity. The third fundamental physical quantity is mass.

## Rest mass and relativistic mass

Einstein argued that the mass of an object will be dependent on its relative speed:

$$m = \gamma m_0 = \frac{m_0}{\sqrt{1 - \dfrac{v_0^2}{c^2}}}$$

The term $m_0$ is known as the **rest mass**, as measured when the mass is stationary in an inertial reference frame. $m$ is the measurement of its mass in a reference frame moving in relation to a stationary frame, and is called the **relativistic mass** or **relativistically corrected mass**.

This means that, as the velocity of an object increases and approaches the speed of light, the mass of the object will increase greatly.

> Rest or proper mass, $m_0$: mass as measured when the mass is stationary in an inertial reference frame. Proper mass never changes.
>
> Relativistic mass or relativistically corrected mass: the greater the relative velocity, the greater the relativistic mass.

> **WORKED EXAMPLE 12.7**
>
> a What is the relativistically corrected mass of an electron whose speed is measured to be $v_0 = 1.8 \times 10^8 \text{ m s}^{-1}$?
>
> Rest mass of an electron is $9.109 \times 10^{-31}$.
>
> b At what speed is a particle moving if its relativistic mass is five times larger than its rest mass?

| ANSWERS | LOGIC |
|---|---|
| a Relativistic mass of electron:<br><br>$m = \dfrac{m_0}{\sqrt{1 - \dfrac{v_0^2}{c^2}}}$ | • Write the formula for relativistic mass. |
| $m = \dfrac{9.109 \times 10^{-31} \text{ kg}}{\sqrt{1 - \dfrac{(1.8 \times 10^8 \text{ m s}^{-1})^2}{(3.0 \times 10^8 \text{ m s}^{-1})^2}}}$ | • Substitute the values with correct units and calculate. |
| $= 1.1 \times 10^{-30} \text{ kg}$ | • State the final answer with correct units and appropriate significant figures. |

| | |
|---|---|
| **b** What $v$ do we need to get $\dfrac{m}{m_0} = 5$? <br><br> $m = \dfrac{m_0}{\sqrt{1 - \dfrac{v^2}{c^2}}}$ | ▪ Use the expression for relativistic mass. |
| $\sqrt{1 - \dfrac{v^2}{c^2}} = \dfrac{m_0}{m}$ <br><br> $1 - \dfrac{v^2}{c^2} = \left(\dfrac{m_0}{m}\right)^2$ <br><br> $\dfrac{v^2}{c^2} = 1 - \left(\dfrac{m_0}{m}\right)^2$ <br><br> $v = c\sqrt{1 - \left(\dfrac{m_0}{m}\right)^2}$ | ▪ Rearrange to make $v$ the subject. |
| $v = c\sqrt{1 - \left(\dfrac{1}{5}\right)^2}$ <br><br> $v = 0.9798c$ | ▪ Substitute the correct values and calculate. |
| $v = 2.94 \times 10^8 \, \text{m s}^{-1}$ | ▪ State the final answer with correct units and appropriate significant figures. |

**TRY THESE YOURSELF**

1 What is the relativistically corrected mass of a proton whose speed is $0.75c$?

2 A neutron has a relativistic mass that is 2.5 times greater than its rest mass. What is the speed of the neutron?

The momentum formula can be thought of as $p = mv$ with the relativistic mass.

## Relativistic momentum

Now that we have a definition for relativistic mass, what does this mean for momentum? Special relativity uses the classical equation for momentum:

$$p = mv$$

However, $m$ is now the relativistic mass, and its magnitude depends on the relative velocity of the mass. This means that the magnitude of an object's relativistically corrected momentum, $p$, is given by:

$$p = \frac{m_0 v}{\sqrt{1 - \dfrac{v^2}{c^2}}}; \; p = \frac{p_0}{\sqrt{1 - \dfrac{v^2}{c^2}}}; \; p = \gamma p_0$$

Think about accelerating a relativistic mass. A rocket works by conservation of momentum. It ejects rocket exhaust backwards at a speed of, say, $v_{\text{exhaust}}$ relative to the rocket. That exhaust has momentum, $p_{\text{exhaust}}$, and the momentum is backwards. We neglect the change in mass of the rocket as it burns the fuel.

Using Newton's third law, we can say:

$$F_{\text{exhaust on rocket}} = -F_{\text{rocket on exhaust}}$$

Recall that $F = ma = \dfrac{\Delta p}{\Delta t}$. If the rocket motor performance is constant, then $\dfrac{\Delta p}{\Delta t}$ is constant for the motor and the force it exerts on the rocket is constant; call it $F_R$. Then, when the rocket begins accelerating:

$$m_{\text{rocket}} a_{\text{rocket}} = F_R$$

9780170409131

where $m_{\text{rocket}}$ is the rest mass of the rocket. As we reach relativistic speeds, we have to use the relativistic mass, so:

$$\gamma m_{\text{rocket}} a_{\text{rocket}} = F_{\text{R}}$$

$$a_{\text{rocket}} = \frac{F_{\text{R}}}{\gamma m_{\text{rocket}}}$$

$$a_{\text{rocket}} \propto \gamma^{-1}$$

We know from Figure 12.12 that $\gamma = \dfrac{1}{\sqrt{1 - \dfrac{v_0^2}{c^2}}}$ gets very large very quickly as $v$ approaches $c$. That means $\gamma^{-1} = \sqrt{1 - \dfrac{v_0^2}{c^2}}$ gets very small, and so does acceleration. The closer the rocket gets to $c$, the weaker the acceleration, and the rocket never quite reaches $c$.

Similar arguments apply when trying to accelerate particles in particle accelerators. Two examples are synchrotrons and particle colliders. These are places where scientists observe masses moving at relativistic velocities. As a particle's velocity approaches $c$, it behaves as if it has much more mass, and becomes progressively more difficult to accelerate further. It also becomes more difficult to force the particle to follow a curve – recall that the force towards the centre to give circular motion must have magnitude $\dfrac{mv^2}{r}$ where $v$ is the particle velocity and $r$ the radius of the curve. As the effective mass, $m$, increases (and so does $v$), the force and/or the radius must also increase. That is the reason that very high energy particle rings, like the Tevatron, the Large Hadron Collider (LHC) and the Advanced Photon Source, need to be large.

Relativistic mass and momentum

Subatomic particles can be accelerated in large rings, where they are controlled using electric and magnetic fields. Some rings, like the LHC, are used for particle physics experiments. Others, like the Australian Synchrotron or the Advanced Photon Source, are used to generate powerful beams of electromagnetic radiation, including X-rays, that can be used in many fields of science, engineering and medicine.

**KEY CONCEPTS**

- The mass, $m$, of an object depends on its relative speed and its rest mass, $m_0$: $m = \gamma m_0$.
- Rest mass is measured when the mass is stationary in an inertial reference frame. It is a property of the object and does not change.
- If we apply the relativistic mass to the momentum formula, $p = mv$, we get relativistic momentum, $p$, from

$$p = \frac{m_0 v}{\sqrt{1 - \dfrac{v^2}{c^2}}}; \quad p = \frac{p_0}{\sqrt{1 - \dfrac{v^2}{c^2}}}; \quad p = \gamma p_0$$

where $p_0 = m_0 v$.

- The increase in mass as an object speeds up makes is increasingly difficult to maintain acceleration.

**Australian Synchrotron**

The Australian Synchrotron accelerates electrons to about $299\,792\text{ km s}^{-1}$. This is very close to the speed of light, and relativistic effects are very significant.

1 Write down relativistic equations for:

   a mass.

   b momentum.

2 A 400 kg rocket has a motor that ejects 5 kg of exhaust per second at a velocity of 3.0 km s$^{-1}$.

   a Find the force the exhaust exerts on the rocket.

   b Find the acceleration of the rocket when it is starting from rest.

   c Neglecting the change in mass of the rocket, find the acceleration when the rocket is travelling at:

   i 0.5$c$

   ii 0.9$c$

   iii 0.999$c$

   iv 0.9999$c$

   A spreadsheet might be useful.

3   A proton is accelerated in a synchrotron to 0.999 999*c*.

   **a**   Calculate its relativistically corrected mass and momentum.

   **b**   What mass travelling at 60 km h$^{-1}$ would have the same momentum?

   **c**   Calculate the ratio of the mass found in part **b** to the proton mass. First check that your calculator or computer can handle squaring a number like 0.999 999 and subtracting the result from one.

4   The Australian Synchrotron accelerates electrons to about 299 792 km s$^{-1}$. Calculate their relativistic mass and momentum.

5   A large particle accelerator ring, like the Australian Synchrotron, needs to be of substantial radius. Use your answers to the questions above, plus your knowledge of circular motion, to explain why this might be so. How does this support the theory of relativity?

# 12.6 Einstein's mass–energy equivalence relationship: $E = mc^2$

To explore the relationship between matter and energy, let us discuss a thought experiment involving the momentum, $p$, of a photon. We have $E = pc$ and so $p = \dfrac{E}{c}$.

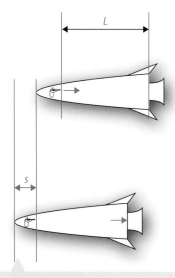

**FIGURE 12.17** A spaceship isolated in space. The pilot is playing with a photon gun and $L$ = distance from the gun to the rear of the spacecraft.

Imagine a stationary spacecraft in distant space, where we can ignore gravity.

Figure 12.17 shows that a pilot of the spacecraft is practising her aim with a laser gun that is capable of firing a single photon at a time. She fires at the rear of the ship, which is distance $L$ away. The photon will take time $t = \dfrac{L}{c}$ to get to the back of the ship. Since photons have momentum, conservation of momentum requires that there will be some recoil. The pilot, spaceship (she is seated) and gun recoil with velocity $v$, and in time $t$ travel distance $s = vt$.

The magnitude of the momentum the spacecraft (of mass $M$) receives will be $Mv$. Of course, the pilot will probably not notice this movement because of the small size of the momentum of a photon. However, as small as it is, the movement will occur. Magnitude of photon momentum must be equal to the magnitude of spaceship momentum. Therefore:

$$\frac{E}{c} = Mv$$

so

$$v = \frac{E}{Mc} = \frac{s}{t}$$

$$s = \frac{Et}{Mc}$$

but we also have $t = \dfrac{L}{c}$, so:

$$s = \frac{EL}{Mc^2}$$

When the photon collides with the rear of the spacecraft, its momentum will be transferred, causing the craft to stop. There is no net external force applied to the system; the total momentum of the system has not changed, yet the ship has moved. It is the photon that has changed position and caused this redistribution. In this interaction with the photon, the spacecraft has behaved exactly as would be expected to behave if there had been a redistribution of mass, even though the photon is massless.

The photon has no rest mass, but can be thought of as behaving as if it has a relativistically corrected mass, $m$. If so, its momentum would be $p = mc$ where $c = \dfrac{L}{t}$, and must be equal in magnitude to the spacecraft's momentum $Mv$, where $v = \dfrac{s}{t}$. Therefore, again equating momenta:

$$\frac{Ms}{t} = \frac{mL}{t} \rightarrow s = \frac{mL}{M}$$

This is a second equation for $s$. We can put it equal to our first equation and simplify:

$$s = \frac{EL}{Mc^2} = \frac{mL}{M}$$

$$E = mc^2$$

This says that mass and energy are equivalent and mass is a manifestation of energy, in this case the energy of the photon. The energy associated with mass at rest is called the **rest energy** of the mass, and is given by the equation:

$$E = m_0 c^2$$

The mass of an electron is approximately $9.1 \times 10^{-31}$ kg. Using $E = mc^2$, the rest energy of an electron is $8.187 \times 10^{-14}$ J. This can be converted to an energy in electron-volts. $1\,\text{eV} = 1.602 \times 10^{-19}$ J, so we get $5.11 \times 10^5$ eV, or 0.511 MeV.

Scientists working in particle physics often express particle masses in eV rather than kg. When they are smashing particles together in an accelerator like the LHC, it is the energy of the particles that matters, not their mass. To make a new particle in a collider, the energy of the colliding particles (say, protons) needs to be converted into the mass of the new particle (say, a Higgs boson). The more energy the particles have (whether rest energy or kinetic), the more massive the new particles that can be made.

## Nuclear reactions

Nuclear reactors, nuclear bombs, and the stars, including our Sun, work by converting rest mass–energy into other forms. Typically, some fraction of the rest energy of the particles entering the reaction is converted into kinetic energy of particles leaving the reaction, some into high energy photons (gamma rays, not to be confused with the quantity $\gamma$ used in relativity!). As in other energy transformations, energy cannot be created or destroyed.

We can use Einstein's formula and conservation of energy to explore nuclear reactions. When a very large atom splits (**fissions**), the total mass of the fragments is less than that of the original atom. That is to say, there is a **mass defect**. The mass has been converted into energy, generally the kinetic energy of the fragments and some gamma rays. This is what happens in nuclear power stations and atomic bombs of the kind used in World War II.

In the Sun, the reverse happens – two or more light atoms are combined (fused) to make one heavier atom, whose mass is less than that of the ingredients. This is called nuclear fusion. **Fusion** power stations have not yet been realised, though the international fusion experiment named ITER is getting close.

A good example of a fission reactor is the OPAL reactor at the Australian Nuclear Science and Technology Organisation in Sydney. A typical reaction occurs when a neutron is absorbed by a $^{235}_{92}\text{U}$ atom. The extra neutron in the centre (**nucleus**) of the atom causes it to split (fission) into pieces. This is illustrated in Figure 12.18.

You might know that the number of protons in the nucleus of an atom defines which element it is. The number of **neutrons** defines which isotope of that element it is. Uranium has 92 protons in the nucleus. The isotope used in fission reactors has 235 nucleons (nuclear particles) in total – 92 protons and $(235 - 92 =) 143$ neutrons. If a single neutron of the right energy hits that nucleus, it can get right into it and create a new atom – $^{236}_{92}\text{U}$, with 144 neutrons and 92 protons.

Mass defect is discussed further in chapter 13.

**ITER**
Thirty-five nations, including Australia, are working together to make generation of electricity using fusion possible.

**OPAL**
Australia has a single working nuclear reactor: OPAL, in Sydney. It is used for scientific experiments and the manufacture of medical radioactive isotopes, not for energy generation.

Before the event, a slow neutron approaches a $^{235}$U nucleus.

**Before fission**

After the event, there are two lighter nuclei and three neutrons.

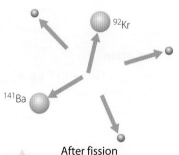

**After fission**

**FIGURE 12.18** A $^{235}$U nucleus fissions after absorbing a neutron.

We write an isotope like this: ${}^{235}_{92}\text{U}$, where the bottom number gives the protons and the top is the total protons plus neutrons. A neutron would then be written ${}^{1}_{0}\text{n}$. So we might write:

$${}^{1}_{0}\text{n} + {}^{235}_{92}\text{U} \rightarrow {}^{236}_{92}\text{U}$$

But ${}^{236}_{92}\text{U}$ is unstable, and it soon splits apart. It can do so in several ways. One of the more common ones can be written like this:

$${}^{1}_{0}\text{n} + {}^{235}_{92}\text{U} \rightarrow {}^{236}_{92}\text{U} \rightarrow {}^{144}_{56}\text{Ba} + {}^{89}_{36}\text{Kr} + 3{}^{1}_{0}\text{n} + \text{energy}$$

which is to say it splits into a barium atom, a krypton atom and three neutrons, and gives off energy. The energy is emitted as a combination of the kinetic energies of the particles and some gamma rays.

How much energy is given off?

To find this, we compare the mass after fission with the mass before.

The masses of nuclei are often given in unified atomic mass units, or u, where $1\,\text{u} \approx 1.660\,54 \times 10^{-27}$ kg. This is roughly the mass of a proton or neutron, but not exactly the same.

**TABLE 12.2** Masses of some nuclei (atoms without electrons)

| NUCLEUS | MASS (UNIFIED ATOMIC MASS UNITS, u) |
|---|---|
| ${}^{235}_{92}\text{U}$ | 235.044 |
| ${}^{1}_{0}\text{n}$ | 1.009 |
| ${}^{144}_{56}\text{Ba}$ | 140.908 |
| ${}^{89}_{36}\text{Kr}$ | 91.905 |

Mass of reactants = 235.055 u + 1.009 u = 236.053 u

Mass of products = 140.908 u + 91.905 u + 3 × 1.009 u = 235.840 u

So we can see that we have less mass at the end than at the start – as we would expect if energy is being given off. The change in mass is:

236.053 u – 235.840 u = 0.213 u

Now, we can convert that to kilograms and use $E = mc^2$ to get the energy.

Change in mass is $0.213\,\text{u} \times 1.660\,54 \times 10^{-27}\,\text{kg}\,\text{u}^{-1}$, so the energy released is:

$$E = (0.213\,\text{u} \times 1.660\,54 \times 10^{-27}\,\text{kg}\,\text{u}^{-1}) \times (3.00 \times 108\,\text{m}\,\text{s}^{-1})^2 = 3.18 \times 10^{-11}\,\text{J} = 198\,\text{MeV}$$

That is a lot from a single atom.

Fusion is the other example of a nuclear reaction delivering energy. It is the mechanism at work in stars and in the most devastating weapon ever invented by human beings – the hydrogen bomb, or H-bomb.

## WORKED EXAMPLE (12.8)

Inside a star such as our Sun, four protons are forced together at enormous pressure and temperature, and fuse to form a helium nucleus (an alpha particle, α) plus two positrons (see below). How much energy is released?

$m_{\text{proton}} = 1.673 \times 10^{-27}$ kg, $m_{\text{positron}} = 9.108 \times 10^{-31}$ kg, $m_{\alpha} = 6.646 \times 10^{-27}$ kg.

| SOLUTION | LOGIC |
|---|---|
| $m_{\text{proton}} = 1.673 \times 10^{-27}$ kg, $m_{\text{positron}} = 9.108 \times 10^{-31}$ kg, $m_{\alpha} = 6.646 \times 10^{-27}$ kg | ▪ Identify the relevant data from the question. |
| $E_R = mc^2$ | ▪ Find the expression linking mass and energy (where in this case *m* is the change in mass). |
| $\Delta m = m_b - m_a$ | ▪ Find the change in mass due to reaction (a = after, b = before). |



9780170409131

| | |
|---|---|
| $m_b = 4m_{proton}$ | ▪ Find the mass before reaction, $m_b$. |
| $m_a = m_\alpha + 2m_{positron}$ | ▪ Find the mass after reaction, $m_a$. |
| $\Delta m = 4m_{proton} - (m_\alpha + 2m_{positron})$ | ▪ Find $\Delta m$. |
| $E_R = (4m_{proton} - (m_\alpha + 2m_{positron}))c^2$ | ▪ Find the energy released. |
| $E_R = (4(1.673 \times 10^{-27})$ $- (6.646 \times 10^{-27}\,kg + 2(9.108 \times 10^{-31}\,kg))(2.998 \times 10^8)^2$ $= 3.97 \times 10^{-12}\,J$ | ▪ Substitute in and calculate the final answer with correct units and appropriate significant figures. |
| This is an order of magnitude smaller than the fission result, but the reactants are two orders of magnitude lighter so the fusion reaction is in fact even more powerful than fission. | |

**TRY THIS YOURSELF**

Express the result above in eV and MeV.

The worked example above shows that the fusion reaction taking place in the Sun gives off $3.97 \times 10^{-12}$ J of energy each time it occurs. This energy is mostly given off as the kinetic energies of the product particles. These particles collide and share their energy with others and the whole Sun heats up and glows. The Sun gives off energy at a rate of about $3.846 \times 10^{26}$ W, or $3.846 \times 10^{26}$ J s$^{-1}$.

Using $E = mc^2$, we can work out how much mass the Sun uses up in one second:

$$m = \frac{E}{c^2} = \frac{3.846 \times 10^{26}\ \text{J}}{\left(3.0 \times 10^8\ \text{m s}^{-1}\right)^2} = 4.3 \times 10^9\ \text{kg}$$

That's about 4 million tonnes per second. In reality, more than one kind of nuclear reaction occurs. However, if they were all of the kind in Worked example 12.8, then that would be about $10^{38}$ reactions per second.

## Positron emission tomography

In the 1920s, British physicist Paul Dirac was working on quantum mechanics. His calculations led him to suggest that every fundamental particle that has mass has an antiparticle. Photons are massless and do not have an antiparticle. The antiparticle of a charged particle has the same mass but opposite charge. For the electron, the antiparticle therefore has positive charge. It is known as the positron. Its mass is equal to $m_e$ and it has a charge of $+1.9 \times 10^{-19}$ C. When a particle interacts with its antiparticle, both are annihilated. That is, they cease to exist. But both had rest energy, so by conservation of energy something has to come out of the process – in this case we get two gamma rays (high energy photons). We saw in chapter 11 that photons have momentum. If the positron and electron were stationary when they annihilated, total system momentum was zero. Therefore, the two gamma rays must carry equal and opposite momentum, so the sum is still zero. That means they head off in exactly opposite directions.

This forms the basis of the medical imaging technique positron emission tomography (PET). Positrons are created when certain radioactive atoms decay. Electrons are very common, so a positron soon runs into an electron and the two particles undergo pair annihilation. If the radioactive atoms are attached to glucose molecules and injected into the body, the gamma rays will be more intensely emitted from parts of the body that are very active. That's because the active parts will be using more glucose. If we put the patient in a ring-shaped detector, we can use the fact that the gamma rays come in pairs going in opposite directions to work back and find out where in the body they originated. Each pair defines a line, and if we measure lots of pairs we can see where the lines tend to cross. Some tumours metabolise (use) a lot of glucose. The brain uses a lot of energy too. So this method helps medical physicists and doctors work out where in the body a tumour might be, or what parts of the brain are most active (Figure 12.19).

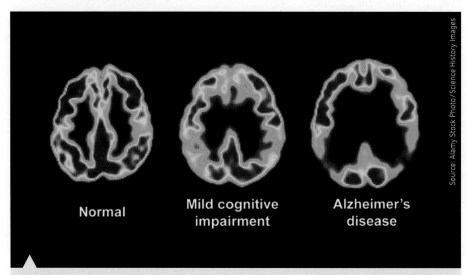

**FIGURE 12.19** PET scans of three brains, one healthy (left), one with slight damage (centre) and one showing signs of Alzheimer's disease (right). The darker colours on the right indicate lower glucose use and therefore lower activity.

## WORKED EXAMPLE (12.9)

An electron and a positron collide. If they have negligible kinetic energy and momentum, calculate the energy of the emitted photons.

| ANSWER | LOGIC |
|---|---|
| $E = mc^2$ | ■ We are converting masses to energies, so we use Einstein's formula. |
| First, the two photons must have equal and opposite momenta. If they have momenta of equal magnitude, they must have the same energy. Therefore, each photon will have energy equal to half the combined mass of the particles. This is just the mass–energy of one electron. So for each photon, $E_{photon} = m_{electron}c^2$. | ■ Think about the situation. |
| $E_{photon} = (9.11 \times 10^{-31} \, \text{kg})(2.998 \times 10^8)^2$ | ■ Substitute in the values with correct units. |
| $E_{photon} = 8.19 \times 10^{-14} \, \text{J} = 0.511 \, \text{MeV}$ | ■ Calculate the final answer with correct units and appropriate significant figures. |

**TRY THIS YOURSELF**

What are the momenta of the photons? Recall that $E = pc$ for a massless particle.

## Chemical reactions

Conventional combustion is a chemical reaction. That means it involves exchange of electrons between atoms. However, chemical reactions do not involve the nuclei of atoms. This means the energies associated with chemical reactions are very low compared with nuclear reactions.

## ► WORKED EXAMPLE (12.10)

One barrel of crude oil, if burned, releases about $6 \times 10^9$ J of energy. The oil in the barrel has density of about $800 \, \text{kg} \, \text{m}^{-3}$, and there are about 160 L of oil in a barrel. 1 kg of $^{235}$U, if completely fissioned, releases about $24 \times 10^6$ kW h of energy.

**a** What mass of $^{235}$U needs to be fissioned to generate the same energy as burning the barrel of oil?

**b** Compare the mass of $^{235}$U and the mass of oil.

| ANSWER | LOGIC |
|---|---|
| **a** $6 \times 10^9$ J per barrel | ▪ Identify the relevant data from the question.<br>▪ Convert both energies to the same units – joule.<br>▪ Energy from barrel of oil in joule |
| $24 \times 10^6 \ \text{kW h kg}^{-1}$<br>$= 24 \times 10^9 \ \text{J s}^{-1} \times 3600 \ \text{s kg}^{-1}$<br>$= 8.64 \times 10^{13} \ \text{J kg}^{-1}$<br><br>$24 \times 10^6 \ \text{kW h kg}^{-1}$<br>$= 24 \times 10^9 \ \text{J s}^{-1} \times 3600 \ \text{s kg}^{-1}$<br>$= 8.64 \times 10^{13} \ \text{J kg}^{-1}$ | ▪ Energy from 1 kg $^{235}$U in joule |
| $\dfrac{6 \times 10^9 \ \text{J barrel}^{-1}}{8.64 \times 10^{13} \ \text{J kg}^{-1}} \approx 7 \times 10^{-5} \ \text{kg barrel}^{-1}$ | ▪ Calculate the ratio. |
| 70 micrograms of $^{235}$U gives the same energy as a barrel of oil. | ▪ State the result. |
| **b** $m_{\text{oil}} = \rho V$ | ▪ Find the mass of the oil. Use the expression relating mass, volume and density: mass = density × volume |
| $\rho = 800 \, \text{kg} \, \text{m}^{-3}$<br>$V = 160 \, \text{L} = 0.160 \, \text{m}^3$ | ▪ Identify the relevant data from the question. |
| $m_{\text{oil}} = (800 \, \text{kg} \, \text{m}^{-3})(0.160 \, \text{m}^3) = 128 \, \text{kg}$ | ▪ Substitute in the values and correct units. |
| $\dfrac{128 \ \text{kg}}{70 \ \mu\text{g}} \approx 18 \times 10^6$ | ▪ Calculate the ratio $\dfrac{\text{mass oil}}{\text{mass } ^{235}\text{U}}$. |
| If we take the ratio of the masses, we get about 1.8 million. That is, we would need to burn 1.8 million kilograms of oil to get the same energy as we get from fissioning 1 kg of $^{235}$U. | ▪ State the comparison. |

**TRY THIS YOURSELF**

Estimate how many joules of energy would be released if one of your fingers were totally converted into energy.

Note that pure $^{235}$U is not used in reactors, and not all the $^{235}$U is fissioned.

- Einstein's mass–energy equivalence relationship is $E = mc^2$. It calculates the energy equivalent to the rest mass of a body.
- We can use it to calculate the energy released during nuclear reactions, fission and fusion.
- Nuclear reactions involve much greater energies than chemical reactions.
- When a particle meets its antiparticle, all the mass is converted into massless photons of high energy.

**CHECK YOUR UNDERSTANDING**

**12.6**

1 Define 'mass defect'. How is it related to energy?

2 What is the importance of mass defect to nuclear physics?

3 $^{214}$Po (mass 213.995 u) is radioactive and decays into an alpha particle (mass 4.003 u) and $^{210}$Pb (mass 209.984 u).

 a What is the mass defect?

 b Is this consistent with the alpha particle having an energy of 7.68 MeV? Explain your answer.

4 Show that fission of 190 kg of pure uranium-235 could produce approximately $10^{16}$ J of energy.

5 Outline the important of relativity to nuclear physics.

6 We get $6 \times 10^9$ J of energy by burning a barrel of oil. This energy comes from the small mass change that happens when energy is released by breaking and forming chemical bonds. How much mass is converted to energy in this case?

- The speed of light in a vacuum has a value of $299\,792\,458\,\text{m s}^{-1}$ or an approximate value of $3.0 \times 10^8\,\text{m s}^{-1}$.

- An inertial frame of reference is one that is not accelerating; that is, it is stationary or moving at constant velocity.

- The principle of relativity states that the laws of motion are the same in all inertial frames of reference.

- The proper length of an object is the length measured by an observer at rest relative to the object.

- Classical or Galilean relativity says that the velocity of A relative to B is the velocity of A relative to C plus (vector addition) the velocity of C relative to B, or $v_{AB} = v_{AC} + v_{CB}$.

- Michelson and Morley found no evidence for the existence of the aether.

- First postulate of special relativity: The laws of physics are the same in all inertial frames of reference.

- Second postulate of special relativity: The speed of light has the same value in all inertial frames.

- The Lorentz factor ($\gamma$) is defined as: $\gamma = \dfrac{1}{\sqrt{1 - \dfrac{v^2}{c^2}}}$.

- The time interval between two events occurring at one place in an inertial frame, as measured by an observer in that inertial frame, is called proper time, $t_0$ (sometimes called local time).

- Time dilation: The time interval measured between two events occurring at different locations in an inertial frame will be greater than the proper time; $t = \dfrac{t_0}{\sqrt{1 - \dfrac{v^2}{c^2}}} = \gamma t_0$.

- Length contraction: All observers will measure a moving object as being shorter or contracted in the direction of relative motion than when the object is at rest.

- $l = l_0 \sqrt{1 - \dfrac{v^2}{c^2}} = \dfrac{l_0}{\gamma}$, where $l_0$ is the length of an object at rest, and $l$ is the length of the object measured by an observer who is moving at relative velocity $v$ to the object's inertial frame.

- The predictions of the theory of special relativity can be seen in experiments performed on Earth and in astronomical observations.

- The behaviour of light from distant, powerful sources like supernovas is what we would expect if Einstein's theory were correct.

- The time subatomic particles like muons apparently take to decay depends on the relative speeds of the muon and the observer, as Einsteinian relativity predicts.

- Atomic clocks, which are extraordinarily precise, show the effects of relativity even when moving at the speed of a commercial jetliner. These speeds are quite slow compared to $c$.

- Rest mass or proper mass is mass as measured in an inertial reference frame in which the object is stationary.

- Relativistic mass increases as the velocity of the object increases: $m = \gamma m_0$.

- Relativistic momentum, $p$, is given by $p = \gamma m_0 v$, or
$$p = \frac{m_0 v}{\sqrt{1 - \dfrac{v^2}{c^2}}}\,;\quad p = \frac{p_0}{\sqrt{1 - \dfrac{v^2}{c^2}}}\,;\quad p = \gamma p_0\,.$$

- The rest energy of a mass is the energy associated with a mass at rest and is given by $E_{\text{rest}} = m_0 c^2$, where $m_0$ is the rest mass and $c$ is the speed of light.

- In a nuclear reaction that gives off energy, the mass of the products is less than the mass of the reactants. The 'lost' mass, $\Delta m$, is released as energy, given by $E = \Delta m c^2$.

- This energy can take the form of the kinetic energy of the reaction products and gamma rays.

- The amount of energy released in a nuclear reaction is many orders of magnitude larger, per unit mass of reactants, than in a chemical reaction.

1   State Einstein's two postulates for relativistic motion.

2   Define these terms.
    a   Inertial frame of reference
    b   Relative motion
    c   Proper time
    d   Proper motion
    e   Time dilation
    f   Length contraction
    g   Mass defect

3   State all the relativistic equations presented in this chapter.

4   What is the importance of Maxwell's equations for relativity?

5   Starting at the front, a passenger walks quickly (at $3.0\,\text{m s}^{-1}$) towards the rear of a train carriage that is travelling forwards at $16\,\text{m s}^{-1}$. Relative to an observer on the ground nearby:
    a   what is the speed of the passenger?
    b   where is the passenger after $5.0\,\text{s}$?

6   Describe the main features of the Michelson–Morley experiment. Why is the null result in this experiment significant?

7   Outline the key equations of Galilean relativity, defining all terms.

8   A space probe travels into deep space at an average velocity of $0.6c$. When it is $1.8 \times 10^9\,\text{m}$ away as measured from Earth, it sends back to Earth a pulse of light of duration $2.00\,\text{s}$ in the space probe's frame of reference.
    a   Find the speed of the light pulse:
        i   according to the space probe.
        ii  according to Earth-based observers.
    b   According to Earth-based observers:
        i   how long does the pulse take to reach Earth?
        ii  what is the duration of the pulse?

9   Electrons in an electron gun are about 0.5% heavier than electrons at rest. What is the velocity of the electrons in the electron gun?

10  Two identical clocks are synchronised. One clock is sent off in a spaceship travelling with a speed $v = 0.70c$. After 49 years on the Earth clock, calculate the time on the spaceship, as observed from Earth.

11  A pion has a mean lifetime of $26\,\text{ns}$ in Earth's frame of reference. Calculate its mean lifetime in its own frame of reference if it is travelling at $0.5c$.

12  Muons are created by the collision of cosmic rays with air molecules high in Earth's atmosphere ($10\,000\,\text{m}$). They approach the ground with a velocity of $0.999c$. From the viewpoint of an observer travelling at the velocity of the muons, how high do they measure Earth's atmosphere to be?

13  The main nuclear fusion reaction that creates the energy in a hydrogen bomb is: $4\,{}^1_1\text{H} \rightarrow {}^4_2\text{He} + 2e^+ + \text{energy}$.
    The mass of H is $1.673 \times 10^{-27}\,\text{kg}$, the mass of He is $6.644 \times 10^{-27}\,\text{kg}$ and the mass of a positron is $9.109 \times 10^{-31}\,\text{kg}$.
    Calculate the amount of energy, in MeV and joule, produced from $10\,\text{kg}$ of hydrogen.

14  The Twin Paradox illustrates the effects of time dilation. Fred and Pierre are twins. Fred takes an extraterrestrial journey that takes him to a distant stellar system and then returns. The average speed of Fred's spacecraft is $0.6c$. Pierre remains on Earth. When Fred returns, Pierre is 20 years older. Calculate how much Fred has aged according to his space watch.

15  Xavier, Iris and Maxine are triplets. Xavier pilots a spacecraft away from Earth towards the centre of the Milky Way for a distance of 7 light years relative to Earth. Maxine pilots a similar spacecraft in the opposite direction, away from the centre of the Milky Way. Her craft travels outwards for a distance of 7 light years. Their average speed is $0.61c$. Iris remains at home. Xavier and Maxine return at the same time.
    a   Will Xavier and Maxine appear to be the same age on their return?
    b   Calculate how much older than the travellers Iris will be when the travellers return.

16  The pilot of a non-accelerating spacecraft, moving away from Earth at great speed, celebrates the passing of six birthdays. Earth-bound observers measure this elapsed time to be 10 years. Calculate, relative to Earth:
    a   the speed of the spacecraft.
    b   how far the spacecraft travels over these six birthdays.

17  A short pulse of particles from a linear accelerator passes through two sensors that are spaced $85\,\text{m}$ apart (Figure 12.20). The particles are timed to travel the distance in $0.377\,\text{ms}$.

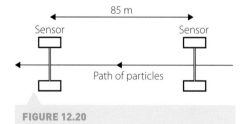

**FIGURE 12.20**

    a   Calculate the velocity of the particles as a proportion of the speed of light.
    b   In the reference frame of the particles, calculate the spacing of the detectors.

**18** A positron is the antimatter equivalent of an electron. Both have a rest mass of $9.109 \times 10^{-31}$ kg. When an electron and a positron collide, they annihilate each other and produce energy in the form of electromagnetic radiation. Calculate how much energy is released by the collision.

**19** Muons have a speed of $0.995c$ when they are recorded at 5000 m above Earth's surface. They have a mean lifetime of 2.2 µs in their rest frame.

In the muon reference frame:

  **a**  **i** How far do muons travel in a mean lifetime?

   **ii** Is it possible for muons to reach Earth? Explain your answer.

  **b** How high above Earth's surface are they created?

  **c** Is it likely that muons will reach Earth?

**20** The ultimate speed is $c$, but there does not appear to be an ultimate kinetic energy. Explain this apparent contradiction.

**21** In a synchrotron, 2.5 GeV electrons travelling at $0.9996c$ are subjected to a 2.8 T magnetic field (see chapter 5).

  **a** Calculate the momentum of the electrons.

  **b** Calculate the radius of their path.

**22** PET relies on fundamental physics to help health professionals look after patients. Can you think of some other examples of fundamental physical inventions contributing to health and wellbeing?

**23** How have atomic clocks been used to test Einstein's ideas?

# MODULE (7): THE NATURE OF LIGHT

**Answer the following questions.**

1   What impact did the work of Christiaan Huygens have on that of James Clerk Maxwell?

2   Why did Maxwell deduce that light was electromagnetic radiation?

3   According to Maxwell, why can electromagnetic radiation travel through space without there being a medium to carry the wave?

4   How can a microwave oven be used to determine the speed of light?

5   How can we distinguish between red dwarf and red giant stars?

6   How can we determine the rotational velocity of a star?

7   A monochromatic light is projected through two slits on an otherwise blackened microscope slide, separated by a distance of 1.05 mm, on to a viewing screen 1.8 m away. If the wavelength of the light is 500 nm, how far from the central bright spot should the first interference fringe appear?

8   What differences might we expect to see when shining a light through a double slit slide on to a screen if the light source were:

    a   monochromatic?

    b   incandescent?

9   Why do we see bright and dark areas on a screen when we shine a light through a narrow single slit?

10  What property of a single, thin strand of copper wire allows it to produce interference fringes when illuminated by a monochromatic light source?

11  You see a car in the distance, at night, with its headlights on. Why don't you see an interference pattern?

12  William Lawrence Bragg and his father William Henry Bragg illuminated crystalline solids with X-rays and discovered intense patterns of radiation peaks and troughs at certain angles and wavelengths. What might these results suggest?

13  A beam of plane-polarised light is passed through a filter oriented at 60° to the direction of the light's plane of polarisation. What is the intensity of the filtered beam compared to the original beam?

14  In a black body spectrum, the wavelength at which the intensity is highest, $\lambda_{max}$, is given by Wien's Law.

    a   State Wien's law and write a sentence explaining what happens to $\lambda_{max}$ as $T$ increases.

    b   Sketch a typical plot of intensity versus wavelength for radiation from a black body. Label the axes and note the position of $\lambda_{max}$.

    c   Qualitatively, what happens to an incandescent light bulb if the voltage across it is increased? What happens to the colour of a bar of metal as it gets hotter and hotter? Does this agree with Wien's law?

    d   A black body is at 4000 K. What is $\lambda_{max}$?

15  The energy of a photon of frequency $f$ is given by $E = hf$.

    a   Calculate the energy of a photon of wavelength 500 nm.

    b   Sketch a graph illustrating the dependence of photon energy on:

        i   frequency.

        ii  wavelength.

    c   With this is mind, discuss why is it important that the plot of intensity versus wavelength for black body radiation goes to zero at very small wavelengths.

    d   What was the 'ultraviolet catastrophe'?

16  The equation for the photoelectric effect is $K_{max} = hf - \phi$.

    a   Define all the symbols used in this equation.

    b   Give an expression for the maximum possible energy for an electron that leaves the surface of the metal.

    c   What is the value of the minimum possible energy for an electron that leaves the surface of the metal?

    d   Define 'stopping voltage'.

    e   Does the stopping voltage depend on the intensity of the light, the frequency of the light, or both? Use the equation to help you explain.

    f   A metal has a work function of 4.50 eV. Photons of wavelength 250 nm fall on the metal. Calculate the stopping voltage.

17  A red ball and a green ball are rolling along the floor, going in the same direction but at different speeds. At time $t_0$, the green ball is 4 m from the start, in the positive direction. The red ball is 2 m from the start, also in the positive direction.

    a   Sketch the situation. Include a set of labelled axes and some sensible notation for the velocities and positions of the balls.

    b   The green ball is travelling at $2\,m\,s^{-1}$, the red at $3\,m\,s^{-1}$. Calculate the velocity of the green ball relative to the red. Remember, velocity is a vector.

    c   Calculate the position of the red ball relative to the start at time 10 seconds after $t_0$.

    d   Calculate the position of the red ball relative to the green ball at time 10 seconds after $t_0$.

**18** Time dilation is one phenomenon predicted by the theory of special relativity.

**a** In words, define 'time dilation'.

**b** Write down the key equation we use to relate proper time and relativistic time. Define the symbols.

**c** A spaceship is travelling at 0.94c relative to an observer on Earth. If 5 years pass for the observer on Earth, how many pass for an observer on the spaceship?

**d** State the two postulates of special relativity.

**19** Various experiments have been done to demonstrate special relativity.

**a** What phenomenon predicted by special relativity was demonstrated by the Hafele–Keating experiment? Briefly outline the experiment, including the key pieces of equipment needed.

**b** Another prediction of special relativity is the dependence of mass on velocity. Discuss how this effect is important at circular particle accelerators like the Australian Synchrotron or the Large Hadron Collider. Use your knowledge of special relativity and circular motion.

**20** Einstein's formula $E = m_0 c^2$ says that mass is a manifestation of energy.

**a** One of the first atomic bombs was 'Little Boy'. It exploded with an energy equivalent to 15 000 tonnes of TNT – about 63 TJ (TJ = terajoule, or $10^{12}$ J). Calculate the mass, in kilograms, that must be converted to energy to obtain such an explosion.

**b** The mass of a Higgs boson is equivalent to about 126 GeV. Calculate the mass in kilograms.

**c** When $^{234}$Th (thorium-234) undergoes radioactive decay, it emits an electron and transmutes into $^{234}$Pa (protactinium-234). Each decay also releases about 270 keV of energy. Calculate the amount of mass lost in each decay, in kilograms.

---

**DEPTH STUDY SUGGESTIONS**

→ Investigate applications of spectroscopy to cosmology.

→ Investigate applications of diffraction in industry.

→ Investigate polarisation of light in astronomy.

→ Sending a spacecraft to investigate a body in the solar system involves a number of remote sensing techniques. What are some of these techniques, how do they work and when are they applicable? Include ground penetrating radar (GPR) and X-ray fluorescence in your investigation.

→ How does the GPS system work and what are its limitations?

→ Lasers have many applications. What are the basic principles and how are they implemented? Include examples of their uses from science and industry.

→ Do a literature study to explore how black body radiation relates to climate change.

→ Find out about Millikan's photoelectric experiments. If you have access to a photoelectric experiment laboratory demonstration, try to demonstrate the effect yourself.

→ Find several different types of light globes – incandescent, plus one or two of halogen, diode and fluorescent. Using a spectroscope or diffraction grating – even reflecting the light off a CD and onto a screen is useful – show how the incandescent lamp gives a continuous spectrum, as would be expected from a black body, but the others do not. See if you can find out why.

Real Time Relativity

→ Explore the Real Time Relativity software using the weblink. It simulates travelling at relativistic speeds and can be downloaded for free.

→ The electrons within heavy atoms – like gold – move so fast that they have to be thought of as relativistic particles. This is one reason gold looks yellow and silver does not. Write a brief introduction to the principles of relativistic quantum chemistry. There is no need to include equations.

→ Explore the 'twin paradox'. Set up a spreadsheet to compare the ages of two twins if one stays on Earth and the other comes back after a 10 light year voyage at relativistic speed. Do the comparison for different speeds. Plot some results and summarise them.

# FROM THE UNIVERSE TO THE ATOM

Science Photo Library / M. Kornmesser / European Southern Observatory

9780170409131

# 13 Origins of the elements

**OUTCOMES**

**Students:**

- investigate the processes that led to the transformation of radiation into matter that followed the 'Big Bang' CCT, ICT
- investigate the evidence that led to the discovery of the expansion of the Universe by Hubble (ACSPH138) ICT, N
- analyse and apply Einstein's description of the equivalence of energy and mass and relate this to the nuclear reactions that occur in stars (ACSPH031) CCT
- account for the production of emission and absorption spectra and compare these with a continuous black body spectrum (ACSPH137) CCT, ICT
- investigate the key features of stellar spectra and describe how these are used to classify stars N
- investigate the Hertzsprung–Russell diagram and how it can be used to determine the following about a star: CCT, ICT, N
  - characteristics and evolutionary stage
  - surface temperature
  - colour
  - luminosity
- investigate the types of nucleosynthesis reactions involved in Main Sequence and Post-Main Sequence stars, including but not limited to: CCT, ICT
  - proton–proton chain
  - CNO (carbon–nitrogen–oxygen) cycle

Physics Stage 6 Syllabus © 2017 NSW Education Standards Authority (NESA) for and on behalf of the Crown in right of the State of New South Wales.

Science Photo Library/Roger Harris

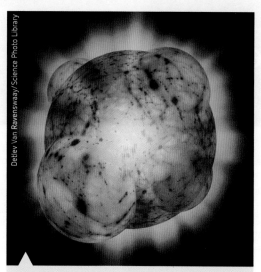

FIGURE 13.1 An artist's impression of the Universe just after the Big Bang as inflation occurred

Theories are used to explain a range of related phenomena. For example, the **Big Bang** Theory (BBT) is used to explain the formation of the elements of the periodic table over the lifetime of the Universe (approx. 13.8 billion years). The BBT was developed and modified by the collection of evidence from the Universe. The Universe is so vast that we cannot collect chemical elements and compounds directly. Instead, evidence of their existence is deduced from their signature radiations. Each element has its own characteristic spectrum of radiations, which can be identified by instruments that are sensitive to electromagnetic radiations such as microwaves, infrared, visible light, ultraviolet, X-rays and γ-rays.

Evidence is used to develop, support and change scientific theories. Significant international scientific cooperation has led to the collection of a great deal of evidence for the BBT. In the process, the original theory has been modified. Originally, it was assumed that the Universe just started to expand uniformly in all directions from a single point. Later, new evidence was used to modify this to include a 'slow' start, followed by a period of 'rapid' inflation (Figure 13.1). In the last 10 years, further evidence has suggested that the Universe is not expanding uniformly in all directions but that its expansion is accelerating.

Observations of stars and the use of technology have enabled us to learn much about stars and the mechanisms by which they use their fuel. Analysis of a star's spectrum reveals much about the star's nature, composition, age and luminosity.

## 13.1 Radiation transforms into matter

Critical and creative thinking

The BBT is an attempt to understand the origin and development of the Universe and everything in it. The idea of energy is crucial. According to the BBT, energy is all there was at the beginning. This energy spread out from an incredibly concentrated point called a singularity. The temperature was incredibly high. As the energy spread out, it was distributed over vaster and vaster distances. In any one region, the temperature became much less than in the singularity.

Particles can be formed from energy. Forces, mediated by field particles, also form out of the energy. At very high temperatures, particles that form are rapidly destroyed. But at lower temperatures, the particles become more stable and longer lived. Consequently, they can start to combine. Eventually, the conditions become possible for atoms and molecules to form.

The energy from the singularity is all the energy in the Universe. Over the life of the Universe, approximately 14 billion years, this energy has been redistributed into the matter and radiation that can be observed everywhere. But the total amount of energy is still the same as at the beginning.

### Elementary particles

An **elementary particle** is a particle that cannot be further divided. The search for elementary particles has revealed a number of things. Early on, it was thought that atoms were indivisible. Then electrons and radioactivity were discovered, and alpha particle experiments showed that atoms were mainly empty space between a central, positive nucleus and surrounding electrons. Protons were used to finalise the basis of the periodic table in 1914. The neutron, which was posited to account for mass problems associated with atomic nuclei, was discovered in 1932. Later experiments have shown that protons and neutrons are divisible but that electrons are not divisible. Thus, electrons are elementary particles, but protons and neutrons are not.

There is an antiparticle for every particle. Slightly more matter than antimatter was formed overall, so that most of what we detect now is matter, not antimatter. As yet there is no explanation for this imbalance. When matter and antimatter encounter each other, they return to radiant energy. For example, the antiparticle of an electron is a positron (Figure 13.2). When an electron and a positron collide, two characteristic gamma rays are released in opposite directions. The particles have been annihilated.

There are two types of elementary particle: quarks and leptons.

» Quarks are small, elementary particles, which can combine to produce other particles, such as protons and neutrons. There are six quarks: up, down, strange, charm, bottom, top. Each has its own antiquark.

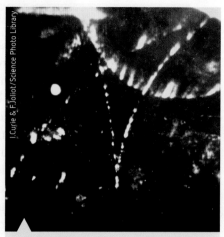

FIGURE 13.2 A gamma ray photon decays into two oppositely charged particles that leave behind vapour trails, visible in this cloud chamber image.

**TABLE 13.1 Quarks and antiquarks**

| QUARK | up | u | down | d | strange | s | charm | c | bottom | b | top | t |
|---|---|---|---|---|---|---|---|---|---|---|---|---|
| ANTIQUARK | anti-up | $\bar{u}$ | anti-down | $\bar{d}$ | anti-strange | $\bar{s}$ | anti-charm | $\bar{c}$ | anti-bottom | $\bar{b}$ | anti-top | $\bar{t}$ |

» Leptons are small elementary particles: electron, electron neutrino, muon, muon neutrino, tau and tau neutrino. Leptons interact via the gravitational force, electromagnetic force and weak force, but not the strong force.

**TABLE 13.2 Leptons and antileptons**

| LEPTON | electron | electron neutrino | muon | muon neutrino | tau τ | tau neutrino |
|---|---|---|---|---|---|---|
| ANTIPARTICLE | positron | electron anti-neutrino | anti-muon | muon anti-neutrino | anti-tau $\bar{\tau}$ | tau anti-neutrino |

## Hadrons and field particles

Hadrons are relatively large particles that are composed of elementary particles. Hadrons can be divided into middle-sized mesons and heavy baryons. Mesons are composed of one quark and one antiquark. Baryons comprise three quarks. See Figure 13.3.

» Middle-sized mesons decay to electrons, positrons, neutrinos and photons.

» The heavier baryons always decay to protons and sometimes decay to other particles such as neutrons.

Field particles are the particles that mediate force through field particle exchange mechanisms. For example, the electromagnetic force acts when charged particles exchange photons. The four forces are:

» electromagnetic (mediated by the photon)

» weak (mediated by W and Z bosons)

» gravitational (mediated by the graviton; not yet detected)

» strong (mediated by gluons and pions).

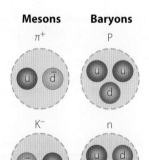

FIGURE 13.3 The quark compositions of two mesons ($\pi^+$ and $K^-$) and two baryons (proton and neutron)

Forces and particles are discussed in more detail in chapter 17.

## BBT timeline for the formation of particles

According to the BBT, all four forces separated and quarks and leptons came into existence in the first microsecond (Figure 13.4). This corresponded to a decrease in temperature from $10^{32}$ K to $10^{15}$ K. Within 3 minutes, hadrons such as protons and neutrons were formed. Over the next 380 000 years, nuclei, mainly hydrogen, helium and some lithium, formed as the temperature dropped to 3000 K. These nuclei were charged ions because atoms were not yet able to form. During this time, radiation was produced by collisions between matter and antimatter. The radiation was strongly scattered from the ions so that the Universe was opaque in the same way that light, being scattered from water droplets, makes a fog opaque.

After around 380 000 years, conditions were ripe for the formation of hydrogen atoms. Neutral atoms do not scatter radiation, so this could now escape. The Universe became transparent to radiation. Clumps of neutral matter – atoms and molecules – formed. Under the influence of gravity, gas clouds, stars and galaxies clumped together.

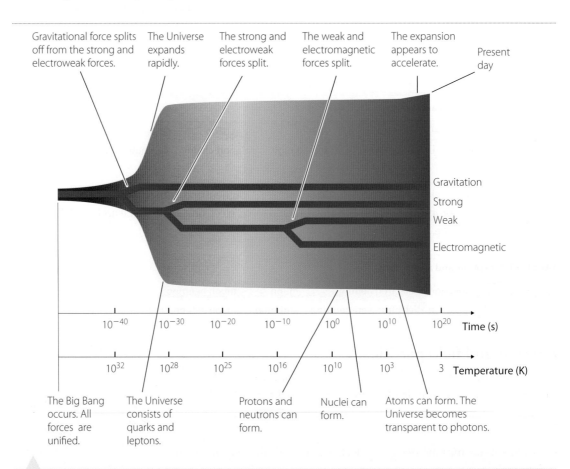

**FIGURE 13.4** The history of the Universe. The four forces separated and quarks and leptons formed in about a microsecond. Hadrons appeared by 3 minutes and charged nuclei in the next 700 000 years, at which time atoms began to form. The Universe cooled to 3000 K and became transparent. Molecules, then gas clouds and galaxies, began forming.

## Evidence for the BBT

Radiation transforms into matter

The residual radiant energy that did not form into matter can be detected today as microwaves. Its signature wavelength is consistent with the predictions of the BBT.

According to the predictions, any residual gamma radiation from the Big Bang would be seen from every direction. The wavelengths would be extremely stretched after so much expansion of the Universe. They should be seen as microwaves, corresponding to a temperature of approximately 3 K.

In 1965, Robert Wilson and Arno Penzias measured this radiation for the first time. Satellites launched by NASA (COBE from 1989 to 1993 and WMAP from 2001 to 2011) and the European Space Agency (Planck from 2009 to 2013) have measured the cosmic microwave background radiation (CMBR) to very high levels of accuracy. This is the strongest evidence for the BBT.

A computer-enhanced map of the cosmic microwave background radiation is shown in Figure 13.5. Figures 13.6 and 13.7 summarise the postulated history of the Universe.

© ESA and the Planck Collaboration

**FIGURE 13.5** The remnant radiation from the Big Bang has been stretched by the expansion of the Universe. It is now observed in the microwave region of the electromagnetic spectrum.

Information and communication technology capability

ESA Planck Mission

Cosmic microwave background radiation

An explanation of the origins of the CMBR

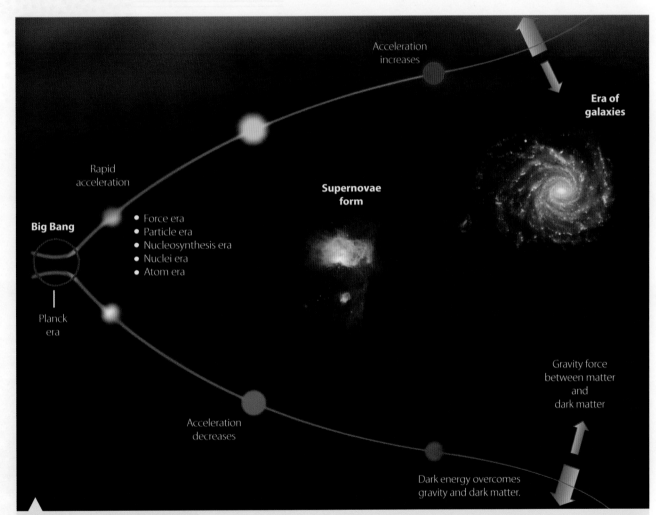

Acceleration increases

Era of galaxies

Rapid acceleration

Supernovae form

**Big Bang**

- Force era
- Particle era
- Nucleosynthesis era
- Nuclei era
- Atom era

Planck era

Acceleration decreases

Gravity force between matter and dark matter

Dark energy overcomes gravity and dark matter.

**FIGURE 13.6** Artist's impression of the stages of the Universe from the Big Bang to the present. Notice the accelerating expansion of the Universe today, which was discovered by, among others, Nobel Laureate Brian Schmidt and colleagues at the Australian National University.

| Time since Big Bang | Era | Temperature (K) |
|---|---|---|
| $10^{-43}$ s | Planck | $>10^{32}$ |
| $10^{-35} - 10^{-32}$ s | Inflation | $10^{14}$ |
| $10^{-10}$ s | Force separation | $10^{13}$ |
| $10^{-3}$ s (1 ms) | Particles | $10^{12}$ |
| 3–20 min | • Nuclear fusion<br>• Nucleosynthesis | $10^9$ |
| Up to 380 000 years | Atoms | Falls to 3000 |
| 380 000 to 200 million years | Dark ages | Falls to 100 |
| 200 million to 380 million years | First stars | Falls below 100 |
| 400 million years to present | Galaxies | Falls to 2.7 |

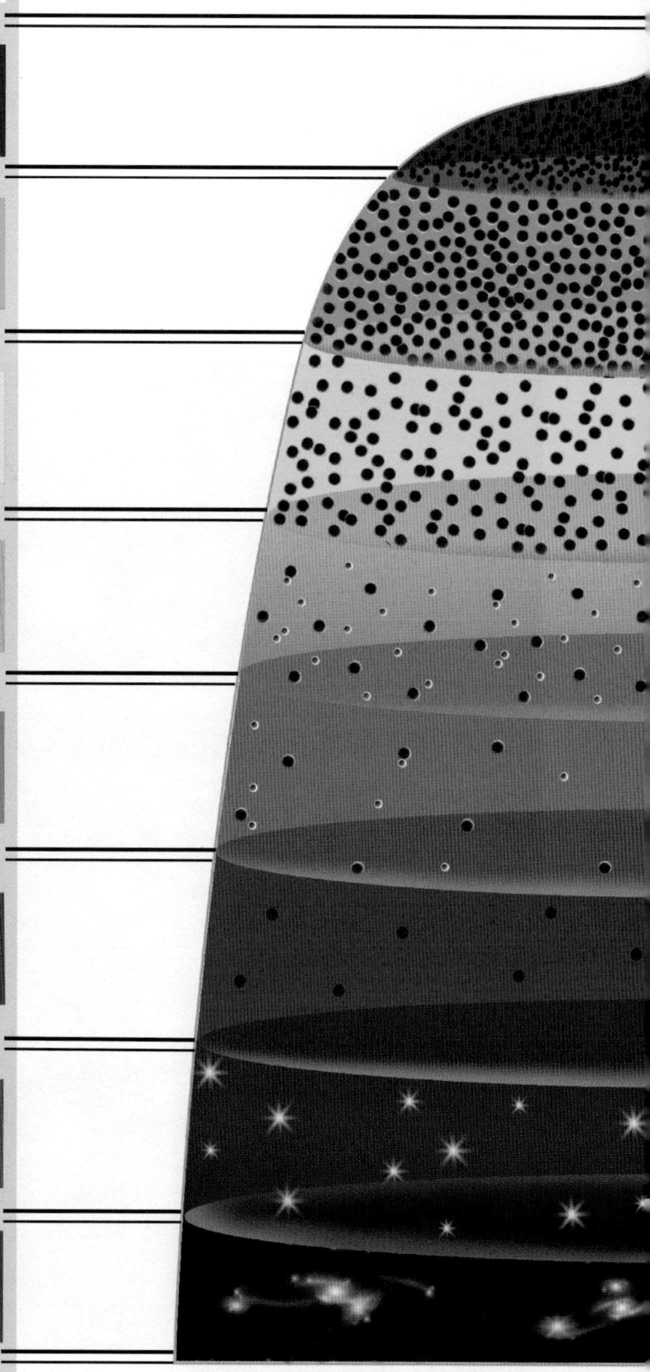

FIGURE 13.7 The history of the Universe

| Description | Major events in the era |
|---|---|
| Incredibly high temperature | Space and time established |
| Fundamental particles form and unform | Strong force becomes distinct; cosmic inflation |
| Fundamental particles last long enough to interact; explosion of photons as matter and antimatter collide | Separation of strong force, weak force and electromagnetism |
| Fundamental particles | Fusion begins |
| Protons, neutrons, electrons and neutrinos stabilise | Fusion ceases; primordial element ratios established |
| Plasma of hydrogen, helium and electrons | Photons are trapped, but freed when atoms are formed |
| Universe is transparent but there are no stars to shine | Cosmic background radiation is unimpeded; only other light is faint hydrogen glow |
| First stars, 50–500 solar masses, appear | Light from first stars begins to re-ionise gas clouds |
| Galaxies appear and evolve | Universe becomes as we know it today |

Information and communication technology capability

Origin and use of the term 'Big Bang'

**KEY CONCEPTS**

- The BBT is widely accepted as a theory for explaining the origins of energy and matter in the Universe.
- The Universe started in an extraordinarily high energy singularity.
- The singularity began to expand, then inflated very rapidly.
- Four forces separated within $10^{-32}$ s: electromagnetic, weak, strong and gravitation.
- Elementary particles – quarks and leptons – formed in the first microsecond, along with their antiparticles.
- Hadrons, including mesons, protons and neutrons, were produced, along with their antiparticles, within 3 minutes.
- Matter and antimatter annihilated; eventually, slightly more matter than antimatter remained.
- Charged nuclei (ions) formed over the next 380 000 years.
- Radiation was scattered from ions in all directions – the Universe was opaque.
- The Universe cooled sufficiently for stable atoms, such as hydrogen and helium, to form – the Universe became transparent to radiation.
- By 700 000 years, atoms, molecules, then giant gas clouds and galaxies began to form.

**CHECK YOUR UNDERSTANDING**

13.1

1  **a**  What is the purpose of the BBT?

   **b**  Describe the main ideas of the BBT.

2  Identify the major events that occurred in the first 3 minutes of the Universe.

3  **a**  Define 'elementary particle'.

   **b**  Name the elementary particles and their antiparticles.

4  In quark theory, what is the difference between a meson and a baryon?

5  What is the significance of the CMBR?

## 13.2  The expanding Universe

The origins of the Universe have long been the topic of thought and conjecture. Once the enormous size of galaxies and the distances between them came to be realised in the early 20th century, astronomers had to explain how the Universe had not been contracted back into itself by the force of gravity long ago.

### Distances in the Universe

There are other commonly used astronomical distance units: the **parsec** (about 3.26 light years) and the **astronomical unit**, or AU, the distance between the Sun and Earth.

Distances in space are so huge that using normal units such as metres or kilometres becomes very difficult. The distance to Proxima Centauri, the next nearest star to Earth, is about $4 \times 10^{16}$ m, or 40 000 000 000 000 km. A **light year** is the distance that light will travel in a vacuum in 1 Earth year: $9.46 \times 10^{15}$ m. Light travels at $3.0 \times 10^8$ m s$^{-1}$ in a vacuum. At this speed, it takes a little over 1 second to reach the Moon. Light from the Sun takes 8 minutes and 20 seconds to reach Earth. The light from Proxima Centauri takes 4.3 years to get to us. Using the Hubble Space Telescope, light that has taken nearly 14 billion years to travel across the Universe has been collected and photographed.

### First predictions about the expanding Universe

Einstein published the General Theory of Relativity in 1916. He was convinced that the Universe was static. However, in 1922 Russian mathematician Alexander Friedmann (Figure 13.8a) found solutions to the equations of general relativity that suggested that the Universe could either be expanding or contracting. Independently, in 1927, Georges Lemaître, a Belgian priest, mathematician and astronomer

(Figure 13.8b), came to similar conclusions. He went further, to propose that the early Universe was like a 'primeval atom' full of all the mass in the Universe. This primeval atom expanded, spreading out all the mass into the Universe. Lemaître used empirical cosmological data to decide that the Universe must be expanding. He correctly deduced that the Universe was expanding at a speed $v$ that was proportional to the distance $D$ from Earth:

$$\frac{v}{D} = \text{constant}$$

The value of Lemaître's proposed rate of expansion was similar to the rate measured in 1929 by Edwin Hubble. Hubble's contribution to cosmology is honoured in Hubble's Law, which is now written as:

$$\frac{v}{D} = H_0$$
$$\Rightarrow v = H_0 D$$

Information and communication technology capability

Who discovered the expanding Universe?

where $H_0$ = Hubble's constant.

Einstein did not readily accept the work of Friedmann. He saw it as mathematical but without physical reality. Nor did Einstein initially accept Lemaître's expanding Universe, the original Big Bang Theory, despite Lemaître briefing him on the evidence from cosmological measurements. Eventually, in the early 1930s, Einstein accepted the expanding Universe solution, writing that Friedmann was correct. By this time Friedmann had died of typhoid.

## Finding Hubble's constant

According to Friedmann and Lemaître, the expansion rate of the Universe is greatest at greatest distance. Thus, finding ways to measure these distances became important. The speed of expansion should also have an effect on the observed radiation (spectra) emitted from celestial objects.

## Measuring astronomically large distances

Stars radiate energy in every direction. The radiation observed on Earth is from the surface of the stars. Measurements of the radiation can be used to find distances to stars.

FIGURE 13.8 **a** Alexander Friedmann; **b** Einstein and Lemaître

### Luminosity and distance

Luminosity, or absolute brightness, is the rate of energy output from the surface of a star. The difference between what is observed on Earth (apparent brightness) and what is happening at the star (absolute brightness) can then be used to measure the distance to the star. Luminosity, $L$ (watt; W; $\text{J s}^{-1}$), is related to the surface temperature, $T$ (absolute temperature, K), of a star and its surface area, $A$ (square metre; $\text{m}^2$). The relationship is called the **Stefan–Boltzmann Law**, and the constant of proportionality, $\sigma$, is called the Stefan–Boltzmann constant:

$$L = \sigma A T^4$$

where $\sigma = 5.7 \times 10^{-8}\,\text{W m}^{-2}\,\text{T}^{-4}$.

The intensity of a star's output is the luminosity per unit area of the surface:

$$I = \frac{L}{A} = \sigma T^4\,\text{W m}^2$$

The temperature of a star can be determined from its colour, and its area can be found by measuring the diameter. Thus, the absolute brightness of a star can be computed.

FIGURE 13.9 Henrietta Swan Leavitt at work

FIGURE 13.10 M31, the Andromeda galaxy, now measured as being 2.5 million light years away

FIGURE 13.11 Edwin Hubble at the 100-inch telescope on Mt Wilson, California, that he used to observe the spectrum of other galaxies

The energy radiates spherically from a star so that, at a distance $d$ (m) from a star, the energy is distributed over a surface of area $4\pi d^2$; hence, the apparent brightness $m$ (W) of a star of luminosity $L$ (W) is:

$$m = \frac{L}{4\pi d^2}$$

If you measure the apparent brightness and have found the absolute brightness, then it is possible to find the distance to the star.

## Cepheids – cosmic distance indicators

In 1912, Henrietta Leavitt analysed a group of stars, called Cepheid variables, whose apparent brightness changes periodically. Leavitt was one of many 'girls' who were employed on low wages to analyse the photographs produced by the men who 'manned' the telescopes (Figure 13.9). She showed that the period was directly related to the star's luminosity.

## Hubble's Law

Leavitt's result was well-known to Hubble when he began a series of observations of stars in the nearby Andromeda galaxy (M31) (see Figures 13.10, 13.11). In 1923, Hubble identified a Cepheid variable, which he later used to measure the distance to M31. In 1929, Hubble determined the distance to M31 to be around 900 000 light years. In 1953, Walter Baade, who worked with Hubble at the Mt Wilson telescope from 1931, showed that M31 was 2.5 million light years away.

## The expanding Universe

Hubble's calculation of the enormous distance to M31 prompted further observations of the spectra of similar objects. By carefully comparing the wavelengths of spectral lines of hydrogen from other nebulae to the wavelengths of the same spectral lines observed in the laboratory, the relative speeds of these nebulae to Earth could be found. This was done using the Doppler effect – a shortening of the wavelengths (shift towards the blue end of the spectrum) indicates an approaching nebula, while a lengthening (a shift towards the red end of the spectrum) is due to a receding source.

Hubble measured the relative speed of a small number of galaxies. He quickly realised that they are moving much faster than any known object within our own galaxy. He concluded that they were indeed separate galaxies in their own right, a long distance away from our own. He assumed that their apparent brightness was an indication of their distance. Putting the data together, he was able to show a simple relationship: in general, the further away the galaxy is, the faster it is moving away from us. Surprisingly, M31, the Andromeda galaxy, which is the closest

The Doppler effect, as applied to sound, was discussed in *Physics in Focus Year 11* chapter 10.

major galaxy to our own Milky Way galaxy, is moving towards us. This is likely due to the mutual gravitational attraction between the two galaxies.

The explanation for the expanding Universe (Figure 13.12) is that it is the space between the galaxies that is expanding, in the same way that dots drawn on the surface of a balloon move further apart when the balloon is inflated.

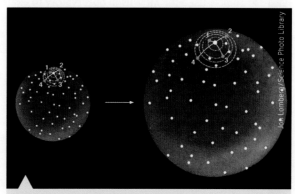

**FIGURE 13.12** As the Universe expands, the distance between galaxies increases like the distance between dots on the surface of an expanding balloon.

Data similar to Hubble's data are plotted in Figure 13.13. The recession speed is plotted against distance. The equation is in the form of Hubble's Law:

$$v = H_0 D$$

The value of the gradient is the Hubble constant $H_0$. The unit for the Hubble constant is $s^{-1}$. Thus, the inverse value of $H_0$, or $\frac{1}{H_0}$, gives the age of the Universe. Many corrections to the value of $H_0$ have been made over recent years and continue to be made. Each time the value of $H_0$ is revised, the estimated age of the Universe is revised. Currently, it is about 13.7 billion years.

If the Universe is expanding, from where did it originate? The nature of the expansion means that from wherever the Universe is being observed, it seems that it is expanding away from that point. The 'centre' of the Universe cannot be found. However, if time is wound back to the instant that everything came together, what was it that caused the Universe to form? The expanding Universe gives further support to the Big Bang Theory.

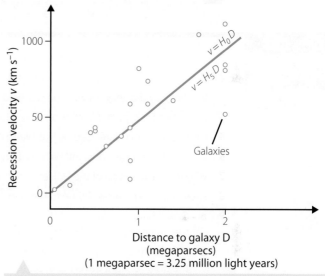

**FIGURE 13.13** Hubble's data appeared similar to that shown on this graph of the speed of recession of galaxies versus their distance away.

- Friedmann (1922) and Lemaître (1927) theoretically predicted the possibility of either an expanding or contracting Universe as a consequence of Einstein's Theory of General Relativity.
- Lemaître used evidence to support the expanding Universe hypothesis and deduced Hubble's Law, including the value of the Hubble constant, $H$, two years before Hubble:
  $v = H_0 D$
  where $v$ = recession speed
  $D$ = distance to a distant celestial object, such as a galaxy
  $H_0$ = Hubble's constant, $s^{-1}$; so $\frac{1}{H_0}$ is the age of the Universe.
- Leavitt used Cepheid variables to calculate luminosity and hence measure the distance to galaxies.

WS

The expanding Universe

»

- Hubble used calculations from Cepheid variables and apparent brightness to measure the distances to other galaxies.
- Hubble used the Mt Wilson observatory to examine the spectra of nebulae that he soon realised were distinct, separate galaxies.
- Hubble and others measured the wavelengths of hydrogen from galaxies, which showed redshifted spectra (distant galaxies moving away from Earth) and blueshifted spectra ('local' galaxies moving towards Earth)
- The speed of galaxies moving away from Earth, the recession speed of these galaxies, was found to be greater with increasing distance.
- All regions of space are moving away from each other, just like the distance between dots on the surface of an expanding balloon.
- Hubble's Law, which is based on theory and observations, is strong evidence in favour of the BBT.
- The luminosity, or absolute brightness, of a star is the rate at which energy is emitted from its surface:

$$L = \sigma A T^4$$

where $\sigma = 5.7 \times 10^{-8}\,\mathrm{W\,m^{-2}\,T^{-4}}$.

- The apparent brightness of a star follows an inverse-square law:

$$m = \frac{L}{4\pi d^2}$$

**CHECK YOUR UNDERSTANDING**

**13.2**

1  Why was a static Universe deemed not possible?

2  Compare Friedmann's prediction about the Universe with Lemaître's prediction.

3  In what way did Leavitt's work on Cepheid variables influence Hubble?

4  Describe the Doppler effect as observed by Hubble.

5  Outline the procedure used by Hubble to estimate the speed of other galaxies that he observed.

6  What is the 'Hubble constant'? Why is it important?

7  Why is the expansion of the Universe not observed in everyday life?

8  How does the reversal of time to the origin of the Universe support the Big Bang Theory?

 Numeracy

9  **a**  Given that the value of the Hubble constant $H_0$ is the inverse of the age of the Universe in seconds, find the numerical value of $H_0$. Hint: The age of the Universe is approximately 13.7 billion years.

   **b**  How fast would a galaxy that is 5 billion light years away be moving away from us? Hint: 1 light year is the distance light travelling at $3.0 \times 10^8\,\mathrm{m\,s^{-1}}$ travels in 1 Earth year.

10  The Sun is $1.496 \times 10^8\,\mathrm{km}$ from Earth. The energy reaching Earth's upper atmosphere is measured as $1373\,\mathrm{W\,m^{-2}}$. Calculate the luminosity of the Sun.

11  The luminosity of Rigil Kentaurus is $8.36 \times 10^{26}\,\mathrm{W}$. Its apparent brightness is very much smaller, at $3.99 \times 10^{-8}\,\mathrm{W\,m^{-2}}$. How far is Rigil Kentaurus from Earth?

# 13.3 The production of spectra

Information about celestial objects comes from the radiation they emit. Leavitt and Hubble used the periodicity of the light from Cepheid variables to calculate the luminosity and hence the distance to other galaxies. Hubble used spectral analysis to show how fast the Universe is expanding.

A spectrum is observed by allowing the light from a source to pass through a device that spreads the wavelengths apart. A triangular prism used for the dispersion of white light into the colours of the rainbow – and indeed raindrops, which produce rainbows – are examples of the production of a spectrum

9780170409131

(Figure 13.14). The human brain interprets a combination of colours and wavelengths as one resultant colour, or white if the right combination of colours is present. This is why spectra cannot be observed with the human eye alone.

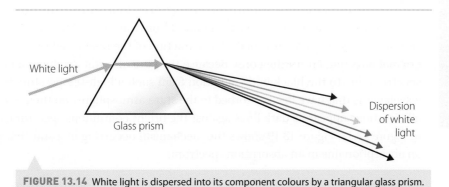

**FIGURE 13.14** White light is dispersed into its component colours by a triangular glass prism.

## Emission spectra

An **emission spectrum** is produced when low pressure gas atoms are heated or 'excited' by other means such as a strong electric field. Electrons in the atoms absorb the energy and 'jump' to a higher energy level, as shown in Figure 13.15.

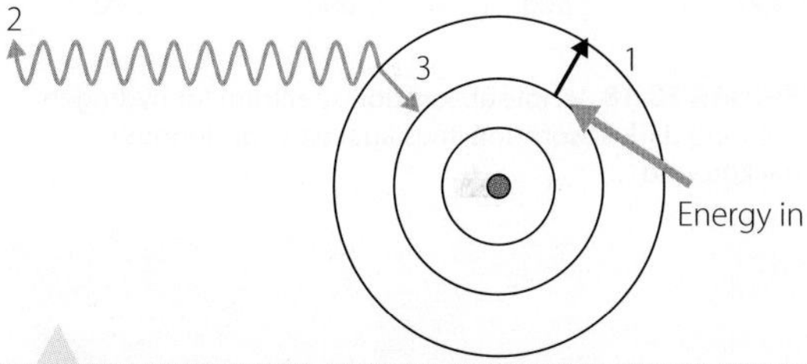

Niels Bohr, in his model of the atom, described the allowable orbits of electrons in terms of energy levels that electrons could jump between. When moving up to higher energy levels, the electron would absorb an amount of energy equal to the energy difference between the two levels. The excited atom returns to its normal or ground state when the electron loses energy by emitting a photon of light and 'falling' to a lower allowed energy level. The frequency of the emitted photon is determined by the equation $E = hf$ (see chapter 11), where $E$ is the energy difference between the two allowed energy levels the electron moves between, and $h$ is Planck's constant. As the allowed energy levels are fixed for a particular element, only certain frequencies, characteristic of that element, can be emitted.

The release of the absorbed energy by an electron occurs only at certain frequencies, so the observed spectrum has bright lines against a dark background (Figure 13.16). Useful emission spectra sources include gas discharge tubes, fluorescent light tubes, and sodium or mercury vapour street lights such as the light shown in Figure 13.17.

**FIGURE 13.15** An electron jumps up an energy level (1) when it absorbs energy. It releases the energy as a photon of light (2) with a set frequency when it returns to its original energy level (3).

This concept involving the absorption and emission of spectral lines is also discussed in chapters 9 and 15.

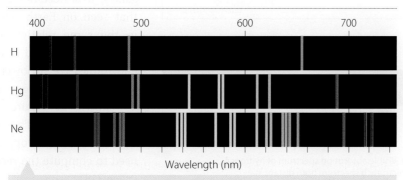

**FIGURE 13.16** Emission spectra for hydrogen, mercury and neon showing the bright emission lines against a dark background.

FIGURE 13.17 A low pressure sodium vapour street light

## Absorption spectra

An **absorption spectrum** is produced when electrons in an atom, ion or molecule in the atmosphere of a star absorb radiation at set wavelengths. The absorbed wavelengths are determined by the differences in the energy levels that the electrons jump between. The absorbed wavelengths, originally emitted from the core of the star, are re-emitted very soon after they are absorbed. Only a fraction of the re-emitted radiation is in the original direction (from the core). Because the core of a star produces a continuous spectrum due to the black body radiation from such a high-temperature source, the wavelengths that have been absorbed in the star's atmosphere and then re-emitted in all directions appear as dark lines against the bright continuous spectrum, as shown in Figure 13.18. Figure 13.19 shows the mechanism occurring in a star that produces an absorption line in an absorption spectrum.

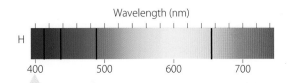

FIGURE 13.18 Visible absorption spectrum for hydrogen showing dark absorption lines against a continuous background

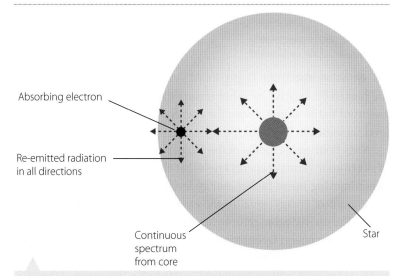

FIGURE 13.19 The production of an absorption line within a star

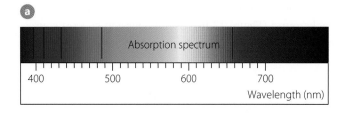

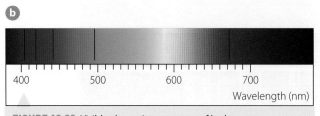

FIGURE 13.20 Visible absorption spectrum of hydrogen: **a** laboratory spectrum; **b** spectrum from distant star

## Hubble and redshifted spectra

When Hubble viewed the hydrogen spectrum emitted from a distant galaxy, he noticed that it differed from that seen on Earth. The lines were in the same relative positions but had all been shifted towards the red end, indicative of lower frequencies and therefore longer wavelengths, as shown in Figure 13.20. They were redshifted. The **redshift** was attributed to the Doppler effect and used to compute the recession speed (page 318).

9780170409131

## Continuous black body spectra

Continuous spectra are produced from hot bodies, like the tungsten filament in an incandescent light globe. The peak temperature is the energy emitted by most of the surface. But there are other emissions from the surface at higher and at lower energies. There is, therefore, a spread of energies being emitted. The typical spread of energies on a graph of intensity vs wavelength is known as a Planck curve (Figure 13.21). As the temperature of the body increases, the peak wavelength of the radiation becomes shorter. This causes the colour of the object to change from red through to orange, yellow and then white. This is expressed quantitatively in Wien's Displacement Law:

$$\lambda_{max} = \frac{b}{T_{max}}$$

where $b = 2.9 \times 10^{-3}\,\text{mK}$.

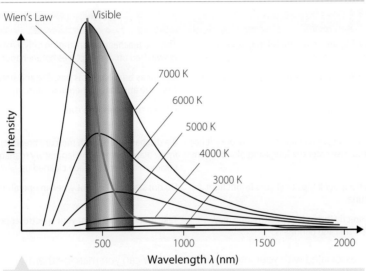

FIGURE 13.21 Planck curve for hot bodies at different peak temperatures. The red line maps the peak wavelengths according to Wien's Law.

The core of a star, a region of dense nuclei heated to many million degrees kelvin, is a source of a continuous spectrum, like that shown in Figure 13.22.

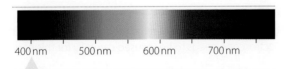

FIGURE 13.22 An example of a continuous spectrum

When the peak wavelength of a continuous spectrum corresponds to green, the object appears white, not green. This is because at such a temperature there is a significant amount of blue being emitted. Our brain interprets this colour mixture as white.

Wien's Law is discussed in more depth in chapter 11.

The production of spectra

## Teacher demonstration: Investigating spectra

Critical and creative thinking

### AIM

To observe and compare examples of emission, absorption and continuous spectra

### MATERIALS

- Hand-held spectroscopes
- 12V power pack
- Induction coil
- Discharge tubes with various gas contents
- High intensity sodium vapour discharge lamp
- Incandescent globe
- Alligator leads

RISK ASSESSMENT

| WHAT ARE THE RISKS IN DOING THIS INVESTIGATION? | HOW CAN YOU MANAGE THESE RISKS TO STAY SAFE? |
|---|---|
| High voltage produced by an induction coil may cause a severe electric shock. | This is a teacher demonstration only. Ensure the power source is switched off at two places before handling. |
| The spark from a high voltage induction coil produces X-ray radiation. | Use for as brief a time as possible and stand back at least 3 m from the apparatus when in operation. |
| Glass discharge tubes are delicate and may break easily, releasing small shards of sharp glass. | Handle the tubes with care. |
| Glass discharge tubes may contain mercury which, once it has escaped, vaporises and may enter the lungs and bloodstream. | If a discharge tube with mercury breaks or cracks, immediately inform the laboratory managers so that it can be removed and any spillage cleaned up. |
| High intensity sodium vapour lamps emit very bright light that can harm the retina. | Do not stare at the light when in operation. |
| Observing sunlight directly through a spectroscope may cause burning of the retina and blindness. | Take great care to point your spectroscope away from the Sun. |

What other risks are associated with your investigation and how can you manage them?

### METHOD

1 With teacher guidance, connect the power pack to the induction coil and test its operation.
2 Darken the room to a level where the apparatus can still be operated safely.
3 Connect a gas discharge tube to the high voltage output of the induction coil and, through the spectroscope, observe the spectrum of several different gases produced.
4 Observe the spectrum emitted from a sodium vapour lamp.
5 Connect an incandescent globe and observe the spectrum it produces.
6 Finally, go outside and observe the spectrum of reflected sunlight without directing the spectroscope towards the Sun. The spectroscope needs to be focused carefully for this part.

### RESULTS

Sketch what you observe through the spectroscope or, if possible, photograph the spectra observed.

### DISCUSSION

1 Compare the spectra produced by each of the different gas discharge tubes. How are they similar and how are they different?
2 Compare the three different types of spectra observed.
3 Discuss how the way each spectrum is produced determines the type of spectrum produced.

### CONCLUSION

With reference to the data obtained and its analysis, write a conclusion based on the aim of this investigation.

- Light from a source can be dispersed into a spectrum in much the same way that sunlight is dispersed into a rainbow.
- Emission spectra are produced by the excitation of a low density gas and contain bright lines against a dark background.
- Absorption spectra are produced by stars and show dark lines against a continuous background spectrum.
- A continuous spectrum, or Planck curve, is produced by a hot body such as the tungsten wire in an incandescent globe.
- Wien's Displacement Law:

$$\lambda_{max} = \frac{b}{T_{max}}$$

where $b = 2.9 \times 10^{-3} \, \text{m K}$

1 When we see the colours in a rainbow, what are we observing?

2 Why do we not see a spectrum when we see sunlight directly?

3 Name the three different types of spectrum.

4 What conditions give rise to an emission spectrum?

5 Why do most stars have an absorption spectrum?

6 What type of spectrum does a glowing hot object produce?

7 A star has a peak temperature of $3.0 \times 10^4$ K. What is its peak wavelength?

# 13.4 Features of stellar spectra – classifying stars

The classification of stars by their spectra began before astronomers fully understood the link between the patterns within the spectrum and the surface temperature of the star (see chapter 9). In 1814, Joseph Fraunhofer carefully studied the hundreds of absorption lines present in the Sun's spectrum.

## Spectral class and surface temperature

The key features of a star's spectrum used by astronomers when classifying a star include the appearance and intensity of spectral lines, the relative width of certain absorption lines, and the wavelength at which peak intensity occurs. The apparent colour of the star is determined by its surface temperature, as discussed in chapter 9 and shown in Figure 13.23.

The strongest Balmer series absorption lines occur for a surface temperature of about 10 000 K (class A). Cooler red stars exhibit very weak Balmer series lines in their spectra (type M). Very hot stars have no discernable Balmer series lines (type O). The reason for the lack of hydrogen lines in stars above 20 000 K is that hydrogen is completely ionised. That is, atomic electrons responsible for the production of the hydrogen lines in the spectrum have been stripped from the proton/nucleus.

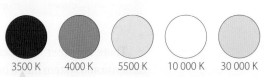

3500 K   4000 K   5500 K   10 000 K   30 000 K

FIGURE 13.23 The colour of a star is determined by its surface temperature.

Originally when stars were classified the spectral types were simply placed in the order hottest to coolest and simplified to eliminate overlapping and confusing spectra. This work was mainly done at Harvard University from 1918 to about 1924, most significantly by Annie Jump Cannon (1863–1941). Cannon was a university-educated astronomer and physicist who was employed at Harvard University, along with a number of other women, to catalogue stars from photographs. Over her 43-year career, Cannon became a well-respected astronomer whose stellar classification work was published each year. She classified over 225 000 stars and discovered 300 variable stars and five novae. She could classify stars at a rate of three per minute.

The classification system was originally based on the strength of the Balmer lines within the hydrogen absorption spectrum from the surface. The spectral types were given letter names:

O, B, A, F, G, K, M

Table 13.3 shows Cannon's Harvard system of spectral types, along with their characteristic colours and temperatures. It includes later additions of L and T spectral types. Some examples of stars for each spectral type are also listed.

**TABLE 13.3** Colour and temperature of stars using the spectral type classification

| SPECTRAL TYPE | COLOUR | PEAK WAVELENGTH (nm) | SURFACE TEMPERATURE (K) | BALMER LINE FEATURES | OTHER SPECTRAL FEATURES | EXAMPLES |
|---|---|---|---|---|---|---|
| O | blue | 72 | >30 000 | weak | Ionised $He^+$ lines, strong UV continuum | Orionis C |
| B | light blue | 145 | 11 000–30 000 | medium | Neutral He lines | Achernar, Rigel, Spica |
| A | white | 290 | 7500–11 000 | strong | Strong H lines, ionised metal lines | Sirius, Vega |
| F | yellow-white | 380 | 6000–7500 | medium | Weak ionised $Ca^+$ | Procyon, Canopus |
| G | yellow | 530 | 5000–6000 | medium | Ionised $Ca^+$, metal lines | Sun, Capella |
| K | orange | 725 | 3500–5000 | very weak | $Ca^+$, Fe, strong molecules, CH, CN | Arcturus, Aldebaran |
| M | red | 960 | <3500 | very weak | Molecular lines, e.g. TiO, neutral metals | Betelgeuse, Antares |
| L | orange-brown | 1200 | <2700 | (–) | Ionised $He^+$ lines, strong UV continuum | GD 165B |
| T | brown | 4000 | <900 | (–) | Neutral He lines | Gleise 229B |

There are other classification systems. For example, the Morgan–Keenan, or MK, system is a multi-dimensional system based on the Harvard classification. The dimensions are temperature and luminosity. Within a class, temperatures are given numbers from 0 (hottest) to 9 (coolest). Luminosities are denoted by Roman numerals: Ia, Ib, II, III, IV and V. In MK, for example, the Sun's classification is G2V.

The luminosity of stars is plotted against the re-organised order, O B A F G K M, on the Hertzsprung–Russell diagram (Figure 13.24).

The advent of infrared astronomy, possible with satellite-based telescopes, has led to recent modification of the spectral types classification. Stars previously too cool to classify (because they

9780170409131

were not detectable by light telescopes) are now assigned the spectral types R, N or S, depending on the elements present in their spectra. Other additions include the WR (Wolf-Rayet) and the T (T Tauri) categories.

The thickness of the absorption lines visible in a star's spectrum gives an indication of the pressure of the gases in the star's atmosphere. Small, compact stars have a relatively dense atmosphere with higher pressure due to the higher gravitational acceleration near the surface of the star. Larger, giant stars have lower pressure and therefore lower density of gas near their surfaces. Despite their greater mass, the much greater radius of the star means that the surface of the star is a greater distance from the centre. Gravitational force obeys an inverse-square relationship with the distance from the centre of the star and is much weaker at the surface for giant stars compared to stars of smaller radius.

A 'white dwarf' is the remnant of a star in the final stages of cooling down after its nuclear fuel has been depleted. Despite their relatively high surface temperature, they are very dim due to their size, about the same as that of Earth. These stars are no longer fusing nuclei in their core, unlike 'dwarf' and 'sub-dwarf' stars, which are so named due to their comparatively small size.

Features of stellar spectra – classifying stars

**CHECK YOUR UNDERSTANDING**

**13.4**

1   Why are the letters assigned to spectral class not in alphabetical order?
2   From hottest to coolest, what is the order of spectral classes?
3   Approximately what is the colour of a star with a surface temperature of 10 000 K?
4   Why do larger stars such as red giants have lower pressure in their atmospheres?
5   How does the width of spectral lines give astronomers information regarding the nature of a star?

## 13.5  The Hertzsprung–Russell diagram

Naked-eye astronomers invented a scale where the brightest stars were assigned the magnitude value 1. Other stars that were less bright were assigned positive values: magnitudes 2 and 3 were the second and third brightest. Brighter stars have negative magnitudes; fainter stars have higher positive magnitudes. Magnitude and luminosity are different ways of reporting the actual and observed brightness of stars.

The Hertzsprung–Russell diagram, or H–R diagram, is named after Ejnar Hertzsprung, a Danish astronomer who, in 1911, plotted stars' absolute magnitude against their colour. Two years later an American astronomer, Henry Russell, plotted the absolute magnitude of stars against their spectral class. Both astronomers found that stars tended to fall into distinct groupings on their diagrams. An example of an H–R diagram is shown in Figure 13.24.

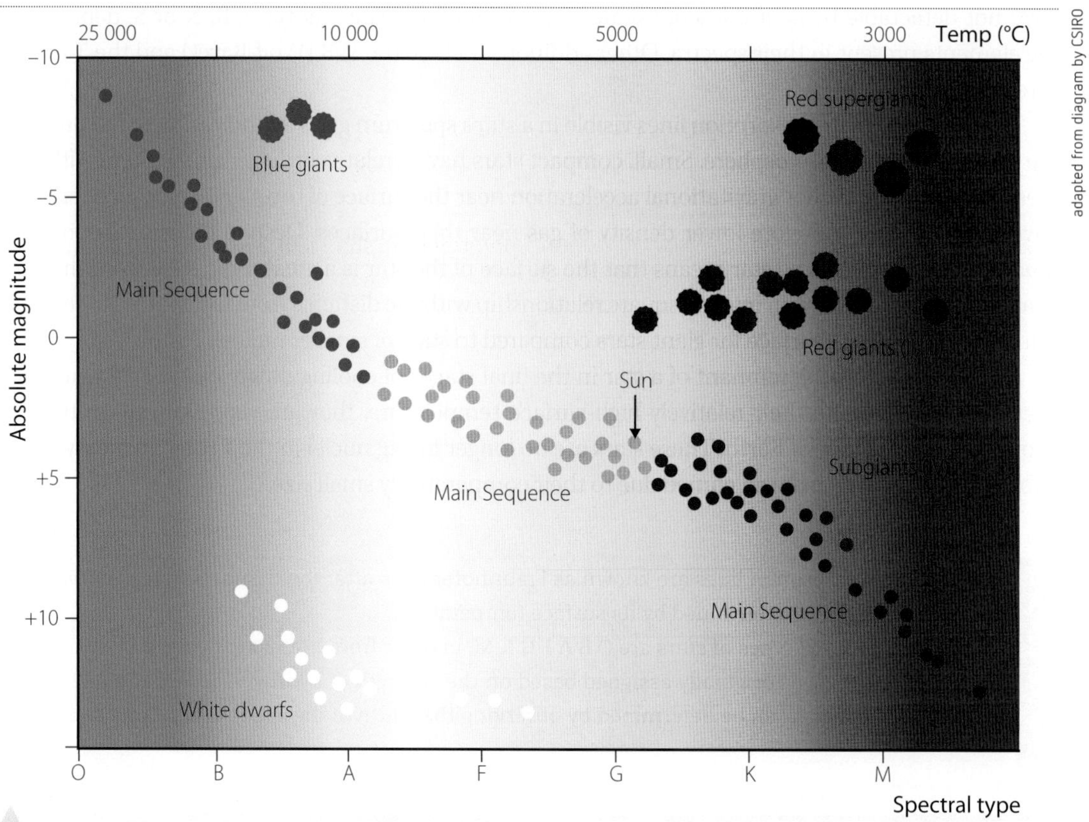

adapted from diagram by CSIRO

**FIGURE 13.24** A Hertzsprung–Russell diagram with luminosity classes shown. The regions where most stars lie are clearly evident. The spectral type scale shows peak wavelengths increasing from left to right but peak energy from right to left.

## Star groups in the H–R diagram

Most of the stars plotted on an H–R diagram lie within certain regions.

### Main sequence

Most stars observed fall into a group known as the main sequence, with hotter stars being more luminous. This groups extends from the upper left to the lower right of an H–R diagram. It contains stars such as our Sun that are fusing hydrogen to helium in their cores.

### Red giants

Other groups of stars include the red giants, so named because, despite being relatively cool, they are very luminous. This means that they must be very large, an inference confirmed by the luminosity class assigned by observation of the thickness of their spectral lines. Red super giants are even less common and are more luminous than red giant stars.

### White dwarfs

White dwarf stars have low luminosity despite the relatively high surface temperatures that make them white. These stars are very small and are believed to be the collapsed remnants of stars like our Sun that have depleted their reserves of elements that can undergo fusion.

9780170409131

## Blue giants

These rare, short-lived stars are very luminous and hot, but are not main sequence stars because they are fusing heavier elements in their cores.

## Key stages in a star's life

Stars go through an evolutionary life cycle. Before conditions are sufficient for fusion in their cores, a hot ball of gas collapses under gravity.

The stages of a star's life can be summarised in a flow chart (Figure 13.25).

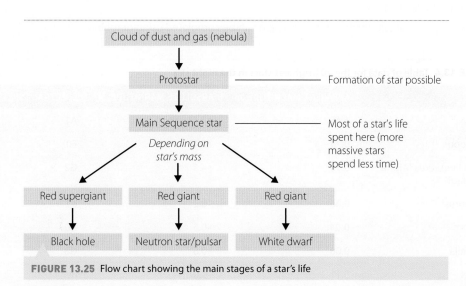

**FIGURE 13.25** Flow chart showing the main stages of a star's life

The protostar stage, in which gravity is pulling in more material from the surrounding gas and dust cloud, does not involve nuclear reactions. The source of energy is the transformation of lost gravitational potential energy from the material. It is not until the commencement of nuclear fusion of hydrogen into helium within its core that the star is truly 'born'. For the star to remain stable on the main sequence, there must be an equilibrium between the outward radiative force and the pressure of the gas in the star against the inward force of gravity. The greater the mass of the star, the greater the force of gravity, allowing greater density and temperature in the core. This in turn results in the rate of the nuclear fusion reactions being greater, so that the surface temperature of the star is higher. Stars with the same mass as the Sun have surface temperatures of about 5850 K, while stars with about ten solar masses are very hot blue-white stars of around 20 000 K or more. These stars may consume their nuclear fuel in only a few tens of millions of years, which is why they are rare. The least massive stars of about 0.1 solar masses are small red main sequence stars, consuming their nuclear fuel so slowly that it is believed they may be as old as the Universe itself.

The following stage of a star's life occurs when the hydrogen fuel has been depleted to the extent that the core of the star collapses. Gravity takes over, elevating the core's density and temperature. The layer of helium that has formed around the core is compressed to such an extent that the fusion of helium into carbon begins – the 'helium flash'. Such processes, known as post-main sequence nuclear reactions, are discussed in more detail in Section 13.6.

ws

The Hertzsprung–Russell diagram

## Using the H–R diagram to classify stars

Plotting the 20 brightest stars in our night sky yields a pattern on an H–R diagram that is very different from the result when the 20 closest stars are plotted.

# INVESTIGATION 13.2

## Plotting stars on an H–R diagram

Critical and creative thinking

Numeracy

Information and communication technology capability

**AIM**

To compare the results of plotting the 20 brightest and the 20 closest stars on an H–R diagram

**MATERIALS**

- Data in Tables 13.4 and 13.5
- Pen, paper, ruler (although this investigation can be completed using graphing software)

**TABLE 13.4** Table of data for the 20 brightest stars in the night sky

| STAR | APPARENT MAGNITUDE, *m* | ABSOLUTE MAGNITUDE, *M* | SPECTRAL TYPE |
|---|---|---|---|
| Sirius | 1.45 | +1.41 | A1 |
| Canopus | 0.73 | +0.16 | F0 |
| Rigil Kentaurus (Alpha Centauri A) | 0.10 | +4.3 | G2 |
| Arcturus | 0.06 | −0.2 | K2 |
| Vega | 0.04 | +0.5 | A0 |
| Capella | 0.08 | −0.6 | G8 |
| Rigel | 0.11 | −7.0 | B8 |
| Procyon | 0.35 | +2.65 | F5 |
| Achernar | 0.48 | −2.2 | B5 |
| Hadar | 0.60 | −5.0 | B1 |
| Altair | 0.77 | +2.3 | A7 |
| Betelgeuse | 0.80 | −6.0 | M2 |
| Aldebaran | 0.85 | −0.7 | K5 |
| Acrux | 0.9 | −3.50 | B2 |
| Spica | 0.96 | −3.4 | B1 |
| Antares | 1.0 | −4.7 | M1 |
| Pollux | 1.15 | +0.95 | K0 |
| Fomalhaut | 1.16 | +0.08 | A3 |
| Deneb | 1.25 | −7.3 | A2 |
| Mimosa | 1.26 | −4.7 | B0 |

**TABLE 13.5** Table of data for the 20 closest stars in the night sky

| STAR | DISTANCE (ly) | APPARENT MAGNITUDE | ABSOLUTE MAGNITUDE | SPECTRAL TYPE |
|---|---|---|---|---|
| Proxima Centauri | 4.2 | 11.1 | 15.5 | M5 |
| Rigil Kentaurus | 4.3 | −0.01 | 4.4 | G2 |
| Alpha Centauri B | 4.3 | 1.33 | 5.7 | K1 |

»



9780170409131

| STAR | DISTANCE (ly) | APPARENT MAGNITUDE | ABSOLUTE MAGNITUDE | SPECTRAL TYPE |
|---|---|---|---|---|
| Barnard's Star | 6.0 | 9.54 | 13.2 | M4 |
| Wolf 359 | 7.7 | 13.5 | 16.7 | M6 |
| BD +362147 | 8.2 | 7.5 | 10.5 | M2 |
| Luyten 726-8A | 8.4 | 12.5 | 15.5 | M6 |
| Luyten 726-8B | 8.4 | 13.0 | 16.0 | M6 |
| Sirius A | 8.6 | −1.46 | 1.4 | A1 |
| Sirius B | 8.6 | 8.3 | 11.2 | A |
| Ross 154 | 9.4 | 10.45 | 13.1 | M4 |
| Ross 248 | 10.4 | 12.29 | 14.8 | M5 |
| Epsilon Eridani | 10.8 | 3.73 | 6.1 | K2 |
| Ross 128 | 10.9 | 11.1 | 13.5 | M4 |
| 61 Cygnus A | 11.1 | 5.2 | 7.6 | K4 |
| 61 Cygnus B | 11.1 | 6.0 | 8.4 | K5 |
| Epsilon Indi | 11.2 | 4.7 | 7.0 | K3 |
| BD +4344A | 11.2 | 8.1 | 10.4 | M1 |
| BD +4344B | 11.2 | 11.1 | 13.4 | M4 |
| Procyon A | 11.4 | 0.4 | 2.6 | F5 |

## METHOD

1  Draw axes for your H–R diagram. These should be redrawn from the example shown in Figure 13.26, using about half an A4 page.

2  Use the data in Table 13.4 to plot the positions of the 20 brightest stars in the night sky. Use the absolute magnitude column and the spectral type columns. Note that each spectral type is divided into ten sub-types, for example A0, A1, A2 … A9.

3  On the same axes but using a different colour for the data points, use the data in Table 13.5 to plot the positions of the 20 closest stars in the night sky.

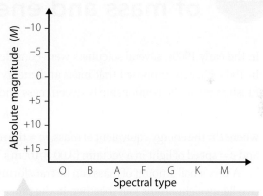

**FIGURE 13.26** An H–R diagram showing axes labelled as absolute magnitude (M) and spectral type

## RESULTS

1  Retain the resulting plots for your notes.

2  Identify the main sequence, red giant and white dwarf stars by drawing a line around each region on the H–R diagram.

## DISCUSSION

1  Compare and contrast the plots obtained from the two data sets.

2  Which data set would be a more valid random sample of stars in our night sky? Justify your response.

3  Suggest reasons why the data set from Table 13.5 contains a predominance of very luminous stars and very few small main sequence stars.

## CONCLUSION

With reference to your results, write a conclusion based on the aim of this investigation.

- An H–R diagram is used to plot stars' absolute magnitude against spectral class.
- The main regions on an H–R diagram are main sequence, red giants and supergiants, and white dwarfs.
- Luminosity is the amount of light being given off by a star. Astronomers represent luminosity by a number called absolute magnitude.
- Main sequence stars fuse hydrogen in their cores and have a luminosity class of V.
- Three main regions of stars are found on H–R diagrams: main sequence, red giants and white dwarfs.
  - Red giant stars are fusing heavier elements and have a luminosity class of I.
  - White dwarf stars are the hot, condensed remnants of stars that have depleted their nuclear fuel.

## CHECK YOUR UNDERSTANDING

### 13.5

1 Explain why red giant stars are so luminous.

2 In which region of an H–R diagram are most stars found?

3 Explain why a red giant is found in the upper right region on an H–R diagram.

4 White dwarf stars are found in the lower left of an H–R diagram. Given that they are very hot, what feature causes these stars to have low luminosity?

5 What can be said about the evolutionary age of a star found in the lower left of an H–R diagram?

6 Why are luminous stars over-represented in a plot of the 20 brightest stars in the night sky?

7 Why are red main sequence stars relatively dim?

## 13.6 Einstein's equivalence of mass and energy

Einstein explains the equivalence of energy and mass

The American Museum of Natural History articles on $E = mc^2$

The concept of mass defect was introduced in chapter 12.

Nuclear fission interactive
Use this to deepen your understanding.

$E = mc^2$ is treated quantitatively in chapters 12 and 16.

In the early 1900s, several scientists were considering a possible connection between mass and energy. In 1905, Einstein proposed that mass and energy could be transformed from one to the other. In 1905, Einstein made the relationship between mass and energy quantifiable in the equation:

$$E = mc^2$$

where $E$ = the energy equivalent of mass, $m$ = the mass of an object at rest in an inertial frame of reference, and $c$ = speed of light in a vacuum ($3.00 \times 10^8 \, \text{m s}^{-1}$).

A very small amount of mass can be transformed into a very large amount of energy. In nuclear fusion, two lightweight atoms join together. In the process, a small loss of mass, called the mass defect, occurs. Multiplying this mass defect by very large numbers of fusion events can produce enormous amounts of energy. Similarly, and as it turns out, more readily, when a heavy atom splits into two or more fragments, there is a mass defect. Again, the large number of atoms in even quite small masses can multiply this fission effect enormously. The enormity of an uncontrolled fission explosion was demonstrated at Hiroshima and Nagasaki in 1945.

The most precise confirmation of the change in the mass of a radioactive nucleus when it emits a gamma ray was made in 2005. It showed Einstein's equation $E = mc^2$ to be accurate to within $\frac{1}{400\,000}$ %.

### The energy source of stars

In the 19th century, scientists believed that the Sun was powered by chemical reactions similar to combustion reactions observed on Earth. Simple calculations based on the amount of solar energy striking each square metre of the Earth's surface and the distance from the Sun to Earth showed that a huge amount of energy was being released from the Sun every second. If this energy were indeed from chemical reactions, the fuel, or reactants, would be consumed in a very short time. It was not until the 1920s that nuclear fusion was proposed as the Sun's energy source. The energy equivalent of the mass defect was found to be sufficient to power the Sun for many billions of years.

On Earth, uncontrolled fusion reactions have been produced in thermonuclear bombs by using a fission reaction to cause hydrogen nuclei to collide with sufficient energy to undergo fusion. Controlled fusion reactors, which are predicted to be much safer than fission reactors, are still the subject of ongoing research and development.

Critical and creative thinking

Nuclear fusion, mass defect and energy transformations will be treated quantitatively in chapter 16.

**KEY CONCEPTS**

- Einstein's equation $E = mc^2$:
  - predicted that mass could be converted into energy
  - showed that mass and energy are equivalent.
- Mass defect occurs in nuclear reactions, including fusion and fission, and is evident as energy.
- A very small mass defect can be converted into a large energy release because of the enormous numbers of possible reactions in a very small amount of material.
- Nuclear fusion occurs when two lightweight nuclei stick together to form a new nuclide.
  - Nuclear fusion is the energy source for stars in the core of stars.
- Nuclear fission occurs when a heavy nucleus is split into smaller nuclides.
  - Nuclear fission is the source of controlled energy for nuclear reactors, such as power stations, and uncontrolled energy in bombs.

WS

Einstein's equivalence of mass and energy

**CHECK YOUR UNDERSTANDING**

**13.6**

1 Write Einstein's mass energy equation in symbols. Define each term.
2 Define 'mass defect'.
3 Outline the reason why nuclear fusion and nuclear fission can produce so much energy.
4 Compare the processes of nuclear fusion and nuclear fission.
5 Why were chemical reactions ruled out as being the energy source for stars such as the Sun?

## 13.7 Nucleosynthesis in stars

According to the BBT, favourable conditions for nuclear fusion occurred from about 3 minutes after the start of the Universe. This process, nucleosynthesis, caused protons to combine to form a small range of lightweight nuclides, mainly isotopes of hydrogen, helium and some lithium, the three lightest elements. Matter began to clump together under the influence of the gravitational force, which causes masses to be attracted to each other. Given the vast time scales involved, these clumps became giant clouds of matter inside which stars were born. In the core of stars, new nuclides were synthesised by nuclear fusion. Nucleosynthesis in stars is possible because of the enormous pressure and very high temperatures.

Every element has been synthesised in the core of stars. Fusion reactions that produce elements up to and including iron, element 26 in the periodic table, are self-sustaining because they release more energy than is required to ignite them. The synthesis of elements heavier than iron is not self-sustaining, and is believed to occur only within the core of supernovas, that is, large stars that collapse and then explode, and in even more extreme events such as collisions between neutron stars.

### Main Sequence stars

Most of the stars in the Universe are Main Sequence stars, as is our Sun. These stars are in the stable period of their life cycles and are fusing hydrogen into helium in their cores. They can vary in size from red dwarf stars, about one-tenth the diameter of the Sun, to very large, blue-white stars, which are very short lived. Main Sequence stars, when plotted on an H–R diagram (Figure 13.24, page 328), lie within a band stretching from the upper left to the lower right. It is not a sequence as such, but the region can easily be wrongly interpreted as one. It is believed that, due to the relationship between mass, luminosity and size of all Main Sequence stars, they have a common nuclear energy source – the fusion of hydrogen nuclei into helium nuclei.

A star is composed initially of hydrogen nuclei, which are protons $\left(^1_1H\right)$. A helium nucleus comprises two protons and two neutrons $\left(^4_2He\right)$. Mostly, when protons approach each other they experience a glancing collision and continue on as protons. In order to synthesise a helium nucleus, four hydrogen nuclei undergo fusion. Two of these protons must then decay to neutrons. The probability of a simultaneous fusion of four hydrogen nuclei is so small that such a mechanism can be discounted. The energy required for this collision to be successful is also too high for it to be considered as contributing to a star's energy output.

Fusion is believed to occur via two main processes, involving a series of steps.

## The proton–proton chain

At the temperatures and pressures present in the core of stars, the most likely pathway is the proton–proton chain. This involves the collision of only two particles at a time, an event with much greater probability than the collision of four particles and therefore occurring far more frequently.

Two separate fusions of protons produce deuterium $\left(^2_1H\right)$, which has one proton and one neutron. The neutron arises because one of the protons decays by positron emission to a neutron. For energy conservation reasons, an electron neutrino is also emitted:

$$^1_1H + ^1_1H \rightarrow ^2_1H + e^+ + \nu$$

Each of the two deuterium nuclides now undergoes fusion to helium-3 by incorporating another proton. Most of the energy is released from this reaction as a gamma ray:

$$^2_1H + ^1_1H \rightarrow ^3_2He + \gamma$$

Finally, the two helium-3 nuclides undergo fusion to helium-4, which is stable; two protons are released back into the star.

$$^3_2He + ^3_2He \rightarrow ^4_2He + ^1_1H + ^1_1H$$

Figure 13.27 outlines the reactions involved in the proton–proton chain.

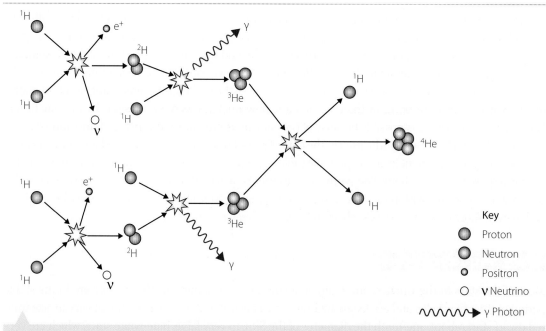

**FIGURE 13.27** Steps in the proton–proton chain

Along the way, six protons are involved in fusion reactions. Two protons are subsequently released (final fusion event), along with two positrons and two neutrinos (after the first fusion events) and two gamma rays (after the second fusion events).

The net equation for this form of the proton–proton chain is:

$$4^1_1\text{H} \rightarrow {}^4_2\text{He} + 2e^+ + 2\nu + 2\gamma$$

Neutrinos interact with matter so rarely that it is estimated that billions pass through us every second. Only a few neutrinos are detected each day in experimental neutrino capture systems. One such system involves a huge underground water tank. When a neutrino interacts with a water molecule, a small flash of light is emitted and recorded.

While there are believed to be other forms of the proton–proton chain, the above process accounts for an estimated 85% of the energy produced in the Sun. Stars with masses up to approximately 1.5 solar masses with core temperatures of up to 18 MK also produce most of their energy in this way.

## The CNO cycle

Another pathway leads to a nearly identical overall reaction, but requires the presence of carbon-12 and temperatures over $18 \times 10^6$ K to proceed. It is known as the CNO cycle.

The carbon–nitrogen–oxygen (CNO) cycle is a pathway for nuclear fusion that commences with the fusion of one proton with a carbon-12 nucleus. The carbon-12 undergoes transmutation to nitrogen-13, but it re-emerges after several more steps in the process, acting in a similar way to a catalyst in a chemical reaction. Figure 13.28 shows the steps involved in the CNO cycle.

Information and communication technology capability

The Sudbury Neutrino Observatory

The Ice Cube Neutrino Observatory in Antarctica

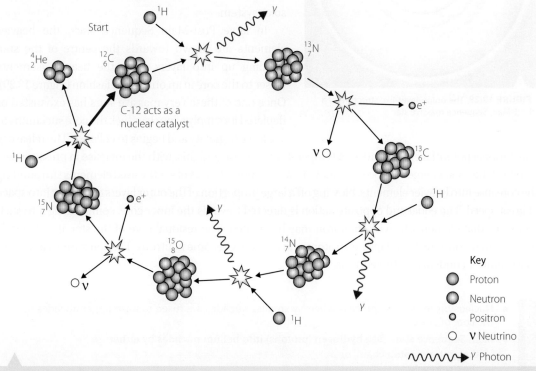

**FIGURE 13.28** The CNO cycle, showing the steps in the fusion of hydrogen to helium nuclei

It can be seen that at no stage in the CNO cycle is a collision between more than two particles required. The carbon-12 is transmutated several times until the last step when, by alpha decay, a nitrogen-15 nucleus decays back to the carbon-12 nucleus.

The net equation for the CNO cycle is:

$$4^1_1\text{H} \rightarrow {}^4_2\text{He} + 2e^+ + 2\nu + 3\gamma$$

The overall equations for the proton–proton chain and the CNO cycle are nearly identical. Four protons produce one helium nucleus, with energy being released as motion (kinetic energy) of the helium, as mass and kinetic energy of electron neutrinos and as gamma rays. The source of the energy is the slight decrease in the mass of the products, the mass defect, when compared with the reactants.

Nucleosynthesis in stars

## Post-Main Sequence stars

Once a star has consumed most of its hydrogen fuel, the core will begin to collapse and a layer of extremely hot helium nuclei, which has built up during the main sequence stage of the star's life, will surround the core. The mass of the star will determine what happens next. With sufficient mass, a star's gravitational force will be able to sustain the density necessary for helium to fuse into heavier elements such as carbon and oxygen. This process commences as the 'helium flash', resulting in the star becoming a red giant. Very massive stars are able to continue fusing elements all the way to the formation of iron. These fusion reactions are all exothermic; that is, they release energy and thus provide the outward forces of radiation, preventing the star from collapsing under its own gravity. The formation of elements heavier than iron is unsustainable, as such fusion requires a net input of energy. The existence of these heavier elements on Earth is due to supernovas or neutron star collisions that must have occurred before the formation of our solar system.

In large Post-Main Sequence stars, the heavier elements are drawn towards the centre of the star, building up into layers, with the heavier elements closer to the core in an onion-like fashion (Figure 13.29). Once one of these very massive stars has exhausted or depleted its supply of nuclear fuel, the core succumbs to the force of gravity and begins to collapse. The release of gravitational potential energy causes a temperature increase and this, with the increase in pressure, sets off one of the most energetic processes known – a supernova. Layers of synthesised elements surrounding the core fuse into heavier elements, blowing off a large proportion of the outer layers of the star into space at great speed. The equal and opposite action is directed towards the inner core, compressing it to such an extent that elements heavier than iron may be created. The residual core of the star, if sufficiently massive, may continue to collapse as protons and electrons become neutrons. The star shrinks as matter as we know it condenses into neutrons.

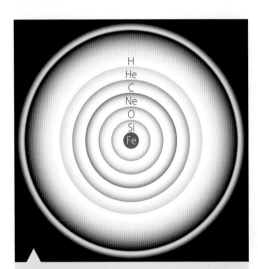

**FIGURE 13.29** The onion-layer-like structure of a Post-Main Sequence massive star

KEY CONCEPTS

- Nucleosynthesis is the process whereby smaller nuclides are fused to form larger nuclides in the core of stars.
- Main Sequence stars fuse hydrogen (protons) into helium nuclides by either:
  - the proton–proton chain, or
  - the CNO cycle.
- Post-Main Sequence stars produce nuclei of elements up to and including iron in a self-sustaining process.
- Elements heavier than iron are produced in the core of supernovas – giant stars at the end of their life cycles – or in collisions between neutron stars.

9780170409131

1 Under which conditions would a star be most likely to fuse hydrogen into helium using the proton–proton chain?

2 Why do both the proton–proton chain and the CNO cycle involve several steps in the fusion process rather than the processes being single-step?

3 Compare the elements synthesised in the cores of Main Sequence stars and Post-Main Sequence stars.

4 Why can stars synthesise elements up to and including iron, but not heavier elements?

5 What evidence do we have on Earth for the theory that a supernova must have occurred in our region of the galaxy at some time in the past?

6 Describe how our Universe would be different today if nucleosynthesis in stars did not occur.

7 Explain why it is so difficult to achieve nuclear fusion on Earth.

- The Big Bang Theory is widely accepted as explaining the origins of the Universe.

- At first the temperature of the Universe was too high for matter to exist – there was only energy.

- Within a very short period of time, the Universe cooled sufficiently for matter and antimatter to form.

- Matter and antimatter collided and annihilated each other, reverting back into energy.

- Eventually more matter was left than antimatter.

- Friedmann and Lemaître used Einstein's Theory of General Relativity to predict an unstable Universe. Lemaître used data to show that the expanding Universe is physically correct.

- Leavitt used Cepheid variable periodicity to measure astronomical distances.

- Hubble used the Mt Wilson observatory to examine very distant Cepheid variables and the spectra of nebulae.

- Hubble realised the nebula were distinct, separate galaxies.

- Hubble used spectral data (spectral redshift) to confirm Lemaître's prediction that the recession speed of distant galaxies would be greater with increasing distance.

- The space between the galaxies is increasing like the distance between dots on the surface of an expanding balloon.

- Light from a source can be dispersed into a spectrum, like sunlight is dispersed into a rainbow.

- Emission spectra are produced by the excitation of a low density gas and contain bright lines against a dark background.

- Absorption spectra are produced by stars and show dark lines against a continuous background spectrum.

- The Sun's absorption lines are known as Fraunhofer lines after the first person to observe them.

- A continuous spectrum is produced by a hot body such as a tungsten wire in an incandescent globe.

- An H-R diagram is used to plot stars' absolute magnitude versus spectral type.

- The main regions on an H–R diagram are Main Sequence, red giants and supergiants, and white dwarfs.

- Luminosity is the amount of light being given off by a star and is called absolute magnitude by astronomers.

- Spectral type as ordered from hottest to coolest is O B A F G K M.

- Spectral type is determined by the spectral lines present in the star's spectrum, which in turn is determined by the star's surface temperature.

- Surface temperature determines the colour of the star.

- Einstein's equation $E = mc^2$ predicted that mass could be converted into energy: mass and energy are equivalent.

- In nuclear reactions, the mass defect is the difference between the mass of reactants and mass of products.

- A very small amount of mass converted into energy in each reaction can be multiplied by enormous numbers of possible reactions to release a large amount of energy.

- Nuclear fission occurs when a large nucleus is split into smaller nuclei and some mass is lost, as in atom bombs:
    - Nuclear reactors utilise fission to heat water into steam to drive turbines and in turn, electrical generators.

- Nuclear fusion occurs in the core of stars.

- Nucleosynthesis is the process whereby small nuclei are fused to form larger nuclides in the core of stars.

- Main Sequence stars fuse hydrogen nuclei (protons) into helium nuclei by either the proton–proton chain or by the CNO cycle.

- Post-Main Sequence stars produce nuclei of elements up to and including iron in a self-sustaining process.

- Elements heavier than iron are produced in the core of supernovas – giant stars at the end of their life cycles.

1  Explain why matter could not form immediately after the Big Bang.

2  How do we know that more matter than antimatter formed after the Big Bang?

3  What elements were the first formed after the Big Bang?

4  Outline the process that has led to the formation of heavier elements up to and including iron.

5  Describe the extreme process in which elements heavier than iron can form.

6  What observations, made by Sir Edwin Hubble, led to the discovery of the expanding Universe?

7  Outline how cosmic microwave background radiation formed.

8  What evidence other than Hubble's observations do we have for the expanding Universe?

9  Explain how the Hubble constant, $H_0$, relates the speed of recession of a galaxy to its distance from Earth.

10  Why do we not notice the expansion of the Universe is making objects on Earth further apart?

11  The Sun emits huge amounts of energy as radiation and is believed to have been doing this for billions of years. How did this observation help give rise to the idea of the equivalence of mass and energy?

12  Compare the proton–proton chain to the CNO cycle in terms of:

   a  the overall reactions involved.

   b  the steps involved in the production of helium nuclei.

13  Why is it thought that a multi-step process such as the proton–proton chain or the CNO cycle must be the mechanism for the fusion of hydrogen in stars?

14  Outline the reason why stars produce absorption spectra.

15  What do the presence of certain spectral lines in a star's spectrum tell us about the star?

16  Describe how a star's luminosity class is determined.

17  Explain why a red giant star has a luminosity class of $I$ and the Sun has a luminosity class of V.

18  How can the evolutionary stage of a star be determined by plotting its position on an H–R diagram?

19  Describe the relationship between the surface temperature and colour of stars.

20  Account for the production of emission spectra and compare these to absorption spectra.

21  Account for the fact that the fresh fuel for a nuclear power station has greater mass than the spent fuel.

22  Elements heavier than iron are present in the Earth's crust. How does this fact suggest that before the solar system formed, a supernova event occurred in our region of the galaxy?

# 14 Structure of the atom

## INQUIRY QUESTION

How is it known that atoms are made up of protons, neutrons and electrons?

**OUTCOMES**

Students:

- investigate, assess and model the experimental evidence supporting the existence and properties of the electron, including: ICT
  - early experiments examining the nature of cathode rays
  - Thomson's charge-to-mass experiment
  - Millikan's oil drop experiment (ACSPH026)
- investigate, assess and model the experimental evidence supporting the nuclear model of the atom, including: ICT
  - the Geiger–Marsden experiment
  - Rutherford's atomic model
  - Chadwick's discovery of the neutron (ACSPH026)

Physics Stage 6 Syllabus © 2017 NSW Education Standards Authority (NESA) for and on behalf of the Crown in right of the State of New South Wales.

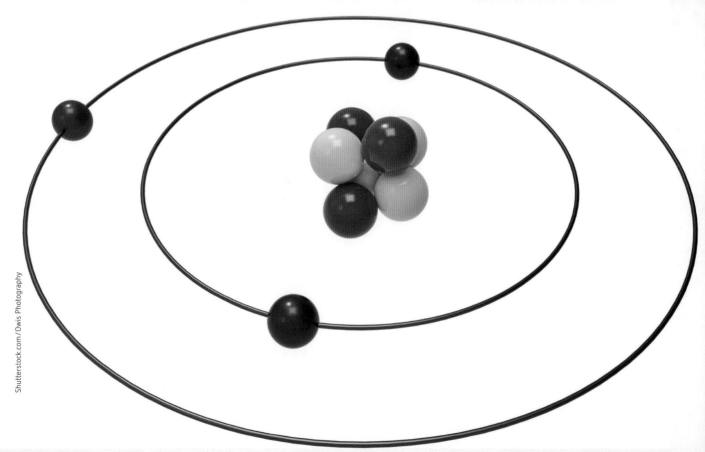

Shutterstock.com/Owis Photography

9780170409131

Our knowledge of the structure of the atom has been modified and updated as technological advances allowed new observations to be made (Figure 14.1). From the time of the Greek philosopher Democritus (460–370 BCE) to the modern era of multi-billion dollar particle accelerators, the development of the model of the atom is one of science's big stories.

The discovery of electrons, protons and finally neutrons came about from the perseverance of many scientists – both theorists and experimenters. It was

**FIGURE 14.1** Early work using cathode ray tubes has led to sweeping changes in society through the use of electronics.

not until around 90 years ago that Chadwick discovered the neutron. This subatomic particle is now the tool used by facilities such as ANSTO's nuclear reactor, OPAL, at Lucas Heights near Sydney. The unique properties of the neutron can be used to probe matter to further our understanding of atoms, nanoparticles and pharmaceuticals.

# 14.1 The electron

The discovery of the electron was made after decades of observation, inference, models of the atom and finally meticulous measurements of the smallest unit of charge, which it carries. Technological advances in the 19th century, in this case the development of a better vacuum pump, opened the door to experiments and observations that led to the electron's discovery.

Until recently, TVs used a large cathode ray tube to produce the images on the screen.

## The nature of cathode rays

In 1855 Heinrich Geissler invented the first glass vacuum tube, using his mercury air pump. In the 1870s the tube was modified by William Crookes to observe cathode rays.

A cathode ray tube (CRT), as shown in Figure 14.2, consists of an evacuated glass tube (almost all gas removed) and two metal electrodes, one embedded at each end of the glass tube. The two electrodes are connected to a power source, usually via an induction coil.

The electrode connected to the negative terminal of the power source is named the cathode, while the electrode connected to the positive terminal is named the anode. When the power is on, the cathode rays are observed to flow from the negative cathode to the positive anode inside the tube, just like current flowing in an electric circuit.

### How cathode ray tubes operate

Figure 14.3 is a schematic diagram of a cathode ray tube.

There are several requirements for the operation of a cathode ray tube.

*Low pressure:* A cathode ray tube must have its glass tube evacuated to a very low gas pressure, preferably close to vacuum. This is because the cathode and anode are separated by a large distance inside the tube. The low pressure inside the tube ensures minimal collisions between the air molecules inside the tube and the electrons (cathode rays) as they make their way from the cathode to the anode.

**FIGURE 14.2** An example of a cathode ray tube with a fluorescent screen inside the tube

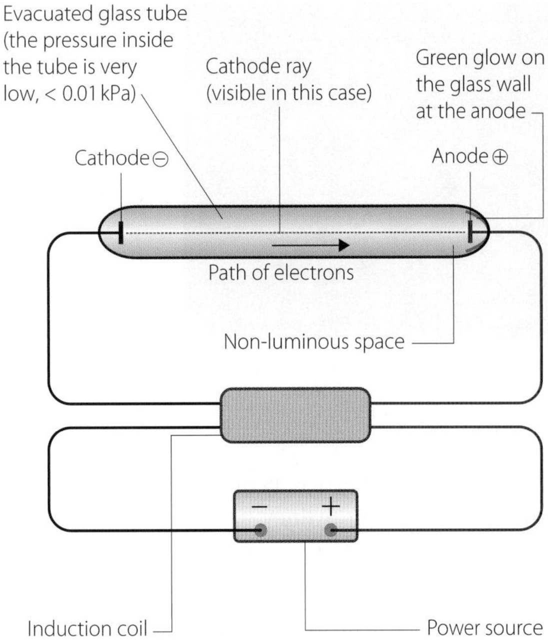

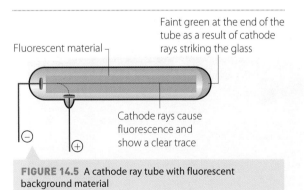

Evacuated glass tube
(the pressure inside
the tube is very
low, < 0.01 kPa)

Cathode ray
(visible in this case)

Green glow on
the glass wall
at the anode

Cathode⊖

Anode⊕

Path of electrons

Non-luminous space

Induction coil

Power source

**FIGURE 14.3** Diagram showing the basic operation of a
cathode ray tube

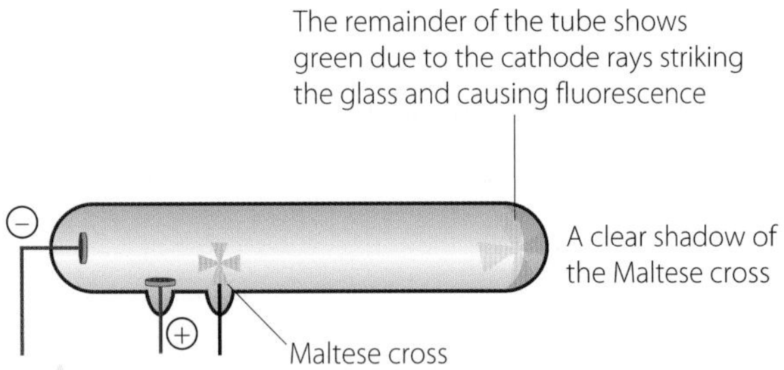

The remainder of the tube shows
green due to the cathode rays striking
the glass and causing fluorescence

A clear shadow of
the Maltese cross

Maltese cross

**FIGURE 14.4** A cathode ray tube containing a Maltese cross

Faint green at the end of the
tube as a result of cathode
rays striking the glass

Fluorescent material

Cathode rays cause
fluorescence and
show a clear trace

**FIGURE 14.5** A cathode ray tube with fluorescent
background material

*High voltage:* Low pressure alone is not enough to ensure the electricity can 'jump' across such a big gap. Extremely high voltage is required to pull the electrons off the cathode and give them enough kinetic energy to make their way to the anode. It is important to note that cathode ray tubes only work on DC, or direct current. Transformers cannot be used to step up the voltage to the required value because these only operate on AC, or alternating current. An induction coil is therefore used to step up the voltage.

## Early observations

With the development of better vacuum pumps, Crookes and others were able to make careful observations of cathode rays using a variety of tubes.

These observations were:

1   Cathode rays are emitted at the cathode and travel in straight lines. This is shown by using a CRT containing a Maltese cross (Figure 14.4). The cathode rays illuminate the Maltese cross and cast a clearly defined shadow of it at the other end of the tube.

2   Cathode rays can cause fluorescence. This is shown by using a cathode ray tube containing a background fluorescent material (Figure 14.5). As the cathode ray passes from the cathode to the anode, it causes the material to fluoresce, leaving a clear trace. Cathode rays are also able to cause the wall of the glass tube to glow, as shown in many other scenarios.

3   Cathode rays can be deflected by magnetic fields. When a pair of bar magnets is placed next to the cathode ray tube, the cathode rays are deflected as predicted for negatively charged particles by the right-hand rule (Figure 14.6).

4   Cathode rays can be deflected by electric fields. When oppositely charged plates are placed one each side of the tube (Figure 14.7), the cathode rays are deflected in the direction opposite to that of the electric field.

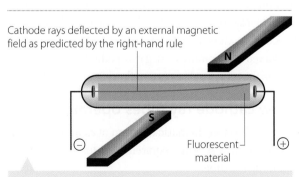

Cathode rays deflected by an external magnetic
field as predicted by the right-hand rule

Fluorescent
material

**FIGURE 14.6** Cathode rays deflected by a magnetic field

It is interesting to note here that, in initial investigations, cathode rays could not be deflected by the electric fields applied to the tubes. This led to debate about their nature and discussion as to whether they were particles or waves. It turned out that the electric charge used to try to deflect the cathode rays was discharging too quickly. Improvements in the vacuum used solved this issue.

9780170409131

**5** Cathode rays carry and are able to transfer momentum. This is shown by using a cathode ray tube containing a paddle wheel (Figure 14.8). As the cathode rays strike the paddle wheel, some of their momentum is transferred to the paddle, which makes the paddle wheel roll in the same direction as the cathode rays are travelling.

**6** Cathode rays are identical regardless of the type of material used as the cathode.

Cathode rays can also facilitate some chemical reactions and expose photographic films.

FIGURE 14.7 A cathode ray is deflected by an electric field

## Thomson's plum pudding model of the atom

Collectively, these observations pointed to a particulate nature for cathode rays, or cathode ray particles. They seemed to be common to all types of elements and therefore atoms. For the first time it was suggested that atoms were themselves composed of smaller particles.

Prior to these observations, the model of the atom was a basic, structureless sphere, rather like a billiard ball. Dalton's 'billiard ball' model of the atom clearly needed to be replaced. Thomson proposed his 'plum pudding' model, as shown in Figure 14.9. This model had discrete negatively charged particles within a sea of positive charge. The total positive charge was cancelled by the total negative charge in each atom.

FIGURE 14.8 A cathode ray tube containing a paddle wheel, which is forced to rotate by the cathode rays

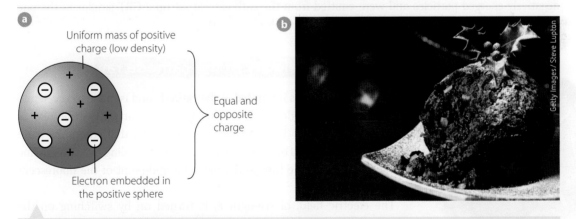

FIGURE 14.9 **a** Thomson's 'plum pudding' model of the atom; **b** In the model, the negative charges were distributed like the fruit in a matrix of positively charged pudding.

In Thomson's model the atom was still assumed to be spherical in shape. It had the electrons embedded randomly within the atom. Since the atom needed to be neutral overall, there must be positive charge to balance the negative electron charges. Thomson proposed that the rest of the atom was uniformly positively charged, with its mass evenly distributed but low in density. This model was analogous to a plum pudding, where the electrons were like fruit scattered throughout the 'pudding-like' atom, as shown in Figure 14.9.

The electron

● Cathode rays:
  – travel from the cathode to the anode in straight lines
  – can cause fluorescence in glass
  – can be deflected in magnetic fields
  – can be deflected in electric fields
  – can transfer momentum to a paddle wheel device in a cathode ray tube
  – are identical regardless of the cathode material used
  – behave as negatively charged particles
● Thomson's plum pudding model of the atom has discrete negative particles embedded in a sea of positive charge.

CHECK YOUR UNDERSTANDING

14.1

1  Which observation of cathode rays is consistent with them being negatively charged?

2  Why is a high voltage between the cathode and anode necessary to produce cathode rays?

3  What is the significance of the paddle wheel in the cathode ray tube?

4  Why did Thomson modify Dalton's 'billiard ball' model of the atom?

5  Explain why the plum pudding model of the atom has positive charge spread throughout the atom rather than as discrete, small particles.

6  How does Thomson's model of the atom help explain observations made in cathode ray tubes?

# 14.2 Thomson's charge-to-mass ratio experiment

The aim of Thomson's experiment was to measure the charge-to-mass ($q/m$) ratio of cathode rays.

Before conducting the experiment, Thomson (Figure 14.10) assumed that cathode rays were negatively charged particles and were emitted from the cathode. He set up a CRT similar to the one shown in Figures 14.11 and 14.12. The experiment involves two parts.

FIGURE 14.10 J J Thomson in his laboratory

## Selecting the velocity of the cathode ray

A beam of cathode rays is emitted at the cathode and is made to accelerate towards the multi-anode collimators to enter the main part of the tube, as shown in Figure 14.11. This ensures the cathode ray that enters the main tube is narrow and well defined. The beam keeps travelling in a straight line to reach the end of the tube and strikes the mid-point of the fluorescent screen.

The electric field, of strength $E$, is turned on by switching on the voltage supply to the electric plates. The beam is deflected in the direction opposite to that of the electric field, say for this case up (Figure 14.11).

The magnetic field, of strength $B$, is turned on by supplying a current to the coil. The current is directed so that the magnetic field produced by the coil deflects the cathode ray in the opposite direction to that imposed by the electric field, say for this case down (Figure 14.11).

The strengths of the electric and magnetic fields are adjusted so that the deflections created by each field exactly balance out. Consequently the beam will travel to the end of the tube undeflected, and will again hit the middle of the fluorescent screen.

Thus the force acting on the cathode ray from the electric field can be equated to the force acting from the magnetic field; hence:

$$F_E = F_B$$

$$F_E = qE$$

and since θ is 90°, because the cathode ray is always perpendicular to the magnetic field:

$$F_B = qvB \sin\theta = qvB$$

Therefore:

$$qE = qvB$$

$$v = \frac{E}{B}$$

## Finding the charge-to-mass ratio of the cathode ray

The electric field is then turned off and the magnetic field is left on. The cathode ray is deflected by the magnetic field only, and thus curves down in an arc of a circle as shown in Figure 14.12.

Since the magnetic force ($F_B = qvB$) provides the electron with the centripetal force:

$$F_c = F_B$$

$$F_c = \frac{mv^2}{r}$$

where $r$ is the radius of the arc described by the cathode ray, and

$$F_B = qvB$$

Therefore:

$$\frac{mv^2}{r} = qvB$$

$$mv = qrB$$

$$\frac{q}{m} = \frac{v}{rB}$$

since $v = \frac{E}{B}$. Therefore:

$$\frac{q}{m} = \frac{E}{rB^2}$$

The strength of the electric field ($E$) and magnetic field ($B$) can be determined (by measuring the size of the applied voltage and current) and the radius $r$ of the arc described by the cathode ray can be measured. Thus the charge-to-mass ratio ($q/m$) can be calculated.

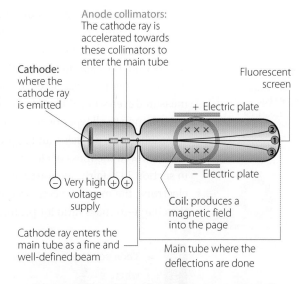

Anode collimators:
The cathode ray is accelerated towards these collimators to enter the main tube

Cathode: where the cathode ray is emitted

Fluorescent screen

+ Electric plate

− Electric plate

Very high voltage supply

Coil: produces a magnetic field into the page

Cathode ray enters the main tube as a fine and well-defined beam

Main tube where the deflections are done

**1** When undeflected, the cathode ray travels straight to the mid-point of the fluorescent screen. This happens again when the deflection along the electric field is balanced out by the deflection due to the magnetic field

**2** Deflection of the cathode ray due to the electric field

**3** Deflection due to the magnetic field

**FIGURE 14.11** The CRT setup used by Thomson showing how the magnetic field deflects the cathode ray in the opposite direction to the electric field. When the two forces are balanced, no deflection occurs

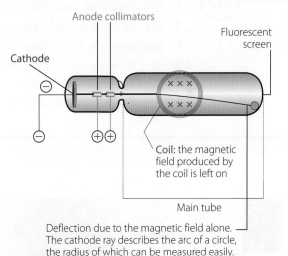

Anode collimators

Cathode

Fluorescent screen

Coil: the magnetic field produced by the coil is left on

Main tube

Deflection due to the magnetic field alone. The cathode ray describes the arc of a circle, the radius of which can be measured easily.

**FIGURE 14.12** The CRT setup used by Thomson showing the deflection caused by the magnetic field alone

## Conclusions from the *q/m* ratio experiment and their implications

The experiment proved cathode rays were indeed particles with a negative charge. The fact that the charge-to-mass ratio of cathode rays was successfully measured indicated that cathode rays had measurable mass. Because waves do not possess mass, this in turn provided definitive evidence for the particle nature of cathode rays. This effectively ended the debate over the nature of cathode rays.

**Thomson's charge-to-mass ratio experimental apparatus**

Work through the simulation and write up the investigation for your records.

Thomson's charge-to-mass ratio experiment

Thomson's experiment showed that the particles had a large (negative) charge with very little mass (especially compared to alpha particles).

It contributed to the discovery of electrons and the development of further models of the atom. The results from the experiment laid the foundation for Thomson's discovery that cathode rays were in fact a new class of particles, later to be called electrons. The fact that the same charge-to-mass ratio was measured even when different materials were used as the cathode indicated that cathode rays (electrons) are common to all types of atoms. This was one piece of evidence that led Thomson to believe electrons were subatomic particles, and to propose the plum pudding model of atoms.

It allowed the mass of electrons to be calculated. Millikan's famous oil drop experiment, discussed in section 14.3, fairly accurately determined the charge of electrons. Using the charge-to-mass ratio of electrons, the mass of electrons could be easily calculated. Later, a similar idea was used to measure the charge-to-mass ratio for protons, from which other useful information was deduced.

**KEY CONCEPTS**

- Thomson used crossed electric ($E$) and magnetic ($B$) fields as a velocity selector for cathode rays, where $v = \dfrac{E}{B}$.
- The cathode rays with known velocity were then deflected in an arc with a radius $r$ within a magnetic field only.
- The ratio of the charge $q$ to the mass $m$ of the cathode rays was calculated as $\dfrac{E}{rB^2}$.
- The resulting very high $q/m$ ratio meant that the particles had a large (negative) charge with a small mass.

**CHECK YOUR UNDERSTANDING**

**14.2**

1 Why were the electric and magnetic fields perpendicular to each other in Thomson's cathode ray velocity selector?

2 Why are cathode rays deflected towards the positively charged electric plate?

3 Why did Thomson not find a value for the mass of cathode ray particles?

4 Explain why the $q/m$ ratio for cathode ray particles has a much larger value than for protons.

5 What inference can be drawn from the observation that the $q/m$ ratio for cathode rays has the same value regardless of the cathode material used?

6 How does the fact that cathode rays have a $q/m$ ratio support the inference that they are particles and not waves?

## 14.3 Millikan's oil drop experiment

FIGURE 14.13 Robert Millikan

In 1909, Robert Millikan (Figure 14.13) performed an experiment that showed that electric charge was quantised (occurred in discrete amounts) rather than continuous. He was awarded the Nobel Prize in Physics in 1923 for his work.

The experiment involved tiny charged droplets of oil being squirted into a chamber with electric plates above and below the oil droplets. When a voltage was applied across the plates, some of the oil drops which had been charged by friction with the spray nozzle could be made to stop falling. Millikan could calculate the electric field between the two plates because $E = \dfrac{V}{d}$ where $V$ = voltage applied and $d$ = separation between the two charged plates. His experimental setup is shown schematically in Figure 14.14.

Having calculated the mass of the oil droplets, Millikan used a simple calculation equating the weight of the oil drop with the force applied to it due to the electric field.

$$F_{\text{weight}} = F_E$$
$$mg = E \times q$$
$$mg = \frac{V}{d} \times q$$
$$q = mg \div \frac{V}{d}$$
$$q = \frac{mgd}{V}$$

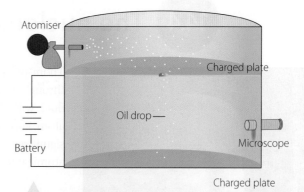

FIGURE 14.14 A diagram of the apparatus Millikan used to perform his oil drop experiment

Millikan found that, to make other oil drops stop falling, the voltage across the charged plates had to be stepped up by the same difference. These oil drops had picked up multiple charges when they were squirted into the chamber. The minimum charge an oil drop could have was found by Millikan to be $1.59 \times 10^{-19}$ C, or about 6% lower than the currently accepted value of the charge on an electron. Any oil drop was found to have only integer multiples of this amount of charge.

## Charge is quantised

The results of Millikan's experiment, which was published in 1913, helped convince the scientific community that charge was indeed quantised. This was despite the fact that electric circuits and the operation of electric devices could be explained using a model of electricity being a continuous quantity like a flowing fluid through wires. Thomson's results were reinforced and explained by Millikan's. Cathode ray particles were indeed tiny negatively charged particles, electrons.

With Millikan able to calculate the magnitude of the smallest unit of charge and Thomson having found the $q/m$ ratio, the mass of electrons finally became known. A value close to $9.1 \times 10^{-31}$ kg, or around $\dfrac{1}{1800}$ times the mass of a proton, meant that these electrons were indeed very small particles carrying the unit of electric charge.

Millikan's oil drop experiment

**Millikan and his oil drop experiment**
Work through this investigation and write it up for your records.

---

- Millikan observed the motion of charged oil drops between charged plates.
- He found the smallest unit of charge to be about $1.59 \times 10^{-19}$ C.
- Electric charge on other oil drops only existed as multiples of this base unit of charge.
- The observations led to the knowledge that charge is quantised and not continuous.

---

1  What is the purpose of the charged plates in Millikan's experiment?
2  How did the oil drops acquire their charge?
3  What did Millikan adjust so that an oil drop could be made to stop falling due to gravity?
4  What significance did the results of Millikan's experiment have on the understanding of the nature of electricity?
5  If electric charge was continuous rather than quantised, how would Millikan's results have been different?
6  Were Millikan's results consistent with Thomson's plum pudding model of the atom? Explain your answer.

## 14.4 The nucleus

**FIGURE 14.15** Ernest Rutherford

Thomson's plum pudding model of the atom had the mass of the atom spread throughout its volume. In the model the 'pudding' was like a sea of positive fluid, with the fruit in the pudding being the electrons embedded within the atom. Further investigation and observations proved that this model required significant modification as evidence built pointing to the existence of an atomic nucleus.

### The Geiger–Marsden experiment

The most significant piece of evidence came from an experiment performed by Hans Geiger and Ernest Marsden, under the direction of Ernest Rutherford (Figure 14.15).

In 1911, Ernest Rutherford directed his assistants Geiger and Marsden to perform an experiment using the newly discovered alpha particles with the aim of confirming Thomson's model of the atom. Alpha particles are emitted by certain radioactive isotopes and are now known to comprise two protons and two neutrons. They are, in fact, nuclei of helium atoms. Thomson's model showed electrons occupying only a very small space and the rest of the atom filled with positive charge. As this positive charge filled the entire volume of the atom it must have had a relatively low density. Rutherford proposed that if this was the case, when alpha particles were fired at a thin foil made of gold, the alpha particles should either go straight through or go through with very minimal deflections. He set up the apparatus as shown in Figure 14.16.

However, the results of the experiment were not what Rutherford had expected. Although most of the alpha particles went through the atoms with either no deflection or very small deflections as predicted, 1 in 8000 alpha particles were deflected back at an angle greater than 90°, as shown in Figure 14.16. This was totally surprising, because it suggested that there must exist a sufficiently dense positively charged mass inside the atoms to cause the alpha particles to rebound. Repetition of the experiment achieved the same result.

Information and communication technology capability

**Alpha particle scattering**

The interactive Phet simulation for the alpha particle scattering experiment will help deepen your understanding of this important work.

A simulation of the alpha particle scattering experiment with further information.

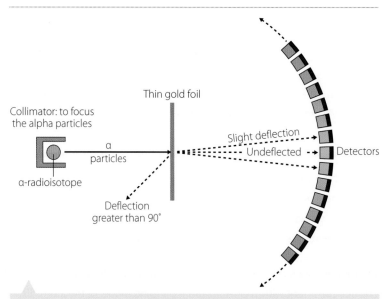

**FIGURE 14.16** Rutherford's alpha particle scattering experiment

Rutherford's surprise at the rebounding alpha particles was expressed in his statement 'It was almost as incredible as if you had fired a 15-inch shell at a piece of tissue paper and it came back and hit you.'

## Rutherford's model of the atom

From the analysis of the results of his experiments, Rutherford proposed that Thomson's model of the atom needed to be modified to account for his observations. He stated that the only way the alpha particles could be deflected through such a large angle was if all of the atom's positive charge and nearly all of its mass were concentrated in a very small region. This region was later named the nucleus. The electrons, first proposed by Thomson, were to be placed around the nucleus in a circular fashion. The rest of the atom consisted of empty space, as shown in Figure 14.17.

This model was adequate in explaining the deflection of the alpha particles. Usually these alpha particles would actually pass through the empty space between the nucleus and electrons, and hence would not have their path altered. If the alpha particles skimmed past the nucleus or collided with the electrons, then their path would be altered slightly. Deflection back at an angle greater than 90° must be due to the alpha particles colliding with the positively charged nucleus. Since the nucleus was proposed to be very small compared to the size of the atom, the chance of this happening was remote.

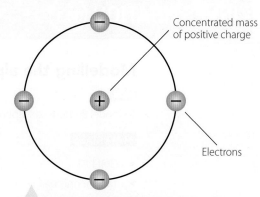

**FIGURE 14.17** Rutherford's model of the atom showing the positive nucleus with electrons around the outside.

### What was new in Rutherford's model?

The concept of the indivisible, structureless model of the atom, as proposed by the chemist John Dalton (the 'billiard ball' model) was the accepted view of the atom for over 50 years in the 1800s. Dalton could not envisage 'empty space'; his atomic theory stated that atoms occupy all of the space in matter.

Thomson's work in 1897 suggested that the atom may indeed be divisible, because electrons were thought to be a part of any atom. He proposed the plum pudding model to explain his findings; another scientist, Philipp Lenard, proposed a model in which positive and negative pairs were found throughout the atom. Rutherford's challenge was to devise a way of probing the atom in an attempt to find its structure.

Rutherford's experimental results, in which some alpha particles actually rebounded back off the thin gold foil, and his careful analysis, led him to propose his 'planetary' model. The nucleus of the atom, with almost all of the mass and all of the positive charge, was orbited by the electrons, like planets around the Sun.

Rutherford's model of the atom was the first to propose a nucleus with the electrons in separate motion. It led to insights in the field of chemistry, which deals with the interaction between the electrons of different atoms.

### How did Rutherford's model change the direction or nature of scientific thinking?

The motion of the electrons in Rutherford's model of the atom violated the laws of classical physics. However, rather than being disregarded, Rutherford's model triggered the further work of Bohr and others on their journey to develop quantum physics. The first step along this journey was to suggest that Rutherford's electrons could exist in a stable state and not emit radiation. These ideas are presented in chapter 15.

**Atomic theory**
Background information on atomic theory around the time of Rutherford's experiment.

**Ernest Rutherford**

## Evaluation of Rutherford's work

Rutherford's work paved the way for major changes in scientific thinking – the proposals of electrons orbiting a positive nucleus and the idea that much of the volume of atoms was empty space. The answers to Rutherford's puzzles led to the development of **quantum theory** and changes to the way in which matter was explained.

## Modelling the alpha particle scattering experiment

### AIM

To model Rutherford's alpha particle scattering experiment

### MATERIALS

- Golf ball
- Table tennis ball
- Metre ruler
- Device to record the motion of the balls

RISK ASSESSMENT

| WHAT ARE THE RISKS IN DOING THIS INVESTIGATION? | HOW CAN YOU MANAGE THESE RISKS TO STAY SAFE? |
| --- | --- |
| The balls may roll onto the floor and cause a person to slide or fall. | Account for all balls and make sure they do not roll away on the floor. |

What other risks are associated with your investigation, and how can you manage them?

### METHOD

1 On an even surface, roll the table tennis ball towards the golf ball from approximately 2 m away.
2 Observe the motion of both balls when the table tennis ball collides with the golf ball.
3 Repeat steps 1 and 2 about 20 times.
4 Record how many times the table tennis ball:
   **a** missed the golf ball and passed straight by.
   **b** collided head-on with the golf ball and rebounded back the way it came.

### RESULTS

Enter your results in a table.

### DISCUSSION

1 Describe how the balls represent the alpha particles and gold atoms in the foil in Rutherford's alpha particle scattering experiment.
2 Suggest why the table tennis ball was rolled from a significant distance away from the golf ball.
3 Suggest modifications to the design of this investigation which may make the analogy with Rutherford's experiment more accurate.

### CONCLUSION

With reference to the data obtained and its analysis, write a conclusion based on the aim of this investigation.

 9780170409131

## Limitations of Rutherford's model of the atom

Rutherford's model was successful in accounting for the surprising results of his experiment. However, there were still a few aspects that he was unable to explain.

First, he could not explain the composition of what he called the nucleus. Although he proposed that most of the atom's mass and positive charges were concentrated into the very small and dense area called the nucleus, he could not explain what was in the nucleus. The existence of protons and neutrons was not known at the time.

Second, although he proposed that the electrons should be placed around the nucleus, he did not know how to arrange the electrons around the nucleus, except 'like planets around the Sun'.

The biggest problem that Rutherford failed to explain was how the negative electrons could stay away from the positive nucleus without collapsing into it. The only way to overcome the attractive force between the positive nucleus and the negative electrons was to have the electrons orbiting around the nucleus. However, electrons, when circulating around the nucleus, would have centripetal acceleration, because centripetal acceleration applies to all circular motion. It was already known that accelerating charges produce electromagnetic radiation, a form of energy. This meant that electrons would release electromagnetic radiation as they were orbiting, and this would result in a loss of energy. This loss of energy would have to be derived from the kinetic energy of the electrons, resulting in the electrons slowing down. With the loss of kinetic energy, they would be unable to maintain their orbit around the nucleus and they would spiral back into the nucleus. Every atom would therefore be unstable and short lived. Obviously, this does not happen – but Rutherford's model failed to provide a reason for this.

The nucleus

**CHECK YOUR UNDERSTANDING**

**14.4**

1  Why were alpha particles expected to pass through a thin gold foil?

2  What specific observation from the alpha particle scattering experiment led Rutherford to propose the existence of the nucleus?

3  Where are the electrons in Rutherford's model of the atom?

4  Describe the motion of the electrons in Rutherford's atomic model.

5  Why were electrons in Rutherford's model expected to emit electromagnetic radiation?

6  What observation of atoms, and matter in general, shows that Rutherford's atomic model has certain limitations?

7  What did Rutherford propose as being the contents of the nucleus?

8  How was it possible for a few alpha particles to rebound off the gold foil?

# Chadwick's discovery of neutrons

Rutherford was the first person to use the term 'nucleus' as part of the conclusion of his alpha particle scattering experiment. It is now known that there are two types of particles that exist inside the nucleus: protons and neutrons. Protons and neutrons are collectively called nucleons. The properties of the types of nucleons are summarised in Table 14.1.

**TABLE 14.1** Properties of nucleons

| PROPERTIES | PROTON | NEUTRON |
|---|---|---|
| Location | Inside the nucleus | Inside the nucleus |
| Mass | $1.673 \times 10^{-27}$ kg | $1.675 \times 10^{-27}$ kg |
| Charge | $1.602 \times 10^{-19}$ C | 0 |

## The contents of the nucleus

The nucleus as described by Rutherford was simply a concentrated mass of positive charge. No one at the time knew what it was composed of and what its internal structure was. Again, as with the electron models, it took the work of many scientists, over several decades, to determine the structure and components of the nucleus.

Protons were the first nucleons to be discovered, in a similar fashion to the way electrons were discovered. The charge-to-mass ratio of the proton was measured in a discharge tube containing hydrogen ions. Hydrogen ions are hydrogen atoms with the single electron removed, leaving only a proton. Neutrons were discovered later.

### Early theories concerning the nucleus

Soon after learning about the existence of the nucleus and protons, some scientists started to speculate that the nucleus should possess $A$ protons and $A - Z$ electrons, where $A$ is the mass number and $Z$ the atomic number. For example, for sodium atoms, with $A = 23$ and $Z = 11$, there should be 23 protons and $23 - 11 = 12$ electrons in the nucleus. This hypothesis worked well for two main reasons.

1 It successfully explained why atoms had a mass number that was larger than the actual number of positive charges (protons). Note that for our example of sodium the 12 electrons in the nucleus cancelled out the positive charges of 12 protons, leaving a net charge of positive 11 rather than 23.

2 It explained beta particle emission, a radioactive decay in which electrons are ejected from the nucleus. Although incorrect, scientists at the time thought the only way to account for this phenomenon was if there were electrons inside the nucleus.

However, other scientists believed that another type of particle existed inside the nucleus. They hypothesised that these particles had a similar mass to protons and were neutral in charge. These particles were neutrons.

### Chadwick's discovery of neutrons

In 1930, the German scientist Walther Bothe noted that when the element beryllium was bombarded with alpha particles (helium nuclei), a neutral but highly penetrative 'radiation' could be obtained. However, he could not explain the nature of this 'radiation'.

**FIGURE 14.18** James Chadwick

Source: Alamy Stock Photo/Science History Images

In 1932, the English physicist James Chadwick (Figure 14.18) demonstrated that the unknown radiation obtained from the alpha particle bombardment of beryllium was in fact particles and not electromagnetic radiation. The arrangement used for this bombardment of beryllium is shown in Figure 14.19. Chadwick used the data from this experiment to try to quantitatively study this unknown 'radiation'. The neutral charge of this 'radiation' was easily demonstrated since it was not deflected by electric fields or magnetic fields. To measure the mass of the proposed neutrons, Chadwick modified the design of the experiment using beryllium as the first target.

Paraffin wax is rich in hydrogen atoms, and therefore protons. The proposed neutrons, when directed towards the paraffin wax, should have a good chance of colliding with those protons and knocking them out. As a result, protons would be ejected from the paraffin wax and could be measured by the detector, which allowed the energy and the velocity of the ejected protons to be assessed. By applying the law of conservation of momentum and the law of conservation of energy, Chadwick was able to work backwards to calculate that the mass of the neutron was approximately equal to that of the proton. The existence of the neutron had been experimentally shown. Chadwick demonstrated the existence of neutrons without directly observing them. Rather, it was achieved through observing the properties of neutrons.

Neutrons are difficult to assess directly because their lack of any charge means they cannot be manipulated easily. The clever part of Chadwick's experiment is that it translates the difficult-to-measure neutrons to the easily measured protons. Since protons have charges, they can be manipulated easily, just as electrons can be manipulated in cathode ray tubes to allow their properties to be assessed.

Chadwick's discovery of neutrons

A nuclear equation can be used to summarise the reaction taking place in Chadwick's experiment:

$$^4_2\text{He} + {}^9_4\text{Be} \rightarrow {}^{12}_6\text{C} + {}^1_0\text{n}$$

Note that when writing nuclear equations, the total mass number on the left-hand side of the equation should equal to the total mass number on the right; here, $4 + 9 = 12 + 1$; similarly the atomic numbers should also be equal on both sides: $2 + 4 = 6 + 0$.

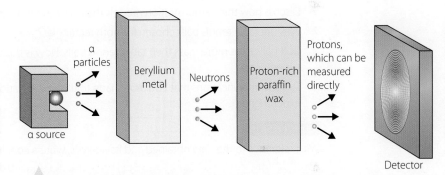

**FIGURE 14.19** Chadwick's experimental setup

---

# INVESTIGATION (14.2)

## Modelling Chadwick's experiment

### AIM

To model the experimental observations that led to the discovery of the neutron

### MATERIALS

- Golf ball
- 6 white table tennis balls
- 6 orange table tennis balls
- Means of making a visual record of the investigation, such as a mobile phone camera

RISK
ASSESSMENT

| WHAT ARE THE RISKS IN DOING THIS INVESTIGATION? | HOW CAN YOU MANAGE THESE RISKS TO STAY SAFE? |
| --- | --- |
| The balls may roll onto the floor and cause a person to slide or fall. | Account for all balls and make sure they do not roll away on the floor. |

What other risks are associated with your investigation, and how can you manage them?

## METHOD

1 Arrange the table tennis balls on a suitable flat surface in two rectangles of 3 balls by 2, with the rectangle of white balls first and then the rectangle for the orange balls, as shown in Figure 14.20.

2 Gently roll a golf ball aimed at the white table tennis balls.

3 Carefully observe and record the resulting motion of the balls.

4 Repeat steps 1–3 several times.

Golf ball

White table
tennis balls

Orange table
tennis balls

**FIGURE 14.20**

## RESULTS

Analyse the speeds of the table tennis balls when the white balls collide with the orange balls.

## DISCUSSION

1 Discuss how this demonstration models the experiment used by Chadwick.

2 Why are table tennis balls chosen for both rectangles?

3 Golf balls have more mass then table tennis balls. How is this significant in this modelling of the experiment?

4 Suggest modifications that could be made to this investigation to improve how it models Chadwick's experiment.

## CONCLUSION

With reference to the data obtained and its analysis, write a conclusion based on the aim of this investigation.

**ANSTO**

Further reading from ANSTO, Australia's only nuclear research facility, on the outskirts of Sydney

KEY CONCEPTS

- Bothe was the first to observe neutrons but did not know what this 'penetrative radiation' was.
- Chadwick's experiment in 1932 showed the existence of a neutral particle with a mass approximately equal to a proton.
- Chadwick used the data from an experiment where alpha particles were directed at beryllium, which caused the emission of neutrons towards a paraffin block.
- The paraffin block had protons dislodged and emitted from it, and these were detected.
- This process involves the conservation of momentum as neutrons collide with the protons in the paraffin.
- Neutrons themselves are hard to detect because they have zero charge.

**CHECK YOUR UNDERSTANDING**

14.5

1 Why were electrons believed to exist in the nucleus?

2 What is a nucleon?

3 How was it known that neutrons did not have a charge?

4 What is the role of the paraffin wax in Chadwick's experiment?

5 In the symbol $^{12}_{6}C$, what does the number 6 mean?

6 Explain why protons are easier to manipulate than neutrons.

- Cathode rays:
  - travel from the cathode to the anode in straight lines
  - can cause fluorescence in glass
  - can be deflected in magnetic fields
  - can be deflected in electric fields
  - can transfer momentum to a paddle wheel device in a cathode ray tube
  - are identical regardless of the cathode material used
  - behave as negatively charged particles.

- Thomson's plum pudding model of the atom has discrete negative particles embedded in a sea of positive charge.

- Thomson used crossed $E$ and $B$ fields as a velocity selector for cathode rays, where $v = \dfrac{E}{B}$.
  - The cathode rays with known velocity were then deflected in an arc with a radius $r$ within a magnetic field only.
  - The ratio of the charge $q$ to the mass $m$ of the cathode rays was calculated as $\dfrac{E}{rB^2}$.
  - The resulting very high $q/m$ ratio meant that the particles have a large (negative) charge with a small mass.

- Millikan observed the motion of charged oil drops between charged plates.

- The smallest unit of charge was found to be about $1.59 \times 10^{-19}$ C.

- Electric charge on other oil drops only existed as multiples of this base unit of charge. These observations led to the knowledge that charge is quantised and not continuous.

- Rutherford's assistants, Geiger and Marsden, performed the alpha particle scattering experiment.
  - Alpha particles were fired from a radioactive source at a thin gold foil.
  - As expected, most alpha particles went straight through; however, 1 in 8000 rebounded.
  - Rutherford interpreted these results as the atom having a nucleus.
  - The nucleus was proposed to contain nearly all the mass and all of the positive charge of an atom.

- The electrons in Rutherford's model orbited the nucleus like planets orbiting the Sun.

- Rutherford could not explain why electrons, although they were accelerating, would not radiate energy and fall into the nucleus.

- Bothe was the first to observe neutrons but did not know what this 'penetrative radiation' was.

- Chadwick's experiment in 1932 showed the existence of neutral particles with a mass approximately equal to a proton.

- Chadwick used the data from an experiment where alpha particles were directed at beryllium, causing the emission of neutrons towards a paraffin block'

- The paraffin block had protons dislodged and emitted from it, and these were detected.

- This process involves the conservation of momentum as neutrons collide with the protons in the paraffin.

- Neutrons themselves are hard to detect because they have zero charge.

Review quiz

1   Suggest reasons why the electron was the first of the subatomic particles to be discovered.

2   Suggest why the neutron was the last subatomic particle to be discovered.

3   How does modelling, as performed in some of the investigations in this chapter, assist in the understanding of Rutherford's and Chadwick's work?

4   Why was an induction coil required for cathode ray tubes rather than a transformer?

5   From which electrode do cathode rays originate? Explain your answer.

6   State three observations that are consistent with cathode rays being negatively charged.

7   What did the cathode ray tube with the Maltese cross show the early experimenters?

8   Which cathode ray tube experiment showed beyond doubt that cathode rays carry momentum?

9   Explain why a high voltage is necessary to generate cathode rays.

10  Why is a very good vacuum necessary for cathode rays to be observed?

11 How did the fact that cathode rays were identical regardless of the material used for the cathode support the idea that electrons were common to all atoms?

12 Outline the differences between Dalton's and Thomson's models of the atom.

13 In a diagram of Thomson's model of the atom, as shown in Figure 14.9, why are the positive signs not enclosed in circles?

14 Explain how Thomson's model of the atom became known as the plum pudding model.

15 What two forces on the cathode ray particles must be balanced in order for their speed to be calculated?

16 Outline how Thomson was able to find the charge-to-mass ratio of electrons.

17 Why did Millikan's calculation of the charge on electrons allow their mass to be calculated?

18 Explain why the charge-to-mass ratio for electrons is much larger than for protons.

19 How was Millikan able to show that electric charge was quantised rather than being continuous?

20 Explain how the very surprising observations made in the alpha particle scattering experiment provided strong evidence for the existence of an atomic nucleus.

21 Explain why Rutherford's model of the atom would have electrons losing energy and spiralling into the nucleus.

22 Explain the role of the conservation of momentum in Chadwick's calculations that led to the discovery of the neutron.

23 Assess the roles of Thomson, Rutherford and Chadwick in the development of our understanding of the structure of the atom.

24 Why is it important for us to pursue endeavours such as improving our understanding of the structure of the atom?

# Quantum mechanical nature of the atom

## INQUIRY QUESTION

How is it known that classical physics cannot explain the properties of the atom?

**Students:**

- assess the limitations of the Rutherford and Bohr atomic models ICT
- investigate the line emission spectra to examine the Balmer series in hydrogen (ACSPH138) ICT
- relate qualitatively and quantitatively the quantised energy levels of the hydrogen atom and the law of conservation of energy to the line emission spectrum of hydrogen using:

  - $E = hf$

  - $E = \dfrac{hc}{\lambda}$

  - $\dfrac{1}{\lambda} = R\left[\dfrac{1}{n_{f}^{2}} - \dfrac{1}{n^{2}}\right]$ (ACSPH136) ICT, N

- investigate de Broglie's matter waves, and the experimental evidence that developed the following formula:

  - $\lambda = \dfrac{h}{mv}$ (ACSPH140) ICT, N

- analyse the contribution of Schrödinger to the current model of the atom

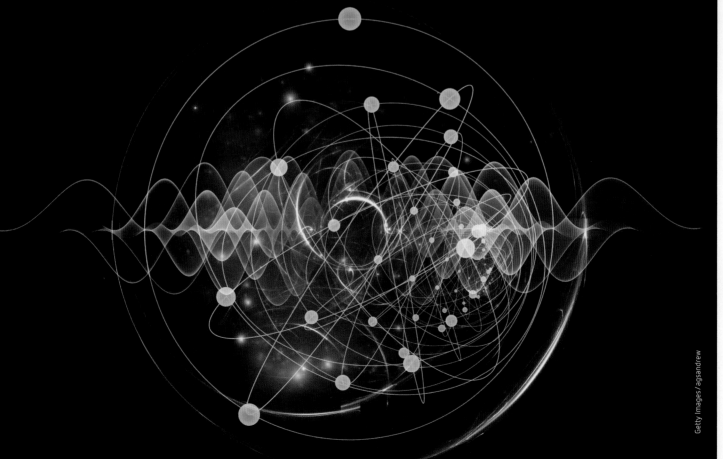

**FIGURE 15.1** A hydrogen emission discharge tube used as the source of a hydrogen spectrum

In chapter 14 we saw how the model of the atom was dramatically revised from that of Dalton's billiard ball model to Rutherford's nuclear model. It was clear at the time that Rutherford's planetary model of the atom had severe limitations. Using Planck's idea of quantised energy, Neils Bohr built on Rutherford's model so that it could explain easily observable phenomena, including the emission spectrum of hydrogen (Figure 15.1).

Bohr proposed a model that allowed electrons to orbit the nucleus in a stable state. With the emergence of quantum theory, scientists and mathematicians derived some very elegant and simple equations that were able to explain observations. Models of the atom became more complex, yet in many cases they still had limitations. The work of Bohr, de Broglie and Schrödinger progressed our understanding of the atom beyond the world of classical physics.

# 15.1 Bohr's atomic model

Neils Bohr, a Danish physicist, developed a model of the atom in 1913 that was based on Rutherford's model. Bohr also used quantum theory, specifically Planck's hypothesis that stated that electromagnetic radiation consisted of discrete packets of energy, or quanta, which are now referred to as photons. Bohr observed the spectral lines of hydrogen in the visible region. His model, illustrated in Figure 15.2, was based on four postulates.

## Bohr's postulates

Bohr's postulates are:

Postulate 1: An electron in an atom moves in a circular orbit about the nucleus under the influence of the electrostatic attraction of the positive nucleus.

Postulate 2: Electrons can only exist in certain, stable energy levels and orbits surrounding the nucleus. These levels are known as energy shells. When in these stable energy shells, electrons do not lose energy so they do not emit radiation.

This postulate quantises the energies electrons are allowed. An electron could not possess an 'in-between' energy when in the atom.

Postulate 3: When electrons move from one energy level to another, they either absorb or release a quantum of energy in the form of electromagnetic radiation. If they absorb energy, they jump to a higher allowed energy level with a greater radius. When falling down from a higher energy level to a lower orbit with a smaller radius, they emit energy. The energy of this quantum is related to the frequency of the radiation by $E = hf$. Here, $h$ is Planck's constant and $f$ is the frequency of the radiation. The energy, $E$, is measured in joules. The energy released is:

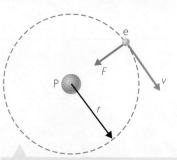

**FIGURE 15.2** The motion and forces involved in the Bohr model of the hydrogen atom

$$E = E_i - E_f = hf$$

Postulate 4: The angular momentum of electrons in orbit about the nucleus is quantised. Angular momentum, $L$, is given by $mvr$, so we have:

$$mvr = \frac{nh}{2\pi}$$

$$r = \frac{nh}{2\pi mv}$$

where $n$ is the principal quantum number.

The innermost energy shell is assigned $n = 1$, with the next shell $n = 2$, and so on.

Postulate 4 is an **empirical formula** derived from observations of the hydrogen emission spectrum.

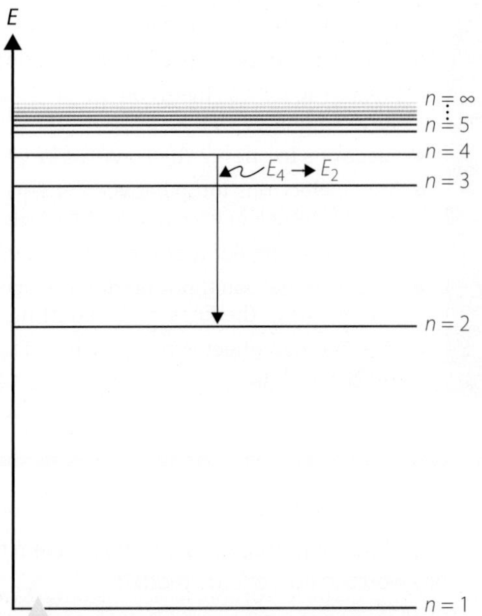

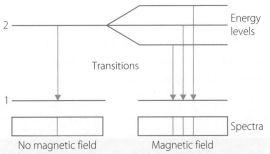

**FIGURE 15.3** The energy levels in Bohr's model of the atom showing an electron falling from the 4th to the 2nd energy shell. This would result in the emission of radiation with a frequency $f$ where $E = hf$.

Information and communication technology capability

## Limitations of the Bohr model

Bohr was able to explain qualitatively and quantitatively the existence and positions of the spectral lines of a hydrogen atom (Figure 15.3). His estimate of the size of the largest stable radius also agreed closely with the measured size of the hydrogen atom.

However, the Bohr model could not predict the spectra of multi-electron atoms, even one as simple as two-electron helium. It also could not explain the different intensities of lines or why some lines split into multiple, closely spaced lines – fine and hyperfine structure – or the magnetically induced **Zeeman effect**.

**Models of the atom**

Phet's simulation showing models of the atom, including the Bohr model

### Zeeman effect

In 1896, Dutch physicist Pieter Zeeman was using laboratory equipment to observe the effect of strong magnetic fields on spectral lines. He observed that the individual lines were split into multiple closely spaced lines (Figure 15.4). Unfortunately, he was supposed to be working on something else, and had been explicitly told by his supervisor not to use the equipment for his own research. When his supervisor found out he was fired.

Zeeman and Hendrik Lorentz later showed that the splitting was due to the electron's intrinsic magnetic field (spin) and the magnetic field due to its orbital motion interacting with the applied magnetic fields. For this they were awarded the 1902 Nobel Prize in Physics.

Finally, Bohr's model introduced the idea of quantised atomic energy levels, but it did not offer any explanation for why they should be quantised. Successful models have both predictive power and explanatory power. Bohr's model lacked explanatory power, and had limited predictive power. It was superseded by a more fully quantum mechanical model developed by Schrödinger and Heisenberg. This modern quantum model is built on the ideas of de Broglie, which are described in section 15.4.

**FIGURE 15.4** The Zeeman effect. Spectral lines are split by the presence of the magnetic field.

WS

Bohr's atomic model

● Rutherford's atomic model was modified by Bohr in an attempt to overcome many of its limitations.
● Bohr's model could predict the positions of the spectral lines of hydrogen.
● Bohr's model was based on his postulates, stating electrons could orbit the nucleus only in allowed energy shells.
● Electrons in energy shells with a larger radius have more energy.
● When electrons jump between energy shells they emit or absorb a quantum of energy equal to the difference in energies of the shells.
● Electrons' angular momentum is quantised.
● Bohr's model could not predict the spectra of multi-electron atoms, the differences in intensities of the lines or the splitting of lines (the Zeeman effect).
● The Zeeman effect is the splitting of spectral lines when hydrogen atoms are subjected to a magnetic field.

1  State the limitations of Rutherford's atomic model with respect to electrons orbiting the nucleus.
2  What did Bohr propose to solve the problem that orbiting electrons did not continuously emit radiation, as they would in Rutherford's model?
3  What quantity is quantised in Bohr's model of the atom?
4  What mechanism was proposed by Bohr that gave rise to an emission line?
5  How did Bohr explain the presence of several emission lines in the hydrogen spectrum?
6  Identify the specific limitations of Bohr's model of the atom in terms of the observations it cannot explain.
7  Outline the Zeeman effect.

# 15.2 Spectra

The different types of spectra – emission, absorption and continuous black body spectra – were introduced in chapter 13. In this chapter, we shall examine the spectrum produced by hydrogen gas. Hydrogen is the simplest of the elements and has only one electron. Early observations of the spectrum produced by hydrogen using hydrogen gas in a glass tube, excited by applying a high voltage across electrodes embedded in the tube at either end, led to further understanding of the nature of the structure of the atom.

## Observing spectra

Different elements were observed to produce characteristic colours in a gas discharge tube, as illustrated in Figures 15.5 and 15.6. Gustav Kirchhoff and Robert Bunsen had recognised as early as the 1860s that line spectra can be used to identify elements. In 1861, they discovered two new elements, caesium and rubidium, using spectroscopy.

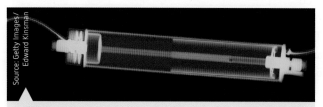

FIGURE 15.5 Neon lights are an example of a gas discharge tube. The electrical energy is transformed into light.

FIGURE 15.6 Colours produced by helium, neon, argon, krypton and xenon in discharge tubes.

Although Kirchhoff, Bunsen and others had observed and used spectra, there was no theory that explained why they existed. However, it was presumed that the characteristic spectra were related to the internal structure of the atom. To solve this puzzle, the simplest atom, hydrogen, was the subject of intense theoretical and experimental investigation.

# INVESTIGATION (15.1)

## Teacher demonstration: Observing spectra

### AIM

To observe and photograph the emission spectrum of hydrogen

### MATERIALS

- Hand-held spectroscope
- Hydrogen gas discharge tube
- Induction coil or similar high-voltage apparatus
- 12V power pack
- Alligator leads
- Darkened room/laboratory
- Suitable camera (e.g. mobile phone)

RISK ASSESSMENT

| WHAT ARE THE RISKS IN DOING THIS INVESTIGATION? | HOW CAN YOU MANAGE THESE RISKS TO STAY SAFE? |
| --- | --- |
| High voltage can cause electric shock. | Take extreme care when handling the induction coil and discharge tube by first disconnecting the apparatus before handling. |

What other risks are associated with your investigation, and how can you manage them?

### METHOD

1 Connect the hydrogen discharge tube to the induction coil and 12V power pack, taking care to ensure the power is disconnected while handling the apparatus.
2 Observe the spectrum produced by the hydrogen through the spectroscope.
3 Align the camera lens with the eyepiece of the spectroscope and record the spectrum.

### RESULTS

Describe the pattern of emission lines observed.

### DISCUSSION

1 Explain why this spectrum is classified as an emission spectrum.
2 Describe the mechanism in the hydrogen that is producing this spectrum.

### CONCLUSION

With reference to the data obtained and its analysis, write a conclusion based on the aim of this investigation.

## The hydrogen spectrum

Hydrogen produces infrared, visible and ultraviolet emission spectra. The emission and absorption spectra of hydrogen are shown in Figure 15.7. Dark lines in the absorption spectrum of any element coincide with the bright lines in its emission spectrum.

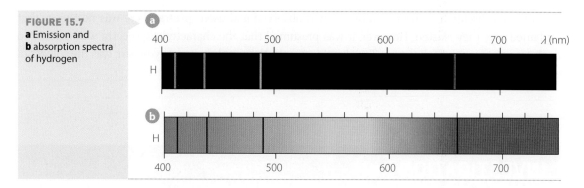

**FIGURE 15.7**
**a** Emission and
**b** absorption spectra of hydrogen

In 1855, Johann Balmer derived an empirical formula for the visible series that now bears his name (Figure 15.8).

Balmer showed that the observed wavelengths were proportional to $\dfrac{m^2}{m^2 - n^2}$, with $n = 2$ and $m$ greater than $n$.

The other spectral series of hydrogen are named after their discoverers. These are represented in Figures 15.9 and 15.10. The Lyman series is in the ultraviolet and the Paschen series is in the infrared part of the spectrum. Other series of even longer wavelength are the Brackett series and the Pfund series. These lines have a similar pattern of separation, but with different values for $n$ and $m$.

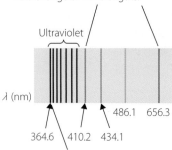

The lines shown in colour are in the visible range of wavelengths.

Ultraviolet

$\lambda$ (nm)

486.1   656.3

364.6   410.2   434.1

This line is the shortest wavelength line and is in the ultraviolet region of the electromagnetic spectrum.

**FIGURE 15.8** The Balmer series of spectral lines for atomic hydrogen

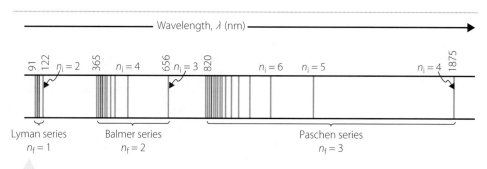

**FIGURE 15.9** The hydrogen spectrum in the UV (Lyman series), visible (Balmer series) and IR (Paschen series) wavebands.

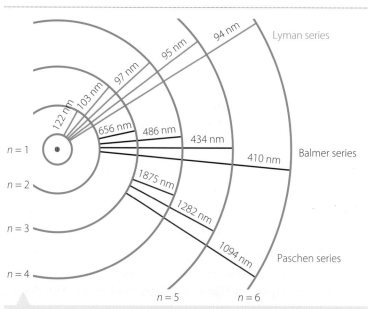

**FIGURE 15.10** The electron transitions between energy shells that correspond to lines in the hydrogen spectrum. Wavelengths are shown in nanometres, nm.

**TABLE 15.1** Lines in the spectral series of hydrogen

| SERIES NAME | PART OF SPECTRUM | SHORTEST WAVELENGTH (nm) | LONGEST WAVELENGTH (nm) |
|---|---|---|---|
| Lyman | ultraviolet | 91.1 | 121.6 |
| Balmer | visible | 364.5 | 656.3 |
| Paschen | infrared | 820.1 | 1870 |

- A spectroscope can be used to observe spectra.
- The three main types of spectra are emission, absorption and continuous (black body).
- The dark lines in the solar spectrum are an example of an absorption spectrum.
- The series of lines in the visible region of a hydrogen spectrum is called the Balmer series.
- The Balmer series occurs when electrons move from a higher energy shell to shell $n = 2$ for emission lines, or from $n = 2$ to higher energy shells for absorption lines.
- The wavelengths of lines in the emission spectrum are equal to the wavelengths of the lines in absorption spectrum for hydrogen.
- Other hydrogen spectrum series are the Lyman (ultraviolet) and Paschen (infrared) series.

1  Explain how a spectroscope enables the viewing of a spectrum, whereas we normally cannot see a spectrum with the unaided eye.
2  What is the number of the final energy shell for the Balmer series emission lines?
3  Outline the difference between an emission and an absorption spectrum.
4  Outline how a hydrogen emission spectrum can be observed.
5  Suggest reasons why the Lyman series gives rise to emission lines in the ultraviolet region of the electromagnetic spectrum.
6  State the main differences between the Balmer, Lyman and Paschen series in terms of the energy shells and electron transitions.
7  Why are the wavelengths of an emission series equal to those of the corresponding absorption lines?

# 15.3 Energy, frequency, wavelength and the hydrogen spectrum

The observed spectrum of the Balmer series for hydrogen is caused by the transition of electrons from higher energy shells to the second highest energy shell, $n = 2$. The transition with the smallest energy difference occurs between shells $n = 3$ to $n = 2$. The relationship between the energy difference between energy shells and the frequency of the hydrogen emission lines is described by $E = hf$.

## The energy of a photon

Bohr's model explained the observed line spectra as resulting from transitions between energy levels in atoms. The lowest wavelength or highest energy line corresponds to the ionisation energy of electrons in the lowest possible energy level. This level is called the ground level of the atom and corresponds to the electron being in the orbit with the smallest radius. Ionisation is the removal of an electron from the atom to infinitely far away, or at least so far that the electrostatic attraction is negligible.

Energy levels can be represented in two different ways:

1  with the ionisation energy being taken as zero (all the energy states then have a negative potential energy) (Figure 15.11a)

2  with the ground state level being taken as zero (Figure 15.11b).

In both representations, the energy *difference* between levels is the same. We shall generally use the first representation, because this corresponds to the usual convention for choosing the zero of potential energy in electromagnetism.

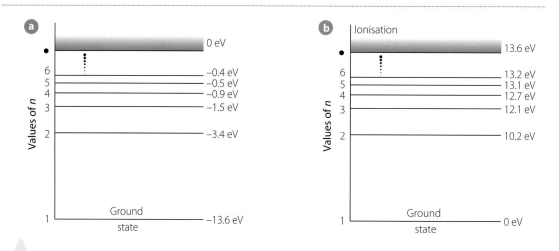

**FIGURE 15.11**  The energy levels of the hydrogen atom have either **a** the ionisation energy as zero or **b** the ground state as zero.

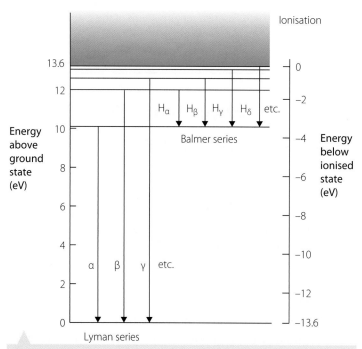

**FIGURE 15.12**  This energy level model for hydrogen shows the transitions corresponding to the Lyman and Balmer emission spectra series.

The energy of each level can be deduced from the wavelengths, and hence energies, of the lines in the emission spectrum. The highest energy lines for hydrogen are the Lyman series in the ultraviolet. These lines correspond to transitions to the ground state from higher energy levels, hence for these lines $n_f = 1$ and $n_i = 2, 3, 4 \ldots$ The Balmer series, in the visible region, corresponds to transitions to the $n = 2$ level, so $n_f = 2$ and $n_i = 3, 4, 5 \ldots$

Figure 15.12 shows the energy levels for hydrogen and the transitions corresponding to these two series of spectral lines. The labels $\alpha$, $\beta$, $\gamma$ and so on through the Greek alphabet are used to identify the lines in each series starting with the line with the longest wavelength.

**a** Construct an energy level diagram like that shown in Figure 15.11a showing the transitions corresponding to the Paschen series, for which $n_f = 3$.

**b** Calculate the frequency of a photon released in a transition from $n_i = 6$ to $n_f = 3$.

| ANSWER | LOGIC |
|---|---|
| 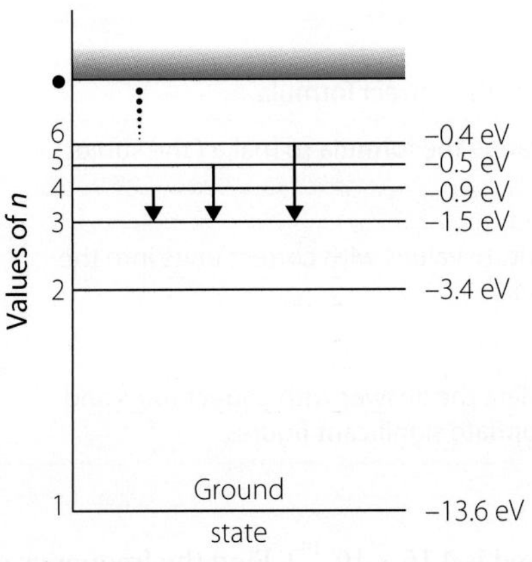 <br>**FIGURE 15.13** | ▪ Use the energies given in Figure 15.11a to draw an energy level diagram showing the energy levels up to $n = 6$. On this diagram (Figure 15.13) the Paschen series corresponds to transitions to $n_f = 3$ from levels $n_i = 4$, 5 and 6. |
| **b** $E_f = E_i - hf$ | ▪ Relate the change in energy of the electron to the photon frequency. |
| $f = \dfrac{E_i - E_f}{h}$ | ▪ Rearrange for frequency. |
| $f = \dfrac{-0.4 \text{ eV} - (-1.5 \text{ eV})}{4.14 \times 10^{-15} \text{ eV s}}$ | ▪ Insert values with correct units, noting that we must use $h$ in units of eV s or convert our energies to J. |
| $f = 2.6 \times 10^{14} \text{ Hz}$ | ▪ Calculate the answer with correct units and appropriate significant figures. |

**TRY THESE YOURSELF**

**1** Calculate the frequency of the lowest energy line in the Lyman series.

**2** Calculate the wavelength of the highest energy line in the Balmer series.

As discussed earlier, the relationship between the energy of a photon and its frequency is given by:

$$E = hf$$

where

$E$ = energy in joules (J)

$h$ = Planck's constant = $4.14 \times 10^{-15}$ eV s (or $6.626 \times 10^{-34}$ J s)

$f$ = the frequency of the photon of light in hertz (Hz)

When an electron changes energy shells and releases energy, we have:

$$E = E_i - E_f = hf$$

> The electron-volt, eV, is a commonly used non-SI unit of energy.
> 1 eV = $1.6 \times 10^{-19}$ J

## WORKED EXAMPLE (15.2)

The energy difference between the energy shells $n_i = 3$ and $n_f = 2$ is $3.04 \times 10^{-19}$ J. When an electron falls from $n_i = 3$ to $n_f = 2$, what is the frequency of the light emitted?

| ANSWER | LOGIC |
|---|---|
| $E = 3.04 \times 10^{-19}$ J <br> $h = 6.626 \times 10^{-34}$ J s | ▪ Identify the relevant data from the question. |
| $E = hf$ | ▪ Identify the correct formula. |
| $f = \dfrac{E}{h}$ | ▪ Rearrange the formula to make $f$ the subject. |
| $f = \dfrac{3.04 \times 10^{-19} \text{ J}}{6.626 \times 10^{-34} \text{ J s}}$ | ▪ Substitute values with correct units into the formula. |
| $f = 4.59 \times 10^{14}$ Hz | ▪ Calculate the answer with correct units and appropriate significant figures. |

### TRY THESE YOURSELF

1 When an electron falls from $n_i = 4$ to $n_f = 2$, the energy released is $4.16 \times 10^{-19}$ J. Find the frequency of the light emitted.

2 When an electron falls from $n_i = 5$ to $n_f = 2$, 2.9 eV of energy is released as a photon of light. What is the frequency of this light?

## The relationship between the energy and wavelength of a photon

As discussed in Year 11, the relationship for any wave between its speed $v$, frequency $f$ and wavelength $\lambda$ is

$$v = f\lambda$$

If we apply this relationship to $E = hf$, and we rearrange $v = f\lambda$ so that $f = \dfrac{v}{\lambda}$, we have:

$$E = \frac{hc}{\lambda}$$

## WORKED EXAMPLE (15.3)

An electron transitioning between energy shells emits a photon with a wavelength of 1094 nm. What is the difference in energy levels between the two shells?

| ANSWER | LOGIC |
|---|---|
| $\lambda = 1094$ nm, $h = 6.626 \times 10^{-34}$ J s, $c = 3.00 \times 10^8$ m s$^{-1}$ | ▪ Identify the relevant data from the question. |
| $E = \dfrac{hc}{\lambda}$ | ▪ Identify the correct formula. |
| $E = \dfrac{6.626 \times 10^{-34} \text{ J s} \times 3.00 \times 10^8 \text{ m s}^{-1}}{1094 \times 10^{-9} \text{ m}}$ | ▪ Substitute values with correct units into the formula. |
| $= 1.82 \times 10^{-19}$ J | ▪ Calculate the answer with correct units and appropriate significant figures. |

1  What is the difference in energy levels for an electron that emits a photon with a wavelength of 122 nm?

2  If the difference between the energy levels of electron shells was $4.55 \times 10^{-19}$ J, what wavelength would a photon of light have when an electron falls from the higher to the lower energy shell?

## Rydberg's equation

In the 1880s Johannes Rydberg was working on finding a mathematical description of the line spectra of alkali metals (group 1 metals such as lithium and sodium). He read Balmer's work on hydrogen and realised that his own mathematical model and Balmer's were equivalent. Rydberg expressed the relationship as:

$$\frac{1}{\lambda} = R\left(\frac{1}{n_f^{\,2}} - \frac{1}{n_i^{\,2}}\right)$$

where $\lambda$ is the wavelength of the line, $n_1$ and $n_2$ are integers. $R$ is a constant known as the Rydberg constant, $R = 1.097 \times 10^7\,\text{m}^{-1}$.

Rydberg arrived at his formula empirically. That is, he derived the equation by fitting it to the observed data. At the time there was no theoretical model of the atom that could predict the relationship between positions of spectral lines, or even the existence of spectral lines. His work was important in that it led to the development of the first quantum mechanical model of the atom – the Bohr model.

The Rydberg equation predicts the wavelength of absorption and emission lines for the different series of hydrogen.

For emission lines, the Lyman series corresponds to $n_f = 1$, the Balmer series to $n_f = 2$ and the Paschen series to $n_f = 3$. However, the equation does not work when applied to more complicated atoms with multiple electrons.

In the following examples, we shall see how simple it is to use the Rydberg formula to predict the wavelengths of the emission line produced when an electron in a hydrogen atom transitions between any two allowed energy shells.

Energy, frequency, wavelength and the hydrogen spectrum

▶ **WORKED EXAMPLE** (15.4)

Find the wavelength of a photon emitted when an electron in a hydrogen atom makes the smallest energy transition for the Paschen series.

Numeracy

| ANSWER | LOGIC |
|---|---|
| $n_2 = 3$ (Paschen series) <br><br> $n_1 = 4$ (next energy shell) <br><br> $R = 1.097 \times 10^7\,\text{m}^{-1}$ | ▪ Identify the relevant data from the question. |
| $\dfrac{1}{\lambda} = R\left(\dfrac{1}{n_f^{\,2}} - \dfrac{1}{n_i^{\,2}}\right)$ | ▪ Identify the correct formula. |
| $\lambda = \dfrac{1}{R\left(\dfrac{1}{n_f^{\,2}} - \dfrac{1}{n_i^{\,2}}\right)}$ | ▪ Rearrange the formula to make $\lambda$ the subject. |

| | |
|---|---|
| $= \dfrac{1}{1.097 \times 10^7\, \text{m}^{-1}\left(\dfrac{1}{3^2} - \dfrac{1}{4^2}\right)}$ | ▪ Substitute values with correct units. |
| $= 1.875 \times 10^{-6}\, \text{m}$ <br> $= 1875\, \text{nm}$ | ▪ Calculate the answer with correct units and appropriate significant figures. The energy shell levels are exact numbers, so here 4 significant figures is appropriate as given in Rydberg's constant. <br> ▪ Units of the metre and nm are both acceptable |

**TRY THESE YOURSELF**

1 Calculate the wavelength corresponding to the smallest energy transition for an electron that gives an emission line in the Balmer series for hydrogen.

2 Find the emission wavelength that occurs when an electron moves from energy level $n_i = 5$ for the Lyman series.

▶ **WORKED EXAMPLE** (15.5)

A hydrogen emission line has a wavelength of $1.88 \times 10^{-6}$ m, which is in the infrared. If the final energy shell $n_f = 3$, calculate the energy shell from which the electron fell.

| ANSWER | LOGIC |
|---|---|
| $R = 1.097 \times 10^7\, \text{m}^{-1}$ <br> $n_f = 3$ <br> $\lambda = 1.88 \times 10^{-6}\, \text{m}$ <br> $n_i = ?$ | ▪ Identify the relevant data from the question. |
| $\dfrac{1}{\lambda} = R\left(\dfrac{1}{n_f^{\,2}} - \dfrac{1}{n_i^{\,2}}\right)$ | ▪ Identify the correct formula. |
| $\left(\dfrac{1}{n_f^{\,2}} - \dfrac{1}{n_i^{\,2}}\right) = \dfrac{1}{\lambda R}$ <br> $-\dfrac{1}{n_i^{\,2}} = \dfrac{1}{\lambda R} - \dfrac{1}{n_f^{\,2}}$ <br> $\dfrac{1}{n_i^{\,2}} = \dfrac{1}{n_f^{\,2}} - \dfrac{1}{\lambda R}$ | ▪ Rearrange the formula. |
| $\dfrac{1}{n_i^{\,2}} = \dfrac{1}{3^2} - \dfrac{1}{1.88 \times 10^{-6}\, \text{m} \times 1.097 \times 10^7\, \text{m}^{-1}}$ | ▪ Substitute values with correct units. |
| $\dfrac{1}{n_i^{\,2}} = \dfrac{1}{0.0626}$ <br> $n_i^{\,2} = 15.97$ <br> $n_i = 4$ | ▪ Calculate the answer with appropriate significant figures. |

**TRY THESE YOURSELF**

1 From which energy shell would an electron have fallen in a hydrogen atom if, when falling to the second energy shell, it emitted a photon with a wavelength of 486 nm?

2 From which energy shell would an electron have fallen when it fell to the first energy shell if the wavelength emitted is the longest possible?

## Conservation of energy

Information and communication technology capability

The law of conservation of energy states simply that 'energy cannot be created nor destroyed, but can be transformed into other forms'. We have seen how mass can be considered as another form of energy in $E = mc^2$, and that there exists an equivalence between mass and energy when they are transformed from one to the other.

When an electron loses energy, the lost energy must be accounted for. When an electron transitions between energy shells in an atom by falling from a higher energy shell to a lower shell, the law of conservation of energy tells us that this lost energy must be transformed into another form. We have seen that this lost energy appears in the form of a photon of electromagnetic radiation with a frequency $f$. The energy of the photon is given by $E = hf$. Indeed, if this photon was absorbed by another electron in the lower energy shell the electron would jump back to the higher energy shell, and energy would be conserved.

Electron transitions in atoms, along with the production of emission or absorption lines, are yet another example of the law of conservation of energy in the natural world.

**Online calculator for Rydberg equation solutions**

Enter the energy shell numbers. Extra information is also found here.

Students studying Chemistry for the HSC will discuss the conservation of energy along with another quantity, entropy, in their course. This is not a requirement of the Physics course.

> **KEY CONCEPTS**
> - The Balmer series for hydrogen occurs when electrons fall to the second energy shell and emit a photon, or jump up from the second energy shell when absorbing a photon.
> - Other hydrogen spectrum series occur, for example when electrons fall to the first energy shell or third energy shell.
> - The Rydberg equation predicts the wavelengths emitted when an electron moves between specified energy shells in a hydrogen atom.
> - The Rydberg equation does not work for other elements.
> - The process of photon emission or absorption is governed by the law of conservation of energy.

**CHECK YOUR UNDERSTANDING**

**15.3**

1. What prevents an electron from emitting a photon when it is in the ground state?
2. Explain why there are a number of spectral lines with different frequencies in the Balmer series.
3. Why are the spectral lines caused when an electron falls to the first energy shell not visible to the naked eye?
4. Use the information in Figure 15.11b on page 364 to find the frequency of a photon emitted when an electron in a hydrogen atom falls from the fifth to the third energy shell.
5. Use your answer to the previous question to find the wavelength of a photon that would be absorbed by an electron in the third energy shell that would allow it to jump to the fifth energy shell.
6. Write down the Rydberg equation and give a definition of $n_i$ and $n_f$.
7. Why is Rydberg's equation considered to be an empirical equation?
8. How does the law of conservation of energy play a role in the production of emission and absorption spectra?

## 15.4 de Broglie's matter waves

For many centuries, light had always been thought of as waves. Light can undergo reflection, refraction, diffraction and interference, all of which are fundamental wave properties. However, as shown in chapter 11, when Einstein was trying to explain the photoelectric effect theoretically in 1905, he had to assume that light behaved like particles, where each particle had the energy equal to Planck's constant multiplied by the light frequency, $E = hf$. This particle model of light successfully explained the photoelectric effect. The phenomenon that light can behave as both a wave and a particle at the same time is known as **wave–particle duality**.

So, how can light possess both wave and particle characteristics at the same time? The answer lies in the fact that the wave model and particle model should be seen as complementing each other to give a more complete description of light. When dealing with properties such as reflection, deflection, refraction, diffraction and interference, the wave model of light applies. On the other hand, when explaining phenomena like the photoelectric effect, the particle model of light is more suitable.

## Relationship between wave characteristics and particle properties

To describe wave–particle duality quantitatively, it is possible to derive a formula using Einstein's mass–energy equivalence and Planck's hypothesis (see chapter 11).

Einstein's equation:

$$E = mc^2$$

and then Planck's hypothesis:

$$E = hf$$

Equating the two equations gives:

$$mc^2 = hf$$

$$mc^2 = \frac{hc}{\lambda}$$

and

$$mc = \frac{h}{\lambda}$$

so:

$$\lambda = \frac{h}{mc}$$

where:

$\lambda$ = wavelength of the light (m)
$h$ = Planck's constant = $6.62 \times 10^{-34}$ J s
$m$ = mass of the photon (kg)
$c$ = speed of the photon (light) = $3.00 \times 10^8$ m s$^{-1}$
$mc$ = momentum of the photon, measured in kg m s$^{-1}$

It is important to note that the left-hand side of the equation is 'wavelength', which is one feature of a wave, whereas the right-hand side contains 'momentum', which is an exclusive particle property.

**FIGURE 15.14** Louis de Broglie

### Matter waves

The wave–particle duality is not limited to light or other forms of electromagnetic radiation. In fact, it can be generalised to all other substances. This principle was first proposed by Louis de Broglie in 1924. de Broglie (Figure 15.14) 'borrowed' the equation $\lambda = \dfrac{h}{mc}$ and, believing that particles have properties similar to waves, generalised the equation by replacing the $c$ (speed of light) by $v$, the speed of any particle. The consequence of this transformation is that any particle that has a momentum can have a wavelength, and therefore can behave like a wave called the matter wave. This is particularly true for small particles like electrons, as we shall see in the examples.

With the formula $\lambda = \dfrac{h}{mv}$, the wavelength of any particle with momentum, that is, any object with mass that is also moving, can be found. The following examples show why normal, everyday objects, do not appear to have a wavelength.

Recall that for everyday objects of mass $m$, where $v =$ speed of the particle ($\text{m s}^{-1}$), momentum of the particle ($\text{kg m s}^{-1}$) $= mv$.

> **WORKED EXAMPLE (15.6)**

Numeracy

What is the de Broglie wavelength of an object with a mass of $1.00\,\text{kg}$ moving at a velocity of $1.00\,\text{m s}^{-1}$?

| ANSWER | LOGIC |
|---|---|
| $m = 1.00\,\text{kg}$ <br><br> $v = 1.00\,\text{m s}^{-1}$ <br><br> $h = 6.626 \times 10^{-34}\,\text{J s}$ | ▪ Identify the relevant data from the question. |
| $\lambda = \dfrac{h}{mv}$ | ▪ Identify the correct formula. |
| $= \dfrac{6.626 \times 10^{-34}\ \text{J s}}{1.00\ \text{kg} \times 1.00\ \text{m s}^{-1}}$ | ▪ Substitute values with correct units. |
| $\lambda = 6.626 \times 10^{-34}\,\text{m}$ | ▪ Calculate the answer. |
| $\lambda = 6.63 \times 10^{-34}\,\text{m}$ | ▪ Express the answer with the correct units and appropriate significant figures. |

**TRY THESE YOURSELF**

1   An electron (mass $= 9.1 \times 10^{-31}\,\text{kg}$) is moving with a speed of $4.0 \times 10^{3}\,\text{m s}^{-1}$. Find its de Broglie wavelength.

2   A tennis ball is served with a speed of $50.0\,\text{m s}^{-1}$. Its mass is $58.5\,\text{g}$. Find its de Broglie wavelength and explain why the wave nature of the ball is not observed during the tennis match.

The wavelengths calculated in Worked example 15.6 are far too small to be observed. Although all objects with momentum possess wave characteristics, most of the wavelengths formed by massive objects are too small to be observed. We usually neglect their wave nature.

## Experimental evidence for matter waves

The proposition of the wave nature of matter by de Broglie needed to be supported through observations. The experiment to confirm the existence of matter waves was performed by Clinton J Davisson and Lester H Germer in 1927. In order to understand the experiment, we need to review the wave property known as diffraction. Diffraction is the bending of waves as they pass around the end of an obstacle. It is best observed as waves pass through a narrow slit.

Consider the situation where a wave is allowed to pass through two slits that are adjacent to each other, as shown in Figure 15.15. As we would expect, the waves will undergo diffraction. Furthermore, the diffracted waves will now make contact with each other; therefore, they will interact with each other to cause interference. Recall that when the crest of one wave meets the crest of another wave, they

Diffraction is discussed in detail in *Physics in Focus Year 11* chapter 8 and in chapter 10 of this volume. It is also recapped in Figure 15.15.

will combine to give an even bigger crest. A similar principle applies to two troughs; this is known as constructive interference. On the other hand, when a crest meets a trough, they will cancel each other out – this is known as destructive interference. Because waves have crests and troughs that alternate, there will be alternating constructive and destructive interferences throughout the region where the two waves are in contact, with the exact pattern determined by the wavelength of the two waves (Figure 15.16). Furthermore, if the waves are visible light, then a series of dark and bright lines can be seen. In the case of sound waves, alternating loud and soft sounds can be heard. For any other waves, an alternating maximal and minimal signal intensity can be detected by instruments. Remember also that diffraction and interference are exclusively wave properties.

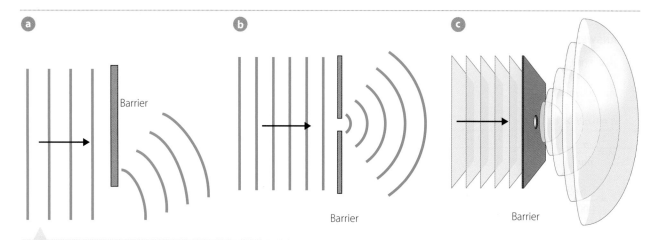

**FIGURE 15.15** **a** Diffraction of a wave as it passes around a corner **b** Diffraction of a wave as it passes through a slit **c** Diffraction of a wave as it passes through a small circular hole

**FIGURE 15.16** Interference pattern formed by two adjacent diffracted waves. Note that the brighter regions represent constructive interference whereas the darker regions represent destructive interference.

## The experiment

In 1927 Davisson and Germer set up an experiment as shown in Figure 15.17. They fired electrons towards a nickel crystal and studied the behaviour of these electrons as they scattered off the nickel surface.

The electrons were accelerated using a potential difference of 54 V to achieve a high velocity, and were then directed towards the nickel crystal. The electrons, on reaching the nickel crystal, would be scattered off different planes of the nickel crystal. Some of the returning (scattered) electrons would pass through the gaps between the nickel atoms, which act as many 'slits', so diffraction would occur. The situation was similar to that shown in Figure 15.16. Consequently, interference patterns would be formed by the returning electrons. If a detector was run alongside the nickel crystal, a series of maxima and minima of electron intensity should be detected if diffraction were occurring.

In their experiment, Davisson and Germer were able to observe the predicted series of maxima and minima of the scattered electrons, thus proving the wave nature of the electrons and hence the existence of matter waves. Furthermore, from the interference pattern they were able to measure the wavelength of the electron waves that resulted from this diffraction pattern. The value they found agreed with the wavelength calculated using de Broglie's equation $\lambda = \dfrac{h}{mv}$. In summary, the experiment was

Information and communication technology capability

Quantum wave interference simulation – visualising

9780170409131

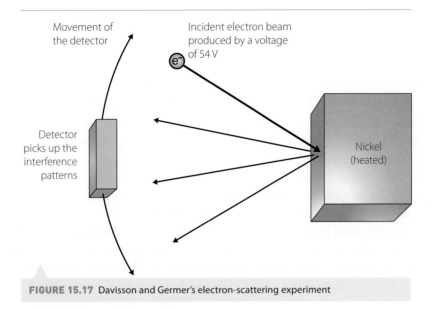

**FIGURE 15.17** Davisson and Germer's electron-scattering experiment

successful not only in determining the existence of electron waves, and therefore matter waves, but also in confirming the validity of de Broglie's equation to describe these matter waves, $\lambda = \dfrac{h}{mv}$.

## GP Thomson's electron diffraction experiment

Another important experimental observation confirmed the existence of de Broglie's matter waves. In 1928, GP Thomson, son of JJ Thomson, passed an electron beam through a thin foil of gold. The electrons went through the thin foil and were scattered to land on the photographic film behind the foil, where they created an interference pattern. He compared the pattern to that obtained from using X-rays, which had previously been shown to have wave characteristics. He observed that the two patterns were remarkably similar. These patterns are shown in Figure 15.18. From this, he was able to confirm the wave nature of the electrons. JJ Thomson was awarded the Nobel Prize for showing the particle nature of electrons, then a couple of decades later his son GP Thomson discovered the wave characteristics of electrons and was also awarded a Nobel Prize.

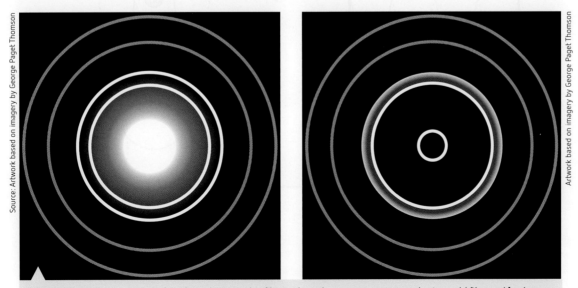

Source: Artwork based on imagery by George Paget Thomson

Artwork based on imagery by George Paget Thomson

**FIGURE 15.18** The images produced on photographic film **a** when electrons were scattered using gold film; and **b** when X-rays, known to have wave characteristics, were scattered from a zirconium oxide crystal

## Applying matter waves to the electrons in an atom

de Broglie stated that electrons can behave as waves, and this is true for all electrons including those found in atoms. de Broglie went on to propose that:

'The electrons in atoms behave like standing waves, which encircle the nucleus in an integral number of whole wavelengths. These are known as electron waves.'

The concept of standing waves is shown in Figure 15.19. Standing waves are waves that do not propagate but instead vibrate between two boundaries. The points that do not vibrate are called nodes, and the points that vibrate between maximum and minimum positions are known as anti-nodes. A faster vibration will result in a higher frequency and hence more waves. If we pick any of the standing waves from Figure 15.19a–c and join the waves from the head to the tail so that they form a closed loop, they resemble the electron waves that wrap around the nucleus.

Furthermore, as pointed out by de Broglie, the number of wavelengths for the electron waves wrapping around the nucleus must be an integer. This is because in order for the standing wave to wrap around the nucleus, the beginning point of the wave must be in phase with the end point of the wave, and this only occurs if the wave finishes with a complete wavelength. For non-integral

Note: It is not correct to think that the sketches in Figure 15.20 represent the pathways for the electrons to move around the nucleus. *The entire wave is actually one electron.*

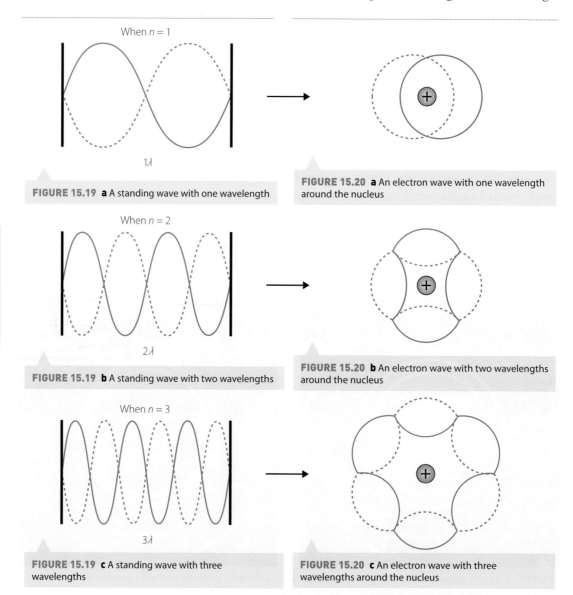

When $n = 1$

$1\lambda$

**FIGURE 15.19 a** A standing wave with one wavelength

**FIGURE 15.20 a** An electron wave with one wavelength around the nucleus

When $n = 2$

$2\lambda$

**FIGURE 15.19 b** A standing wave with two wavelengths

**FIGURE 15.20 b** An electron wave with two wavelengths around the nucleus

When $n = 3$

$3\lambda$

**FIGURE 15.19 c** A standing wave with three wavelengths

**FIGURE 15.20 c** An electron wave with three wavelengths around the nucleus

9780170409131

wavelengths, the beginning and end position of the wave will be out of phase and consequently result in destructive interference, which diminishes the wave. Hence, electrons in the first energy level have one wavelength, as shown in Figure 15.20a, electrons in the second energy level have two wavelengths, as shown in Figure 15.20b, and so on.

## The implications of de Broglie's electron wave model

With the electron wave model of the atom it is possible to explain why electrons, when in their own energy level, are stable and do not emit electromagnetic radiation. Using de Broglie's model, Bohr's first postulate can be explained. Electrons are now understood to be standing waves. They are no longer thought of as moving charges and hence do not emit any radiation. Furthermore, standing waves do not propagate, and therefore they are stable and will not lose any energy.

Second, de Broglie's electron wave model enables a mathematical derivation for Bohr's third postulate. This postulate proposes the quantisation of angular momentum, which Bohr made without any theoretical support:

$$2\pi r_n = n\lambda$$

and

$$\lambda = \frac{h}{mv}$$

so

$$2\pi r_n = \frac{nh}{mv}$$
$$mv(2\pi r_n) = nh$$
$$mvr_n = \frac{nh}{2\pi}$$

Using de Broglie's theory of matter waves and the matter wave equation, we are able to theoretically derive Bohr's third postulate, thereby adding rationality to this postulate.

Historically, de Broglie's model and his matter wave equation $\lambda = \dfrac{h}{mv}$ also formed the foundation of a new area of physics known as **quantum mechanics**. Such physics was later expanded and advanced by the work of many physicists. Quantum mechanics involves complicated physics theory and mathematics.

It is not only electrons that can show wave characteristics such as interference patterns. Atoms including helium, sodium and even large molecules such as carbon-60 'buckyballs' have been used to produce interference patterns. However, the more mass an object has, the smaller its de Broglie wavelength becomes and the harder it is to observe the de Broglie wavelength.

The wave–particle theory originally put forward by de Broglie is now the basis for several important technologies. Electron microscopes such as the one in Figure 15.21a use the wave nature of electrons to create images of objects too small to be resolved using light (Figure 15.21b). It is not possible to 'see' or resolve objects that are smaller than the wavelength used to observe them. Electron microscopy, using the very short wavelengths of electrons, has been enormously valuable to medical and biological sciences. It has allowed organisms such as bacteria and viruses to be resolved, imaged and studied. Neutron diffraction is carried out at the OPAL nuclear reactor at Lucas Heights near Sydney. The wave nature of neutrons is used to produce diffraction patterns. These are used to investigate the properties of materials and test for stress damage in machine parts, among other applications.

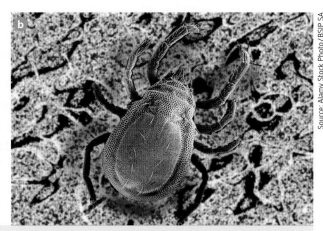

**FIGURE 15.21** **a** An electron microscope; **b** An image taken with an electron microscope of a mite that is less than 1 mm long

## Single particle interference experiments

We have still not answered the question: what is it that is 'waving'? *How* do electrons, which show particle behaviour, produce an interference pattern? What exactly is distributed over space such that interference can take place?

Experiments using electrons and double slits also showed interference patterns, with a pattern of maxima and minima predicted by the de Broglie wavelength. Initially, it seemed that perhaps electrons passing through each slit were interfering with each other. Lots of experiments were done to try to determine what exactly was passing through the slits, and lots of competing theories were developed.

One experimental result that was troubling to physicists was the single photon interference demonstrated by G I Taylor in 1909. The apparatus used for this experiment is shown schematically in Figure 15.22.

Taylor demonstrated that an interference pattern occurred when the light was so dim that only a single photon could be present in the apparatus at a time. This wasn't a problem if light was a wave. Regardless of the intensity, which is related to the amplitude, the wave still spreads out in space and hence passes through both slits and recombines on the other side. But how could this result be explained by the particle theory of light?

Physicists wanted to see if the same thing happened with other particles. In 1974, Pier Giorgio Merli, Giulio Pozzi and GianFranco Missiroli performed a double-slit experiment with electrons. In their experiment, as in Taylor's, only a single particle was in the apparatus and incident on the slits at a time. Again, interference was observed. The most famous single-electron interference experiment was performed by Akira Tonomura and colleagues in Japan in 1989. Their experimental set-up is shown in Figure 15.23.

Their results clearly showed the build-up of an interference pattern, one electron at a time, as shown in Figure 15.24.

This experiment can be analysed in the same way as the double-slit experiment for light in chapter 10. The electron biprism acts as a double slit for the incident electrons. Interference maxima are observed at angles given by $d\sin\theta = m\lambda$, $m = 0, 1, 2, \ldots$ where $d$ is the effective slit spacing due to the biprism and $\lambda$ is the de Broglie wavelength of the electron. The angle $\theta$ is the angle between the normal line joining the biprism and the detector screen and the line from the biprism to the point of interest on the screen.

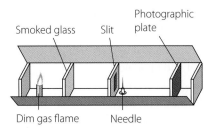

**FIGURE 15.22** Taylor's experiment (1909). There was never more than one photon in the box at any one time, yet an interference pattern developed on the photographic plate.

9780170409131

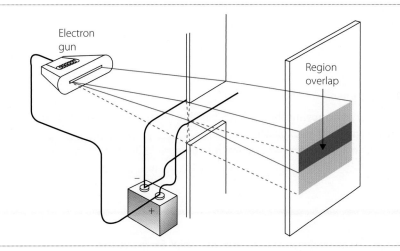

FIGURE 15.23
Tonomura's electron biprism experiment. Electrons diffract around the charged central wire and recombine to produce an interference pattern.

The results were now clear and indisputable. Single electrons can produce interference patterns. So it cannot be that the electrons passing through one slit interfere with electrons passing through the other. Each electron in some way interacts with and passes through both slits simultaneously.

It is hard to imagine how a particle could do this. We need to use the wave model here. In the modern interpretation of quantum mechanics we think of the particle as being built up of waves that can pass through both slits at the same time. What exactly these waves represent has been a matter of dispute among physicists ever since. The most accepted theory currently is that the waves represent the *probability* of the electron passing through each slit. The theory states that the probability of the particle passing through each slit is represented by a wave. It is these probability waves that interfere, creating the pattern seen in Figure 15.24. The pattern appears even though the electrons themselves are detected as particles at single distinct points.

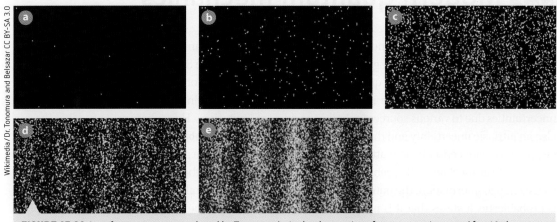

**FIGURE 15.24** Interference pattern produced by Tonomura's single-electron interference experiment. **a** After 10 electrons and **b** after 100 electrons the pattern looks random, but **c** after 3000 electrons have arrived, the interference pattern is starting to emerge. In **d** and **e** the interference pattern has become even clearer.

So what happens if we know which slit the electron passes through? In this case the probability of the wave passing through one slit is zero, and through the other slit is one. Surely if this interpretation of the wavelength is correct, the pattern should disappear. When the experiment was done, the pattern *did* disappear. There was argument over the validity of the experiment, over whether the detector prevented the interference, and so on. But many versions of the experiment have been performed, with the same results.

de Broglie's matter waves

The electron double-slit experiment can be analysed in exactly the same way as Young's double-slit experiment for light, as described in chapter 10.

- de Broglie proposed the existence of matter waves where $\lambda = \dfrac{h}{mv}$.
- Davisson and Germer observed the diffraction of electrons and confirmed the existence of matter waves.
- The wave nature of electrons can be applied to their orbits around the nucleus to explain the existence of stable electron orbits, where $mv r_n = \dfrac{nh}{2\pi}$.
- The wave nature of everyday moving objects is too small to observe or measure due to the extremely short wavelengths of such objects.
- Matter waves are used in such devices as electron microscopes and for neutron scattering to observe the structure of materials.

## CHECK YOUR UNDERSTANDING

### 15.4

1 Show how the formula $\lambda = \dfrac{h}{mv}$ can be derived from the relationships $E = hf$ and $v = f\lambda$.

2 Use an example of a 1000 kg car moving at $25\,\text{m s}^{-1}$ to show that the car's wavelength is too small to be observed.

3 Outline the observations made by Davisson and Germer that confirmed the existence of matter waves.

4 Outline one other experiment that also confirmed the existence of matter waves.

5 Describe an application of matter waves.

6 What relationship must exist between the circumference of an electron's orbit and its wavelength in order for a stable orbit to exist?

7 Why are the properties of diffraction and interference so important in the experiments that proved the existence of matter waves?

## 15.5 Modern quantum mechanics

From the 1920s, what is now known as modern quantum mechanics was developed by physicists including Niels Bohr, Louis de Broglie, Werner Heisenberg and Erwin Schrödinger. The term 'modern' is used to distinguish it from earlier quantum mechanics, such as the Bohr model.

The new model included the idea of uncertainty. In any experiment, as you know, there will be uncertainties due to various sources including equipment limitations. Heisenberg proposed that there is also an intrinsic uncertainty and that the behaviour of particles is probabilistic. In other words, it cannot be predicted with certainty no matter how much you know about the particles.

In classical mechanics the behaviour of all objects including subatomic particles is deterministic. Given enough information, the outcome of any experiment is predictable. Modern quantum mechanics is probabilistic. It states that it is not possible to predict the outcome of an experiment with complete certainty. This is a fundamental difference between quantum mechanics and classical mechanics.

### Schrödinger's contribution

In 1926, Erwin Schrödinger applied mathematical equations involving probability to the electrons in the Bohr atom. Solutions to what is known as Schrödinger's equation give solutions to any quantum mechanical model. The solution to Schrödinger's equation when applied to electrons can give the likelihood that an electron will be found in a particular position around the nucleus. This gave rise to the idea of an electron cloud or orbital around the nucleus, with different orbitals having a different cloud shape. The density of the electron cloud gave the probability of the electron being found at that position. Figure 15.25 shows the contrast between the Bohr model and Schrödinger's quantum model in visualising the position of electrons around the nucleus in an atom. The quantum model can only predict the probability of finding an electron at a certain position.

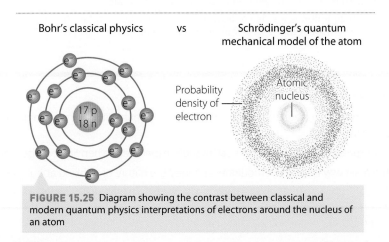

**FIGURE 15.25** Diagram showing the contrast between classical and modern quantum physics interpretations of electrons around the nucleus of an atom

## Schrödinger's cat

In an attempt to show the absurdity of applying quantum mechanical solutions to large-scale systems, Schrödinger proposed a thought experiment, illustrated in Figure 15.26. A cat was in a closed box. A small amount of a radioactive substance that had exactly 50% chance of decaying was placed inside the box. If it decayed, the radioactivity would be detected by a Geiger counter in the box, which could then release a hammer that would fall and smash a vial of poison, killing the cat. The interpretation of a quantum mechanical solution to this situation is that the cat would be both alive and dead until the box is opened and the cat observed. The act of observing the cat would cause the wavefunction to collapse, resulting in the cat being either alive or dead, but no longer both. Such an interpretation of quantum mechanics is known as the Copenhagen interpretation, where a system can be in all states of existence until it is observed. Even though it was believed that Schrödinger's thought experiment was his attempt to support the application of quantum mechanics to larger systems, Schrödinger was not content with such an interpretation.

Modern quantum mechanics

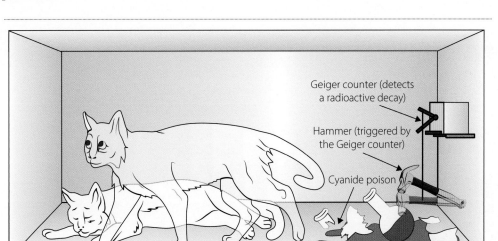

**FIGURE 15.26** Schrödinger's cat thought experiment was proposed to show the absurdity of applying quantum mechanics to large-scale systems.

● Schrödinger developed a mathematical model and equation that could give the probability of a quantum event or position in quantum mechanical systems.

● The model of the atom was advanced from Bohr's interpretation, which had electrons in known positions, to that of Schrödinger's electron probability clouds.

● Schrödinger's equation gives rise to absurd situations when applied to large-scale systems, as illustrated by the cat in the box thought experiment.

**CHECK YOUR UNDERSTANDING**

**1** What does an electron density cloud represent?

**2** Outline the ways in which a diagram of Bohr's atomic model would differ from a quantum mechanical diagram of an atom.

**3** What was the purpose of the 'Schrödinger's cat' thought experiment when Schrödinger first proposed it?

**4** Summarise the main way in which Schrödinger changed the nature of thinking about the model of the atom.

- Rutherford's atomic model was modified by Bohr in an attempt to overcome many of its limitations.

- Bohr's model could predict the positions of the spectral lines of hydrogen.

- Bohr's model was based on his postulates, stating electrons could orbit the nucleus only in allowed energy shells.

- Electrons in energy shells with a larger radius have more energy.

- When electrons jump between energy shells they emit or absorb a quantum of energy equal to the difference in energies of the shells.

- Electrons' angular momentum is quantised.

- Bohr's model could not predict the spectra of multi-electron atoms, the differences in intensities of the lines or the splitting of lines (the Zeeman effect).

- The Zeeman effect is the splitting of spectral lines when hydrogen atoms are subjected to a magnetic field.

- A spectroscope can be used to observe spectra.

- The three main types of spectra are emission, absorption and continuous.

- Incandescent lights emit a continuous spectrum.

- Fluorescent lights emit an emission spectrum because they are an excited low pressure gas.

- The dark lines in the solar spectrum are an example of an absorption spectrum.

- The series of lines in a hydrogen spectrum called the Balmer series occurs when electrons move from a higher energy shell to shell $n = 2$ for emission lines, or from $n = 2$ to higher energy shells for absorption lines.

- The wavelengths of lines in the emission spectrum are equal to the wavelengths of the lines in absorption spectrum for hydrogen.

- Other hydrogen spectrum series are the Lyman (ultraviolet) and Paschen (infrared) series.

- The Balmer series for hydrogen occurs when electrons fall to the second energy shell and emit a photon, or jump up from the second energy shell when absorbing a photon.

- Other hydrogen spectrum series occur when electrons fall to the first energy shell or third energy shell, and so on.

- The Rydberg equation predicts the wavelengths emitted when an electron moves between specified energy shells in a hydrogen atom.

- The Rydberg equation does not work for elements other than hydrogen.

- The process of photon emission or absorption is governed by the law of conservation of energy.

- de Broglie proposed the existence of matter waves where $\lambda = \dfrac{h}{mv}$.

- Davisson and Germer observed the diffraction of electrons and confirmed the existence of matter waves.

- The wave nature of electrons can be applied to their orbits around the nucleus to explain the existence of stable electron orbits, where $mvr_n = \dfrac{nh}{2\pi}$.

- The wave nature of everyday moving objects is too small to observe or measure due to the extremely short wavelengths of such objects.

- Matter waves are used in such devices as electron microscopes and for neutron scattering to observe the structure of materials.

- Schrödinger developed a mathematical model and equation that could give the probability of a quantum event or position in quantum mechanical systems.

- The model of the atom was advanced from Bohr's interpretation, which had electrons in known positions, to that of Schrödinger's electron probability clouds.

- Schrödinger's equation gives rise to absurd situations when applied to large-scale systems, as illustrated by the cat in the box thought experiment.

1   Draw a diagram to show a model of the atom as proposed by Rutherford.

2   Why were electrons in Rutherford's atom expected to collapse into the nucleus?

3   What force within the atom kept electrons in their orbits in the Rutherford model?

4   Describe Bohr's postulates and how they helped to explain the limitations of Rutherford's model of the atom.

5   Identify the observations that could not be explained by Bohr's atomic model.

6   Describe the Zeeman effect, using diagrams in your response.

7   How did Bohr's model explain that spectral lines in absorption and emission spectra occurred at the same frequencies?

8   What is the purpose of a spectroscope?

9   Give examples of sources for each of the three types of spectra, absorption, emission and continuous.

10  List an important risk and a control when using a spectroscope outdoors.

11  Identify three spectral series evident for hydrogen and state in which part of the electromagnetic spectrum they can be observed.

12  For the three spectral series identified in Question **11**, explain why they form separate series in terms of electron transitions between energy shells.

13  Describe the relationship between the energy of a photon and its frequency.

14  How is the energy of a photon emitted when an electron falls from one energy shell to another related to the energies of the shells?

15  Using the Rydberg equation, find the wavelength of a photon emitted when an electron in a hydrogen atom falls from the seventh to the second energy shells.

16  What are the limitations of the Rydberg equation?

17  Calculate the de Broglie wavelength of an electron moving at half the speed of light. Use $m_e = 9.1 \times 10^{-31}$ kg.

18  Why did the diffraction of electrons show that they had wave properties?

19  Outline how G P Thomson showed the wave nature of electrons.

20  Describe how the wave nature of electrons can be used to determine stable electron orbits.

21  Describe two examples of the application of the wave nature of neutrons or electrons.

22  Analyse the significance of Schrödinger's contribution to advances in the model of the atom.

23  Compare and contrast two models of the atom and describe why one is more accurate than the other.

INQUIRY QUESTION

How can the energy of the atomic nucleus be harnessed?

OUTCOMES

**Students:**

- analyse the spontaneous decay of unstable nuclei, and the properties of the alpha, beta and gamma radiation emitted (ACSPH028, ACSPH030) ICT
- examine the model of half-life in radioactive decay and make quantitative predictions about the activity or amount of a radioactive sample using the following relationships:

  – $N_t = N_0 e^{-\lambda t}$
  – $\lambda = \dfrac{\ln 2}{t_{1/2}}$

  where $N_t$ = number of particles at time $t$, $N_0$ = number of particles present at $t = 0$, $\lambda$ = decay constant, $t_{1/2}$ = time for half the radioactive amount to decay (ACSPH029) ICT, N
- model and explain the process of nuclear fission, including the concepts of controlled and uncontrolled chain reactions, and account for the release of energy in the process (ACSPH033, ACSPH034) ICT
- analyse relationships that represent conservation of mass-energy in spontaneous and artificial nuclear transmutations, including alpha decay, beta decay, nuclear fission and nuclear fusion (ACSPH032) ICT, N
- account for the release of energy in the process of nuclear fusion (ACSPH035, ACSPH036) ICT
- predict quantitatively the energy released in nuclear decays or transmutations, including nuclear fission and nuclear fusion, by applying: (ACSPH031, ACSPH035, ACSPH036) ICT, N
  – the law of conservation of energy
  – mass defect
  – binding energy
  – Einstein's mass–energy equivalence relationship $E = mc^2$

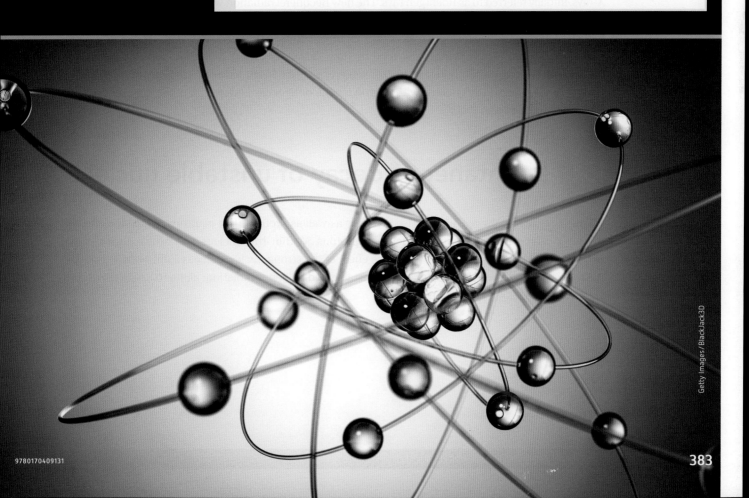

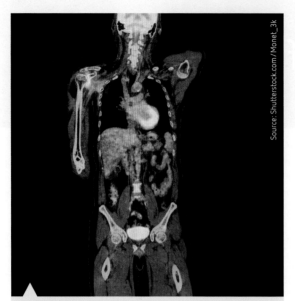

**FIGURE 16.1** Radiolabelled tracers introduced into the body by intravenous injection travel to specific organs or tissues and produce emissions that can be seen by PET scanners

Since the discovery of radioactivity by Henri Becquerel in 1896, the stability of the nucleus of the atom has been the subject of intense scrutiny and study. Einstein's prediction of the possibility of an atomic bomb led to its development during World War II's Manhattan Project. The resources of much of the western world's scientific community were directed at developing the bombs that helped bring an end to the conflict. Now there are over 300 nuclear power stations operating in dozens of countries around the world. We have utilised radioactivity to develop medical techniques that detect and treat cancer (Figure 16.1), and industrial applications such as checking for cracks in the turbines of jet engines, to name just two of the many applications beneficial to our society. The study of radioactivity to increase and broaden our understanding of matter will continue to yield many exciting and useful discoveries in the years to come.

Becquerel called the radiation 'metal phosphorescence' – his training was in studies of phosphorescence – and he thought of these emanations as an invisible type of light.

Becquerel showed that atoms were capable of emitting smaller particles, meaning that atoms must be divisible. This was the beginning of investigations into radioactivity and the use of nuclear radiations to investigate the structure of the atom. Just prior to Becquerel's discovery, Roentgen had discovered X-rays, initially referred to as Roentgen rays. The new Becquerel radiation was soon shown to be different from X-rays. The first of these radiations to be identified was called alpha particles, the second **beta particles**, and the third gamma rays, after the first three letters in the Greek alphabet – α, β and γ. Alpha particles were extensively studied and used by New Zealander Ernest Rutherford after he discovered them while working at Cambridge in 1899. Becquerel identified beta particles as electrons in 1900. Paul Villard, however, is widely credited with identifying the differences between gamma rays, other nuclear radiations and X-rays in 1900.

## 16.1 Spontaneous decay of unstable nuclei

All nuclei, except hydrogen-1, comprise neutrons as well as protons. The neutrons reduce the effect of the electrostatic force of repulsion between protons. Within the nucleus the strong nuclear force overcomes the electrostatic force. When the effect of the strong nuclear force is sufficiently strong, the nucleus is stable. Otherwise the nucleus is unstable and will emit radiation: alpha (α), beta (β) and/or gamma (γ) rays.

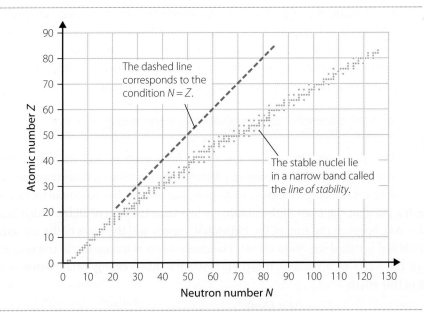

FIGURE 16.2 The stability curve for nuclides

Nuclides with nuclei comprising up to about 40 nucleons (neutrons plus protons) are stable when they have equal, or nearly equal, numbers of protons and neutrons. Heavier nuclei are more stable when the balance is in favour of neutrons.

In the stability curve shown in Figure 16.2, unstable nuclides below the line of stability are most likely to decay by $\beta^-$ decay. This decay increases the proton number, $Z$, by +1, while reducing the neutron number, $N$, by one. The resulting daughter nuclide will then be on or closer to the stability line. These decay processes are discussed in detail later in this chapter.

Uranium and thorium are examples of naturally occurring radioactive substances. Artificial radioisotopes are routinely made for medical and industrial applications.

## Features of nucleons

Protons and neutrons are collectively called nucleons. The atomic mass number (A) is the nucleon number. The atomic number (Z) is the proton number. The number of neutrons (N) in the nucleus of an atom is the difference between the mass number (A) and the atomic number (Z):

$$N = A - Z$$

**TABLE 16.1**

| SYMBOL | NAME | DESCRIPTION |
|--------|------|-------------|
| A | Mass number (nucleon number) | Number of protons and neutrons in the atom |
| Z | Atomic number (proton number) | Number of protons |
| N | Number of neutrons | $N = A - Z$ |

The peninsula of nuclear stability

Stability curve

This link provides nuclide information and allows you to predict decay of nuclides.

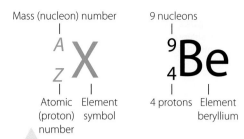

Mass (nucleon) number — $A$ — $X$ — $Z$ — Atomic (proton) number — Element symbol

9 nucleons — $^{9}_{4}$Be — 4 protons — Element beryllium

**FIGURE 16.3** International standard notation for representing a nuclide

The number of protons (atomic number) is used to name atoms, as illustrated in Figure 16.3. For example, all atoms with 8 protons are called oxygen. All atoms with 79 protons are called gold. The periodic table found on the inside front cover of this book shows the names and symbols for each of the elements, sorted according to their proton number.

## Energy stored in, and released from, nuclei

We have seen that nuclei are made from protons and neutrons. The energy that would be needed to disassemble a nucleus into its component nucleons is the nuclear binding energy. Each nucleon, on its own, has a mass. But when nucleons are brought together to form a nucleus, the mass of the nucleus is less than the sum of all the individual nucleons. The difference, $\Delta m$, between the sum of the individual masses and the mass of the nucleus into which they are combined is called the mass defect. The mass defect is a measure of the energy, $\Delta E$, needed to bring all the parts of a nucleus together. Einstein's mass–energy relationship is a quantitative statement of this effect:

$$\Delta E = (\Delta m)c^2$$

where $c$ = the speed of light, $3.00 \times 10^8\,\mathrm{m\,s^{-1}}$.

Some of the mass of the individual nucleons appears as the binding energy of the nucleus. In this sense, it is best to consider mass and energy as interchangeable:

$$\text{mass} \Leftrightarrow \text{energy}$$

Mass and energy are equivalent, as has been explored in chapters 12 and 13. Mass is a manifestation of energy.

To understand what is happening when unstable nuclei decay and emit radiation, we need to explore some commonly used terms that apply to particular characteristics of nuclei.

### Nuclides, 'the element' and isotopes

A nuclide is a species of atom or nucleus classified according to the number of protons and neutrons as well as its energy state. As with electrons, the energy level shell model is used to explain energy transitions in the nucleus. The most stable nuclides exist in their lowest or ground state. Nuclides that are in excited energy states above the ground state are unstable and decay radioactively until a stable nuclide is formed. The energies involved are far greater than for electron energy level transitions.

A substance that has only nuclides with the same number of protons is called an element. Unfortunately, there is some ambiguity about the use of the word 'element'. When you take a naturally occurring sample of an element, you typically find several different nuclides or isotopes. Each has the same number of protons but different isotopes have different numbers of neutrons. The nuclide that is most common is sometimes called 'the element', but this is not the standard meaning of element. All nuclides with the same number of protons are isotopes of the same element.

## Radioactive decay

Radioactive atoms can emit α or β particles that have mass, and massless gamma rays, which are pure energy with no mass. As a result of α and β emission, the radioactive atom is transformed into an atom of a different chemical element. An example of this is when radioactive carbon-14, with 6 protons and 8 neutrons, undergoes β⁻ decay to form nitrogen-14, with 7 protons and 7 neutrons. Changes in radioactive decay are different from chemical changes. Chemical reactions involve a rearrangement of electrons, while nuclear reactions involve changes in the nucleus. Chemical changes may release energy, but nuclear processes release far, far more.

When the energy conditions are right in a nucleus, it can enter a transitional state with a higher energy than is stable. It then releases the energy as a particle ($\alpha$ decay or $\beta$ decay) or as electromagnetic radiation ($\gamma$-ray). Figure 16.4 shows the decay of iron-59 to the unstable transition state cobalt-59*. This then $\gamma$-decays to stable cobalt-59. Depending on the radioactive decay sequence, neutrinos or antineutrinos, tiny particles with almost no mass, are also emitted.

Radioactive decay is a random, spontaneous and uncontrollable process involving the nucleus of an atom. When a radioactive nucleus emits an alpha or beta particle, it breaks into two parts – the lighter, emitted particle and a new nucleus of a different element. This is because, in radioactive decay, it is the nucleus of the atom that changes. In a gamma emitter, the same nucleus emerges, but at a lower energy level. When a nucleus of one element (the parent nuclide) decays, it becomes a nucleus of a different element (the daughter nuclide). The daughter nuclide and the emitted particles are the decay products.

Radioactive decay is also called nuclear transformation, disintegration and transmutation.

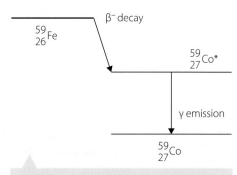

**FIGURE 16.4** For the radioactive nuclide $^{59}_{26}$Fe, a beta decay followed by gamma emission results in a stable nuclide in its ground state.

## Types of radioactivity

Alpha particles are helium nuclei. Beta particles may be electrons or their anti-particle equivalent, positrons. They are assigned the charge −1 for electrons and +1 for positrons, but they have no nucleon number because they have negligible mass. Gamma rays have neither nucleon number nor charge. In all nuclear reactions, the nucleon number remains constant. The net charge is also conserved. The nucleus never takes part in chemical reactions; thus, the nucleus goes wherever the atom goes. It is indestructible by means of chemical reactions. Table 16.2 summarises the types of radiation.

Before and after a nuclear reaction there is the same:

▹ number of nucleons

▹ net charge.

**TABLE 16.2** Summary of terms

| RADIATION TYPE | SYMBOL | DESCRIPTION |
|---|---|---|
| Alpha particle | $\alpha$, or $^4_2$He | Helium-4 nuclide |
| Beta particle | $\beta^-$, or $^0_{-1}$e | Electron |
| | $\beta^+$, or $^0_{+1}$e | Positron |
| Gamma ray | $\gamma$, or $^0_0\gamma$ | Electromagnetic radiation |
| Neutrino | $\nu_e$, or $^0_0\nu_e$ | Energy carrier |
| Antineutrino | $\bar{\nu}_e$, or $^0_0\bar{\nu}_e$ | Energy carrier |

## Alpha decay: $^4_2$He or $\alpha$

The $\alpha$ particle is a positively charged helium nucleus, which contains two protons and two neutrons. It has been stripped of its two electrons, so it carries a +2 charge. The most common nuclide of uranium, $^{238}_{92}$U, undergoes $\alpha$-decay, resulting in 90 protons and 234 nucleons in the daughter nucleus. This is thorium, $^{234}_{90}$Th. The nuclear reaction equation is:

$$^{238}_{92}\text{U} \rightarrow {}^{234}_{90}\text{Th} + {}^4_2\text{He}$$

Simulation of
alpha decay

In all radioactive decays, energy is released. In this decay reaction, almost all the energy released is taken away by the α particle. The mass numbers balance ($238 - 234 = 4$) and the atomic numbers balance ($92 - 90 = 2$).

In general, α-decay can be written as:

$$^{A}_{Z}X \rightarrow \,^{A-4}_{Z-2}Y + \,^{4}_{2}He$$

Use the periodic table in Appendix 6 for all worked examples.

In a nuclear reaction equation involving alpha emission, different atomic symbols appear on either side of the equation. This is because new nuclides are formed as a result of the emission of alpha particles. In a nuclear reaction equation the same number of nucleons appear on both sides of the equation. That is, the sum of the top row of numbers, the mass numbers, is equal on either side of the reaction.

## WORKED EXAMPLE (16.1)

Neptunium-237 decays by emitting an α particle and changes to a completely different element.

**a** Write a complete nuclear reaction equation that includes the symbol for the daughter nuclide.

**b** What is the name of the daughter nuclide?

| ANSWERS | LOGIC |
|---|---|
| **a** $^{237}_{93}Np \rightarrow \,^{233}_{91}Pa + \,^{4}_{2}He$ | ▪ Use the correct symbols for all particles, and conservation of nucleon number and proton number. |
| **b** Pa = protactinium | ▪ Locate the symbol Pa on the periodic table and state the correct name. |

### TRY THESE YOURSELF

1 Francium-211 decays by emitting an α particle and changes to a completely different element.

    **a** Write a complete nuclear reaction equation that includes the symbol for the daughter nuclide.

    **b** What is the name of the daughter nuclide?

2 Polonium-213 decays by emitting an α particle and changes to a completely different element.

    **a** Write a complete nuclear reaction equation that includes the symbol for the daughter nuclide.

    **b** What is the name of the daughter nuclide?

## Beta decay

There are two forms of beta decay: electron emission and positron emission. An electron has opposite charge to a proton, but it is not a nucleon because it cannot reside in the nucleus. Its symbol is written $^{0}_{-1}e$ or $\beta^{-}$.

A positron is an anti-electron. It has the same charge as a proton, but it is not a nucleon. Its symbol is written $^{0}_{+1}e$ or $\beta^{+}$.

The mass of each of the two beta particles, positrons and electrons, is very small compared with the masses of nucleons. They have a mass of $9.11 \times 10^{-31}$ kg, which is tiny compared with the mass of a proton ($1.6726 \times 10^{-27}$ kg) or a neutron ($1.6749 \times 10^{-27}$ kg). This difference in mass is good evidence that, although these particles could be emitted from the nucleus, they are not nucleons.

## Electron emission or β⁻ decay

The ejection of an electron from the nucleus, β⁻ decay, can be modelled, for the time being, by regarding a neutron as capable of changing into a proton and an electron. Another particle, an uncharged and almost undetectable antineutrino, $\bar{\nu}$, is also released in this nuclear reaction.

This model of β⁻ decay can be written as:

$$^1_0n \rightarrow ^1_1p + ^0_{-1}e + \bar{\nu} \quad \text{or} \quad ^1_0n \rightarrow ^1_1H + ^0_{-1}e + \bar{\nu}$$

The electron formed is ejected from the nucleus immediately as a β⁻ particle with a very high speed, around two-thirds the speed of light.

When thorium undergoes β⁻ decay, it produces the nuclide with 91 protons but an unchanged mass number of 234. The new element is protactinium, Pa:

$$^{234}_{90}Th \rightarrow ^{234}_{91}Pa + ^0_{-1}e + \bar{\nu}$$

It is important to recognise that the electron (β⁻ particle) comes from the nucleus and not from the atomic shell electrons.

When a nucleus emits a β⁻ particle, the atomic mass (nucleon) number does not change, but the atomic number ($Z$) of the daughter nuclide increases by 1. That is, the resulting atom belongs to an element one place along in the periodic table.

In general, β⁻ decay can be represented as:

$$^A_Z X \rightarrow ^A_{Z+1} Y + ^0_{-1}e + \bar{\nu}$$

Several beta-emitters are used as radiopharmaceuticals to treat a range of diseases. The choice of radiopharmaceutical depends on the problem and location. For example, iridium-192 is used to treat prostate cancer while iodine-131 is used for thyroid cancer.

## Positron emission or β⁺ decay

The ejection of a positron from the nucleus, β⁺ decay, can be modelled by regarding a proton, or hydrogen nuclide, as capable of changing into a neutron and a positron. Another particle, an uncharged and almost undetectable neutrino, $\nu$, is also released in this nuclear reaction.

This model of β⁺ decay can be written as:

$$^1_1p \rightarrow ^1_0n + ^0_{+1}e + \nu \quad \text{or} \quad ^1_1H \rightarrow ^1_0n + ^0_{+1}e + \nu$$

When thallium-195 undergoes β⁺ decay, it produces a nuclide with 80 protons but an unchanged mass number of 195. The new element is mercury, Hg:

$$^{195}_{81}Tl \rightarrow ^{195}_{80}Hg + ^0_{+1}e + \nu$$

When a nucleus emits a positron, the mass (nucleon) number does not change, but the atomic number ($Z$) of the daughter nuclide decreases by 1; that is, the resulting atom belongs to an element one place back in the periodic table.

In general, positron decay can be represented as:

$$^A_Z X \rightarrow ^A_{Z-1} Y + ^0_{+1}e + \nu$$

During positron emission tomography (PET), a patient is given a positron emitter, such as $^{18}_9F$-fluorodeoxyglucose (FDG) or oxygen-15 $\left(^{15}_8O\right)$. The selected positron emitter is known to accumulate in an organ or other tissue of interest. The positrons interact with electrons to produce gamma rays of known energy and direction. Collectors identify these gamma rays and compute an image (Figure 16.5).

**Radioisotopes in medicine**
Information regarding the many uses of radioisotopes in medical applications around the world.

Positron emission tomography is described in more detail on page 299.

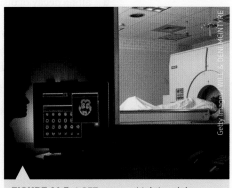

**FIGURE 16.5** A PET scanner (right) and the image it can produce (left)

Spontaneous decay
of unstable nuclei

## A note on models of beta emission

We have been careful to say that a proton can be *converted into* a positron and a neutron, and that a neutron can be *converted into* an electron and a proton. A neutron *does not contain* an electron and a proton, nor does a proton contain a neutron and a positron.

When β radiation from nuclei was first observed it seemed to physicists that this was the case, as it explained how a neutron could emit an electron. However, it later turned out that this could not be correct. More than just an electron is emitted in $\beta^-$ decay, and there are other reactions involving protons and neutrons that could not be explained by them containing electrons or positrons. Chapter 17 discusses some of these issues in the context of the Standard Model of matter.

▶ **WORKED EXAMPLE**

What daughter nuclide is produced after:

**a** alpha decay of a polonium-214 nuclide?

**b** beta-minus decay of carbon-14?

**c** positron emission of sodium-20?

**d** gamma emission from cerium-139?

| ANSWERS | LOGIC |
|---|---|
| **a** $^{214}_{84}\text{Po} \rightarrow \text{X} + {}^{4}_{2}\text{He}$ <br><br> $^{214}_{84}\text{Po} \rightarrow {}^{210}_{82}\text{Pb} + {}^{4}_{2}\text{He}$ | ▪ Use correct nomenclature for alpha particle. <br><br> ▪ Use correct symbol for daughter nuclide. |
| **b** $^{14}_{6}\text{C} \rightarrow \text{X} + {}^{0}_{-1}\text{e}$ <br><br> $^{14}_{6}\text{C} \rightarrow {}^{14}_{7}\text{N} + {}^{0}_{-1}\text{e}$ | ▪ Use correct nomenclature for electron. <br><br> ▪ Use correct symbol for daughter nuclide. |
| **c** $^{20}_{11}\text{Na} \rightarrow \text{X} + {}^{0}_{+1}\text{e}$ <br><br> $^{20}_{11}\text{Na} \rightarrow {}^{20}_{10}\text{Ne} + {}^{0}_{+1}\text{e}$ | ▪ Use correct nomenclature for positron particle. <br><br> ▪ Use correct symbol for daughter nuclide. |
| **d** $^{139}_{58}\text{Ce} \rightarrow \text{X} + {}^{0}_{0}\gamma$ <br><br> $^{139}_{58}\text{Ce} \rightarrow {}^{139}_{58}\text{Ce} + {}^{0}_{0}\gamma$ | ▪ Use correct nomenclature for gamma ray. <br><br> ▪ Use correct symbol for daughter nuclide. |

**TRY THESE YOURSELF**

**1** What daughter nuclide is produced after:

   **a** alpha decay of bismuth-211?

   **b** beta-minus decay of bromine-82?

   **c** positron emission from gold-190?

   **d** gamma emission from samarium-145?

**2** What daughter nuclide is produced after:

   **a** alpha decay of lead-204?

   **b** beta-minus decay of lead-209?

   **c** positron emission of lead-199?

   **d** gamma emission from lead-203?

- A nucleon is a particle in the nucleus – a proton or neutron.
- The atomic number, $Z$, is the number of protons in the nucleus.
- The mass number, $A$, is the number of nucleons (protons plus neutrons) in the nucleus.
- Nuclear radiation originates from unstable nuclei.
- Alpha radiation is the emission of an alpha particle, which is identical to a helium nucleus.
- Beta-minus emission occurs when a neutron transforms into a proton and a high speed electron, a beta particle, is emitted from the nucleus.
- Positron, or beta-plus, emission occurs when a proton transforms into a neutron and a positron, a positive electron, which is emitted from the nucleus.
- Gamma emission is the release of a photon of energy from the nucleus.

1 Identify two numbers that must have the same total on each side of a nuclear equation.

2 Write the general equation for:

   a alpha decay.            c positron decay.

   b beta-minus decay.      d gamma radiation.

3 How are nuclear reactions different from chemical reactions? Consider the two sides of the equation.

4 Fluorine-21 is a beta-minus emitter. What is the daughter nuclide?

5 Holmium-151 decays by emitting an α particle and changes to a different element.

   a Write a complete nuclear reaction equation that includes the symbol for the daughter nuclide.

   b What is the name of the daughter nuclide?

6 Radon-210 decays to polonium-206. What type of radioactive particle is emitted in this decay? Write the decay equation.

7 Write the decay equation for $^{15}$O.

8 Use correct symbols to show the decay of terbium-158 by alpha emission followed by gamma emission.

9 Summarise the conditions that produce stable nuclei and those that produce unstable nuclei.

# 16.2 Properties of alpha, beta and gamma radiation

Alpha, beta and gamma rays affect materials in different ways according to their power to ionise and penetrate materials.

## Ionising power

Atoms become ions by losing or gaining electrons. When electrons are removed from a neutral atom, it becomes a positive ion, and when a neutral atom gains electrons, it becomes a negative ion.

Slow-moving, positively charged α particles attract electrons from atoms and ionise them. With each ionisation, α particles lose energy.

Negatively charged β⁻ particles are repelled by electrons in atoms. This causes particles to be bounced between atoms. These collisions may cause some electrons to be ejected from atoms, which become ionised. These collisions transfer less energy than the interactions between α particles and atoms. Positively charged β⁺ particles interact with electrons in atoms. The effect of these interactions is that the atoms become ionised. These interactions transfer less energy than the interactions between α particles and atoms.

The ionising power of a gamma ray depends upon its energy. High-energy rays can ionise atoms and molecules by colliding with them and transferring energy to their electrons. If an electron gains enough energy it can leave the atom or molecule, leaving behind a positive ion. The electron may then remain free for some time before binding to another atom or molecule to form a negative ion. Low-energy rays are more likely to result in heating of a material than ionisation.

α particles have far more ionising ability than β particles or gamma rays. The ionising power is inversely proportional to the penetrating power. This is to be expected because the particles expend their energy in causing ionisation, and so do not penetrate as far. Consequently, the range of α particles in air is much less than that of either β particles or gamma rays.

Neutrinos and antineutrinos are weakly interacting particles that do not ionise atoms.

## Penetrating power

α particles are the most easily absorbed particles and, therefore, the least penetrating. They are absorbed by a thin sheet of paper or outer skin (Figure 16.6). β particles can be stopped by a few millimetres of aluminium. Gamma rays can penetrate up to 30 cm of steel. A 1 cm thick sheet of lead reduces the intensity of γ radiation to about half of the original intensity. A comparison of penetrating power is illustrated in Figure 16.7.

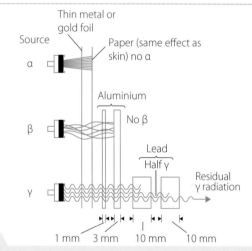

**FIGURE 16.6** Gamma rays are much more penetrating than α or β particles. In turn, β particles are more penetrating than α particles.

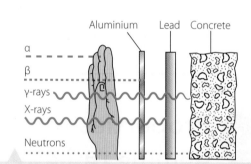

**FIGURE 16.7** Penetrating power for different radiations is indicated by their relative absorptions in materials.

Neutrons are highly penetrating in air and other materials. They interact with and are absorbed strongly by materials containing a lot of hydrogen. Hence materials such as water and concrete are good neutron absorbers and are used as shielding at nuclear reactors. The core of the Open Pool Australian Lightwater reactor (OPAL) at Lucas Heights in Sydney's south is contained in a large pool of water.

### Range in air

Information and communication technology capability

Explore the penetrating power of radiation

There are many ions and other charged particles in air. The greater the charge on a radioactive particle, the more likely it is that it will pick up these charged particles and become neutral. For this reason, α particles penetrate only a few centimetres in air. The range of β particles is difficult to determine, partly because they can have widely different energies, but they can certainly travel a few metres in air before stopping, as shown in Figure 16.8. γ radiation is so energetic it can penetrate hundreds of metres through air. Luckily for life on Earth, most gamma radiation from space is absorbed by the Earth's atmosphere.

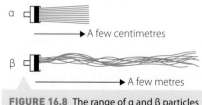

**FIGURE 16.8** The range of α and β particles in air. Gamma rays pass easily through air.

9780170409131

## The penetrating power of radiation

### AIM

To observe the penetrating power of alpha, beta and gamma radiation. Write an inquiry question for this investigation.

### MATERIALS

- Approved samples of alpha, beta and gamma emitters
- Sheets of paper, cardboard, aluminium, lead
- Gloves
- Ruler
- Stopwatch
- Geiger–Muller tube and counter

| WHAT ARE THE RISKS IN DOING THIS INVESTIGATION? | HOW CAN YOU MANAGE THESE RISKS TO STAY SAFE? |
|---|---|
| Radiation is potentially harmful. | Radioactive samples should only be handled by the teacher. |
| | The samples should be checked for cracks or damage prior to use. |
| | Keep back a safe distance from the samples. |
| | Wear gloves whenever handling the samples. |
| | Use the samples expediently and then place them back in the metal box, wrapped in lead and moved away from the laboratory. |
| Handling lead sheets may leave residue of lead on your skin. | Wash your hands after the samples and the lead have been packed away. |

What other risks are associated with your investigation, and how can you manage them?

### METHOD

1 Place the Geiger–Muller tube 1 cm from each radioactive sample in turn and record the radiation count over a 15 second interval for each.
2 Repeat step 1 but with the sheet of paper placed over the face of the Geiger–Muller tube.
3 Repeat step 2 with, in turn and separately, the cardboard, aluminium and lead sheets.
4 Record your results in a suitable table.
5 For each radioactive sample in turn, record the radiation counts over a 15 second period at distances of 1 cm, 2 cm, 3 cm, 4 cm, 5 cm and 10 cm.

### RESULTS

Display the results obtained from step 5 in a suitable graph of radiation count versus distance.

### DISCUSSION

1 Relate the results obtained to the nature of the three types of radiation tested.
2 Give an answer to your inquiry question.

### CONCLUSION

With reference to the data obtained and its analysis, write a conclusion based on the aim of this investigation.

RISK ASSESSMENT

Australian guidelines

Australian guidelines for the use of radioactive samples in schools can be found at the weblink *Australian guidelines* and must be read prior to conducting this investigation.

## Effects of electric and magnetic fields on radiation

Charged α and β particles moving in straight lines can be deflected by electric and magnetic fields. Gamma rays, being a form of electromagnetic radiation, have no mass or charge; therefore, they are not deflected by either electric or magnetic fields.

Charged particles emitted from nuclei are generally travelling very fast. We can use electric and magnetic fields to distinguish between the different types of radiation.

## α, β and γ radiation in electric fields

α and β⁺ particles will be accelerated in the direction of an applied electric field, while β⁻ particles will be accelerated in the opposite direction. This can be observed as a curving of the path of a moving charged particle when it enters an electric field. α and β⁺ particles will be deflected in one direction, β⁻ particles are deflected in the opposite direction. The force depends on the charge of the particle, but the acceleration depends on both the force and the mass, and it decreases with increasing mass. Hence, even though an α particle in an electric field experiences twice the force of a β particle, the curvature of its path is much less because its mass is about 7000 times greater than that of a β particle. A gamma ray has no charge, and so will pass through a region of electric field without any deflection.

## α, β and γ radiation in magnetic fields

A magnetic field applies a force to any moving charged particle, causing it to follow a curved path.

The magnitude of the force depends on the speed at which the particle is moving and the magnitude of its charge. The direction of the force depends on the sign of the charge. Hence the force experienced by a β⁺ particle is the same size but the opposite direction to that experienced by a β⁻ particle, if they are moving at the same speed. An α particle at the same speed experiences a force in the same direction as a β⁺ particle, but the force is twice as large because it has twice the charge.

How this magnetic force affects the path of the particle depends on the mass of the particle. An α particle has a mass about 7000 times that of a β particle. Hence an α and a β⁺ particle will be deflected in the same direction by a magnetic field, but the β⁺ particle will be deflected far more. A β⁻ particle will be deflected just as much as a β⁺ particle, but in the opposite direction. A gamma ray, which has no charge, is not deflected by a magnetic field. This is shown in Figure 16.9.

Properties of alpha, beta and gamma radiation

**FIGURE 16.9**
Deflections of α, β and γ radiation in a magnetic field. There is no deflection in the absence of a magnetic field.

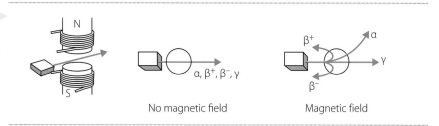

No magnetic field

Magnetic field

**TABLE 16.3** Summary of the properties of α, β and γ radiation.

|  | α PARTICLES | β PARTICLES | γ-RAY |
|---|---|---|---|
| Composition | A helium nucleus (i.e. 2 protons and 2 neutrons) | A fast-moving electron or positron | High-frequency (short wavelength) electromagnetic radiation (i.e. a high-energy photon) |
| Charge | +2 elementary charges | −1 (electron) +1 (positron) elementary charge | Uncharged |
| Mass | 4 atomic mass units (i.e. 4 u); $4 \times 1.66 \times 10^{-27}$ kg | 0.0005 u; $9.11 \times 10^{-31}$ kg | No mass |
| Ionising effect | Strong | Weak | Very weak |
| Penetrating power | Few centimetres in air | Few metres in air | Very weakly absorbed in air (most radiation absorbed by a few centimetres of lead) |
| Effect of electric and magnetic fields | Small deflection | Large deflection | No deflection |
| Typical emission velocity | 5–7% of speed of light | 30–90% of speed of light | Speed of light ($3 \times 10^8$ m s⁻¹) |

- Alpha particles have the least penetrating power; gamma rays have the greatest.
- Alpha particles have the greatest ionising ability due to their charge (+2) and mass; gamma rays have the least.
- Alpha particles are slightly deflected in electric or magnetic fields, beta particles have a large deflection while gamma rays are not deflected.
- Like all nuclear radiation, beta particles emanate from the nucleus. When a neutron transforms into a proton a $\beta^-$ is released; when a proton transforms into a neutron, a $\beta^+$ is released.

1  For α, β and γ rays, which is:

   **a**  most penetrating?      **b**  most ionising?      **c**  least likely to cause damage?

2  Suggest why α particles travel the smallest distance through air.

3  Give a reason for gamma rays' low ionising ability.

4  Three identical radioactive sources, A, B and C, are placed in a vacuum in front of identical Geiger–Muller tubes. Their count rates are measured simultaneously. Sources A, B and C are at effective distances of 25 cm, 50 cm and 1.50 m respectively from the Geiger–Muller tube. Relative to B, is the count rate likely to be greater than, the same as or less than the count rate at B at:

   **a**  C?      **b**  A?

5  Which type of nuclear radiation has the most mass? Explain.

6  1600 counts are recorded in a Geiger–Muller tube in 20.00 s. What is the count rate?

7  Why do alpha particles have a smaller deflection in electric or magnetic fields than beta particles?

8  Why do gamma rays remain undeflected when passing through regions with an electric or magnetic field?

# 16.3 Decay and half-life

Radioactive decay is a random event. For any given nucleus of a radioactive (unstable) isotope it is impossible to predict when it will decay, or even if it will decay at all in some given time period. This is very much like throwing dice or tossing coins. The result of a single coin toss or throw of a die cannot be predicted.

The half-life is the average time taken for half the nuclei in a sample to decay. If there is a large number of radioactive nuclei, we can say that some fraction of them will decay in a given time. The fraction that decays in any given time period depends upon the **half-life** of the particular nuclide. The shorter the half-life, the faster the rate of decay.

For example, in 1 g of uranium, there may be $10^{20}$ unstable nuclei. In one half-life, half of these nuclei will decay. In the next half-life, half of the remaining nuclei will decay. In these two half-lives, three-quarters of the original nuclei will have decayed, leaving a quarter of the original unstable nuclei.

When tossing coins, we cannot say which will show heads and which will show tails, but for a large sample we will get 50% heads and 50% tails.

In general, for a sample of $N_0$ particles, the number, $N_t$, remaining after time $t$ is given by the equation:

$$N_t = N_0 e^{-\lambda t}$$

where $\lambda$ = the decay constant (with units s$^{-1}$)

and $\lambda = \dfrac{\ln 2}{t_{1/2}}$ where $t_{1/2}$ = time for half the radioactive particles to decay.

After 300 years, the number of remaining unstable nuclei in a radioactive sample is 10% of the original number. Find the decay constant, $\lambda$, for this radioactive sample.

| ANSWER | LOGIC |
|---|---|
| $t = 300$ years | ■ Identify the relevant data from the question. |
| $= 300 \times 365 \times 24 \times 60 \times 60$ <br> $= 9.46 \times 10^9\,\text{s}$ | ■ Convert to SI units. |
| $\dfrac{N_t}{N_0} = 10\%$ <br> $= 0.1$ | ■ Identify the relevant data from the question. |
| $N_t = N_0 e^{-\lambda t}$ | ■ Identify the correct formula. |
| $\dfrac{N_t}{N_0} = e^{-\lambda t}$ | ■ Rearrange the formula. |
| $0.1 = e^{-\lambda \times (9.46 \times 10^9)}$ | ■ Substitute the values with correct units. |
| $\ln 0.1 = -\lambda \times (9.46 \times 10^9)$ <br> $\lambda = -\dfrac{\ln 0.1}{9.46 \times 10^9}$ <br> $= 2.43 \times 10^{-10}\,\text{s}^{-1}$ | ■ Calculate the answer and express with correct units and appropriate significant figures. |

**TRY THESE YOURSELF**

1 Calculate the decay constant, $\lambda$, if three-quarters of the unstable nuclei in a radioactive sample decay in 80 hours.

2 Find the decay constant, $\lambda$, for a radioactive sample with a half-life of 6.7 hours.

The time required for the decay of half the original sample of atoms is called the half-life ($t_{1/2}$) of the radioactive material. Each radioactive isotope has a unique half-life.

Radioactive decay is a random event. Half-life is the time taken for half the nuclei to decay.

$$N_t = N_0 \left(\frac{1}{2}\right)^n$$

> This formula is included here because it is very helpful, but it is not included in the syllabus.

where $n$ = whole number of half-lives.

The letter $N$ is used here to represent the number of nuclei at any time when radioactive decay is being measured. As we have seen, $N$ is also used to represent the number of neutrons in a nucleus. You need to use the context to distinguish which $N$ is meant.

## WORKED EXAMPLE (16.4)

Calculate the decay constant, $\lambda$, for a radioactive sample that has a half-life of 6.00 hours.

| ANSWER | LOGIC |
|---|---|
| $t_{1/2} = 6.00$ hours <br> $= 6.00 \times 60 \times 60$ s <br> $= 2.16 \times 10^4$ s | • Identify the relevant data from the question |
| $\lambda = \dfrac{\ln 2}{t_{1/2}}$ | • Identify the correct formula. |
| $= \dfrac{\ln 2}{21\,600 \text{ s}}$ | • Substitute the values with correct units. |
| $= 3.21 \times 10^{-5} \text{ s}^{-1}$ | • Calculate the answer and express with correct units and appropriate significant figures, |

### TRY THESE YOURSELF

Find the decay constant for:

a   a sample of radon-224 with a half-life of 55 seconds.

b   a sample of carbon-14 with a half-life of 5730 years.

## WORKED EXAMPLE (16.5)

Calculate the half-life of a radioactive sample that has a decay constant of $3.4 \times 10^{-9} \text{ s}^{-1}$.

| ANSWER | LOGIC |
|---|---|
| $\lambda = 3.4 \times 10^{-9} \text{ s}^{-1}$ | • Identify the relevant data from the question. |
| $\lambda = \dfrac{\ln 2}{t_{1/2}}$ | • Identify the correct formula. |
| $t_{1/2} = \dfrac{\ln 2}{\lambda}$ | • Rearrange the formula. |
| $= \dfrac{\ln 2}{3.4 \times 10^{-9} \text{ s}^{-1}}$ | • Substitute the values with correct units. |
| $= 2.039 \times 10^7$ s | • Calculate the answer. |
| $= 2.0 \times 10^7$ s | • Express the answer with the correct units and appropriate significant figures. |

### TRY THESE YOURSELF

1   Find the half-life of a radioactive sample that has a decay constant of $4.5 \times 10^{-6} \text{ s}^{-1}$.

2   Find the half-life of a radioactive sample that has a decay constant of $7.8 \times 10^{-8} \text{ s}^{-1}$.

Let us look at another example. Polonium-218 has a half-life of 3 minutes and decays to Pb-214. If we start with $1.00 \times 10^{20}$ Po-218 nuclei at some time, then after 3 minutes $0.50 \times 10^{20}$ Po-218 nuclei will have decayed to Pb-214, and we will have $0.50 \times 10^{20}$ remaining Po-218 nuclei. After 6 minutes, two half-lives have passed, and three-quarters of the original Po-218 nuclei will have decayed, leaving only $0.25 \times 10^{20}$ Po-218 nuclei. Table 16.4 shows the number of nuclei remaining as a function of time.

Decay and half-life

**TABLE 16.4** Decay of a particular sample of a radioactive nuclide

| TIME (MINUTES) | 0 | 3 | 6 | 9 | 12 | 15 |
|---|---|---|---|---|---|---|
| Number of half-lives elapsed | 0 | 1 | 2 | 3 | 4 | 5 |
| Proportion of Po-218 nuclei remaining ($\times 10^{20}$) | 1.00 | 0.50 | 0.25 | 0.125 | 0.0625 | 0.03125 |
| Proportion of Po-218 nuclei decayed into Pb-214 ($\times 10^{20}$) | 0 | 0.50 | 0.75 | 0.875 | 0.9375 | 0.96875 |

Numeracy

Our example shows us that, after 5 half-lives, the fraction of radioactive nuclei has dropped to less than 5% of the original number, and hence more than 95% of the nuclei have decayed. The level of radioactivity of a sample is dependent on the number of undecayed unstable nuclei remaining. Radioisotopes used in medicine that are injected into patients to perform a scan must have short half-lives so that the level of radioactivity diminishes quickly. For example, Tc-99m has a half-life of only 6 hours. After 24 hours, only 6.25% of the original radioactivity is present.

## WORKED EXAMPLE (16.6)

What percentage of nuclei will be left undecayed after 40 days in a radioactive sample with a half-life of 8 days?

Information and communication technology capability

A radioactive decay simulation

| ANSWER | LOGIC |
|---|---|
| $n = \dfrac{40}{8} = 5$ <br><br> Let $N_0 = 1$ | ▪ Identify the relevant data from the question. |
| $N_t = N_0 \left(\dfrac{1}{2}\right)^n$ | ▪ Identify the correct formula. |
| $= \left(\dfrac{1}{2}\right)^5$ | ▪ Substitute values into the formula. |
| $= 0.031 \times 100\%$ | ▪ Calculate the answer. |
| $= 3.1\%$ | ▪ Express the answer with the appropriate significant figures. |

### TRY THESE YOURSELF

1 What percentage of nuclei will be left undecayed after 1 year if the radioactive sample has a half-life of 12 hours?

2 What percentage of nuclei will be left undecayed after 1 day if the half-life of the sample is 6 hours?

9780170409131

# INVESTIGATION (16.2)

## A simulation of random decay and half-life

### AIM

To simulate the random decay and half-life of a radioactive nuclide

### MATERIALS

Per group:

- 1 bag or cup
- 80 small counters (or disc-shaped sweets) with an obvious distinction between the two sides
- Gloves
- Dice (for Extension)

| WHAT ARE THE RISKS IN DOING THIS INVESTIGATION? | HOW CAN YOU MANAGE THESE RISKS TO STAY SAFE? |
|---|---|
| | |

In your report, include any risks you can think of, as well as ways to manage them.

### METHOD

1 As a class, decide which side of the counters represents decay (Up) and which side represents not-decay (Down).
2 Shake the bag and pour the counters on to the table.
3 Remove all counters that represent decay.
4 Count and record the counters that represent the remaining nuclides that are yet to decay.
5 Replace these in the bag.
6 Repeat the process until there is one or no counters remaining.

### RESULTS

Combine the results from all groups into a single data table for the class.

### ANALYSIS OF RESULTS

1 Use whole-class data to plot a graph of the number of counters remaining vs the number of trials.
2 From the graph, determine the half-life of the counters.

### DISCUSSION

1 For radioactivity, how did this experiment model:
   a the randomness of decay?
   b half-life?
2 A radioactive nuclide is an unstable nucleus that could decay at any moment. The decay occurs because the daughter nuclide is more stable than the parent. Discuss.

### CONCLUSION

Write a suitable conclusion based on the aim of this investigation.

### EXTENSION

In this experiment the half-life was one 'throw' of the counters. Try modelling a half-life that is less than one 'throw' by repeating this experiment using a large number of dice. Remember to write an inquiry question for your investigation. What is the half-life when you remove all dice with a one showing at each throw? What if you remove all those with a one or a two showing?

- The half-life of a sample of radioactive material is the time taken for half of the nuclei present to decay.
- $N_t$, the number of undecayed particles present at time $t$, is given by $N_t = N_0 e^{-\lambda t}$ where $N_0$ = number of particles present at $t = 0$ and $\lambda$ is the decay constant.
- $\lambda = \dfrac{\ln 2}{t_{1/2}}$ with the units s$^{-1}$.
- The number of undecayed particles present is related to the level of radioactivity of the sample.

1 Explain what the term 'half-life' means.

2 Why should a radioactive substance that is injected into a patient to perform a scan have a short half-life?

3 What is the decay constant, $\lambda$, for a radioactive sample with a half-life of 14 hours?

4 Find the half-life of a radioactive sample with a decay constant of $5.6 \times 10^{-10}$ s$^{-1}$. Express your answer in days.

5 Why is the proportion of undecayed nuclides present in a sample representative of the sample's radioactivity level?

6 What proportion of nuclides remains undecayed in a radioactive sample after 8 half-lives?

7 Would it be possible for a radioactive sample to ever decay so that its level of radiation decreases to zero? Explain.

# 16.4 Energy from the nucleus

The energy released in nuclear reactions, including fission and fusion, is enormous compared to the energy released in the most explosive chemical reaction. The differences are typically of the order of 1–10 billion times more energetic. Fission involves the splitting of a larger nucleus into smaller fission fragments. Fusion, on the other hand, is the process of forming a larger nucleus from smaller nuclei.

## Binding energy

The total energy needed to hold a nucleus together is called the nuclear binding energy. This is shown graphically in Figure 16.10 for nuclides up to uranium, which is the heaviest naturally occurring element.

The strong nuclear force acts most strongly between nearby nucleons. The greater the binding energy per nucleon, the harder it is to pull the nucleus apart. Therefore, a nuclide is more stable when its binding energy per nucleon is greater. Iron-56 is the most stable of the nuclides.

Table 16.5 shows the total binding energy for nuclides as well as the binding energy per nucleon. Consider helium-4, which has a total binding energy of 28.29 MeV. It has four nucleons, so its binding energy per nucleon is $\dfrac{28.29 \text{ MeV}}{4} = 7.07$ MeV per nucleon.

Fusion is the coming together of two nuclides to form a new nucleus with a greater atomic number. The new composite nucleus is more stable because its binding energy per nucleon is greater. This is more likely to occur for nuclides that have an atomic number $Z < 56$.

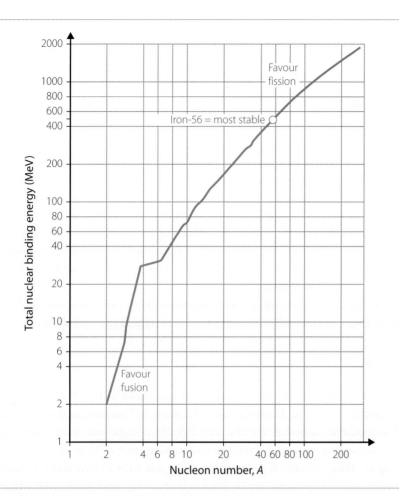

**TABLE 16.5** Binding energy and binding energy per nucleon for some nuclides

| ELEMENT | BINDING ENERGY (MeV) | BINDING ENERGY PER NUCLEON (MeV) |
|---|---|---|
| Deuterium (Hydrogen-2) | 2.23 | 1.12 |
| Helium-4 | 28.29 | 7.07 |
| Lithium-7 | 40.15 | 5.74 |
| Beryllium-9 | 58.13 | 6.46 |
| Iron-56 | 492.24 | 8.79 |
| Silver-107 | 915.23 | 8.55 |
| Iodine-127 | 1072.53 | 8.45 |
| Lead-206 | 1622.27 | 7.88 |
| Polonium-210 | 1645.16 | 7.83 |
| Uranium-235 | 1783.80 | 7.59 |
| Uranium-238 | 1801.63 | 7.57 |

## Fission

Nuclear fission is the process by which a heavy nucleus ($Z > 56$) splits into two fragments. In general, the fragments are rarely the same size, so it is incorrect to say that the atom 'splits in half'. In the process, neutrons are released and energy stored as binding energy is released.

An example of a fission reaction is the splitting of a uranium-235 nucleus into krypton-92 and barium-141 nuclei:

$$_{0}^{1}\text{n} + {}_{92}^{235}\text{U} \rightarrow {}_{36}^{92}\text{Kr} + {}_{56}^{141}\text{Ba} + 3\,{}_{0}^{1}\text{n}$$

Nuclear fission is triggered by the absorption of a neutron. Irène Joliot-Curie, daughter of Marie and Pierre Curie, was the first to identify the products of nuclear fission. The process was first suggested to the German chemist Otto Hahn by Ida Noddack, a Hungarian chemist. Lise Meitner, who worked with Hahn, developed a model to explain the process, which was written up by her nephew Otto Frisch, the first person to use the term 'fission' in the literature. At Meitner's suggestion, Hahn and Fritz Strassmann undertook similar experiments to those of Joliot-Curie. The two men received the Nobel Prize in 1944. Meitner's decisive contributions were not recognised.

Enrico Fermi was the first to control nuclear fission. On 2 December 1942, in a squash court under the stadium at the University of Chicago, the first self-sustaining nuclear reactor began operation.

## Fission chain reaction

Uranium-235 can absorb a slow neutron or thermal neutron that has about 5–10 keV of energy. The neutron is absorbed in the uranium nucleus and forms an unstable nucleus. This nucleus then splits into two fragments, each with lower atomic mass than the uranium-235. On average, between two and three neutrons are also released. The neutrons released from the fission of uranium-235 do not usually get absorbed in uranium-235 nuclei. This is a natural process that rarely amounts to anything substantial. However, if conditions are right, then the process can be used to harness the energy released, which is about 200 MeV per fission event. This process will be discussed in more detail in following sections.

A chain reaction occurs when more than one of the neutrons released from the initial fission event causes new events to occur. In Figure 16.11, each fission event produces two or three neutrons, some of which go on to cause new fission events. This rapidly multiplies to huge numbers of fission events. Unless this is carefully controlled, a runaway explosion will occur.

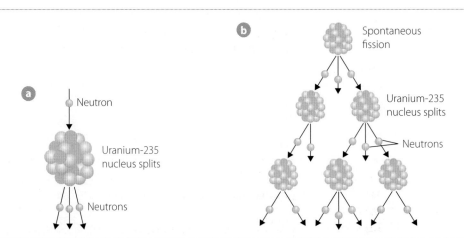

**FIGURE 16.11** Nuclear fission. **a** A slow neutron causes a uranium-235 nucleus to split, releasing three fast neutrons. **b** A chain reaction occurs, if, for example, two of the released neutrons cause further nuclear fission in other uranium nuclei. Vast amounts of energy can be released.

## Controlled fission chain reaction

The chain reaction can be controlled for nuclear power generation. The heat energy released is used to produce steam that powers the turbines and generators of an electricity production plant.

In a controlled chain reaction, one neutron produces one fission neutron. If more than one neutron is produced on average, then a runaway reaction occurs. If the average number of neutrons produced is less than one, then the reaction dies away.

It is difficult to establish and maintain a chain reaction. The proportion of uranium-235 in naturally occurring uranium ore is very small (0.65%) compared to uranium-238 (99.3%). The proportion of the other natural uranium isotope, uranium-234, is negligible (0.05%). Uranium-238 acts almost exclusively as a neutron poison. Uranium-238 beta decays, leading to the production of elements heavier than uranium, or transuranic elements, the most important of which is the highly fissile plutonium.

There are several requirements necessary for a controlled chain reaction to be sustained.

Information and communication technology capability

Simulation of fission chain reaction

### Enrichment

For fission to occur in uranium-235 the initial neutron must be a slow neutron. The neutrons produced in fission events are fast neutrons. The probability of these neutrons causing new fission events in uranium-235 is very small. It is more likely that they will be captured by uranium-238 or some other neutron poison already produced by earlier fission events. These neutron poisons absorb the neutrons and then emit $\alpha$, $\beta$ or $\gamma$ radiation. The neutrons are therefore no longer available to trigger further fission events.

In order to ensure sustained fission, the proportion of uranium-235 must be increased. This enrichment process is achieved by separating uranium-235 out of naturally occurring uranium ore. This separated amount is then put back into a quantity of naturally occurring uranium. This increases, or enriches, the proportion of uranium-235 in the quantity. It is a complex and expensive process. To guarantee an uncontrolled chain reaction, as required for nuclear weapons, enrichment may be as high as 97% uranium-235. For a controlled reaction, 1–4% enrichment is sufficient.

### Moderator

To reduce the energy of the neutrons produced by fission events, a moderator is used. The moderator is a material with nuclides that have slightly larger masses than the neutron, for example hydrogen $\left(^{1}_{1}H\right)$, deuterium $\left(^{2}_{1}H\right)$ and tritium $\left(^{3}_{1}H\right)$. Neutrons share their energy with these nuclides through multiple collisions. They rapidly lose energy, which increases the probability of neutrons entering a uranium-235 nucleus and causing fission.

### Reactor vessel

In a reactor, a controlled chain reaction will not proceed unless most of the neutrons available can be used. Apart from absorption in neutron poisons, some neutrons will escape from the fuel. This is because the absorption of a neutron and subsequent fission occurs only when there is a head-on collision between nucleus and neutron. The reactor vessel is designed to have the right surface area-to-volume ratio, and is made of a high nucleon number material so that the neutrons are reflected back into the sample.

### Control rods

On average, the number of neutrons produced by fission events must be equal to the number of fission-producing neutrons in order to sustain a nuclear reaction. Sometimes the reaction threatens to run away, so control rods containing neutron-absorbing materials, such as boron-10, are moved into the fuel to lower the number of neutrons in the sample. The control rods are removed when the chain reaction starts to produce too few neutrons.

The main components of a nuclear reactor for controlled nuclear fission are depicted in Figure 16.12. They are:

- nuclear fuel, usually 1–4% enriched uranium-235
- a moderator, used to slow down fast-moving neutrons to slower-moving neutrons that can maintain the chain reaction; either graphite or water is used
- control rods, either cadmium or boron steel, that absorb some of the neutrons produced by fission to control the rate of the chain reaction
- a coolant, to absorb the heat generated by the chain reaction and to remove it from the reactor.

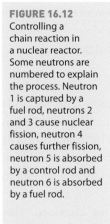

**FIGURE 16.12**
Controlling a chain reaction in a nuclear reactor. Some neutrons are numbered to explain the process. Neutron 1 is captured by a fuel rod, neutrons 2 and 3 cause nuclear fission, neutron 4 causes further fission, neutron 5 is absorbed by a control rod and neutron 6 is absorbed by a fuel rod.

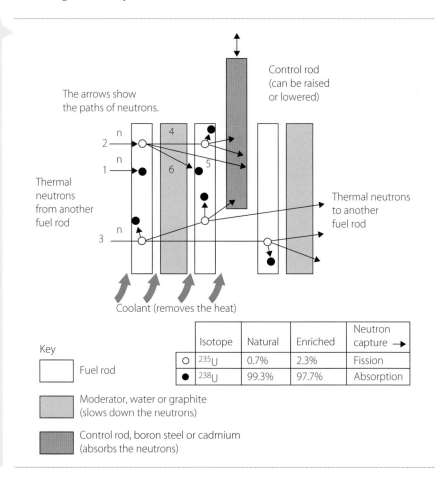

| | Isotope | Natural | Enriched | Neutron capture → |
|---|---|---|---|---|
| ○ | $^{235}U$ | 0.7% | 2.3% | Fission |
| ● | $^{238}U$ | 99.3% | 97.7% | Absorption |

Key

☐ Fuel rod

▨ Moderator, water or graphite (slows down the neutrons)

▧ Control rod, boron steel or cadmium (absorbs the neutrons)

## Mass defect in fission

In nuclear fission of uranium-235, the total mass before fission is greater than the total mass after the fission event. This mass difference is what is converted into energy. This energy is transferred via the fission products. The large daughter nuclei carry most of this energy as kinetic energy. The released neutrons also have kinetic energy.

For example, when uranium-235 undergoes fission to produce krypton-92 and barium-141, three neutrons are released:

$$^{1}_{0}n + {}^{235}_{92}U \rightarrow {}^{92}_{36}Kr + {}^{141}_{56}Ba + 3\,{}^{1}_{0}n$$

There are 236 nucleons before and after this fission event, yet the mass of the products is less than the mass of the original neutron and uranium-235 nuclide. This is the mass defect.

Rather than use the mass of a nucleon, $1.660 \times 10^{-27}$ kg, we shall use its equivalent in unified atomic mass units, u (Table 16.6). 1 u is defined as the mass of $\dfrac{1}{12}$ of the mass of a neutral carbon-12 atom. This is almost the same value as the mass of one nucleon. Why? Because the mass is a common factor for each calculation. If we want to convert back to units of kilogram at the end, we need only multiply our answer by the conversion factor $1\,u = 1.660 \times 10^{-27}$ kg.

**TABLE 16.6** Particles in a fission event involving uranium-235

| PARTICLE | $^{1}_{0}n$ | $^{235}_{92}U$ | $^{92}_{36}Kr$ | $^{141}_{56}Ba$ |
|---|---|---|---|---|
| MASS ($\times 1.660 \times 10^{-27}$ kg) | 1.009 | 235.044 | 91.926 | 140.914 |

In atomic mass units:

Initial mass $= 1.009\,u + 235.044\,u = 236.053\,u$

Final mass $= 91.926\,u + 140.914\,u + 3 \times 1.009\,u = 235.867\,u$

Mass defect $= 236.053\,u - 235.867\,u = 0.186\,u$

It is this mass defect, $\Delta m$, that is converted into energy, $\Delta E$:

$$\Delta E = (\Delta m)c^2$$

At this point, it is necessary to convert back to SI units:

$\Delta m = 0.186\,u \times 1.66 \times 10^{-27}\,kg\,u^{-1} = 3.090 \times 10^{-28}\,kg$

$\Delta E = 3.090 \times 10^{-28}\,kg \times (3.0 \times 10^8\,m\,s^{-1})^2$

$\Delta E = 2.78 \times 10^{-11}\,J$

This is a very small amount of energy; however, 250 g of pure uranium contains more than $10^{23}$ nuclides. If all of these were to undergo fission, the effect would be enormous:

$$\Delta E = 2.78 \times 10^{-11} \times 10^{23}\,J = 2.78 \times 10^{12}\,J \text{ (or 2.78 TJ)}$$

Energy from the nucleus

▶ **WORKED EXAMPLE** (16.7)

A thermal neutron, mass 1.01 u, causes fission of uranium-235 (235.04 u). The fission fragments and their masses, in unified mass units, are rubidium-93 (92.92 u) and caesium-141 (140.92 u).

Numeracy

a How many fast neutrons are released in this fission event?

b Write the nuclear fission reaction using correct nuclide and nucleon symbols.

c What is the mass defect in this event in:

  i unified mass units?

  ii kilogram?

d How much energy is released? Give your answer in joule.

| ANSWERS | LOGIC |
|---|---|
| **a** Nucleon number is conserved.<br>Total nucleons before $= 1 + 235 = 236$<br>Total nucleons after $= 93 + 141 + x = 236$ | ▪ Use the correct conservation rule.<br>▪ Calculate the correct sum.<br>▪ Calculate the correct sum. |
| $x = 236 - 234$<br>$x = 2$ | ▪ Solve for $x$. |
| Two neutrons are released. | ▪ Write the answer. |
| **b** $^1_0 n + {}^{235}_{92}U \rightarrow {}^{93}_{37}Rb + {}^{141}_{55}Cs + 2 {}^1_0 n$ | ▪ Use the correct atomic numbers and symbols.<br>▪ Place fission reactants and products in the correct order. |
| **c** Mass defect $= (1.01 + 235.04) -$<br>$(92.92 + 140.92 + 2 \times 1.01)$ u<br>  **i** mass defect $= 0.19$ u<br>  **ii** mass defect $= 0.19$ u $\times 1.66 \times 10^{-27}$ kg u$^{-1}$<br>      $= 3.15 \times 10^{-28}$ kg | ▪ Substitute the correct values and add to get the answer.<br>▪ Convert u to J. |
| **d** Energy released: $\Delta E = (\Delta m)c^2$ | ▪ Write the equation. |
| $\Delta E = 3.15 \times 10^{-28}$ kg $\times (3.00 \times 10^8 \, \text{m s}^{-1})^2$ | ▪ Substitute the values with correct units. |
| $\Delta E = 2.84 \times 10^{-11}$ J | ▪ Calculate the answer with correct units and appropriate significant figures. |

**TRY THESE YOURSELF**

1   In a nuclear reaction that starts with thorium, a thermal neutron, mass 1.01 u, causes fission of uranium-233 (233.044 u). The fission fragments and their masses, in unified mass units, are molybdenum-104 (103.91 u) and tin-126 (125.91 u).

   **a**   How many neutrons are released in this fission event?

   **b**   Write the nuclear fission reaction using correct nuclide and nucleon symbols.

   **c**   What is the mass defect in this event in:

      **i**   unified mass units?

      **ii**   kilograms?

   **d**   How much energy is released? Give your answer in joules.

2   In a fast breeder reactor, a fast neutron, mass 1.01 u, causes fission of plutonium-239 (239.05 u). One of the two fission fragments is technetium-104 (103.91 u). Three neutrons are released. The mass defect in this event is 0.19 u.

   **a**   What is the nuclide symbol for the second fission fragment?

   **b**   Write the nuclear fission reaction using correct nuclide and nucleon symbols.

   **c**   What is the atomic weight of antimony-133?

   **d**   How much energy is released? Give your answer in joules.

**KEY CONCEPTS**

● Fission of U-235 results in two smaller nuclei and three neutrons emitted.
● Fission of U-235 is caused when it absorbs a slow neutron.
● A controlled fission reaction is sustained using a moderator and control rods.
● An uncontrolled fission reaction can result in huge amounts of energy being released quickly – an atomic bomb.
● The sum of the masses of the products of a fission event is less than the mass present before fission.
● The difference between the mass before and after a fission event is known as the mass defect.
● The mass defect is converted into energy by $E = mc^2$.

1   What is nuclear fission? How does it occur?

2   Why does fission occur for uranium?

3   What are fission fragments?

4   What is 'mass defect' when applied to fission? Why is it important for fission?

5   Each individual fission event produces very little energy. Why then is nuclear energy used for electrical power stations and explosive devices?

6   A thermal neutron, mass 1.01 u, causes fission of uranium-233 (233.044u). The fission fragments and their masses, in unified mass units, are niobium-99 (98.912 u) and antimony-131 (130.912 u).

   **a**   How many neutrons are released in this fission event?

   **b**   What is the mass defect in this event in:

      **i**   unified mass units?                    **ii**   kilograms?

   **c**   How much energy is released? Give your answer in joules.

# 16.5 Nuclear fusion

Nuclear fusion is the process in which two nuclei come together to form a larger nucleus. In the process, energy becomes available. This can occur for elements up to iron-56 (atomic number 26). Making new nuclei by fusion requires an enormous energy input. Conditions similar to those found in the core of stars – extreme pressure and temperatures of millions of kelvin – are needed for fusion to be triggered.

Fusion reactions release much more energy than fission reactions. We have seen that fusion is favoured for elements up to iron-56. For light elements, the curve in Figure 16.13 is steep, which means that any fusion reaction will release a relatively large amount of energy when the new nuclide is formed. For example, Figure 16.13 shows that, for tritium, the binding energy per nucleon is about 2.9 MeV. This is higher than the binding energy per nucleon for the proton (0 MeV) and deuterium (1.1 MeV). Fusion of a proton with deuterium to produce tritium releases about 1.8 MeV of energy per nucleon. This amounts to the release of approximately 62% of the original binding energy per nucleon.

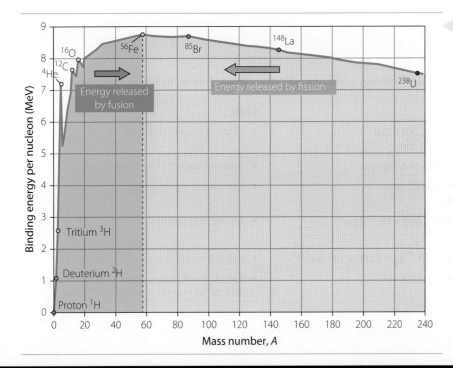

**FIGURE 16.13**
Binding energy per nucleon as a function of nucleon number, $A$

At the other end of the graph, fission is favoured over fusion. The binding energy per nucleon of uranium-235 is about 7.6 MeV. For the two most common fission fragments the binding energy per nucleon is about 8.6 MeV. Taking account of both fission fragments, the difference is about 2.0 MeV. For fission, the release of energy per nucleon is about 26% of the original binding energy per nucleon.

Fusion reactions therefore release a greater proportion of the mass–energy available than do fission reactions.

According to the Big Bang Theory, fusion was possible anywhere in the Universe between the first 1 second and 3 minutes of the Universe's history. During this period, particles began to form through fusion in a process known as nucleosynthesis, which we can still see occurring in stars. Over millions of years a variety of elements can be produced within stars by the fusion-generated process of nucleosynthesis. This topic was covered in chapter 13.

For several decades, scientists have been working on the possibility of using fusion rather than fission to produce electrical energy. Early research into fusion was hampered by a negative energy problem: the energy needed to produce fusion reactions was greater than the energy derived from the reactions. But this is no longer the case. Fusion research has now reached the point where, in the foreseeable future, a fusion reactor will be used to generate electricity, which will feed directly into the electricity grid. The most promising fusion reaction is the deuterium plus tritium reaction, or D–T reaction:

**Fusion as a source of energy**

$$^{2}_{1}\text{H} \quad + \quad ^{3}_{1}\text{H} \quad \rightarrow \quad ^{4}_{2}\text{He} \quad + \quad ^{1}_{0}\text{n}$$
$$2.0141 \text{ u} \quad 3.0160 \text{ u} \quad 4.0026 \text{ u} \quad 1.0086 \text{ u}$$

## Mass defect in fusion

For hydrogen fusion occurring within the core of the Sun, four protons undergo fusion in a multistep process, the proton-proton chain.

$$4^{1}_{1}\text{H} \rightarrow ^{4}_{2}\text{He} + 2\text{e}^{+} + 2\nu + 2\gamma$$

where $\text{e}^{+}$ is a positron, $\nu$ is a neutrino and $\gamma$ is a gamma ray.

Numeracy

The sum of the masses of the protons is greater than the mass of the products. So what becomes of the missing mass, the mass defect? This missing mass is converted into energy by Einstein's equivalence of mass–energy relationship, $E = mc^{2}$.

▶ **WORKED EXAMPLE** (16.8)

How much mass is converted into energy when one helium nucleus is formed by fusion in the Sun?

| ANSWER | LOGIC |
|---|---|
| Mass of four protons $= 4 \times 1.007\,825\,\text{u}$ <br> $= 4.031\,300\,\text{u}$ | ▪ Find the mass of four protons. |
| Mass of one helium nucleus $= 4.00260\,\text{u}$ | ▪ Find the mass of one helium nucleus. |
| Mass defect $= 4.031\,300 - 4.002\,60$ <br> $= 0.028\,70\,\text{u}$ | ▪ Calculate the mass defect. |
| $= 0.028\,70\,\text{u} \times 1.660 \times 10^{-27}\,\text{kg}\,\text{u}^{-1}$ <br> $= 4.764 \times 10^{-29}\,\text{kg}$ | ▪ Convert to SI units. |
| Energy released $= 4.764 \times 10^{-29}\,\text{kg} \times (3.00 \times 10^{8}\,\text{m}\,\text{s}^{-1})^{2}$ | ▪ Apply $E = mc^{2}$. |
| $= 4.288 \times 10^{-12}\,\text{J}$ | ▪ Calculate the answer with correct units and appropriate significant figures. |

1 The fusion of carbon-12 with an alpha particle, $^{12}_{6}C + {}^{4}_{2}He \rightarrow {}^{16}_{8}O + \gamma$, involves the release of energy. Given that the masses are $^{12}_{6}C = 12.0000\,u$, $^{4}_{2}He = 4.0013\,u$, $^{16}_{8}O = 15.9949\,u$, find the energy released each time a carbon-12 and helium-4 nucleus fuse to produce an oxygen-16 nucleus.

2 A proton fuses with a deuterium $\left( {}^{2}_{1}H \right)$ nucleus to form a helium-3 nucleus, $^{3}_{2}He$. Using the data: mass of $^{3}_{2}He = 3.016\,029\,3\,u$; $^{1}_{1}H = 1.007\,825\,u$ and $^{2}_{1}H = 2.014\,101\,78\,u$, calculate:

   **a** the mass defect for this reaction.

   **b** the energy released for this reaction.

Other fusion reactions can and do occur. For example, the production of nitrogen-14 from carbon-12 proceeds by way of two proton additions and positron emission:

$$^{12}_{6}C + {}^{1}_{1}H \rightarrow {}^{13}_{7}N + \gamma$$

$$^{13}_{7}N \rightarrow {}^{13}_{6}C + {}^{0}_{+1}e + \nu$$

$$^{13}_{6}C + {}^{1}_{1}H \rightarrow {}^{14}_{7}N$$

Nuclear fusion

Fusion enables the synthesis of many other nuclides. For such reactions, the mass defect involved results in the release of energy, up to and including the production of iron-56 nuclei. For the synthesis of nuclei heavier than iron-56, a net input of energy is required.

**KEY CONCEPTS**

- Fusion results in the production of heavier nuclei by the joining of lighter nuclei.
- For the production of nuclei up to iron-56, fusion releases energy $E$ where $E = mc^2$ and $m$ is the difference between the masses of the reactant nuclei and product nucleus.
- Mass defect is the difference between the mass of the product nucleus and the mass of the individual constituent nucleons.
- Binding energy is found by multiplying the mass defect by $c^2$.
- Fusion occurs in the core of stars under extreme conditions of pressure and temperature.
- The fusion of deuterium and tritium is the most promising pathway for the future harnessing of fusion for electricity generation.

**CHECK YOUR UNDERSTANDING**

**16.5**

1 Define 'fusion'.

2 Draw a diagram to show fusion with two protons.

3 Outline the process whereby mass defect is converted into energy when fusion occurs.

4 Why is fusion more common than fission for light elements?

5 Outline the conditions required for fusion to occur in stars.

6 What is the binding energy in a deuterium nucleus?

7 During a fusion reaction in a star, lithium-6 takes in a neutron. A new composite nucleus is formed that decays by alpha emission. Write the nuclear equations for the:

   **a** production of the composite nucleus.    **b** subsequent alpha decay.

8 Define 'binding energy per nucleon' and describe why this is an important consideration when assessing the stability of a nucleus.

9 For the fusion reaction $^{12}_{6}C + {}^{1}_{1}H \rightarrow {}^{13}_{7}N + Y$, what is:

   **a** the mass defect?    **b** the energy released?

| PARTICLE | ($^{1}_{1}$H) HYDROGEN | CARBON-12 | NITROGEN-13 |
|---|---|---|---|
| Mass (u) | 1.0078 | 12.0000 | 13.0057 |

10 Why do you think fusion might be a 'cleaner' source of energy than fission?

- A nucleon is a particle in the nucleus – a proton or neutron.

- The atomic number, $Z$, is the number of protons in the nucleus.

- The mass number, $A$, is the number of nucleons (protons and neutrons) in the nucleus.

- Nuclear radiation originates from unstable nuclei.

- Alpha radiation is the emission of an alpha particle; this is identical to a helium nucleus.

- Beta-minus emission occurs when a neutron transforms into a proton and a high speed electron, a beta particle or $\beta^-$, is emitted from the nucleus.

- Positron emission occurs when a proton transforms into a neutron and a positron, a positive electron $\beta^+$, which is emitted from the nucleus.

- Gamma emission is the release of a photon of energy from the nucleus.

- Alpha particles have the least penetrating power, gamma rays have the greatest.

- Alpha particles have the greatest ionising ability due to their charge (+2) and mass; gamma rays have the least.

- Alpha particles are slightly deflected in electric or magnetic fields; beta particles have a large deflection and gamma rays are not deflected.

- The half-life of a sample of radioactive material is the time taken for half of the nuclei in it to decay.

- $N_t$, the number of undecayed particles present at time $t$, is given by $N_t = N_0 e^{-\lambda t}$, where $N_0$ = number of particles present at $t = 0$ and $\lambda$ is the decay constant.

- $\lambda = \dfrac{\ln 2}{t_{1/2}}$, with the units $s^{-1}$.

- The number of undecayed particles present is related to the level of radioactivity of the sample.

- Fission of U-235 results in two smaller nuclei and three neutrons emitted.

- Fission of U-235 is caused when it absorbs a slow neutron.

- A controlled fission reaction is sustained using a moderator and control rods.

- An uncontrolled fission reaction can result in huge amounts of energy being released quickly – an atomic bomb.

- The sum of the masses of the products of a fission event is less than the mass present before fission.

- The difference between the mass before and after a fission event is the mass defect.

- The mass defect is converted into energy by $E = mc^2$.

- Fusion is the production of heavier nuclei by the joining of lighter nuclei.

- For the production of nuclei up to iron-56, fusion releases energy where $E = mc^2$, and $m$ is the difference between the masses of the reactant nuclei and product nucleus.

- Mass defect is the difference between the mass of the product nucleus and the mass of the individual constituent nucleons

- Binding energy is found by multiplying the mass defect by $c^2$.

- Fusion occurs in the core of stars under extreme conditions of pressure and temperature.

- The fusion of deuterium and tritium is the most promising pathway for the future harnessing of fusion for electricity generation.

1   List the three types of radiation that may be emitted from the radioactive decay of a nucleus.

2   Define nucleon.

3   Compare and contrast a $\beta^+$ particle to a $\beta^-$ particle.

4   Compare the penetrating power and ionising ability of alpha particles, beta particles and gamma rays and give an explanation for each.

5   Write a nuclear equation for the fission of a uranium-235 nucleus when it captures a slow neutron.

6   Describe the term 'slow neutron'.

7   With the aid of a diagram, explain how the fission of uranium-235 can be sustained in a controlled chain reaction.

8   What percentage of the original level of radioactivity will a sample have after six half-lives? Show your working.

9   Calculate the decay constant for a radioactive sample that has a half-life of 15 days. Give your answer with the units $s^{-1}$.

10  Compare the processes involved in a controlled chain reaction to an uncontrolled chain reaction for the fission of uranium.

11  Explain why energy is released in the fission of a uranium-235 nucleus. Refer to the concepts of mass defect and mass–energy equivalence.

12  Using appropriate nuclear equations, show why the atomic number of a nucleus increases for beta-minus decay but decreases for alpha decay.

13  Name and write the nuclear symbols for three isotopes of hydrogen.

14  By referring to binding energy and mass defect, explain why nuclear fusion in the core of stars can only be sustained for the production of nuclei up to and including iron-56, but not for heavier nuclei.

15  During gamma decay, a nucleus loses a small quantity of mass. How does the law of conservation of energy apply in this situation?

16  How much energy would be released if 6.0 g of matter was converted into energy?

17  Calculate the energy released when a positron and an electron, each with a mass of $9.1 \times 10^{-31}$ kg, anihilate and are converted into energy.

18  The products of the fission of uranium and the fusion of hydrogen differ in ways that makes one of these processes 'cleaner' than the other. Discuss this statement with reference to the processes' applications to the production of electricity.

19  Calculate the half-life of a radioactive sample with a decay constant of $5.40 \times 10^6 \, s^{-1}$.

20  Give reasons why a radioactive sample will never have its radioactivity level reduced to zero.

21  Compare the properties and deflections in electric and magnetic fields of the various types of nuclear radiation.

22  Discuss the possibility of harnessing fusion power for the future generation of electrical energy.

23  Sketch a graph showing the decrease in radioactivity of a sample with a half-life of 60 years over a period of 60 years. Commence with a radioactivity level of 100%.

24  Discuss the relative merits and difficulties in harnessing the power available from fission and from fusion reactions to generate electrical energy for our society.

25  Discuss how increased knowledge of the structure of the atom has led to applications that have benefits to our society.

# 17 Deep inside the atom

INQUIRY QUESTION

How is it known that human understanding of matter is still incomplete?

OUTCOMES

**Students:**

- analyse the evidence that suggests:
  - that protons and neutrons are not fundamental particles
  - the existence of subatomic particles other than protons, neutrons and electrons
- investigate the Standard Model of matter, including:
  - quarks, and the quark composition hadrons
  - leptons
  - fundamental forces (ACSPH141, ACSPH142) ICT
- investigate the operation and role of particle accelerators in obtaining evidence that tests and/or validates aspects of theories, including the Standard Model of matter (ACSPH120, ACSPH121, ACSPH122, ACSPH146) ICT

Physics Stage 6 Syllabus © 2017 NSW Education Standards Authority (NESA) for and on behalf of the Crown in right of the State of New South Wales.

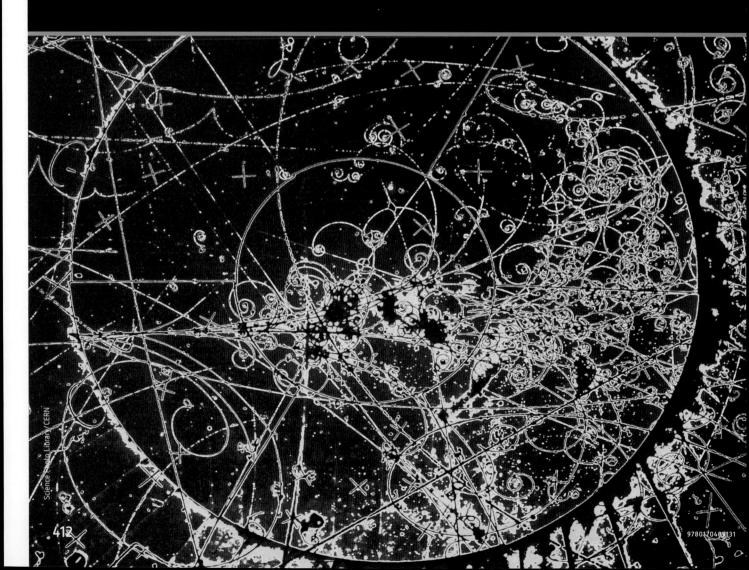

Science Photo Library/CERN

9780170409131

One of the key characteristics of physics as a science is the belief that complex systems can be understood by examining and understanding the motion and interactions of simpler component parts. Underlying this belief is another: that, as we break the system into smaller and smaller parts, we will eventually come to a point where there are no internal components. Then we will be dealing with the basic building blocks out of which everything else is constructed.

The idea that there are fundamental building blocks of matter has persisted through the ages, although it has undergone many changes as new discoveries and observations have been made. The ancient Greeks thought that the Universe was made out of four basic elements: earth, air, fire and water. Then Democritus introduced the concept of the atom. Later it was found that atoms were made of even smaller components. In chapter 14 the development of modern atomic theory was discussed, with advances made by Thomson (plum pudding model), Rutherford (planetary model), Bohr and de Broglie being studied. The particles these models referred to, protons, neutrons and electrons, are now considered to be part of a broader understanding of the nature of matter.

It is now known that the fundamental building blocks are fundamental particles (can also be called elementary particles), and that protons and neutrons are made up from combinations of such particles.

During the 20th century, scientists and engineers working in particle physics made a series of significant discoveries. Their research and discoveries continue today and have led to new theories that describe and predict these fundamental particles and their interactions (Figure 17.1). In this chapter, we describe some of the sequence of observations and discoveries in particle physics. These highlight the important connections between experimental and theoretical progress in physics. We introduce the model that describes our current understanding of particle physics: the Standard Model of matter (the Standard Model for short).

**FIGURE 17.1** Deep inside the Large Hadron Collider (LHC) at CERN, where scientists are endeavouring to discover the nature of matter

## 17.1 Are protons and neutrons fundamental particles?

The model of the atom that you learned about in chapters 14 and 15 was largely developed by 1932. There is a tiny, massive nucleus that contains positively charged protons and neutral neutrons. Surrounding this is a cloud of negatively charged electrons, as depicted in Figure 17.2.

A fourth particle, the photon, was also already known by the 1930s. As we saw in chapter 11, the idea of a particle of light was introduced to explain the photoelectric effect. Recall that the photon has no mass and no charge. It does have quantised energy and momentum, which depend on its frequency. The electron and the photon both appear to be fundamental particles.

When we are considering the gravitational force between the Sun and Earth, we can treat Earth as if all its mass were concentrated at its centre. However, Earth is not really a point mass because it has *internal structure*. When we are close to Earth's surface, the gravitational field no longer appears to be that due to

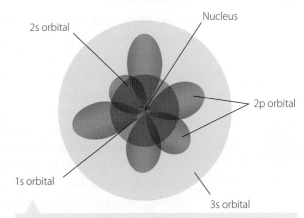

**FIGURE 17.2** In the modern quantum mechanical model of the atom, the protons and neutrons are contained in the nucleus, surrounded by the electron clouds.

a point mass, or a single point. Earth's internal structure becomes important. Similarly, a charged object can be treated as a point charge when it interacts with another charged object a long way away, but close up we must treat it differently.

The electric field of an electron is exactly what we expect for a point charge, suggesting that it has no internal structure and strongly supporting the idea that it is a fundamental particle.

## The evidence – protons and neutrons are not fundamental particles

Experimental evidence suggests that the neutron is *not* a fundamental particle. Although the neutron has no charge, it does have its own intrinsic magnetic field. As we have seen previously, a magnetic field is the result of moving charged particles or a changing electric field. This indicates that the neutron has some internal structure, and contains charges that add to zero total charge. This suggested to physicists that the neutron is *not* a fundamental particle.

Because the proton has a mass very close to that of a neutron and is otherwise a similar particle in its behaviour, the same question was also raised of the proton. Evidence that the proton is not a fundamental particle came from studies of radioactivity. You have seen in the previous chapter that when a nucleus decays, different types of particles may be emitted. These include $\beta^-$ and $\beta^+$ particles. It was observed that these $\beta$ particles had the same mass as an electron, and either a positive or negative charge equal to the electron charge.

It turned out that the $\beta^-$ particle was an electron. When a nucleus emits a $\beta^-$, its proton number increases by one and its neutron number decreases by one. Similarly, when a $\beta^+$ particle is emitted, a proton is converted to a neutron. Hence, it appeared that neutrons could be converted into protons and vice versa by the emission of these negatively or positively charged electrons, suggesting that neither the neutron nor the proton is a fundamental particle.

The positively charged electrons, called positrons, that are involved in $\beta^+$ decay were discovered in cosmic rays in 1932 by Carl Anderson. Positrons were the first antimatter particle to be observed.

By 1932 there were five particles known: three 'normal' matter particles, the electron, proton and neutron; one antimatter particle, the positron; and the photon. Of these, it was already believed that the proton and neutron were not themselves fundamental particles, and so the hunt for their component parts began.

You have already been introduced to these particles in chapter 13.

## Antimatter

In the 1920s, Paul Dirac developed a relativistic version of the Schrödinger equation. Recall from chapter 15 that the Schrödinger equation is the wave equation that describes the behaviour of particles such as electrons. Dirac's new relativistic equation explained the origin of the electron's spin and magnetic moment. This was an important theoretical development in quantum mechanics.

Spin is a quantum number that can take half-integer or integer values. Magnetic moment is a vector; it has direction and can have a positive or negative value. The magnetic moment of a particle depends on the spin, charge and mass of the particle.

However, there appeared to be a difficulty with the theory: the equation that Dirac had developed to describe the electron had two solutions. One of these solutions gave the wave functions for electrons and correctly described their mass, charge and spin. The other solution described a particle with the same mass but the opposite charge and magnetic moment. It also predicted that if these two particles should meet, they would both be destroyed, producing a burst of energy. This is called annihilation.

Hence this second particle was the 'anti-electron'. It was this particle, the antiparticle to the electron, that Anderson observed in 1932. The energy produced when an electron meets a positron is in the form of a pair of gamma rays that move away at 180° to each other. Conservation of energy holds here because the mass is converted into energy according to Einstein's mass–energy relationship $E = mc^2$.

9780170409131

For example, when an electron and a positron annihilate each other, twice the mass of an electron is converted into energy, that is $2 \times (9.1 \times 10^{-31}\,\text{kg})$. The amount of energy released is thus:

$$E = mc^2$$
$$= 2 \times (9.1 \times 10^{-31}\,\text{kg}) \times (3.00 \times 10^8\ \text{m s}^{-1})^2$$
$$= 1.6 \times 10^{-13}\ \text{J}$$

So each gamma ray photon has half this energy, or $8.0 \times 10^{-14}\,\text{J}$.

Using $E = hf$ to find the frequency of the two gamma ray photons, each with equal energy,

$$E = hf$$
$$f = \frac{E}{h}$$
$$= \frac{8.0 \times 10^{-14}\ \text{J}}{6.626 \times 10^{-34}\ \text{J s}}$$
$$= 1.2 \times 10^{20}\ \text{Hz}$$

Anderson used a cloud chamber to study cosmic rays. A cloud chamber is a particle detector that detects ionising radiation, discussed in chapter 16. It uses a supersaturated vapour, like a layer of fog, in a container or chamber. When a charged particle enters the chamber it ionises the vapour. The resulting electrically charged vapour particles act as sites for condensation of the vapour. This produces a visible trail of condensation along the path of the particle.

Anderson placed his cloud chamber in a magnetic field. The magnetic field caused the moving charged particles to follow curved paths. This allowed him to distinguish between positive and negative charges. Recall that the direction in which a particle's path curves depends on its charge, and that the sharpness of the curve (radius of curvature) depends on the mass.

Anderson noted that some of the tracks in his cloud chamber had electron-like curvature but were deflected in a direction corresponding to a positively charged particle. Anderson had discovered the anti-electron, or positron (Figure 17.3). He was awarded a Nobel Prize for Physics in 1936 for this discovery.

Dirac's theory suggested that an antiparticle exists for every particle. It has subsequently been verified that almost every known particle has a distinct antiparticle.

The antiparticle for a charged particle has the same mass as the particle but opposite charge and hence opposite sign for magnetic moment.

Antiparticles for uncharged particles, such as the neutron, are more difficult to describe. For neutral particles with magnetic moments such as the neutron, the antiparticle can be defined by the sign of its magnetic moment. The magnetic moment is a vector quantity that measures the ability to interact with a magnetic field. A few particles, such as photons, are their own antiparticle.

When we write the symbol for a particle, such as n for neutron or $e^-$ for electron, we represent the antiparticle either by placing a bar over the symbol, as in $\bar{n}$, or by showing that the sign is reversed, as in $e^+$.

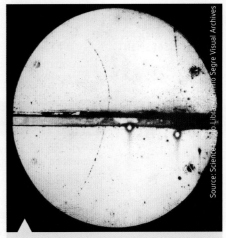

**FIGURE 17.3** This cloud chamber photograph shows the track of the first identified positron.

**TABLE 17.1** Particles and their antimatter equivalent

| PARTICLE | ANTI MATTER EQUIVALENT |
|---|---|
| proton, p | antiproton, $\bar{p}$ |
| neutron, n | antineutron, $\bar{n}$ |
| electron, $e^-$ | positron, $e^+$ |

Are protons and neutrons fundamental particles?

# Build your own cloud chamber and detect cosmic rays

There are very few particle physics experiments that you can do without expensive, and usually high voltage, equipment. However, with some simple equipment you can build your own cloud chamber that will detect cosmic rays. You will need some chemicals and dry ice so you need to be very careful and follow all safety instructions. Typically, it takes about 10 to 20 minutes to detect a high energy cosmic ray, depending on solar flare activity, so you may also need to be a bit patient after you have built your cloud chamber.

## AIM

To build a cloud chamber and detect cosmic rays

## MATERIALS

- Clear glass or clear plastic tank, such as a small fish tank, about 15 cm tall and 15 cm wide by 30 cm long
- Strong source of light such as an overhead or slide projector
- Sheet of metal for a lid for the tank (same size as tank)
- Sheet of cardboard cut to fit the metal lid
- 3 sheets of felt, 30 cm × 30 cm
- Whole roll of black electrical tape
- Foam padding, about 5 cm thick, same dimensions as the tank
- Cardboard box just slightly bigger than the clear tank
- Glue that is not soluble in alcohol, such as silicon sealant
- Isopropyl alcohol (isopropanol) (pure, about 500 mL)
- Dry ice, about 500 g
- Gloves (disposable)
- Gloves (heat-/cold-proof)
- Scissors
- Strong magnet (optional)
- Safety equipment: tongs, lab coats, safety glasses

RISK
ASSESSMENT

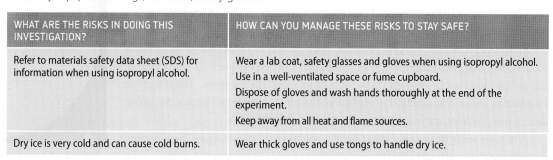

| WHAT ARE THE RISKS IN DOING THIS INVESTIGATION? | HOW CAN YOU MANAGE THESE RISKS TO STAY SAFE? |
|---|---|
| Refer to materials safety data sheet (SDS) for information when using isopropyl alcohol. | Wear a lab coat, safety glasses and gloves when using isopropyl alcohol. Use in a well-ventilated space or fume cupboard. Dispose of gloves and wash hands thoroughly at the end of the experiment. Keep away from all heat and flame sources. |
| Dry ice is very cold and can cause cold burns. | Wear thick gloves and use tongs to handle dry ice. |

What other risks are associated with your investigation, and how can you manage them?

## METHOD

Building the cloud chamber will take some time. You will probably need a whole lab class to build your cloud chamber and then a second class to test it. Stop at step 6 unless you have ample time to proceed and test your cloud chamber.

1 Fold or roll the felt into strips about 5 cm wide and 30 cm long. Attach these to the inside of the clear container with the glue or sealant to form a ring around the bottom, as shown in Figure 17.4. Give the glue or sealant time to dry. (You may want to do this the day before.)

**2** Cover one side of your cardboard sheet with strips of black electrical tape, as neatly as you can. This is to make particle tracks easier to view.

**3** Attach the cardboard, tape side out, to the metal lid so that when the lid is in place the black tape faces into the tank.

**4** Cut the cardboard box down to about 7 cm tall and place the piece of foam in the bottom.

**5** Check that your cloud chamber all fits together: the metal lid sits on top of the foam with the black tape facing upwards. The tank then sits upside-down on top of this so the ring of felt padding is at the top.

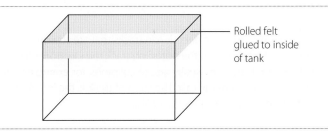
Rolled felt glued to inside of tank

**FIGURE 17.4** Felt folded and attached to tank as a 'soak zone'

It needs to all fit tightly and be well sealed when you are performing your observations. If it is not, air currents will make the particle tracks hard to see.

If it all fits together neatly, then you are ready to add the dry ice and isopropyl alcohol. If not, you will need to make some modifications until it fits neatly together.

**6** Remove the tank and metal lid and, using tongs, place a layer of dry ice on top of the foam. Your teacher may do this step for you.

**7** Soak the felt strips with isopropyl alcohol. Your teacher may do this step for you.

**8** Seal the metal lid to the tank with more electrical tape or duct tape.

**9** Place the tank upside-down on top of the dry ice so the metal lid is in contact with the dry ice.

**10** Arrange the light source so it shines horizontally through the side of the tank. You need a bright light source to illuminate the particle tracks clearly.

**11** Turn on the light, and observe your chamber for at least 10–20 minutes.

You should see a mist-like fog form inside your chamber. This will be most obvious near the bottom; don't worry if you can only see a thin layer, that will be enough.

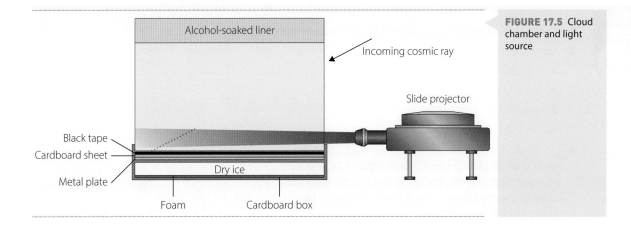
Alcohol-soaked liner

Incoming cosmic ray

Slide projector

Black tape
Cardboard sheet
Metal plate
Dry ice
Foam
Cardboard box

**FIGURE 17.5** Cloud chamber and light source

Step-by-step guide to building a cloud chamber

## RESULTS

1 During the 10–20 minutes you observe your chamber, you should see fine tracks forming in your chamber. The tracks, which are the results of the random passage of cosmic rays through the vapour, can form at any time and will last only a very brief time before they disappear, so you will need to watch carefully for the entire period. Draw these tracks and/or photograph or video them.

2 Once your chamber is working, you can try putting a strong magnet on one side and observe what happens to the new particle tracks. Note that you will need a *very strong* magnet to obtain any significant curvature of the particle paths.

## ANALYSIS OF RESULTS

Record as many tracks as you can. Look for sudden changes of direction in tracks that are straight lines. These could be muon decays, with the incoming track the muon and the outgoing track the electron into which it decays. You may also see tracks that branch like a Y. These are usually due to collisions, for example a muon colliding with an electron and transferring some kinetic energy to it. The stem of the Y is the incoming muon and the two branches are the muon and electron moving off after the collision.

## DISCUSSION

1 Research what cosmic rays are and where they typically originate.

2 Can you identify what is happening in the various tracks that you observe? If not, can you hypothesise what these tracks might be produced by?

## CONCLUSION

Write a suitable conclusion to this investigation based on your observations and the aim.

---

**KEY CONCEPTS**

- Observations of radioactive decay indicated that the proton and neutron are not fundamental particles.
- The electron is a fundamental particle.
- All particles have an antimatter equivalent.
- The positron is the antimatter equivalent of an electron, with opposite charge to an electron.
- Annihilation occurs when a particle collides with its antimatter equivalent.

---

**CHECK YOUR UNDERSTANDING**

1 Outline why electrons are considered to be fundamental particles but protons are not.

2 Identify the charge of:
   a a proton.
   b an antiproton.
   c a positron.
   d an antineutron.

3 An electron enters a region of uniform magnetic field and curves to the right as shown in Figure 17.6. A second charged particle enters the field following the same initial trajectory as the electron, but it curves to the left. How could its path tell us if it was a positron or a proton?

4 What is the symbol for an antineutron?

5 In comparison to an electron, in what ways is a positron:
   a similar?
   b different?

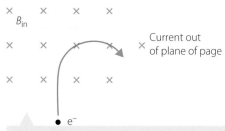

**FIGURE 17.6** An electron's trajectory curves to the right when it enters the magnetic field.

6  What is the process of 'annihilation'? Where does the matter involved go?

7  Outline how the existence of the positron was first predicted.

8  Given that the mass of an electron is $9.1 \times 10^{-31}$ kg, calculate the energy released when a positron meets an electron and they annihilate.

## 17.2 Further evidence for fundamental particles

Following on from the discoveries described above, many 'new' particles were discovered. Some of these were discovered in cosmic-ray experiments such as that carried out by Anderson. Other particles were discovered in nuclear decays.

Cosmic rays pass through Earth's atmosphere at random, and their energies vary widely. Hence it was impossible to design well-controlled experiments that used cosmic rays to create new particles.

The energies of the particles emitted in some nuclear decays are well defined, but they are generally fairly low. The equation $E = mc^2$ suggested that if higher energies could be used, more massive particles might be created.

Thus, physicists looked for ways to produce beams in which there were more particles with high enough energies to produce new particles that they could study. The breakthrough was achieved with the invention of particle accelerators.

From the 1950s onwards, many more particles were discovered in experiments using these accelerators to produce high-energy collisions between known particles. The more energy available in the collisions, the higher the number of new particles produced and the greater their mass.

These new particles are characteristically very unstable and have very short half-lives that range between $10^{-6}$ s and $10^{-23}$ s. Their decays produce lighter particles, some of which are also unstable. So, each collision between just two initial particles may result in many outgoing particles, which need to be detected and identified simultaneously. To enable this, physicists and engineers worked together to develop huge, complex apparatus to use at the new particle accelerators. This is an example of the role that technology plays in allowing new experiments, which in turn leads to the development and refinement of theory.

### Particle accelerators

To detect a particle with a very short half-life, first it is necessary to create it. This is done by causing reactions between stable particles such as protons, electrons and neutrons. However, protons and neutrons have small masses compared to most other particles.

To create a particle of large mass $m$, there needs to be at least an energy of $E = mc^2$ available to convert into mass. If the total mass of the reacting particles is less than $m$, some of this energy must come from the kinetic energy of the reacting particles. Particle accelerators provide this extra energy.

A particle accelerator uses an electric field to accelerate charged particles to very high speeds and hence very high energies. A magnetic field may also be used to contain the charged particles within the accelerators.

A linear accelerator accelerates the particles in a straight line. In a linear accelerator, large electric fields are used to accelerate charged particles such as electrons or protons. This force accelerates the particle to very high speeds.

At many facilities, the linear accelerator is used to feed high-speed particles into a synchrotron ring. The synchrotron ring uses large magnetic fields to contain the charged particles in a circular path.

As the particles are likely to be at very high speeds in a synchrotron, we need to use the relativistic expression for momentum from chapter 12:

$$p_v = \frac{mv}{\sqrt{1 - \dfrac{v^2}{c^2}}}$$

In the synchrotron ring, the particles may be held for storage at constant speed, or they may be further accelerated.

Once the charged particles have been accelerated to the desired speeds they are aimed at a target, again using magnetic and electric fields to steer them. They are made to smash into the target at very high speed, close to the speed of light.

Billiard balls will bounce off each other when they collide at normal speeds. When they are smashed into each other at high speeds, the balls will shatter. This analogy applies to the subatomic particles being smashed into each other in particle accelerators. Reality is far more complex than the billiard ball analogy, however, because new particles are also created in these collisions. Travelling at close to the speed of light and with relativistic momentum many times greater than what they would have otherwise, these particles possess a lot of energy that causes them to break apart into their constituent components – quarks, bosons and other fundamental particles – and new particles also appear.

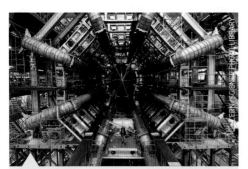

FIGURE 17.7 The 7000 tonne ATLAS detector, one of several at the LHC designed to detect the faintest traces of particles such as the Higgs boson, the particle responsible for attributing mass to matter.

## The Large Hadron Collider

The world's largest and most powerful particle accelerator is the Large Hadron Collider, 100 m underground in Switzerland (Figure 17.7). The result of international collaboration, it is operated by CERN, the European Organisation for Nuclear Research. The LHC is actually a large complex of linear accelerators and synchrotrons, as shown in Figure 17.8. It is capable of accelerating protons to speeds of 0.999 999 99$c$.

**FIGURE 17.8**
The four main LHC experiments. This diagram shows the locations of the four main experiments (ALICE, ATLAS, CMS and LHC-B) that take place at the LHC. Located between 50 m and 150 m underground, huge caverns have been excavated to house the giant detectors. The SPS, the final link in the pre-acceleration chain, and its connection tunnels to the LHC are also shown.

Further evidence for fundamental particles

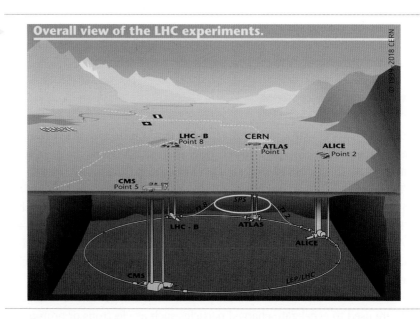

- Particle accelerators give enormous energy to subatomic particles to crash them together and break them into their components.
- Relativistic effects mean that particles moving at close to the speed of light have much more energy than they otherwise would possess.
- Many particles created by colliding high energy particles together have very short half-lives and may therefore be difficult to detect.

CERN's homepage

1 What is a 'cosmic ray'?

2 Why are particle accelerators required for us to further our understanding of matter?

3 What is a 'linear' accelerator?

4 Relativistic kinetic energy is given by $E_k = mc^2 (\gamma - 1)$. Find the relativistic kinetic energy of an electron accelerated to 0.9999c.

5 a Where is the Large Hadron Collider?

   b Outline the principles of its operation.

6 Why is the LHC named the Large Hadron Collider?

# 17.3 The Standard Model of matter

The Standard Model of matter was initially developed to describe all matter and forces in the Universe using fundamental particles. The Standard Model of matter can be summarised by a flow chart as shown in Figure 17.9. You met many of these particles in chapter 13.

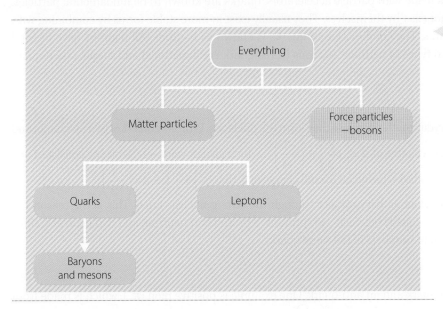

FIGURE 17.9 The Standard Model of matter

Particle adventure

An interactive exploration of the world of subatomic particles

Recent experiments at CERN

Confirm the Standard Model of matter

## Quarks

In 1964, two physicists, Murray Gell-Mann and George Zweig, proposed the existence of particles with charges that were sub-multiples of electron charges, termed quarks. Later, quarks were recognised as fundamental particles – the smallest particles that could not be broken down further. In simple terms, there are six 'flavours' of quarks as well as six corresponding anti-quarks. The term 'flavours' refers to the quantum properties of the different quarks. The names, symbols and charges of the six flavours of quarks

Quarks are also discussed in chapter 13.

are summarised in Table 17.2. Although the six anti-quarks are not shown, they do exist. For instance, an anti-up quark has the symbol $\bar{u}$ and a charge of $-\dfrac{2}{3}$, while an anti-down quark has the symbol $\bar{d}$ and a charge of $+\dfrac{1}{3}$.

**TABLE 17.2** The six 'flavours' of quarks

| GENERATION | QUARKS | SYMBOLS | CHARGE |
|---|---|---|---|
| 1 | Up | u | $+\dfrac{2}{3}$ |
| | Down | d | $-\dfrac{1}{3}$ |
| 2 | Charm | c | $+\dfrac{2}{3}$ |
| | Strange | s | $-\dfrac{1}{3}$ |
| 3 | Top | t | $+\dfrac{2}{3}$ |
| | Bottom | b | $-\dfrac{1}{3}$ |

## Hadrons

Through observations made with particle accelerators, quarks are known to be fundamental particles. However, quarks do not exist by themselves. They usually exist in more stable forms by combining with one or two other quarks. The combination of quarks is known as hadrons. All hadrons have integral charges. There are two types of hadrons: baryons (three-quark combinations) and mesons (two-quark combinations).

## Baryons

Baryons make up everyday matter because they form the nucleons. Protons consist of two up quarks and one down quark, hence a charge of $2 \times \left( +\dfrac{2}{3} \right) + \left( -\dfrac{1}{3} \right)$, which gives a value +1. Neutrons consist of one up quark and two down quarks, giving an overall charge of 0, that is, $\left( +\dfrac{2}{3} \right) + 2 \times \left( -\dfrac{1}{3} \right)$.

All quarks, and therefore all baryons, act through the strong nuclear force. This is the reason the strong nuclear force acts equally between two protons, a proton and neutron, and two neutrons, since all are made from different numbers of the same quarks.

## Mesons

A meson consists of a quark and an anti-quark. One example is positive pion ($\pi^+$), which is made from an up quark $\left( +\dfrac{2}{3} \right)$ and an anti-down quark $\left( +\dfrac{1}{3} \right)$, giving it an overall charge of +1. Mesons are generally unstable and therefore short-lived. This means that detecting and identifying them requires specialised equipment and sophisticated analysis.

## Leptons

Leptons are another type of fundamental particle that have either very little or no mass. There are six 'flavours' of leptons, also grouped into three generations. Within each generation, an electrically charged lepton is coupled with its corresponding neutrino. This is summarised in Table 17.3.

For every lepton, there is a corresponding anti-lepton. All leptons interact through the weak nuclear force and the charged leptons also interact through the electromagnetic force.

**TABLE 17.3** The six 'flavours' of leptons

| GENERATION | LEPTONS | SYMBOL | CHARGE |
|---|---|---|---|
| 1 | Electron | $e^-$ | −1 |
|  | Electron-neutrino | $v_e$ | 0 |
| 2 | Muon | $\mu^-$ | −1 |
|  | Muon-neutrino | $v_\mu$ | 0 |
| 3 | Tau | $\tau^-$ | −1 |
|  | Tau-neutrino | $v_\tau$ | 0 |

## A word about 'generation'

Note that in Tables 17.2 and 17.3, the term 'generation' is used. The first generation particles are the particles that make up ordinary matter. For instance, the generation 1 quarks make up the nucleons, and generation 1 leptons include the electrons and the electron-neutrinos that are released during ordinary $\beta^-$ decays. The second generation particles are less stable and quickly decay to form the first-generation particles; the third generation particles are even less stable and decay rapidly to form the second-generation particles. The fact that the second and third generation particles are unstable and short-lived means they cannot constitute everyday matter and also they are harder to detect. As the generation number increases, the mass of the particles also increases.

## Bosons – the force particles

There are four fundamental forces in the Universe. Using the Standard Model of matter, these four forces are thought to act through the exchange of force particles, called bosons:

1 The electromagnetic force acts through photons.

2 The strong nuclear force acts through gluons.

3 The weak nuclear force acts through Z and W bosons.

4 Gravity is thought by some particle physicists to act through gravitons; however, unlike the other force particles, their existence has not yet been demonstrated. They were originally included in the Standard Model of matter for the sake of mathematical completeness.

This information is summarised in Table 17.4.

Information and communication technology capability

Higgs boson

Other particles predicted by the Standard Model

**TABLE 17.4** The four fundamental forces in nature

| FORCE | PARTICLES INVOLVED | FORCE CARRIER | RANGE | RELATIVE STRENGTH |
|---|---|---|---|---|
| Gravity | all particles with mass | graviton (yet to be observed) | infinity | much weaker |
| Weak nuclear force | quarks and leptons | W and Z bosons | short range | ↓ |
| Electromagnetic force | electrically charged | photon | infinity |  |
| Strong nuclear force | quarks and gluons | gluon | short range | much stronger |

Force particles can be thought to convey attraction forces (between matter) by having the matter pull on the force particles as they are exchanged, whereas repulsion forces are conveyed by having the force particles being pushed away as they are exchanged.

## The scale of the fundamental particles

The fundamental particles themselves are very small compared to the overall size of the nucleus and of the atom. Figure 17.10 has information about the scale of these particles but is not drawn to scale.

**FIGURE 17.10** If the protons and the neutrons in this diagram were 10 cm in diameter, the quarks and electrons would be less than 0.1 mm in size. The atom itself would be 10 km across.

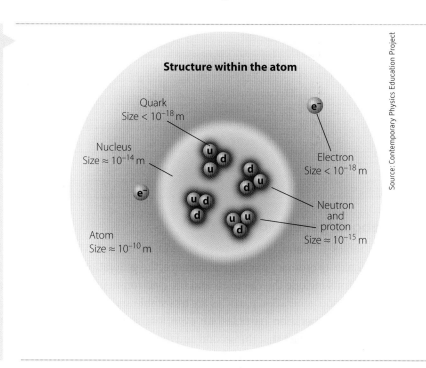

Source: Contemporary Physics Education Project

The Standard Model of matter

Information and communication technology capability

**Large Hadron Collider**

**Problem with the Standard Model**

## Discovery of the Higgs boson

Named after Peter Higgs (Figure 17.11), one of the scientists who first predicted its existence, the Higgs boson is a fundamental particle crucial to the Standard Model. Confirmation of its existence came from data obtained using the ATLAS detector at CERN, using the Large Hadron Collider. On 4 July 2012 it was announced to the world the Higgs boson had been detected with a mass within the range predicted by the Standard Model. Had the Higgs boson remained undetected by the LHC, the Standard Model may have been in doubt. Its detection using the extremely powerful LHC has served to reinforce one of science's most important theories.

Source: Alamy Stock Photo / graham clark

**FIGURE 17.11** Peter W Higgs, jointly awarded the Nobel Prize for Physics in 2013 after confirmation of the existence of the Higgs boson – the force particle that attributes mass - that Higgs and his co-winners had predicted in 1964.

## The future of the Standard Model

The concept of the Standard Model of matter and the existence of quarks, leptons and bosons is an area of physics that has stood for more than 40 years. We may also consider this model as a further advance in our understanding of the atom. Scientists have come a long way from the 'plum pudding' model of the atom suggested by Thomson, to the idea of the quantum and Bohr's model, through the more sophisticated models based on the quantum mechanics of de Broglie, Pauli and Heisenberg and now the Standard Model.

- The Standard Model describes the structure of matter and the forces that hold it together.
- Matter is made of the fundamental particles leptons (e.g. electrons) and quarks.
- Neutrons and protons (baryons) are made from three quarks.
- There are six flavours of quarks and six flavours of leptons.
- Bosons are known as the force carriers.

1  What does the Standard Model attempt to explain?
2  Describe the composition of a neutron according to the Standard Model.
3  Name the six different 'flavours' of quarks.
4  What are the two types of hadron and how do they differ?
5  Outline the role of the strong nuclear force between baryons in the nucleus.
6  Why are mesons difficult to detect? What are they composed of?
7  Discuss the role of particle accelerators in verifying the Standard Model.
8  Outline the three 'generations' of matter.

- Observations of radioactive decay indicated that the proton and neutron are not fundamental particles.
- The electron is a fundamental particle.
- All particles have an antimatter equivalent.
- The positron is the antimatter equivalent of an electron, with opposite charge to an electron.
- Annihilation occurs when a particle collides with its antimatter equivalent.
- Particle accelerators give enormous energy to subatomic particles in order to crash them together and break them into their components and generate new particles.
- Relativistic effects mean that particles moving at close to the speed of light have much more energy than they otherwise would possess.

- Many particles created by colliding high energy particles together have very short half-lives and may therefore be difficult to detect.
- The Standard Model describes the structure of matter and the forces that hold it together.
- Matter is made of the fundamental particles leptons (e.g. electrons) and quarks.
- Neutrons and protons (baryons) are made from three quarks.
- There are six flavours of quarks and six flavours of leptons.
- Bosons are known as the force carriers.

## (17) CHAPTER REVIEW QUESTIONS

Review quiz

1  Define the term 'fundamental particle'.

2  Why was the photon regarded as the 'fourth particle'?

3  The electric field of an electron is exactly what we expect from a point charge. How does this support the idea that it is a fundamental particle?

4  Describe a difference and a similarity between $\beta^-$ and $\beta^+$ particles.

5  Outline what occurs when a particle of matter meets its antimatter equivalent. Refer to an appropriate equation in your response.

6  How does a cloud chamber assist us to detect subatomic particles?

7  Explain how a magnetic field can be used to determine whether a beta particle is an electron or a positron.

8  Explain how particle accelerators were able to reveal the existence of many more particles of subatomic size.

9  Compare a linear accelerator to a synchrotron.

10  Why is it advantageous for particle accelerators to accelerate particles to as close to the speed of light as possible before they are collided?

11  Explain why detecting the particles generated as a result of collisions in particle accelerators requires specialised equipment and sophisticated analysis.

12  Using a table to organise your response, show how quarks are constituents of protons and neutrons.

13  What are the two charges that a quark may possess?

14  Show how the combination of quarks that make up protons and neutrons results in their overall charges.

15  Identify the four types of bosons and the forces they are thought to be responsible for.

16  What is special about the Higgs boson?

17  Construct a chart that shows the various types of particles in the Standard Model and how they are related.

**Answer the following questions.**

1 Spectroscopy, the study of the spectrum of a star, reveals much information about the nature of the star. Describe how characteristics such as evolutionary age, surface temperature, colour, luminosity and chemical composition can be determined using spectroscopy.

2 Discuss the role of nuclear fusion in the nucleosynthesis that occurs in the core of stars, with particular reference to the alternative pathways available to produce helium from hydrogen.

3 Assess the role of cathode ray tubes in the development of the model of the atom from that of an indivisible, featureless particle to the modern model of the atom.

4 With reference to the nature of the radiation emitted, outline the various mechanisms that occur within nuclei that give rise to radioactive decay.

5 Justify the use of particle accelerators such as the Large Hadron Collider to investigate the nature of matter.

6 With reference to binding energy, mass defect and the mass–energy equivalence relationship, explain how energy can be derived from the processes of fission and fusion.

7 Explain how matter waves are relevant to the Bohr model of the atom.

8 Using examples, show that only electron transitions in hydrogen from higher energy shells to $n_f = 2$, that is, the Balmer series, emit photons in the visible part of the electromagnetic spectrum.

9 Assess the role of models in the development of modern scientific theories. Refer to two examples of models in your response.

10 Outline the role of spectroscopy in the discovery of the expansion of the Universe.

## DEPTH STUDY SUGGESTIONS

Standard Model

ANSTO

Cosmic radiation

Pulse@Parkes

Quantum Physics @UNSW

Hubble Space Telescope data

→ A literature review of the discoveries related to the Standard Model over the past 20 years.

→ Work in conjunction with a scientist or scientists at ANSTO reviewing and collating the nature of the research and investigations being undertaken, in particular with the utilisation of the wave nature of neutrons.

→ Research and review secondary-sourced data and undertake a first-hand investigation to explore the nature and source of cosmic radiation that reaches Earth.

→ Undertake further study into the origins of the elements and the nature of interstellar space by participating in the Pulse@Parkes program through the CSIRO's Division of Radiophysics at Marsfield, Sydney.

→ Undertake further research into quantum physics and relate its possible future application to computing.

→ Analyse Hubble Space Telescope data to plot the velocities versus the distances for galaxies to determine the value of $H_0$ and the age of the Universe.

## Appendix 1: SI and non-SI units

### International System of Units (SI)

The international body that decides the appropriate units to be used for the various physical quantities is the Conférence Générale des Poids et Mesures (CGPM). The system of units approved by the CGPM and now widely used by the scientific community throughout the world is known as Système International d'Unités (abbreviated SI).

In your experimental work you should use SI units (or their multiples or submultiples).

The SI consists of seven base units and two supplementary units. All other derived units are based on these nine fundamental units.

The base and supplementary units, together with the derived units with special names that might be relevant to your experimental work, are listed in Tables A1.1 and A1.2.

**TABLE A1.1** SI base units

| PHYSICAL QUANTITY | NAME OF UNIT | ABBREVIATION |
|---|---|---|
| Length | metre | m |
| Mass | kilogram | kg |
| Time | second | s |
| Electric current | ampere | A |
| Thermodynamic temperature | kelvin | K |
| Luminous intensity | candela | cd |
| Amount of substance | mole | mol |

As you become familiar with each new unit you should make a practice of correctly using its abbreviated form.

The internationally recognised prefixes for the SI units together with their abbreviations are given in Table A1.3.

**TABLE A1.2** SI supplementary units derived from SI base units

| NAME | SYMBOL | QUANTITY | EQUIVALENTS | SI BASE UNIT EQUIVALENTS |
|---|---|---|---|---|
| hertz | Hz | frequency | 1/s | $s^{-1}$ |
| radian | rad | angle | m/m | dimensionless |
| steradian | sr | solid angle | $m^2/m^2$ | dimensionless |
| newton | N | force, weight | $kg\,m/s^2$ | $kg\,m\,s^{-2}$ |
| pascal | Pa | pressure, stress | $N/m^2$ | $kg\,m^{-1}\,s^{-2}$ |
| joule | J | energy, work, heat | N m | $kg\,m^2\,s^{-2}$ |
| watt | W | power, radiant flux | J/s | $kg\,m^2\,s^{-3}$ |
| coulomb | C | quantity of electric charge | A s | A s |

9780170409131

| NAME | SYMBOL | QUANTITY | EQUIVALENTS | SI BASE UNIT EQUIVALENTS |
|---|---|---|---|---|
| volt | V | electromotive force, electrical potential difference, electric potential voltage | J/C | $kg\, m^2\, s^{-3}\, A^{-1}$ |
| farad | F | electrical capacitance | C/V <br> s/V | $kg^{-1}\, m^{-2}\, s^4\, A^2$ |
| ohm | V | electrical resistance, impedance, reactance | V/A | $kg\, m^2\, s^{-3}\, A^{-2}$ |
| siemens | S | electrical conductance | 1/V <br> A/V | $kg^{-1}\, m^{-2}\, s^3\, A^2$ |
| weber | Wb | magnetic flux | J/A | $kg\, m^2\, s^{-2}\, A^{-1}$ |
| tesla | T | magnetic field strength, magnetic flux density | $V\, s/m^2$ <br> $Wb/m^2$ <br> N/(A m) | $kg\, s^{-2}\, A^{-1}$ |
| degree Celsius | °C | temperature relative to 273.15 K | K − 273.15 | K − 273.15 |
| lumen | lm | luminous flux | cd sr | cd |
| lux | lx | illuminance | $lm/m^2$ | $m^{-2}\, cd$ |
| becquerel | Bq | radioactivity (decays per unit time) | 1/s | $s^{-1}$ |
| gray | Gy | absorbed dose (of ionising radiation) | J/kg | $m^2\, s^{-2}$ |
| sievert | Sv | equivalent dose (of ionising radiation) | J/kg | $m^2\, s^{-2}$ |

**TABLE A1.3** Prefixes for SI units

| PREFIX | ABBREVIATION | VALUE |
|---|---|---|
| exa | E | $10^{18}$ |
| peta | P | $10^{15}$ |
| tera | T | $10^{12}$ |
| giga | G | $10^9$ |
| mega | M | $10^6$ |
| kilo | k | $10^3$ |
| hecto | h | $10^2$ |
| deka | da | 10 |
| deci | d | $10^{-1}$ |
| centi | c | $10^{-2}$ |
| milli | m | $10^{-3}$ |
| micro | μ | $10^{-6}$ |
| nano | n | $10^{-9}$ |
| pico | p | $10^{-12}$ |
| femto | f | $10^{-15}$ |
| atto | a | $10^{-18}$ |

## Non-SI units

A number of non-SI units are still in use in scientific literature for a variety of reasons. Some of these are being phased out, but others are likely to remain in use. The more common non-SI units that you might come across are listed in Table A1.4.

**TABLE A1.4** Non-SI units

| PHYSICAL QUANTITY | UNIT | ABBREVIATION | CONVERSION TO SI UNITS |
|---|---|---|---|
| time | minute<br>hour<br>day<br>year | min<br>h<br>d<br>y | 60 s<br>$3.6 \times 10^3$ s<br>$8.64 \times 10^4$ s<br>$3.156 \times 10^7$ s |
| mass | unified mass unit<br>tonne | u<br>t | $1.661 \times 10^{-27}$ kg<br>1000 kg |
| angle | degree | ° | dimensionless |
| energy | electron-volt<br>kilowatt hour | eV<br>kW h | $1.602 \times 10^{-19}$ J<br>$3.60 \times 10^3$ J |
| pressure | millimetre of mercury | mmHg | 133.3 Pa |
| charge | elementary or electronic charge | e | $1.602 \times 10^{-19}$ C |
| source activity | curie | Ci | $3.7 \times 10^{10}$ Bq |
| radiation<br>absorbed dose<br>equivalent dose | rad<br>rem | rad<br>rem | 0.01 Gy<br>0.01 Sv |

## Using SI units

There are certain conventions now adopted widely in scientific literature when SI units are being used. Some of the more important ones are given below.

**1** When recording a measurement, write the unit in full or use the recommended abbreviation (e.g. 25 metre or 25 m). Using abbreviations save space and time. Notice the space between the numeral and the unit.

**2** SI units named after scientists:

   **a** If the full word is used, it starts with a lower case letter (e.g. 10 newton, 7 joule, 105 pascal, 50 hertz)

   **b** If the abbreviation is used, it is (or at least commences with) a capital letter (e.g. 10 N, 7 J, 105 Pa, 50 Hz).

   **c** Measurements are written as products. '3 kg' means 'the product of 3 and the mass known as a kilogram', just as '$3x$' in maths means the product of 3 and $x$. Therefore 's' is not added to units (e.g. 5 kg not 5 kgs).

   **d** A full-stop is not placed after the abbreviation of a unit, unless it is at the end of a sentence.

   **e** When units are combined as a quotient (e.g. metre per second), a solidus (/) or negative index may be used. So m/s or m s$^{-1}$ are both acceptable, though the latter is used more widely. Never use more than one solidus in a unit as in m/s/s for acceleration, which should be m/s$^2$ or m s$^{-2}$. It is ambiguous, just as writing 36/6/3 in maths is ambiguous. (This could mean 2 or 18.)

## Converting between units

**Converting between units**

Learn more about converting between units.

Treat the unit as a multiplier. Use the prefixes in Table A1.3 also as multipliers. For example, 4 kg is the same as:

$4 \times k \times g = 4 \times 1000 \times g = 4000$ g

There are 4000 g in 4 kg.

1500 cm is the same as $1500 \times (10^{-2}$ m$) = 15$ m.

9780170409131

# Appendix 2: Some important physical quantities

From time to time you will need to find a value of a physical property from a reputable source. These might include finding:

- the value of a physical constant, such as Newton's universal gravitational constant or the electric constant.
- a physical property, such as boiling point or refractive index, which is characteristic of a particular material.
- a conversion factor such as micrometres to metres, electron-volt to joule, unified mass unit to kilogram.

All physical quantities, including physical constants, are measured to very precise levels of accuracy.

Some important physical quantities, including some physical constants, are listed alphabetically in Table A2.1. They are given to four significant figures. The uncertainty in most of these figures is better than six-figure accuracy. They are taken from sources such as the National Institute of Science and Technology (NIST). NIST is a specialist organisation dedicated to metrology (study of measurement).

**NIST physical reference data**

The National Institute of Standards and Technology (NIST) provides a wide range of data, including Standard reference data (SRF). For example, click on 'Other NIST Data' to enter the NIST Gateway.

**TABLE A2.1** Physical constants, physical measures and conversion factors

| PHYSICAL CONSTANTS | |
| --- | --- |
| Avogadro constant, $N_A$ | $6.022 \times 10^{23}$ mol$^{-1}$ |
| Coulomb law constant, $\dfrac{1}{4\pi\omega_0}$ | $8.988 \times 10^{9}$ N m$^2$ C$^{-2}$ |
| Universal gravitation constant, $G$ | $6.674 \times 10^{-11}$ m$^3$ kg$^{-1}$ s$^{-2}$ |
| Permittivity of free space electric constant, $\varepsilon_0$ | $8.854 \times 10^{-12}$ F m$^{-1}$ |
| Permeability of free space magnetic constant, $\mu_0$ | $4\pi \times 10^{-7}$ H m$^{-1}$ = $12.57 \times 10^{-7}$ H m$^{-1}$ |
| Planck constant, $h$ | $6.626 \times 10^{-34}$ J s = $4.136 \times 10^{-15}$ eV s |
| Speed of electromagnetic radiation in free space, $c$ | $2.998 \times 10^{8}$ m s$^{-1}$ |
| **PHYSICAL MEASURES** | |
| Mass of electron | $9.109 \times 10^{-31}$ kg = $5.486 \times 10^{4}$ u |
| Mass of proton | $1.6726 \times 10^{-27}$ kg |
| Mass of neutron | $1.6749 \times 10^{-27}$ kg |
| Rydberg constant (for hydrogen), $R_H$ | $1.097 \times 10^{7}$ m$^{-1}$ |
| Gravitational field strength at Earth's surface), $g$ | $(9.80 +/- 0.3)$ N kg$^{-1}$ |
| Acceleration due to gravity at Earth's surface, $g$ | $(9.80 +/- 0.3)$ m s$^{-2}$ |
| Mass of Earth | $5.976 \times 10^{24}$ kg |
| Mass of Moon | $7.348 \times 10^{22}$ kg |
| Mass of Sun | $1.989 \times 10^{30}$ kg |
| Period of rotation of Earth | $8.616 \times 10^{4}$ s |
| Radius of Earth (equatorial) | $6.378 \times 10^{6}$ m |
| Radius of Earth (mean) | $6.371 \times 10^{6}$ m |
| Radius of Earth's orbit about Sun (mean) | $1.496 \times 10^{11}$ m |
| Radius of Moon's orbit around Earth (mean) | $3.844 \times 10^{8}$ m |
| Radius of Sun | $6.960 \times 10^{8}$ m |
| Solar constant (mean) | $1.370 \times 10^{3}$ W m$^{-2}$ |

**NIST physical element laboratory**

This website provides atomic and nuclear data for every element.

| PHYSICAL MEASURES | |
|---|---|
| Density of water (pressure and temperature dependent) | $9.982 \times 10^3$ kg m$^{-3}$ |
| Air density (pressure and temperature dependent) | $1.292 \times 10^3$ kg m$^{-3}$ |
| Air pressure (temperature dependent) | $1.013 \times 10^5$ Pa |
| Speed of sound in air at 0°C | 331.4 m s$^{-1}$ |
| **CONVERSION FACTORS** | |
| Absolute zero, 0 K | $-273.15$°C |
| Unified mass unit, u | $1.661 \times 10^{-27}$ kg = 931.5 MeV |
| Electron-volt, eV | $1.602 \times 10^{-19}$ J |
| Elementary electron charge, $e$ | $1.602 \times 10^{-19}$ C |
| Coulomb | $6.242 \times 10^{18}$ elementary charges |

## Appendix 3: Electric and electronic symbols

**TABLE A3.1** Electric and electronic symbols

| COMPONENT GROUP | COMPONENT | SYMBOL |
|---|---|---|
| Sources of emf | Cell | |
| | Battery, DC power supply | |
| | Variable DC power supply | |
| | AC power supply | |
| Resistance | Resistor | |
| | Rheostat, resistor with sliding contact, potentiometer, voltage divider | or |

| COMPONENT GROUP | COMPONENT | SYMBOL |
|---|---|---|
| | Variable resistor | |
| | Light-dependent resistor (LDR) | |
| | Filament globe | |
| | Thermistor | |
| | Fuse | or |
| Capacitance | (Non-polarised) capacitor | |
| | Variable capacitor | |

| COMPONENT GROUP | COMPONENT | SYMBOL |
|---|---|---|
|  | Polarised capacitor, electrolytic capacitor |  |
| Transformer | Iron-cored transformer (one secondary winding) |  |
| Diodes | Junction diode |  |
|  | Zener diode |  |
|  | Photodiode |  |
|  | Light-emitting diode (LED) |  |
|  | Four-diode bridge |  |
| Meters | Ammeter |  |
|  | Voltmeter |  |
|  | Galvanometer |  |

| COMPONENT GROUP | COMPONENT | SYMBOL |
|---|---|---|
|  | Cathode ray oscilloscope (CRO) |  |
| Amplifiers | Voltage amplifier |  |
|  | Operational amplifier (op amp) |  |
| Transducers | Motor |  |
|  | Microphone |  |
|  | Loudspeaker |  |
| External connections | Earth |  |
| Circuit connections | Non-connected leads |  |
|  | Connected leads | or |

# Appendix 4: Resistance codes

A resistor is a physical object or circuit element that has resistance. The resistance of an actual, physical resistor is indicated in one of two ways:

- a colour code: a set of stripes on the resistor to indicate the resistance and the tolerance or uncertainty in the value
- a number–letter code: numbers with a letter to mark the decimal point.

## Resistance colour code

The resistor is marked with three or four colour bands painted on the resistor near one end as shown in Figures A4.1 and A4.2.

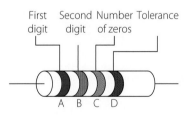

**FIGURE A4.1** A resistor marked with four coloured bands

**FIGURE A4.2** A resistor marked with three coloured bands

The colour of band A nearest the end is the first digit in the resistance value. The colour of band B represents the second digit. The colour of band C gives the number of zeros to follow these two digits (Table A4.1).

The fourth band from the end indicates the tolerance or percentage uncertainty in the resistance value (Table A4.2).

**TABLE A4.1** Colour code for digits

| DIGIT | COLOUR |
|-------|--------|
| 0 | Black |
| 1 | Brown |
| 2 | Red |
| 3 | Orange |
| 4 | Yellow |
| 5 | Green |
| 6 | Blue |
| 7 | Violet |
| 8 | Grey |
| 9 | White |

**TABLE A4.2** Colour code for tolerances of resistors

| TOLERANCE (%) | COLOUR |
|---------------|--------|
| 1 | Brown |
| 2 | Red |
| 5 | Gold |
| 10 | Silver |
| 20 | No band D shown |

If you hold the resistor so the stripes are on the left, you may find it easier to work out the resistance and the tolerance. If there are only three stripes, that is there is no tolerance band, the percentage uncertainty is 20%.

Figure A4.3 summarises the colour code details.

| Colour | Value | Value | Multiply by | Tolerance |
|--------|-------|-------|-------------|-----------|
| Black | 0 | 0 | 1 | Red = +/−2% |
| Brown | 1 | 1 | 10 | Gold = +/−5% |
| Red | 2 | 2 | 100 | Silver = +/−10% |
| Orange | 3 | 3 | 1000 | No band = +/−20% |
| Yellow | 4 | 4 | 10 000 | This gives the |
| Green | 5 | 5 | 100 000 | maximum error in |
| Blue | 6 | 6 | 1 000 000 | the value of the |
| Violet | 7 | 7 | not | resistor. |
| Grey | 8 | 8 | used | |
| White | 9 | 9 | | |

**FIGURE A4.3** Summary of resistance colour code system

*Examples*
Stripe (reading from the stripe nearest to the end):

- A–B–C: yellow (4), violet (7), orange (1000), gold (±5) $\Rightarrow$ (47 000 ± 5%) Ω% or (47 ± 3) kΩ
- A–B–C–D: orange (3), white (9), brown (10), silver (±10%) $\Rightarrow$ (390 ± 10%) Ω or (390 ± 39) Ω
- A–B–C: brown (1), green (5), black (1) and no fourth colour $\Rightarrow$ (15 ± 20%) Ω or (15 ± 3) Ω

## Resistance number–letter codes

In this system, which is often used on circuit diagrams, the numeral may have a letter in front of, behind or between the digits. The resistance is given to two significant figures. The letters R, K and M are used as multipliers: R for '1', K for '×10$^3$' or M for '×10$^6$'. The letters R, K or M are used to show where the decimal point goes.

Tolerances are given letter codes at the end (Table A4.3).

**TABLE A4.3** Letter code for tolerances of resistors

| TOLERANCE (%) | LETTER |
|---|---|
| 1 | F |
| 2 | G |
| 5 | J |
| 10 | K |
| 20 | M |

*Examples*

- $2R5J \Rightarrow 2.5\,\Omega\;(2.5 \pm 5\%)\,\Omega$ or $(2.5 \pm 0.1)\,\Omega$
- $47KM \Rightarrow 47\,k\Omega\;(47 \pm 20\%)\,k\Omega$ or $(47 \pm 10)\,\Omega$
- $M22K \Rightarrow (0.22 \pm 10\%)\,M\Omega$ or $(220 \pm 22)\,k\Omega$

# Appendix 5: Scientific notation and significant figures

When we measure quantities on instruments, the last figure in the measurement is usually uncertain. This is because of the in-built uncertainty in the instrument itself, even if we have avoided errors such as parallax error. Thus, if our electronic balance gives a reading of, say, 10.514 g, then we need to be aware that the last figure (4) will be uncertain. It is likely that the true mass is somewhere between 10.512 g and 10.516 g.

Significant figures show how many digits in the reading are meaningful. The last figure is always deemed to be uncertain. By keeping track of the number of significant figures in all the instrumental measurements used to calculate a quantity, we can determine the extent to which our answer correctly represents the accuracy of our instruments. To reflect this accuracy, we always give our answer to the same number of significant figures as the least accurate data used. (If we use an even lower number of significant figures than this, then we might as well use less accurate instruments!)

## The rules

1 Every non-zero digit is significant. For example, 3.78 and 294 both have three significant figures.

2 Every zero in the middle of a reading is significant. For example, the mass reading of 10.514 g has five significant figures.

3 Every zero to the right of a reading is significant. For example, 31.20 has four significant figures. The exception to this is a number with no decimal point and a trail of zeros, such as 500 kg. This volume may have one, two or three significant figures. To avoid this ambiguity, we must be given more information, stated in standard form. For example, if the volume is provided as $5.00 \times 10^2$ kg, then we know that it has three significant figures. If this is not clarified, we assume that it has the maximum number of significant figures.

4 Every zero before a number is not significant, and only shows the place value. For example, 0.005 only has one significant figure and 0.0090 has two significant figures. Again, rewriting these numbers in standard form clarifies this. (These numbers would be written as $5 \times 10^{-3}$ and $9.0 \times 10^{-3}$ respectively.)

## Calculations

### Rounding off

For rounding off an answer to a given number of significant figures, we examine the next figure on the right only. If it is 5 or more, then we round up.

# Appendix 6: Periodic table of elements

**Key**

element name
**atomic number**
**symbol**
atomic weight*

* standard atomic weight based on 12 C.
( ) indicates mass number of longest-lived isotope
For higher-precision values for atomic masses, visit ciaaw.org

| | |
|---|---|
| **He** | gas at room temperature |
| **Br** | liquid at room temperature |
| **Tc** | synthetic (does not occur naturally) |
| **Li** | solid at room temperature |

s-block
p-block
d-block transition metals
d-block lanthanides and actinides

| 1 | 2 | 3 | 4 | 5 | 6 | 7 | 8 | 9 | 10 | 11 | 12 | 13 | 14 | 15 | 16 | 17 | 18 |
|---|---|---|---|---|---|---|---|---|---|---|---|---|---|---|---|---|---|
| hydrogen 1 **H** 1.008 | | | | | | | | | | | | | | | | | helium 2 **He** 4.003 |
| lithium 3 **Li** 6.941 | beryllium 4 **Be** 9.012 | | | | | | | | | | | boron 5 **B** 10.81 | carbon 6 **C** 12.01 | nitrogen 7 **N** 14.01 | oxygen 8 **O** 16.00 | fluorine 9 **F** 19.00 | neon 10 **Ne** 20.18 |
| sodium 11 **Na** 22.99 | magnesium 12 **Mg** 24.31 | | | | | | | | | | | aluminium 13 **Al** 26.98 | silicon 14 **Si** 28.09 | phosphorus 15 **P** 30.97 | sulfur 16 **S** 32.06 | chlorine 17 **Cl** 35.45 | argon 18 **Ar** 39.95 |
| potassium 19 **K** 39.10 | calcium 20 **Ca** 40.08 | scandium 21 **Sc** 44.96 | titanium 22 **Ti** 47.87 | vanadium 23 **V** 50.94 | chromium 24 **Cr** 52.00 | manganese 25 **Mn** 54.94 | iron 26 **Fe** 55.85 | cobalt 27 **Co** 58.93 | nickel 28 **Ni** 58.69 | copper 29 **Cu** 63.55 | zinc 30 **Zn** 65.38 | gallium 31 **Ga** 69.72 | germanium 32 **Ge** 72.63 | arsenic 33 **As** 74.92 | selenium 34 **Se** 78.97 | bromine 35 **Br** 79.90 | krypton 36 **Kr** 83.80 |
| rubidium 37 **Rb** 85.47 | strontium 38 **Sr** 87.62 | yttrium 39 **Y** 88.91 | zirconium 40 **Zr** 91.22 | niobium 41 **Nb** 92.91 | molybdenum 42 **Mo** 95.95 | technetium 43 **Tc** (98) | ruthenium 44 **Ru** 101.1 | rhodium 45 **Rh** 102.9 | palladium 46 **Pd** 106.4 | silver 47 **Ag** 107.9 | cadmium 48 **Cd** 112.4 | indium 49 **In** 114.8 | tin 50 **Sn** 118.7 | antimony 51 **Sb** 121.8 | tellurium 52 **Te** 127.6 | iodine 53 **I** 126.9 | xenon 54 **Xe** 131.3 |
| caesium 55 **Cs** 132.91 | barium 56 **Ba** 137.33 | 57–71 **lanthanides** | hafnium 72 **Hf** 178.5 | tantalum 73 **Ta** 180.9 | tungsten 74 **W** 183.8 | rhenium 75 **Re** 186.2 | osmium 76 **Os** 190.2 | iridium 77 **Ir** 192.2 | platinum 78 **Pt** 195.1 | gold 79 **Au** 197.0 | mercury 80 **Hg** 200.6 | thallium 81 **Tl** 204.4 | lead 82 **Pb** 207.2 | bismuth 83 **Bi** 209. 0 | polonium 84 **Po** (209) | astatine 85 **At** (210) | radon 86 **Rn** (222) |
| francium 87 **Fr** (223) | radium 88 **Ra** (226) | 89–103 **actinides** | rutherfordium 104 **Rf** (267) | dubnium 105 **Db** (268) | seaborgium 106 **Sg** (271) | bohrium 107 **Bh** (270) | hassium 108 **Hs** (269) | meitnerium 109 **Mt** (278) | darmstadtium 110 **Ds** (281) | roentgenium 111 **Rg** (282) | copernicium 112 **Cn** (285) | nihonium 113 **Nh** (286) | flerovium 114 **Fl** (289) | moscovium 115 **Mc** (289) | livermorium 116 **Lv** (293) | tennessine 117 **Ts** (294) | oganesson 118 **Og** (294) |

| lanthanum 57 **La** 138.9 | cerium 58 **Ce** 140.1 | praseodymium 59 **Pr** 140.9 | neodymium 60 **Nd** 144.2 | promethium 61 **Pm** (145) | samarium 62 **Sm** 150.4 | europium 63 **Eu** 152.0 | gadolinium 64 **Gd** 157.3 | terbium 65 **Tb** 158.9 | dysprosium 66 **Dy** 162.5 | holmium 67 **Ho** 164.9 | erbium 68 **Er** 167.3 | thulium 69 **Tm** 168.9 | ytterbium 70 **Yb** 173.0 | lutetium 71 **Lu** 175.0 |
|---|---|---|---|---|---|---|---|---|---|---|---|---|---|---|
| actinium 89 **Ac** (227) | thorium 90 **Th** 232.0 | protactinium 91 **Pa** 231.0 | uranium 92 **U** 238.0 | neptunium 93 **Np** (237) | plutonium 94 **Pu** (244) | americium 95 **Am** (243) | curium 96 **Cm** (247) | berkelium 97 **Bk** (247) | californium 98 **Cf** (251) | einsteinium 99 **Es** (252) | fermium 100 **Fm** (257) | mendelevium 101 **Md** (258) | nobelium 102 **No** (259) | lawrencium 103 **Lr** (266) |

# NUMERICAL ANSWERS

## CHAPTER 2: PROJECTILE MOTION

### WORKED EXAMPLE 2.1

1 $u_x = 622\,\mathrm{m\,s^{-1}}$

$u_y = 622\,\mathrm{m\,s^{-1}}$

2 $\theta = 90°$

$880\,\mathrm{m\,s^{-1}}$

### WORKED EXAMPLE 2.2

1 $89.80\,\mathrm{s}$

2 $762\,\mathrm{m\,s^{-1}}$

### WORKED EXAMPLE 2.3

1 $34.29\,\mathrm{km}$

2 $t = 77.76\,\mathrm{s}$

$y = 29.6\,\mathrm{km}$

### CHECK YOUR UNDERSTANDING 2.1

2

| | POSITION | VELOCITY | ACCELERATION |
|---|---|---|---|
| a | + | + | − |
| b | + | 0 | − |
| c | + | − | − |

3 a $103.92\,\mathrm{m\,s^{-1}}$

b $60\,\mathrm{m\,s^{-1}}$

4 a $u_x = 19.02\,\mathrm{m\,s^{-1}}$

$u_y = 40.78\,\mathrm{m\,s^{-1}}$

b $u_x = 45.32\,\mathrm{m\,s^{-1}}$

$u_y = -21.13\,\mathrm{m\,s^{-1}}$

5 a $60.81\,\mathrm{m\,s^{-1}}$

b $188.67\,\mathrm{m}$

c $6.25\,\mathrm{s}$

6 $25.56\,\mathrm{m\,s^{-1}}$

### WORKED EXAMPLE 2.4

$9.75\,\mathrm{m}$

### WORKED EXAMPLE 2.5

$1.732\,\mathrm{s}$

### WORKED EXAMPLE 2.6

$12.0\,\mathrm{m\,s^{-1}}, \theta = 45°$

### WORKED EXAMPLE 2.7

$14.70\,\mathrm{m}$

### CHECK YOUR UNDERSTANDING 2.2

3 $4315\,\mathrm{m}$

4 $19.50\,\mathrm{m\,s^{-1}}$

6 a $2.35\,\mathrm{s}$

b $77.12\,\mathrm{m}$

c $47.99\,\mathrm{m\,s^{-1}}$, $46.74°$ below the horizontal

### CHECK YOUR UNDERSTANDING 2.3

2 $\approx 15.18\,\mathrm{m}$, uncertainty of $15.01\,\mathrm{m}$ to $15.35\,\mathrm{m}$

3 $\approx 17.25\,\mathrm{m\,s^{-1}}$, uncertainty of $17.15\,\mathrm{m\,s^{-1}}$ to $17.35\,\mathrm{m}$

4 $18.65\,\mathrm{m\,s^{-1}}$, where $\theta = 67.62°$

6 $12.32 \pm 0.04$

### CHECK YOUR UNDERSTANDING 2.4

1 i mathematically, using equations of motion

ii graphically, including graphs and diagrams

iii numerically, using computer simulations

5 b $45°$

c $15°$ and $75°$

6 b half of a parabola

### CHAPTER REVIEW QUESTIONS

1 $u_x = u\cos\theta$

$u_y = u\sin\theta$

$v = u + at$

$s = ut + \frac{1}{2}at^2$

$x = u_x t$

$y = u_y t + \frac{1}{2}gt^2$

$t = -u_y/g$

$t_{\mathrm{flight}} = -2u_y/g$

$h = y_0 - u_y^2/2g$

$v^2 = u^2 + 2as$

$x_{\max} = 2u_x u_y/g$

$x_{\max} = (2u^2 \sin\theta \cos\theta)/g$

2 0

3 $u_x = u\cos\theta$

4 $851.47\,\mathrm{m\,s^{-1}}, \theta = 49.76°$

5 $32.83\,\mathrm{m\,s^{-1}}$

$-5.79\,\mathrm{m\,s^{-1}}$

7 a $62.5\,\mathrm{m}$

b $3.57\,\mathrm{s}$

11 a $103.32\,\mathrm{m}$

b $76.70$

12 b $87.49\,\mathrm{m}$

c $189.68\,\mathrm{m}$

d $49.16\,\mathrm{m\,s^{-1}}, \theta = 62.76°$

13 a $3.03\,\mathrm{m\,s^{-1}}$

9780170409131

    b  0.49 s

    c  $3.89\,\mathrm{m\,s}^{-1}$

14  a  $26.87\,\mathrm{m\,s}^{-1}$

    b  5.48 s

    c  147.35 m

16  a  $11.65\,\mathrm{m\,s}^{-1}$

    b  Yes. 6.92 m

17  $123.74\,\mathrm{m\,s}^{-1}$

18  21.00 m

20  0.321, all terms same for both cases

21  $\theta = 75.96°$

## CHAPTER 3: CIRCULAR MOTION

### ■ WORKED EXAMPLE 3.1

$14\,\mathrm{m\,s}^{-2}$

### ■ WORKED EXAMPLE 3.2

$v = 8.3\,\mathrm{m\,s}^{-1}$

### ■ WORKED EXAMPLE 3.3

81 revolutions

### ■ CHECK YOUR UNDERSTANDING 3.1

2  $4a$, or a factor of four

4  a  $6.3\,\mathrm{m\,s}^{-1}$

    b  $39\,\mathrm{m\,s}^{-2}$

    c  $2\pi\,\mathrm{rad\,s}^{-1}$

5  $v_{max} = 6\,\mathrm{m\,s}^{-1}$

6  $r = 1.5\,\mathrm{m}$

### ■ WORKED EXAMPLE 3.4

1  a  2.5 N

    b  3.9 N

    c  40°

2  a  12 N

    b  12°

### ■ WORKED EXAMPLE 3.5

$F_T = 0.50\,\mathrm{N}$

### ■ WORKED EXAMPLE 3.6

11°

### ■ CHECK YOUR UNDERSTANDING 3.2

2  a  centre seeking

    b  friction force

    c  $4F_{net}$

3  a  $F_T = 49\,\mathrm{N}, \theta = 17°$

4  a  $2.8\,\mathrm{m\,s}^{-1}$

    b  29 N

5  a  15 kN

    c  231 m

6  $\theta = 15°$

### ■ WORKED EXAMPLE 3.7

$\Delta E_{top\,to\,bottom}$  $-5.9\,\mathrm{J}$

$\Delta E_{bottom\,to\,top}$  $+5.9\,\mathrm{J}$

$\Delta E_{top\,to\,top} = 0$

### ■ CHECK YOUR UNDERSTANDING 3.3

3  a  500 N

    b  50 000 J

4  $E_{total} = 24\,\mathrm{J}$

5  $E_{total\,top} = 26\,\mathrm{J}$

    $E_{total,\,bottom} = 16.5\,\mathrm{J}$

6  a  $-9.8\,\mathrm{J}$

    b  $+9.8\,\mathrm{J}$

    c  0

### ■ WORKED EXAMPLE 3.8

  0.33 m

### ■ CHECK YOUR UNDERSTANDING 3.4

3  61 Nm

4  714 N

6  a  $\tau_{min} = 0$

      $\tau_{max} = 63\,\mathrm{Nm}$

### ■ CHAPTER REVIEW QUESTIONS

6  $a_c = 3.8 \times 10^{15}\,\mathrm{m\,s}^{-2}$

7  a  $15\,\mathrm{m\,s}^{-2}$

    b  18 000 N

    c  friction force

9  $2F_{net}$

10  750 N

11  a  26 m

     b  9.6 s

     c  $0.65\,\mathrm{rad\,s}^{-1}$

12  $16\,\mathrm{m\,s}^{-1}$

13  a  1100 N

     b  2600 N

14  $22\,\mathrm{m\,s}^{-1}$

15  a  350 N

     b  130 N

     c  $F_T = 370\,\mathrm{N}$

     d  130 N

     e  1.1 m

17  34 N, $\theta = 8.4°$ to the horizontal

18 $F_{T, top} = 28\,N$
   $F_{T, bottom} = 38\,N$

19 1.2 s

20 0.10 m

## CHAPTER 4 : MOTION IN GRAVITATIONAL FIELDS

■ WORKED EXAMPLE 4.1

$1.98 \times 10^{20}\,N$, directed towards the moon

■ WORKED EXAMPLE 4.2

$-1.98 \times 10^{20}\,N$, directed towards Earth

■ WORKED EXAMPLE 4.3

1 $g = 0.31\,m\,s^{-2}$

2 $r = 9.0 \times 10^{6}\,m$

■ CHECK YOUR UNDERSTANDING 4.1

5 increasing $r$ by a factor of 4 decreases $F$ by a factor of 16

6 $4.1 \times 10^{3}\,N = 4100\,N$

7 $3.7\,m\,s^{-2}$

8 a $1.0 \times 10^{-47}\,N$

   b $2.3 \times 10^{-8}\,N$

■ WORKED EXAMPLE 4.4

220 days

■ WORKED EXAMPLE 4.5

a $3100\,m\,s^{-1}$

b $g = 0.23\,m\,s^{-2}$

■ WORKED EXAMPLE 4.6

$8.7\,m\,s^{-2}$

88% of Earth's surface gravity

■ CHECK YOUR UNDERSTANDING 4.2

5 $9.6 \times 10^{6}\,m$

6 $r = 2.3 \times 10^{11}\,m$

7 a $T_{Phobos} = 24\,hours$

   $T_{Deimos} = 4\,days$

   b $v_{Phobos} = 670\,m\,s^{-1}$

   $v_{Deimos} = 430\,m\,s^{-1}$

8 17 000 km

9 a i $1.5 \times 10^{11}\,m$

   ii $v_{Earth} = 30\,km\,s^{-1}$

   b $460\,m\,s^{-1}$, < orbital speed of Earth

■ WORKED EXAMPLE 4.7

1.4 hours

■ CHECK YOUR UNDERSTANDING 4.3

4 $2.96 \times 10^{-19}\,s^{2}\,kg^{-1}\,m^{-3}$

5 $5.7 \times 10^{26}\,kg$

6 2.8 years

■ WORKED EXAMPLE 4.8

$2.1 \times 10^{8}\,J$

■ WORKED EXAMPLE 4.9

$-2.4 \times 10^{8}\,J$

■ WORKED EXAMPLE 4.10

$5.0 \times 10^{3}\,m\,s^{-1}$

■ CHECK YOUR UNDERSTANDING 4.4

4 a $-4.7 \times 10^{7}\,J$

   b $2.4 \times 10^{7}\,J$

   c $-2.4 \times 10^{7}\,J$

5 a $1000\,m\,s^{-1}$

   b 28 days

7 60 000 km

■ CHAPTER REVIEW QUESTIONS

7 $890\,m\,s^{-1}$

8 $2.5\,m\,s^{-2}$

9 5.2 years

10 $6.5 \times 10^{34}\,kg$

12 $9.4 \times 10^{3}\,m\,s^{-1}$

13 a 74 N

   b 4.5 N

14 Near-Earth approximation: $6.9 \times 10^{9}\,J$

   Newton's law of gravitation: $-7.2 \times 10^{8}\,J$

16 $\sqrt{3}\ v_{esc}$

17 a $1.5\,m\,s^{-2}$

   b $1.5\,m\,s^{-2}$

   c 75 J (lost)

   d 75 J

18 $7.3 \times 10^{3}\,m\,s^{-1}$

19 $3.1 \times 10^{11}\,J$

20 a $9.0 \times 10^{7}\,m$

   b $3.8 \times 10^{4}\,m\,s^{-1}$

   c $3.96 \times 10^{9}\,J$

21 $6.9 \times 10^{25}\,kg$

■ MODULE 5 REVIEW

1 a $22\,m\,s^{-1}$

   b $v_{x} = 14\,m\,s^{-1}$

   $v_{y} = 17\,m\,s^{-1}$

   d 15 m

   e 3.5 s

   f 49 m

   g 38 J

9780170409131

3  $u = 5.0\,\mathrm{m\,s^{-1}}$, resistance is ignored.

4  d  $r_H > r_K$

   e  $0.5\,\mathrm{rad\,s^{-1}}$

   f  $a_H = 0.63\,\mathrm{m\,s^{-2}}$

      $a_k = 0.50\,\mathrm{m\,s^{-2}}$

   g  $\Delta\theta = 150\,\mathrm{rad}$, same for both girls

      $s_H = 3750\,\mathrm{m}$

      $s_K = 3000\,\mathrm{m}$

5  a  $8.4\,\mathrm{m\,s^{-1}}$

   b  $88\,\mathrm{m\,s^{-2}}$

   c  $7.0\,\mathrm{N}$

   d  $t = 0.45\,\mathrm{s}$

   e  $3.8\,\mathrm{m}$

6  a  $r = 197\,\mathrm{m}$

   b  $\mu_{s,\,min} = 0.09$

7  a  $7.0 \times 10^{25}\,\mathrm{kg}$

   b  As $G$ decreases, the gradient decreases.

8  a  i  $8.7\,\mathrm{m\,s^{-2}}$

      ii  $0.23\,\mathrm{m\,s^{-2}}$

      iii  $0.0027\,\mathrm{m\,s^{-2}}$

   b  $1.98 \times 10^{20}\,\mathrm{N}$ directed towards the moon

   c  $3.5 \times 10^{8}\,\mathrm{m}$

9  a  $2.4 \times 10^{10}\,\mathrm{m}$

   b  $7.4 \times 10^{4}\,\mathrm{m\,s^{-1}}$

   c  $3.3 \times 10^{11}\,\mathrm{J}$

   d  $1.0 \times 10^{5}\,\mathrm{m\,s^{-1}}$

10  a  $1.9 \times 10^{11}\,\mathrm{J}$

    b  $9.4 \times 10^{11}\,\mathrm{J}$

## CHAPTER 5: CHARGED PARTICLES, CONDUCTORS AND ELECTRIC AND MAGNETIC FIELDS

### WORKED EXAMPLE 5.1

ground

### WORKED EXAMPLE 5.2

$-9.6 \times 10^{9}\,\mathrm{m\,s^{-2}}$

down, in the direction of the field

### WORKED EXAMPLE 5.3

$1.1\,\mathrm{mm}$

### CHECK YOUR UNDERSTANDING 5.1

1  $7.5\,\mathrm{V}$

2  $2\,\mathrm{cm}$

4  $1.6 \times 10^{-10}\,\mathrm{m\,s^{-2}}$

   down, in the direction of the field

5  a  $1.6 \times 10^{-14}\,\mathrm{J}$

   b  $1.9 \times 10^{8}\,\mathrm{m\,s^{-1}}$

6  a  $1.9 \times 10^{7}\,\mathrm{m\,s^{-1}}$

   b  $4.4 \times 10^{5}\,\mathrm{m\,s^{-1}}$

   c  $3.1 \times 10^{5}\,\mathrm{m\,s^{-1}}$

### WORKED EXAMPLE 5.4

a  $-0.56\,\mathrm{m\,s^{-2}}$

   left, in the direction opposite to the field

b  $0.49\,\mathrm{m\,s^{-1}}$

c  $0.21\,\mathrm{m}$ left

### CHECK YOUR UNDERSTANDING 5.2

3  b  $7.3 \times 10^{9}\,\mathrm{m\,s^{-1}}$

4  a  $0.38\,\mathrm{m}$ right

   b  $0.91\,\mathrm{m\,s^{-1}}$

5  a  $3.5\,\mathrm{ns}$

   b  $1.8 \times 10^{-6}\,\mathrm{m}$

6  c  $0.77\,\mathrm{J}$

### WORKED EXAMPLE 5.5

$5.0 \times 10^{-16}\,\mathrm{N}$

### WORKED EXAMPLE 5.6

$13\,\mathrm{mT}$

### CHECK YOUR UNDERSTANDING 5.3

4  $1.3 \times 10^{5}\,\mathrm{m\,s^{-2}}$

5  C

6  a  up

   b  down

   c  no force

7  $5.6 \times 10^{-13}\,\mathrm{T}$

   pointing east (right-hand rule)

8  a  $3.1 \times 10^{7}\,\mathrm{m\,s^{-1}}$

   b  $7.1 \times 10^{-7}\,\mathrm{T}$

9  b  $1.1 \times 10^{11}\,\mathrm{m\,s^{-2}}$

   c  $150\,\mathrm{km\,s^{-1}}$

### CHECK YOUR UNDERSTANDING 5.4

1  a  magnetic, electric field and a gravitational field

   b  electric field and gravitational field

   c  gravitational field

3  to the east (right–hand rule) to give a force downwards

5  a  $8.9 \times 10^{-30}\,\mathrm{N}$ down

   b  $1.6 \times 10^{-17}\,\mathrm{N}$ up

   c  $1.2 \times 10^{-16}\,\mathrm{N}$ (don't know the direction of $v$, so don't know the direction of $F$)

### CHAPTER REVIEW QUESTIONS

2  parallel to the field

4  magnetic field: $13\,\mathrm{T}$

   electric field: $6.3 \times 10^{6}\,\mathrm{N\,C^{-1}}$

6 a upper plate

7 a into the page

   b down

   c right

8 a velocity and kinetic energy increase, work is done by the field

   b velocity and kinetic energy decrease, work is done on the field

   c velocity and kinetic energy do not change, work is not done by or on the field

   d velocity and kinetic energy increase, work is done by the field

9 a up

   b out

   c no deflection

10 a no change in magnitude of velocity

   b no change in kinetic energy

   c No work is done.

11 a into page

   b right

   c down

13 $1.6 \times 10^{-10}$ C, negative so the $E$ field pushes it upwards

14 $4.3 \times 10^{16}$ ms$^{-2}$ in the $-z$ direction (right-hand rule, negative charge)

15 $F = 5.0 \times 10^{-8}$ N in the direction of the field

   $a = 5.0 \times 10^{-5}$ ms$^{-2}$, in the direction of the field

16 a $8.8 \times 10^{-17}$ N in the direction opposite the field

   b $3.2 \times 10^{-6}$ m

17 a The force on the electron is down.

   $9.7 \times 10^{13}$ ms$^{-2}$

   b $4.8 \times 10^{7}$ m

   c $4.2 \times 10^{-9}$ J

18 a $-4.8 \times 10^{9}$ ms$^{-2}$

   down, in the direction of the field

   c $0.033$ m

   d $0.13$ m

19 a $1.8 \times 10^{-22}$ J

   b $2.3 \times 10^{-6}$ m

   c $1.1 \times 10^{-3}$ V = 1.1 mV

20 a $8.5 \times 10^{-26}$ J

   b $3.4 \times 10^{-25}$ J

   c $1.6 \times 10^{-6}$ V

   d $3.2 \, 10^{-5}$ Vm$^{-1}$, left

21 $\nu = 2.7 \times 10^{4}$ ms$^{-1}$

   $T = 0.023$ s

22 a $0.63$ m

   b $-0.98$ m, in opposite direction to sodium ion

23 a 9 cm

   b $r = -0.107$ m = 11 cm, circulates in the opposite direction to the sodium ion, but with the radius path

## CHAPTER 6: THE MOTOR EFFECT

### ■ WORKED EXAMPLE 6.1

$30°$

### ■ WORKED EXAMPLE 6.2

down

### ■ CHECK YOUR UNDERSTANDING 6.1

4 down

5 into the page

6 0.09 N

7 c  i  90° and 270°

     ii  0°, 180° and 360°

    iii  30°, 150°, 210° and 330°

8 a $2.5 \times 10^{-4}$ N

   b $3.5 \times 10^{-5}$ T

9 0.098 T, south

### ■ WORKED EXAMPLE 6.3

The force is reduced by a factor of 4.

The forces are still repulsive because the currents are reversed, but still antiparallel.

### ■ CHECK YOUR UNDERSTANDING 6.2

2 No

4 a force is halved

   b force increases by a factor of 4

   c force decreases by a factor of 2

   d $F = 0$

6 a $4 \times 10^{-5}$ N m$^{-1}$, points towards the other wire

   b answer would not change

   c answer would not change

7 2 cm apart

8 2.2 A

### ■ CHAPTER REVIEW QUESTIONS

2 [T] [m][A$^{-1}$] = [kg s$^{-2}$ A$^{-1}$][m][A$^{-1}$] = [kg m s$^{-2}$ A$^{-2}$]

4 a 90° or 270°

   b 0° or 180°

5 upwards

6 a west

   b east

7 0.25 N

11 a $2\pi$ T m A$^{-1}$

   b much larger

12  0.028 T

13  $2.7 \times 10^{-4}$ T

14  39°, 141°, 219° and 321°

15  0.002 A

16  $1.0 \times 10^{-4}$ N m$^{-1}$

17  a  0.013 m

   b  opposite

19  a  $F = IdB$

   b  $a = \dfrac{F}{m} = \dfrac{IdB}{m}$

   c  $v = \sqrt{2aL} = \sqrt{\dfrac{2IdBL}{m}}$

20  a  i  0.02 N

      ii  Side CD is parallel to the field, so experiences no force.

   b  0.34 m

## CHAPTER 7: ELECTROMAGNETIC INDUCTION

■ WORKED EXAMPLE 7.1

60°, 120°, 240° and 300°

■ WORKED EXAMPLE 7.2

$0.6 \times 10^{-3}$ Wb

■ CHECK YOUR UNDERSTANDING 7.1

1  T m$^2$ or Wb (webers)

3  a  $1.1 \times 10^{-8}$ Wb

   b  $7.8 \times 10^{-9}$ Wb

4  0.08 m = 8 cm radius

5  c  i  $\Delta\Phi_{0 \text{ to } 30} = -0.05$ mWb

      ii  $\Delta\Phi_{30 \text{ to } 60} = -0.15$ mWb

      iii  $\Delta\Phi_{60 \text{ to } 90} = -0.2$ mWB

6  The flux has decreased by 0.9 mWb.

■ WORKED EXAMPLE 7.3

30 mT per second

■ CHECK YOUR UNDERSTANDING 7.2

7  0.01 V

8  20 T s$^{-1}$

■ WORKED EXAMPLE 7.4

310 V and 350 V peak

■ WORKED EXAMPLE 7.5

1  The current in the secondary coils is 19 times larger than that in the primary coil.

2  12 V

■ WORKED EXAMPLE 7.6

1  44 A

2  0.023

■ CHECK YOUR UNDERSTANDING 7.3

5  46 V

6  a  step-up transformer

   b  $\dfrac{1}{3}$

9  a  60 000 V

   b  960 W

   c  960 W

   d  0.016 A

10  a  6 V

   b  10 V

   c  i  31

      ii  52

■ CHAPTER REVIEW QUESTIONS

2  $[\text{Wb}] = [\text{T m}^2] = [\text{kg s}^{-2} \text{A}^{-1} \text{m}^2] = [\text{kg m}^2 \text{s}^{-2} \text{A}^{-1}]$

8  a  This is a step-down transformer.

   b  0.3

11  a  $1.4 \times 10^{-7}$ Wb

   b  45°

12  0.24 T

14  $1.0 \times 10^{-5}$ V

15  a  0.02

   b  2 mA

19  0.0048 V

20  a  120 V

   c  4.8 A

21  c  1.2 V

22  a  0.018 A

   b  If the current in the solenoid is constant, the field is constant and no emf or current is induced in the ring after this initial time.

## CHAPTER 8: APPLICATIONS OF THE MOTOR EFFECT

■ WORKED EXAMPLE 8.1

0.002 N m

■ WORKED EXAMPLE 8.2

a  1.9 N

   down (right-hand rule)

b  1.2 N

   (out of page)

d  The coil will not rotate.

■ CHECK YOUR UNDERSTANDING 8.1

4  $1.4 \times 10^{-6}$ N m

6 a $F_{AB} = 0.18$ N in the downwards direction

  b $F_{BC} = 0°$

  c 0.18 N upwards

  d $7.2 \times 10^{-3}$ Nm, out of the page

  e anticlockwise

7 3.1 A

8 30°

■ WORKED EXAMPLE 8.4

156 V

■ CHECK YOUR UNDERSTANDING 8.2

6 a 633 V

  b 448 V

7 0.1 T

8 a $f = 20$ Hz

  b 19 V

  c 13 V

  d 19 V $\cos(40\pi t)$

9 127

■ CHAPTER REVIEW QUESTIONS

8 a 3.0 A

  b 0.60 A, because the signal is a sine curve and symmetric about zero

  c 0 because the signal is a sine curve and symmetric about zero

  d 2.1 A

9 a anti-clockwise

  b $7.7 \times 10^{-2}$ N

  c $6.2 \times 10^{-3}$ Nm

10 a i zero

   ii 54 V

  c 54 V

  d 38 V

11 a doubled

  b doubled

  c doubled

12 b 108

13 between +110 A and −110 A, with a frequency of 50 Hz

17 a $F_{net} = 0$

  b zero

  c It will not rotate.

  d zero

  e 0.048 Nm

  f anticlockwise

18 6 V

■ MODULE 6 REVIEW

1 a −12 V

2 b $1.8 \times 10^{13}$ m s$^{-2}$

  right, opposite the direction of the field

  c $5.6 \times 10^{-12}$ s

  d $2.9 \times 10^{-8}$ V

3 b $1.8 \times 10^{12}$ m s$^{-2}$

  towards you out of the page

  c $8.9 \times 10^{-11}$ s

  d not changed

4 a upwards

  b to the left

  c $F = 8 \times 10^{-6}$ N

  d $4 \times 10^{-6}$ N

5 b $F = IlB \sin\theta$

  c 1 T

  d i 30°

   ii 14°

   iii 0

6 a i will produce an emf across the loop

   ii will not induce an emf

   iii change the flux

7 a step-up transformer

  c no emf is induced

  d $V_{S,rms} = 480$ V

   $V_{S,peak} = 680$ V

8 c anti-clockwise

  e 0.009 Nm

9 a AC generator

  c 216

## CHAPTER 9: ELECTROMAGNETIC SPECTRUM

■ CHECK YOUR UNDERSTANDING 9.2

1 a $3.69 \times 10^{-7}$ N

  b $4.05 \times 10^{23}$ N kg$^{-1}$

  c $2.30 \times 10^{12}$ N C$^{-1}$

2 a $3.18 \times 10^{6}$ m s$^{-1}$

  b $4.94 \times 10^{-17}$ s

  c $2.02 \times 10^{16}$ Hz

  d $1.49 \times 10^{-8}$ m

■ WORKED EXAMPLE 9.1

1 13.9 m

2 1389 rev s$^{-1}$

### WORKED EXAMPLE 9.2

1   0.12 m

2   $6.30 \times 10^{-7}$ m

### CHECK YOUR UNDERSTANDING 9.3

2   0.09 degrees

4   3.23 m

### WORKED EXAMPLE 9.3

1   $1.56 \times 10^{8}\,\text{m s}^{-1}$

2   $1.24 \times 10^{8}\,\text{m s}^{-1}$

### WORKED EXAMPLE 9.4

1   $4.09 \times 10^{-19}$ J

2   $4.85 \times 10^{-19}$ J

### CHECK YOUR UNDERSTANDING 9.4

4   $2.26 \times 10^{8}\,\text{m s}^{-1}$

5   $4.58 \times 10^{-19}$ J

### CHAPTER REVIEW QUESTIONS

2   231 N

3   a   149 million km

4   $1.97 \times 10^{8}\,\text{m s}^{-1}$

5   $2.18 \times 10^{-18}$ J

11   It is moving directly towards us.

## CHAPTER 10: LIGHT: WAVE MODEL

### CHECK YOUR UNDERSTANDING 10.1

2   a   $4.9 \times 10^{-12}$ m

### WORKED EXAMPLE 10.1

a   9.16 mm

b   2.29 mm

### CHECK YOUR UNDERSTANDING 10.2

1   Wave model and particle model

4   a   decreases

b   decreases

c   increases

5   8.4 cm, 16.8 cm, 25.2 cm

6   a   562.5 nm

b   9 cm

7   $\lambda_1$ (630 nm): 25.2 cm

$\lambda_2$ (420 nm): 25.2 cm

### CHECK YOUR UNDERSTANDING 10.3

5   $0.5 \times I_{\text{max}}$

6   21.5°

### CHAPTER REVIEW QUESTIONS

13   0.066°

14   1.43 mm

15   $2.62 \times 10^{-5}$ m

16   $8.83 \times 10^{-6}$ m

17   9

19   a   45 lumens

b   15 lumens

c   No light will pass.

20   45°

## CHAPTER 11: LIGHT: QUANTUM MODEL

### WORKED EXAMPLE 11.1

3410 K

### WORKED EXAMPLE 11.2

1   a   $E = hf$

b   $E = \dfrac{hc}{\lambda}$

2   a   Energy increases as $f$ increases.

b   Energy decreases as $\lambda$ increases

### CHECK YOUR UNDERSTANDING 11.1

1   light as particles; light as waves

3   $E = hf$; $E\,(\text{J})/f(\text{s}^{-1}) = h\,(\text{J/s}^{-1}) = h\,(\text{J s})$

4   Vega is hotter.

5   $6.40 \times 10^{-6}$ m

6   a   $4.04 \times 10^{-21}$ J

b   $2.70 \times 10^{-21}$ J

8   a   380 nm

b   7630 K

9   a   $9.99 \times 10^{-7}$ nm

### WORKED EXAMPLE 11.3

2.11 V

### CHECK YOUR UNDERSTANDING 11.2

5   a   Copper

b   Lithium

6   $5.05 \times 10^{-7}$ m

7   a   2.1 V

b   $8.6 \times 10^{5}\,\text{m s}^{-1}$

### WORKED EXAMPLE 11.4

1   $6.6 \times 10^{-34}$ J s

### WORKED EXAMPLE 11.6

$E_{\text{min}} = 2.88 \times 10^{-19}$ J

$E_{\text{max}} = 4.96 \times 10^{-19}$ J

### CHECK YOUR UNDERSTANDING 11.3

1   The Law of Conservation of Energy

4   $\phi\,(= hf_0) < E \leq K_{\text{max}}\,(= hf)$

5 a $6.74 \pm 0.07 \times 10^{-34}\,\text{Js}$

  b $5.9 \pm 0.05 \times 10^{-19}\,\text{J}$

7 $6.63 \times 10^{-24}\,\text{kg m s}^{-1}$

### CHAPTER REVIEW QUESTIONS

2 1887, Heinrich Hertz

3 dull red, bright red, orange, yellow, white

6 Pt

7 a increase $K_{max}$

  b The photocurrent will double.

9 red

10 107 mm

11 a  i $8.81 \times 10^{-19}\,\text{J}$

    ii $2.26 \times 10^{-7}\,\text{m}$

    iii $1.33 \times 10^{15}\,\text{s}^{-1}$

  b ultraviolet

12 a 505 nm, green

  b 4.14 eV

13 metal = Mn (4.1 eV)

14 a $1.63 \times 10^{-24}\,\text{J} = 1.03 \times 10^{-5}\,\text{eV}$

15 b 5796 K

## CHAPTER 12: LIGHT AND SPECIAL RELATIVITY

### WORKED EXAMPLE 12.1

a 160 m

b 10 m

### WORKED EXAMPLE 12.2

b $1.12\ \text{m s}^{-1}$

  $31.3°$

### CHECK YOUR UNDERSTANDING 12.1

3 $x = x' + v\Delta t;\ y = y' + v\Delta t$

5 Yes – there are no unexplained accelerations.

6 a  i $2\,\text{m s}^{-1}$ (given)

    ii $14\,\text{m s}^{-1}$

  b  i 10 m

    ii 70 m

7 $5\ \text{m s}^{-1}$

  $53.38°$

### WORKED EXAMPLE 12.4

a 4.03 light years

b 0.92 years

### CHECK YOUR UNDERSTANDING 12.3

4 40 m

6 17.20 ns

7 a 22.22 years

  b 9.69 years

### WORKED EXAMPLE 12.5

a  i $8.97 \times 10^{-7}\,\text{ns}$

   ii 0.42

   iii possible but very unlikely

b 0.42

### WORKED EXAMPLE 12.6

$\approx 0.5\ \mu\text{s}$

### CHECK YOUR UNDERSTANDING 12.4

2 8.10 s

3 3.49 m

4 37 s

### WORKED EXAMPLE 12.7

1 $2.52 \times 10^{-27}\,\text{kg}$

2 $2.75 \times 10^{8}\,\text{m s}^{-1}$

### CHECK YOUR UNDERSTANDING 12.5

1 a $m = \dfrac{m_0}{\sqrt{1 - \dfrac{v^2}{c^2}}}$

  b $p = \dfrac{p_0}{\sqrt{1 - \dfrac{v^2}{c^2}}}$ where $p$ is momentum

2 a $1.5 \times 10^{4}\,\text{N}$

  b $37.5\ \text{m s}^{-2}$

3 a $3.54 \times 10^{-16}\,\text{kg}$

  b $1.18 \times 10^{-24}\,\text{kg}$

  c 1:706.59

4 Relativistic $m_e = 5.212 \times 10^{-28}\,\text{kg}$

  Relativistic momentum $p = 1.56 \times 10^{-19}\,\text{kg m s}^{-1}$

### WORKED EXAMPLE 12.8

$2.48 \times 10^{7}\,\text{eV} = 24.8\,\text{MeV}$

### WORKED EXAMPLE 12.9

$1.90 \times 10^{-19}\,\text{eV}$

### WORKED EXAMPLE 12.10

$4.49 \times 10^{15}\,\text{J}$

### CHECK YOUR UNDERSTANDING 12.6

3 a 0.008 u

6 $6.68 \times 10^{-8}\,\text{kg}$

### CHAPTER REVIEW QUESTIONS

3 See chapter 12, page 303.

5 a $13\ \text{m s}^{-1}$

b  passengers change in position = 65 m

8  a  i and ii: $2.998 \times 10^8 \, \text{m s}^{-1}$

   b  i  6.00 s

      ii  1.6 s

9  $9.14 \times 10^7 \, \text{m s}^{-1}$

10  68.61 years

11  22.52 ns

12  447 m

13  $6.202 \times 10^{15} \, \text{J}$

   $3.871 \times 10^{28} \, \text{MeV}$

14  Fred aged 17.32 years while Pierre aged 20 years

15  a  Yes.

   b  2.91 years older

16  a  $2.40 \times 10^8 \, \text{m s}^{-1}$

   b  $1.08 \times 10^9 \, \text{m}$

17  a  $7.52 \times 10^{-4} \, c$

   b  84.99976 m

18  1.022 MeV

19  a  i  656.7 m

      ii  Yes, just barely.

   b  about 10 000 m

   c  Some may make it back.

21  a  $9.652 \times 10^{-21} \, \text{kg m s}^{-1}$

   b  0.0215 m

■ MODULE 7 REVIEW

7  9 cm from the central bright spot

13  25% of the original incident light

14  d  725 nm

15  a  2.48 eV or $3.98 \times 10^{-19} \, \text{J}$

16  b  $K_{\text{max}} = hf - \varphi$

   c  Zero J or zero eV

   f  0.468 V

17  b  $-1 \, \text{m s}^{-1}$

   c  32 m

   d  The red ball will be 12 metres to the right of the green ball after ten seconds.

18  c  1.71 years

20  a  $7 \times 10^{-4} \, \text{kg} = 0.7 \, \text{gram}$

   b  $2.24 \times 10^{-25} \, \text{kg}$

   c  $4.81 \times 10^{-31} \, \text{kg}$

## CHAPTER 13: ORIGINS OF THE ELEMENTS

■ CHECK YOUR UNDERSTANDING 13.2

6  $\dfrac{v}{D} = H_0$

9  a  $2.3 \times 10^{-18} \, \text{s}^{-1}$

   b  $1.1 \times 10^8 \, \text{m s}^{-1}$

10  $3.86 \times 10^{26} \, \text{W}$

11  $4.1 \times 10^{16} \, \text{m}$

■ CHECK YOUR UNDERSTANDING 13.3

7  97 nm

■ CHECK YOUR UNDERSTANDING 13.5

1  Main sequence

7  970 nm

■ CHECK YOUR UNDERSTANDING 13.6

1  $E = mc^2$

■ CHAPTER REVIEW QUESTIONS

3  hydrogen and heliumn mostly hydrogen

## CHAPTER 14: STRUCTURE OF THE ATOM

■ CHECK YOUR UNDERSTANDING 14.3

2  friction with the spray nozzle

3  The voltage on the plates was adjusted.

■ CHECK YOUR UNDERSTANDING 14.5

2  A nucleon is a particle found in a nucleus – protons and neutrons.

5  number of protons in the nucleus

■ CHAPTER REVIEW QUESTIONS

7  cathode rays travelled in a straight line

8  paddle wheel experiment

15  electric and magnetic forces

## CHAPTER 15: QUANTUM MECHANICAL NATURE OF THE ATOM

■ CHECK YOUR UNDERSTANDING 15.1

3  angular momentum of the electrons in their orbits

■ CHECK YOUR UNDERSTANDING 15.2

2  $n = 2$

■ WORKED EXAMPLE 15.1

1  $2.5 \times 10^{15} \, \text{Hz}$

2  370 nm (to two sig. fig.)

■ WORKED EXAMPLE 15.2

1  $6.28 \times 10^{14} \, \text{Hz}$

2  $5.53 \times 10^{14} \, \text{Hz}$

■ WORKED EXAMPLE 15.3

1  10.2 eV

2  436 nm

■ WORKED EXAMPLE 15.4

1  656 nm

2  95 nm

1  4

2  2

■ CHECK YOUR UNDERSTANDING 15.3

4  $2.42 \times 10^{15}$ Hz

5  123 nm

■ WORKED EXAMPLE 15.6

1  180 nm (to two sig. fig.)

2  $2.3 \times 10^{-34}$ m

■ CHAPTER 15 REVIEW QUESTIONS

3  electrostatic attraction between positive and negative charges

8  splits electromagnetic radiation into its components

11  Lyman – ultraviolet

   Balmer – visible

   Paschen – infrared

13  $E = hf$

15  397 nm

17  $4.9 \times 10^{-12}$ m

## CHAPTER 16: PROPERTIES OF THE NUCLEUS

■ WORKED EXAMPLE 16.1

1  a  $^{211}_{87}\text{Fr} \rightarrow ^{207}_{85}\text{X} + ^{4}_{2}\text{He}$, where X is as yet unknown

   b  astatine

2  a  $^{213}_{84}\text{Fr} \rightarrow ^{209}_{82}\text{Pb} + ^{4}_{2}\text{He}$

   b  lead

■ WORKED EXAMPLE 16.2

1  a  thallium, $^{270}_{81}\text{Tl}$

   b  krypton, $^{82}_{36}\text{Kr}$

   c  platinum, $^{190}_{78}\text{Pt}$

   d  samarium-145

2  a  mercury-200

   b  bismuth-209

   c  thallium-199

   d  lead-203

■ CHECK YOUR UNDERSTANDING 16.1

1  Nucleon number and net charge

2  a  $^{A}_{Z}\text{X} \rightarrow ^{A-4}_{Z-2}\text{Y} + ^{4}_{2}\text{He}$

   b  $^{1}_{0}\text{n} \rightarrow ^{1}_{1}\text{p} + ^{0}_{-1}\text{e} + \bar{\nu}$

   c  $^{1}_{1}\text{p} \rightarrow ^{1}_{0}\text{n} + ^{0}_{+1}\text{e} + \nu$

   d  $^{A}_{Z}\text{X} \rightarrow ^{A}_{Z}\text{X} + \gamma$

4  neon-21

5  a  $^{151}_{67}\text{Ho} \rightarrow ^{147}_{65}\text{Tb} + ^{4}_{2}\text{He}$

   b  Tb = terbium

6  $^{210}_{86}\text{Rn} \rightarrow ^{206}_{84}\text{Po} + ^{4}_{2}\text{He}$  (alpha decay)

7  $^{15}_{8}\text{O} \rightarrow ^{15}_{7}\text{N} + ^{0}_{-1}\text{e} + \nu$ (N is nitrogen)

8  $^{158}_{65}\text{Tb} \rightarrow ^{154}_{63}\text{Eu}^{*} + ^{4}_{2}\text{He}$
   $^{154}_{63}\text{Eu}^{*} \rightarrow ^{154}_{63}\text{Eu} + \gamma$

■ CHECK YOUR UNDERSTANDING 16.2

1  a  gamma

   b  alpha

   c  alpha

6  80.00 counts per second

8  Gamma rays are uncharged.

■ WORKED EXAMPLE 16.3

1  $4.8 \times 10^{-5}$ s$^{-1}$

2  $2.9 \times 10^{-5}$ s$^{-1}$

■ WORKED EXAMPLE 16.4

a  $1.3 \times 10^{-2}$ s$^{-1}$

b  $3.8 \times 10^{-12}$ s$^{-1}$

■ WORKED EXAMPLE 16.5

1  43 hr

2  $\approx 100$ days

■ WORKED EXAMPLE 16.6

1  no nuclei left undecayed after a year

2  about 6%

■ CHECK YOUR UNDERSTANDING 16.3

3  $1.4 \times 10^{-5}$ s$^{-1}$

4  $1.4 \times 10^{4}$ day (2 sig. fig.)

6  0.0039

■ WORKED EXAMPLE 16.7

1  a  Three neutrons must be released.

   b  $^{1}_{0}\text{n} + ^{233}_{92}\text{U} \rightarrow ^{104}_{42}\text{Mo} + ^{126}_{50}\text{Sn} + 4\,^{1}_{0}\text{n}$

   c  i  0.194 u

     ii  $3.22 \times 10^{-28}$ kg

   d  $2.90 \times 10^{-11}$ J

2  a  Sb

   b  $^{1}_{0}\text{n} + ^{239}_{94}\text{Pu} \rightarrow ^{104}_{43}\text{Tc} + ^{133}_{51}\text{Sb} + 3\,^{1}_{0}\text{n}$

   c  132.93 u

   d  $2.8 \times 10^{-11}$ J

■ CHECK YOUR UNDERSTANDING 16.4

6  a  Three neutrons must be released from the nucleus.

   b  i  0.19 u

     ii  $3.15 \times 10^{-28}$ kg

   c  $2.84 \times 10^{-11}$ J

1  $9.5482 \times 10^{-13}$ J

2  a  $9.7929957 \times 10^{-30}$ kg

   b  $8.8015056 \times 10^{-13}$ J

## CHECK YOUR UNDERSTANDING 16.5

6  1.1 MeV per nucleon

7  a  $^{1}_{0}n + ^{6}_{3}Li \rightarrow ^{7}_{3}Li$

   b  $^{7}_{3}Li \rightarrow ^{4}_{2}He + ^{3}_{1}H$

9  a  $3.49 \times 10^{-30}$ kg

   b  $3.15 \times 10^{-13}$ J

## CHAPTER 16 REVIEW QUESTIONS

1  alpha (helium-4 nucleus), beta (electron or positron), gamma (photon)

2  particle found in the nucleus

5  $^{1}_{0}n + ^{235}_{92}U \rightarrow ^{A}_{Z}X + ^{235-A-N}_{92-Z}Y + N^{1}_{0}n$

8  about 1.5%

9  $5.3 \times 10^{-7}$ s$^{-1}$ (2 sig. fig. from 15 days)

13  Hydrogen ($^{1}_{1}H$), deuterium ($^{2}_{1}H$), tritium ($^{3}_{1}H$)

16  $5.4 \times 10^{-14}$ J

17  $1.6 \times 10^{-13}$ J

19  $1.28 \times 10^{-7}$ s

## CHAPTER 17: DEEP INSIDE THE ATOM

## CHECK YOUR UNDERSTANDING 17.1

2  a  proton: e

   b  antiproton: −e

   c  positron: e

   d  antineutrino: 0

4  $\bar{n}$

8  $1.6 \times 10^{-13}$ J (mass given to 2 sig. fig.)

## CHECK YOUR UNDERSTANDING 17.2

4  ≈ 36 MeV

5  a  near Geneva (Switzerland)

## CHECK YOUR UNDERSTANDING 17.3

3  up, down, charm, strange, top, bottom

4  Two types of hadrons are baryons and mesons.

6  Mesons are hard to detect because they are very short-lived.

## CHAPTER 17 REVIEW QUESTIONS

13  Quarks: $+\dfrac{2}{3}$ and $-\dfrac{1}{3}$. Antiquarks: $-\dfrac{2}{3}$ and $+\dfrac{1}{3}$

15  ■  graviton (gravity)

    ■  photon (electromagnetism)

    ■  gluon (strong nuclear force)

    ■  W and Z bosons (weak nuclear force)

# GLOSSARY

## A

**absolute frame of reference** a frame of reference against which all others could be measured

**absolute magnitude** a number assigned to a star to represent its luminosity

**absolute uncertainty** the uncertainty in a measurement or derived result expressed in the same units as the measurand

**absorption lines** black lines that appear in an electromagnetic spectrum that indicate an absence of photons of a particular energy or energies

**absorption spectrum** a spectrum showing dark lines at certain wavelengths against a bright, continuous background

**AC generator** a generator that produces an AC (alternating current) output

**AC induction motor** a motor that uses an AC supplied emf and electromagnetic induction to create a current in the squirrel cage and rotate the armature

**accurate** having a mean value close to the 'true value' or accepted value

**aether** a medium through which light waves were thought to propagate

**albedo** the ratio of light reflected by a surface to light incident on it; a surface with an albedo of 1 is perfectly reflective, and a surface with an albedo of 0 is perfectly absorbing

**alpha particle** two protons bound to two neutrons, the nucleus of a helium atom

**alternating current (AC)** a current that varies sinusoidally with time, such that it changes direction periodically

**ampere** the SI unit of current, $1\,A = 1\,C\,s^{-1}$, defined as the constant current which, if maintained in two straight parallel conductors of infinite length, of negligible circular cross-section, and placed 1 m apart in vacuum, would produce between these conductors a force equal to $2 \times 10^{-7}\,N\,m^{-1}$

**angular displacement** the angle, $\Delta\theta$, through which an object has moved, measured between the radial vectors for its initial and final positions

**annihilation** the process that occurs when a matter particle coincides with its antimatter equivalent and their masses are converted into energy

**anode** the positive electrode

## antimatter the opposite of matter; particles with the opposite charge to their matter equivalent

**antineutrino** the antimatter equivalent of a neutrino

**antinode** the point of maximum displacement in a standing wave

**antiparticle** the antimatter partner to a matter particle

**armature** part of a rotor that provides the physical support for the coil

**astronomical unit** the distance between Earth and the Sun

**atomic mass number ($A$)** the number of protons plus neutrons in a nucleus

**atomic number ($Z$)** the number of protons in a nucleus

**axis of rotation** the axis about which an object rotates

## B

**back emf** the induced emf in a motor's coil, created by the change in flux through the coil due to the movement of the coil in a magnetic field

**banked** (of a road) sloped at a curve so that the outside edge is slightly higher than the inside edge

**baryon** a hadron composed of three quarks

**beta particle** a form of radiation originating from the nucleus, either an electron or a positron

**Big Bang** the initial event that formed the Universe

**binding energy** the energy required to break a nucleus apart into its constituent particles

**black body** an object with a perfectly absorbing surface that emits electromagnetic radiation of all wavelengths with a peak intensity characteristic of the temperature of the object

**black body radiation** the electromagnetic radiation emitted by a black body, with a spectrum characteristic of the temperature of the body

**boson** a particle that carries one of the four forces

**brushes** a part of the stator of a motor or generator that acts as a sliding switch connected to the commutator, to disconnect or reverse the direction of current as the rotor turns

## C

**capacitor** a circuit component, such as a pair of parallel electric plates, capable of holding an electric charge

**cathode** the negative electrode

**centripetal acceleration** centre-seeking acceleration; acceleration towards the centre of a circular path

**centripetal force** the net force acting on an object in circular motion, pointing towards the centre of the circular path

**charge** a fundamental property of matter, which creates and is affected by electric fields

**cloud chamber** a device used to detect the paths of radiation as they leave a vapour trail through the chamber

**combustion** a chemical reaction also known as burning or oxidation

**commutator** a part of a rotor on a motor or generator that disconnects or reverses the direction of current as the rotor turns

**conductor** a substance with low electrical resistance that allows current to flow easily

**constructive interference** when the trough of one wave meets the trough of another wave, they will combine to give an even bigger trough

**continuous** able to take any value, sometimes within a fixed range, as distinct from discrete or quantised

**continuous spectrum** a spectrum containing radiation of all wavelengths; for example, a rainbow

**control rods** rods inserted into the core of a nuclear reactor to capture neutrons, typically containing an element such as boron

**controlled chain reaction** where the rate of a fission reaction is kept constant

**controlled variables** variables that are controlled in an experiment and held constant

**corpuscles** tiny particles that make up light according to Newton's theory

**cut-off frequency, $f_0$** the minimum frequency of light needed to eject an electron from a metal surface

9780170409131

## D

**daughter nuclide** a nuclide resulting from a decay event

**DC generator** a generator that produces a DC (direct current) output

**DC motor** a motor that runs on direct current (DC), converting electric potential energy into kinetic energy

**deflection** a change of path or direction

**dependent variable** a variable whose value depends on one or more other variables; the variable that we measure

**destructive interference** when a crest of one wave meets a trough of another wave, they will cancel each other out

**deterministic** where a single solution is found for the outcome of a situation

**diffraction** the bending of a wave as it passes around a corner or obstacle or goes through a gap

**dimensional analysis** the use of measurement units to check a calculation procedure

**direct current (DC)** current that always flows in one direction

**discrete** able to take only specific values, not continuous; for example, a line spectrum is a discrete spectrum

**disintegration** the fission of a large nucleus

**dispersion** the separation of light into its component colours or wavelengths

**Doppler effect** the shortening or lengthening of the wavelength from a source of waves due to relative motion between source and observer

## E

**eddy currents** circulating currents induced by changing magnetic flux

**Einstein's mass–energy relationship** $E = mc^2$

**electric field** the field created by charged objects that exerts a force on charged objects; the force per unit charge acting on a charged object

**electric field lines** visual representation of electric fields. The direction of the field lines shows the direction of force on a positively charged particle, and the density of the field lines is proportional to the field strength

**electric potential** the potential energy per unit charge at a point in space, measured in units of volts, V

**electric potential energy** potential energy arising from the interaction of charged objects; the potential energy stored in an electric field

**electromagnetic induction** the production of an electric field and hence an emf (electromotive force) by a changing magnetic flux

**electromagnetic spectrum** the range of frequencies (or wavelengths) of electromagnetic radiation generated by an incandescent source

**electromotive force, emf, $\varepsilon$** available energy per unit charge, sometimes called a voltage. An emf provides the energy to make a current flow

**electron wave** the quantum mechanical description of an electron, in which it shows wave-like behaviour and surrounds the nucleus of an atom in the form of a standing wave

**electrostatic force** the force exerted by charged objects on other charged objects, when the objects are stationary; the force exerted by the electric field

**element** a substance composed of atoms with the same atomic number (same number of protons)

**elementary particle** a particle that cannot be broken down into further components

**emission lines** photons of a particular frequency (or frequencies) indicating energy emitted by specific energy transitions of electrons in atoms

**emission spectra, emission spectrum** a spectrum showing bright lines at certain wavelengths against a dark background

**empirical formula** a formula derived from observations rather than theory

**enrichment** the process whereby the percentage of uranium-235 in a sample of uranium is increased so that it becomes fissile

**escape velocity** the minimum speed required of an object so that it can escape a gravitational field

**exothermic** releasing heat energy

## F

**falsifiable** able to be proved incorrect or untrue

**fast neutron** a neutron emitted from a fission event with high kinetic energy

**final velocity, $v_{final}$** the velocity of a projectile just before it lands

**fission** a nuclear reaction where a large nucleus splits into smaller nuclei

**fission fragments** the particles resulting from a fission event

**fluorescence** electromagnetic radiation from a substance exposed to energy

**flux linkage** the magnetic connection between coils, the amount of flux from one coil that passes through the other coil

**fractional uncertainty** uncertainty as a fraction of the measurement, often expressed as a percentage

**fusion** the merging of two lightweight nuclei into one

## G

**Galilean transformation** relating coordinates in one inertial frame to those in another inertial frame; classical relativistic transformations

**Gaussian surface** a surface that completely encloses an object such as a charged particle or a piece of magnetic material. The surface does not have to be spherical but it must not double back on itself and must be smooth enough so that a tangent can be taken at any point

**generator** a device that uses electromagnetic induction to convert kinetic energy (usually rotational) into electric potential energy

**geostationary** staying at the same position relative to Earth; for a satellite this is above the same position on the surface of Earth

**geosynchronous** having the same period as Earth's rotation; for a satellite this means an orbital period of one day

**gluon** the elementary particle responsible for the exchange of the strong nuclear force between quarks

**gravitational field, $g$** the force per unit mass acting on a small test mass due to the gravitational force exerted by an object with mass

**gravitational force** the force that any object with mass exerts on any other object with mass via the gravitational field

**gravitational potential energy** potential energy of a system due to the gravitational forces exerted by objects in the system on other objects in the system

**ground level, ground state** the lowest possible energy level for an electron in an atom

## H

**hadron** composite particle composed of quarks; protons and neutrons

**half-life** the time taken for half the radioactive nuclei in a sample to decay

**Hertzsprung–Russell diagram** a graph of luminosity of stars versus spectral type or surface temperature

**horizontal range,** $x_{max}$ the total horizontal distance travelled by a projectile

**Huygens' principle** each point on a wave behaves as a point source for waves in the direction of propagation; the line tangent to these circular waves is the new position of the wave front a short time later

**hypothesis** a testable idea, explanation or answer to a question

## I

**incandescent** light with a range of wavelengths generated by the motion of particles in a heated material; the peak intensity indicates temperature

**independent variable** a variable that does not depend on the value of other variables; the variable that we control and vary

**induced current** a current driven by an induced emf

**induced emf** an emf created by a changing magnetic flux

**induction coil** a device that, using DC, can generate the high voltage necessary for cathode ray tubes

**inertial frame of reference** a non-accelerating frame of reference

**initial velocity** in projectile motion, the velocity at which a projectile is launched; symbol $u$. It is a vector with components $u_x$ and $u_y$

**insulation** material with high resistance, one that does not allow current to flow

**invariant** same in all frames of reference

**ionised** having an electron removed

**ionising ability** the ability to remove an electron from an atom

**isotope** isotopes are atoms with the same number of protons but differing in the number of neutrons in their nucleus

## L

**launch angle** the angle at which a projectile is launched, the angle of the initial velocity, usually given as angle above the horizontal; usual symbol $\theta$

**launch height** the height above ground level, or above landing height, at which a projectile is launched; usually written $y_0$

**length contraction** length appears shorter in a reference frame that is moving relative to a stationary observer frame

**lepton** a light elementary particle, for example the electron

**lever arm** the distance from the pivot point to the point of application of the force, symbol $r$

**Leyden jar** a glass jar with metal on the outside and the inside, not in contact with each other, used to store electric charge; an early form of capacitor

**light year** the distance light travels in a vacuum in one Earth year

**limit of reading** resolution; the smallest increment to which a scale can be read

**linear accelerator** a particle accelerator that accelerates particles in a straight line

**logbook** a record of all planning, experimental works, results and analysis for an investigation

**Lorentz factor** the factor by which time, length and relativistic mass change for an object when that object is moving

$$\gamma = \frac{1}{\sqrt{1 - \frac{v^2}{c^2}}}$$

**lumen** unit of measurement of the brightness of a light source, approximately the absolute brightness of a candle

**luminosity** the amount of light energy being emitted by a star

**lux** unit of measurement of light intensity; one lumen per square metre

## M

**magnetic braking** the braking or slowing of relative motion of a magnet (or electromagnet) and a conductor due to the formation of eddy currents that create a magnetic force opposing the motion

**magnetic field** field created by moving charged particles and currents, which exerts a force on moving charged particles and currents

**magnetic flux** the amount of magnetic field passing through an area, measured in Wb or $T\,m^2$; symbol $\Phi$

**magnetic moment** a vector quantity possessed by an object or particle that measures the tendency to interact with a magnetic field

**Main Sequence** a type of star that is fusing hydrogen in its core

**Malus' Law** the intensity of a beam of plane-polarised light after passing through a rotatable polariser varies as the square of the cosine of the angle through which the polariser is rotated from the position that gives maximum intensity

**mass defect** the difference between the mass of a nucleus and the sum of the masses of the individual nucleons; the difference in mass before and after a nuclear decay or reaction

**matter wave** particles with momentum can have a wavelength and display wave properties

**maximum height** in projectile motion, the highest point on a projectile's trajectory; symbol $h$

**mean lifetime** average lifetime of a radioactive particle, $= 1.443$ half-life

**meson** short-lived particle composed of a quark and an antiquark

**moderator** a substance such as graphite used in a nuclear reactor to slow the fast neutrons to thermal neutrons

**modes of vibration** characteristic patterns of oscillation, usually with a discrete set of allowed frequencies

**motor effect** the force experienced by a current-carrying conductor in a magnetic field

**muon** a meson formed by cosmic rays in the upper atmosphere

## N

**near Earth approximation** also called the 'surface approximation', the approximation that the gravitational force and field are constant; $F = mg$ and $g = 9.8\,m\,s^{-2}$

**nebulae** regions of glowing cosmic gas and dust

**neutrino** a very small, nearly massless, elementary particle with no charge

**neutron** a neutral particle (a hadron) that is a component of most nuclei

**neutron poison** nuclei that absorb neutrons from fission events, thus preventing further fission

**nuclear binding energy** the energy required to break a nucleus apart into its individual nucleons

**nuclear transformation** the changing of a nucleus from one type to another

9780170409131

**nucleons** particles found in the nucleus – protons and neutrons

**nucleosynthesis** the formation of new nuclei

**nucleus** the central part of an atom

**nuclide** a distinct kind of nucleus with a specific number of protons and neutrons

## P

**parent nuclide** a nuclide just before a decay event occurs

**parsec** an astronomical distance approximately equal to 3.26 light years

**particle accelerator** a machine designed to accelerate particles to close to the speed of light

**path difference** the difference in length of two separate paths taken by light rays to reach a certain point

**penetrating power** the ability to move through matter

**period** the time taken to complete one rotation or cycle; symbol $T$

**permeability of free space** $(\mu_0)$ the constant of proportionality which give the strength of a magnetic field due to a current in vacuum, $\mu_0 = 4\pi \times 10^{-7}\,\mathrm{T\,m\,A^{-1}}$

**permittivity of free space** $(\varepsilon_0)$ a constant representing how fast an electric field can propagate. $\varepsilon_0 = 8.854 \times 10^{-12}\,\mathrm{m^{-3}\,kg^{-1}\,s^4\,A^2}$

**photocurrent** the current formed by electrons ejected from a surface by incident photons

**photoelectron** an electron ejected from a metal surface following absorption of a photon of sufficient energy

**photon** a particle or quantum of electromagnetic radiation of a specific energy $E = hf$

**pion** an unstable particle consisting of a quark and an antiquark, the lightest of the mesons and hadrons

**pivot point** the point about which an object rotates or pivots

**Planck constant** the constant of proportionality between energy and frequency for photons; $h = 6.626 \times 10^{-34}\,\mathrm{J\,s}$

**plane polarisation** state of electromagnetic radiation in which the electric fields all oscillate in the same plane

**polarisation** the orientation of light waves based on the direction of the oscillating electric fields

**positron** the antimatter equivalent of an electron; an electron with a positive charge

**positron emission tomography (PET)** a medical visualisation technique that detects the gamma radiation emitted when positrons from a tracer annihilate

**postulate** a proposition or an idea

**potential difference** the difference in electric potential between two points, also called voltage

**precise** having a small spread in values about the mean

**principle of superposition** waves can pass through each other without being disturbed; the wave height at any point is the sum of the individual displacements of the waves present

**probabilistic** having multiple possible outcomes, each with a probability of occurring

**proper length** length measured in an inertial frame of reference in which the object is stationary relative to the observer

**proper time** time measured between two events occurring at the same location in an inertial stationary frame of reference

## Q

**quantised** existing in discrete amounts, not able to be divided into arbitrarily small amounts

**quantum** a discrete unit or amount of some physical property, such as energy, charge, mass or angular momentum

**quantum mechanics** the mathematical description of particles that are atomic scale or smaller

**quantum physics** the study of the smallest units of the physical world in which quantities exist in discrete amounts

**quantum theory** the theory that has as its underlying principle the discrete nature of quantities in the study of the very small

**quark** elementary particles that combine to form hadrons and mesons

## R

**rectilinear motion** motion along a straight line, or in a single direction

**rectilinear propagation** of light, travelling in a straight line

**redshift** the appearance more towards the red end of the specrum than normal of spectral lines of an object such as a star

**refractive index** a number that compares the speed of light in a material with the speed of light in a vacuum

**relativistic mass, relativistically corrected mass** mass measured in a reference frame that is moving relative to the reference frame in which the rest mass was measured

**relativity principle** the laws of physics are the same in all inertial frames of reference

**reliable** trustworthy (of sources); able to be reproduced (of measurements)

**reproducible** able to be repeated, to give the same results

**resistance** the difficulty with which charges flow through a material

**resonant cavity** a physical container that will sustain standing waves with particular frequencies

**rest energy** $E = m_0 c^2$

**rest mass** mass measured in an inertial reference frame in which the object is at rest; symbol $m_0$

**root mean square, rms** for an AC variable, the value obtained by taking the square, averaging and then taking the square root of the variable, giving the single DC quantity that would deliver the same effect

**rotor** the rotating part of a motor, including the armature, the coils and the commutators

## S

**scatter plot** a graph of data shown as points on axes defined by the dependent and independent variables; used to show a mathematical relationship

**slow neutron** a neutron with low kinetic energy that can be absorbed by a uranium-235 nucleus

**spectroscopy** the study of how electromagnetic radiation interacts with matter

**spectrum** the distributed components of light or another wave arranged by frequency or wavelength

**squirrel cage** set of connected loops in an AC induction motor that acts as a current-carrying coil

**stability curve** a curve on a graph of atomic number vs neutron number near where stable nuclei are found

**Standard Model** the scientific model used to classify matter into its fundamental particles

**stator** the part of a motor that does not move, including the casing, the magnets, electrical inputs and brushes

**stellar aberration** apparent change in the position of a star due to relative motion of Earth

**step-down transformer** transformer with fewer turns on the secondary coil than the primary, so that the output voltage (potential difference or emf) is less than the input voltage

**step-up transformer** transformer with more turns on the secondary coil than the primary, so that the output voltage (potential difference or emf) is greater than the input voltage

**stopping voltage** the reverse bias voltage required to stop the flow of photoelectrons in a photoelectric effect experiment

**supplied emf** the potential difference (voltage) supplied to a motor by a battery or other source of emf

**systematic error** a measurement error that gives a constant difference between the measurement and the 'true' value

## T

**tension** a force (a pull) provided by a string, acting in the direction of the string, also the force experienced by the string

**thermal neutron** see 'slow neutron'

**thermonuclear** relating to devices that use fusion to release energy, for example a thermonuclear or hydrogen bomb

**time dilation** the slowing of time in a moving frame of reference as measured by an observer in the stationary frame

**time of flight,** $t_{flight}$ the total time between launch and landing of a projectile

**torque** the rotational equivalent of force, causing rotation; $\tau = rF\sin\theta$

**trajectory** the parabolic path followed by a projectile

**transformer** a pair of magnetically linked coils used to increase or decrease emf and current using electromagnetic induction

**transmutation** the changing of a nucleus from one type to another

**transuranic** relating to elements with nuclei heavier than uranium

## U

**uncertainty** estimate of the range of values within which the 'true' value of a measurement or derived quantity lies; the extent to which the result of an experiment is unknown or unpredictable

**uncontrolled chain reaction** where the rate of a fission reaction is uncontrolled and accelerates rapidly

**uniform** constant in magnitude and direction

**uniform circular motion** motion at constant speed in a circular path

## V

**valid** (of an experiment) reproducible and of good enough accuracy and precision to test the hypothesis

## W

**wave–particle duality** the way in which matter or radiation can be a wave and a particle simultaneously

**work function** the energy required to eject an electron from a metal surface; effectively, it is the ionisation energy for the bulk material

## Z

**Zeeman effect** the splitting of single spectral lines into two when the source atoms are subjected to a magnetic field

9780170409131

# INDEX

fluorescence in CRTs, 342
flux linkage in a transformer, 163–167
force
    on current-carrying wire in magnetic field, 134
    on moving charged particle in magnetic field, 119–120
    on parallel current-carrying wires, 145
    on parallel wires carrying opposite currents, 145
Foucault, Léon
    and speed of light, 209
    testing Newton's particle theory of light, 229
four proton fusion, 334
fractional uncertainty, 20
frames of reference
    inertial and non-inertial, 273
    relative motion and, 277–278
Fraunhofer, *see also* von Fraunhofer, Joseph
Fraunhofer diffraction, 234–236
friction, measuring with turntable, 61–62
Friedmann, Alexander, expansion or contraction of Universe, 316, 317
fundamental forces in nature, 423
fundamental particles, 413
    further evidence for, 419–421
    scale of, 424
fusion
    four proton, 334
    mass defect in, 408–409
    nuclear, 407–409
    in white dwarf stars, 328

galaxies, red shift of, 221
Galilean transformations, 273–275
Galileo, speed of light, 207, 208
gamma rays, 384, 387
Gaussian surface, 204
Gauss's laws for electricity and magnetism, 204–205
Geiger, Hans, particle scattering, 348–349
Geiger-Marsden experiment, 348–349
    modelling, 350
Geissler, Heinrich, glass vacuum tube, 341
Gell-Mann, Murray, named quarks, 421
generations of particles, 423
generators
    alternating current (AC), 182–186
    direct current (DC), 186–187
geostationary satellites, 86–87
geosychronous satellites, 87
Germer, Lester H., matter wave experiment, 371–373
gluons, 311
golf on the moon, 37
Gordon-Smith, A.C., speed of light, 210
graphs, drawing and using, 21–23

gravitation fields, 126–127
    and forces, 80–84
gravitational constant, universal $G$, measuring, 80
gravitational field $g$, 82–84
gravitational fields, energy in, 94–98
gravity and Newton's third law, 81–82
ground level of atom, 363–364
ground state of electrons, 216–217
GUM (Guide to expressions of Uncertainty in Measurement), 20

H–R diagram *see* Hertzsprung–Russell diagram
hadrons, 422
    and field particles, 311
half-life and decay, 395–400
    simulation, 399
half-silvered mirror, 280
Harvard system of spectral types, 326
Hertz, Heinrich, and photoelectric effect, 259
Hertzsprung, Ejnar, and star classification, 327
Hertzsprung–Russell (H–R) diagram of luminosity classes, 326, 328
    classifying stars using, 329
    plotting stars on, 330–331
    star groups in, 328–329
Higgs, Peter, discovers boson, 424
Hooke, Robert, wave model of light, 230–231
horizontal circle, motion in, 63–64
horizontal corner, car turning a, 66
horizontal range of projectiles, 38
Hubble, Edwin
    rate of expansion of Universe, 317
    red-shifted spectra, 322
Hubble Space Telescope orbit, 88
Hubble's constant $H_0$, 317
Hubble's Law, 317, 318
Huygens, Christiaan
    and Hooke's wave model of light, 230–231
    and light, 203, 208
Hyde, William Wallace, and solar spectrum, 216
hydrogen
    ionisation, 325
    spectra, 361–363
hypotheses, 7
    formulating, 8–9
    proposing, 10–11

incandescent light sources, 217
incidence, angles of in light, 228
independent variables, 11
induced current, 155
Induction, Faraday's Law of, 204

induction coils, 341
induction motors, alternating current (AC), 187–188
inertial frames of reference, 273
initial velocity of projectiles, 32
insulation, 132
interference
    and diffraction patterns, 232–242
    experiments with single particles, 376–378
    of monochromatic light, 241
    patterns, 232
interferometry and speed of light, 211
International Space Station (ISS), 88–89
invariance of time, 274
investigations, evaluating, 25
ionisation, 363–364
    of hydrogen, 325
ionising power of radiation, 391–392
ions, 262
isotopes, 386
ITER (international fusion experiment), 297

jumping wire experiment, 138

Kepler, Johannes, three laws of, 91–93
Kirchhoff, Gustav
    and element identification using spectral lines, 360–361
    and speed of electricity, 204
Kohlrausch, Rudolf, and speed of light, 204

Lagrange, Joseph-Louis, 204
Large Hadron Collider, 420–421
launch angles of projectiles, 33
launch height of projectiles, 38
Law of Areas, Kepler's, 92
Law of Orbits, Kepler's, 91–92
Law of Periods, Kepler's, 92
le Verrier, Urbain, and speed of light, 210
Leavitt, Henrietta, and Cepheid variables, 318
Lemaître, Georges, expansion or contraction of Universe, 316–317
Lenard, Philipp, model of atom, 349
length contraction, 284–288
Lenz's Law, 158–160, 179
    and conservation of energy, 189–193
leptons, 311
    flavours of, 423
Leucippus, and nature of light, 227
lever arms, 75
Leyden jars, 203
light
    aberration of, 208
    concept map for, 4–5
    decisive test of Newton's particle theory, 229